Geophysical Monograph Series

Geophysical Monograph Series

190 **The Stratosphere: Dynamics, Transport, and Chemistry** *L. M. Polvani, A. H. Sobel, and D. W. Waugh (Eds.)*
191 **Rainfall: State of the Science** *Firat Y. Testik and Mekonnen Gebremichael (Eds.)*
192 **Antarctic Subglacial Aquatic Environments** *Martin J. Siegert, Mahlon C. Kennicut II, and Robert A. Bindschadler (Eds.)*
193 **Abrupt Climate Change: Mechanisms, Patterns, and Impacts** *Harunur Rashid, Leonid Polyak, and Ellen Mosley-Thompson (Eds.)*
194 **Stream Restoration in Dynamic Fluvial Systems: Scientific Approaches, Analyses, and Tools** *Andrew Simon, Sean J. Bennett, and Janine M. Castro (Eds.)*
195 **Monitoring and Modeling the Deepwater Horizon Oil Spill: A Record-Breaking Enterprise** *Yonggang Liu, Amy MacFadyen, Zhen-Gang Ji, and Robert H. Weisberg (Eds.)*
196 **Extreme Events and Natural Hazards: The Complexity Perspective** *A. Surjalal Sharma, Armin Bunde, Vijay P. Dimri, and Daniel N. Baker (Eds.)*
197 **Auroral Phenomenology and Magnetospheric Processes: Earth and Other Planets** *Andreas Keiling, Eric Donovan, Fran Bagenal, and Tomas Karlsson (Eds.)*
198 **Climates, Landscapes, and Civilizations** *Liviu Giosan, Dorian Q. Fuller, Kathleen Nicoll, Rowan K. Flad, and Peter D. Clift (Eds.)*
199 **Dynamics of the Earth's Radiation Belts and Inner Magnetosphere** *Danny Summers, Ian R. Mann, Daniel N. Baker, and Michael Schulz (Eds.)*
200 **Lagrangian Modeling of the Atmosphere** *John Lin (Ed.)*
201 **Modeling the Ionosphere-Thermosphere** *Jospeh D. Huba, Robert W. Schunk, and George V. Khazanov (Eds.)*
202 **The Mediterranean Sea: Temporal Variability and Spatial Patterns** *Gian Luca Eusebi Borzelli, Miroslav Gacic, Piero Lionello, and Paola Malanotte-Rizzoli (Eds.)*
203 **Future Earth – Advancing Civic Understanding of the Anthropocene** *Diana Dalbotten, Gillian Roehrig, and Patrick Hamilton (Eds.)*
204 **The Galápagos: A Natural Laboratory for the Earth Sciences** *Karen S. Harpp, Eric Mittelstaedt, Noémi d'Ozouville, and David W. Graham (Eds.)*
205 **Modeling Atmospheric and Oceanic Flows: Insightsfrom Laboratory Experiments and Numerical Simulations** *Thomas von Larcher and Paul D. Williams (Eds.)*
206 **Remote Sensing of the Terrestrial Water Cycle** *Venkat Lakshmi (Ed.)*
207 **Magnetotails in the Solar System** *Andreas Keiling, Caitríona Jackman, and Peter Delamere (Eds.)*
208 **Hawaiian Volcanoes: From Source to Surface** Rebecca Carey, *Valerie Cayol, Michael Poland, and Dominique Weis (Eds.)*
209 **Sea Ice: Physics, Mechanics, and Remote Sensing** *Mohammed Shokr and Nirmal Sinha (Eds.)*
210 **Fluid Dynamics in Complex Fractured-Porous Systems** *Boris Faybishenko, Sally M. Benson, and John E. Gale (Eds.)*
211 **Subduction Dynamics: From Mantle Flow to Mega Disasters** *Gabriele Morra, David A. Yuen, Scott King, Sang Mook Lee, and Seth Stein (Eds.)*
212 **The Early Earth: Accretion and Differentiation** *James Badro and Michael Walter (Eds.)*
213 **Global Vegetation Dynamics: Concepts and Applications in the MC1 Model** *Dominique Bachelet and David Turner (Eds.)*
214 **Extreme Events: Observations, Modeling and Economics** *Mario Chavez, Michael Ghil, and Jaime Urrutia-Fucugauchi (Eds.)*
215 **Auroral Dynamics and Space Weather** *Yongliang Zhang and Larry Paxton (Eds.)*
216 **Low-Frequency Waves in Space Plasmas** *Andreas Keiling, Dong-Hun Lee, and Valery Nakariakov (Eds.)*
217 **Deep Earth: Physics and Chemistry of the Lower Mantle and Core** *Hidenori Terasaki and Rebecca A. Fischer (Eds.)*
218 **Integrated Imaging of the Earth: Theory and Applications** *Max Moorkamp, Peter G. Lelievre, Niklas Linde, and Amir Khan (Eds.)*
219 **Plate Boundaries and Natural Hazards** *Joao Duarte and Wouter Schellart (Eds.)*
220 **Ionospheric Space Weather: Longitude and Hemispheric Dependences and Lower Atmosphere Forcing Timothy Fuller-Rowell**, *Endawoke Yizengaw, Patricia H. Doherty, and Sunanda Basu (Eds.)*
221 **Terrestrial Water Cycle and Climate Change Natural and Human-Induced Impacts** *Qiuhong Tang and Taikan Oki (Eds.)*
222 **Magnetosphere-Ionosphere Coupling in the Solar System** *Charles R. Chappell, Robert W. Schunk, Peter M. Banks, James L. Burch, and Richard M. Thorne (Eds.)*
223 **Natural Hazard Uncertainty Assessment: Modeling and Decision Support** *Karin Riley, Peter Webley, and Matthew Thompson (Eds.)*
224 **Hydrodynamics of Time-Periodic Groundwater Flow: Diffusion Waves in Porous Media** *Joe S. Depner and Todd C. Rasmussen (Auth.)*
225 **Active Global Seismology** *Ibrahim Cemen and Yucel Yilmaz (Eds.)*
226 **Climate Extremes** *Simon Wang (Ed.)*
227 **Fault Zone Dynamic Processes** *Marion Thomas (Ed.)*
228 **Flood Damage Survey and Assessment: New Insights from Research and Practice** *Daniela Molinari, Scira Menoni, and Francesco Ballio (Eds.)*
229 **Water-Energy-Food Nexus – Principles and Practices** *P. Abdul Salam, Sangam Shrestha, Vishnu Prasad Pandey, and Anil K Anal (Eds.)*
230 **Dawn–Dusk Asymmetries in Planetary Plasma Environments** *Stein Haaland, Andrei Rounov, and Colin Forsyth (Eds.)*
231 **Bioenergy and Land Use Change** *Zhangcai Qin, Umakant Mishra, and Astley Hastings (Eds.)*
232 **Microstructural Geochronology: Planetary Records Down to Atom Scale** *Desmond Moser, Fernando Corfu, James Darling, Steven Reddy, and Kimberly Tait (Eds.)*
233 **Global Flood Hazard: Applications in Modeling, Mapping and Forecasting** *Guy Schumann, Paul D. Bates, Giuseppe T. Aronica, and Heiko Apel (Eds.)*
234 **Pre-Earthquake Processes: A Multidisciplinary Approach to Earthquake Prediction Studies** *Dimitar Ouzounov, Sergey Pulinets, Katsumi Hattori, and Patrick Taylor (Eds.)*
235 **Electric Currents in Geospace and Beyond** *Andreas Keiling, Octav Marghitu, and Michael Wheatland (Eds.)*
236 **Quantifying Uncertainty in Subsurface Systems** *Céline Scheidt, Lewis Li, and Jef Caers (Eds.)*
237 **Petroleum Engineering** *Moshood Sanni (Ed.)*
238 **Geological Carbon Storage: Subsurface Seals and Caprock Integrity** *Stéphanie Vialle, Jonathan Ajo-Franklin, and J. William Carey (Eds.)*
239 **Lithospheric Discontinuities** *Huaiyu Yuan and Barbara Romanowicz (Eds.)*
240 **Chemostratigraphy Across Major Chronological Eras** *Alcides N.Sial, Claudio Gaucher, Muthuvairavasamy Ramkumar, and Valderez Pinto Ferreira (Eds.)*
241 **Mathematical Geoenergy:Discovery, Depletion, and Renewal** *Paul Pukite, Dennis Coyne, and Daniel Challou (Eds.)*

Geophysical Monograph 242

Ore Deposits
Origin, Exploration, and Exploitation

Sophie Decrée
Laurence Robb
Editors

This Work is a co-publication of the American Geophysical Union and John Wiley and Sons, Inc.

WILEY

This Work is a co-publication between the American Geophysical Union and John Wiley & Sons, Inc.

This edition first published 2019 by John Wiley & Sons, Inc., 111 River Street, Hoboken, NJ 07030, USA and the American Geophysical Union, 2000 Florida Avenue, N.W., Washington, D.C. 20009

Published under the aegis of the AGU Publications Committee

Brooks Hanson, Executive Vice President, Science
Lisa Tauxe, Chair, Publications Committee
For details about the American Geophysical Union visit us at www.agu.org.

Wiley Global Headquarters
111 River Street, Hoboken, NJ 07030, USA
For details of our global editorial offices, customer services, and more information about Wiley products visit us at www.wiley.com.

Library of Congress Cataloging-in-Publication data is available

ISBN: 9781119290537

Cover image: Courtesy of Laurence Robb
Cover design: Wiley

Set in 10/12pt Times New Roman by SPi Global, Pondicherry, India

V10008957_032019

CONTENTS

CONTRIBUTORS

Jean-Marc Baele
Department of Geology and Applied Geology
University of Mons
Mons, Belgium

Shaun L. L. Barker
School of Science
University of Waikato,
Hamilton, New Zealand;
Mineral Deposit Research Unit
University of British Columbia
Vancouver, BC, Canada;
Centre for Ore Deposit and Earth Sciences
University of Tasmania
Hobart, Tasmania

Tamara B. Bayanova
Geological Institute
Kola Science Centre
Russian Academy of Sciences (GI KSC RAS)
Apatity, Russia

Victor V. Chashchin
Geological Institute
Kola Science Centre
Russian Academy of Sciences (GI KSC RAS)
Apatity, Russia

Janine Cole
School of Geosciences
University of the Witwatersrand
Johannesburg, South Africa;
Geophysics and Remote Sensing Unit
Council for Geoscience
Silverton, Pretoria, South Africa

G. R. J. Cooper
School of Geosciences
University of the Witwatersrand
Johannesburg, South Africa

L. Corriveau
Geological Survey of Canada
Natural Resources Canada
Québec, QC, Canada

W. J. Davis
Geological Survey of Canada
Natural Resources Canada
Ottawa, ON, Canada

Sophie Decrée
Geological Survey of Belgium
Royal Belgian Institute of Natural Sciences
Brussels, Belgium

Stijn Dewaele
Department of Geology and Mineralogy
Royal Museum for Central Africa
Tervuren, Belgium;
Department of Geology and Soil Science
Ghent University
Ghent, Belgium

Gregory M. Dipple
Mineral Deposit Research Unit
University of British Columbia
Vancouver, BC, Canada

R. J. Durrheim
University of the Witwatersrand
School of Geosciences
Johannesburg, South Africa

Niels Hulsbosch
KU Leuven
Geodynamics and Geofluids Research Group
Department of Earth and Environmental Sciences
Leuven, Belgium

E. J. Hunt
University of the Witwatersrand
School of Geosciences
Johannesburg, South Africa

Alexey U. Korchagin
Geological Institute, Kola Science Centre
Russian Academy of Sciences (GI KSC RAS),
Apatity, Russia;
JSC "Pana,"
Apatity, Russia

John N. Ludden
British Geological Survey
Keyworth, Nottingham, UK

M. S. Manzi
University of the Witwatersrand
School of Geosciences
Johannesburg, South Africa

S. Master
Economic Geology Research Institute
School of Geosciences
University of the Witwatersrand
Johannesburg, South Africa

Ryan Mathur
Department of Geology
Juniata College
Huntingdon, Pennsylvania, USA

Maarten Minnen
KU Leuven
Geodynamics and Geofluids Research Group
Department of Earth and Environmental Sciences
Leuven, Belgium

Alexander F. Mitrofanov
SRK Consulting
Toronto, Canada

Felix P. Mitrofanov
Geological Institute
Kola Science Centre
Russian Academy of Sciences (GI KSC RAS)
Apatity, Russia

J.-F. Montreuil
Formerly Institut National de la Recherche Scientifique
Québec, QC, Canada;
Red Pine Exploration Inc.
Toronto, ON, Canada

Philippe Muchez
KU Leuven
Geodynamics and Geofluids Research Group
Department of Earth and Environmental Sciences
Leuven, Belgium

N. M. Ndhlovu
School of Geosciences
University of the Witwatersrand
Johannesburg, South Africa

Lyudmila I. Nerovich
Geological Institute
Kola Science Centre
Russian Academy of Sciences (GI KSC RAS)
Apatity, Russia

E. G. Potter
Geological Survey of Canada
Natural Resources Canada
Ottawa, ON, Canada

Brian Rusk
Department of Geology
Western Washington University
Bellingham, Washington, USA

Stephanie E. Scheiber-Enslin
School of Geosciences
University of the Witwatersrand
Johannesburg, South Africa

Pavel A. Serov
Geological Institute
Kola Science Centre
Russian Academy of Sciences (GI KSC RAS)
Apatity, Russia

Da Wang
State Key Laboratory of Geological Processes
and Mineral Resources
School of Earth Sciences and Resources
China University of Geosciences
Beijing, China

Susan J. Webb
School of Geosciences
University of the Witwatersrand
Johannesburg, South Africa

Dmitry V. Zhirov
Geological Institute
Kola Science Centre
Russian Academy of Sciences (GI KSC RAS)
Apatity, Russia

PREFACE

The volatility of financial markets over the past decade has had a major impact on the upstream sector of the global resource industry. Exploration and replenishment of natural resources have not kept pace with consumption, and the declining rate of discovery of new, viable mineral deposits is cause for concern. Coupled with this is the fact that world-class mineral deposits are increasingly difficult to find because large, shallow ores have largely been discovered. A major challenge of the 21st century, therefore, is how to locate buried mineral deposits that do not have a footprint at the surface, and also how to identify new sources of mineral wealth.

Recent trends in exploration and mining have seen a number of amazing innovations, exemplified by technologies that have, for example, enabled the mining of massive sulphide deposits on the ocean floor. Even more astounding are the developments aimed at exploiting asteroids from near-Earth orbits. While many might see these innovations as futuristic, they are nevertheless counterbalanced by the ability of geoscientists to continue pushing the frontiers of mineral exploration and seek new land-bound metallotects, as well as to develop innovative methods for detecting metal anomalies under cover. This book brings together a variety of papers that, in Section I, highlight the features of less conventional mineral deposit styles that offer alternative exploration opportunities, and, in Section II, describe some of the recent technological advances that will assist in the future discovery of mineral deposits.

Whereas most of the world's mineral exploration is still focused on well-trodden metallotects, such as magmatic arcs and stable cratonic blocks, Section I emphasizes the features of atypical ores such as metamorphosed porphyry deposits of Proterozoic age, stratiform copper deposits hosted in sandstone, and fractionation mechanisms in S-type granitoids. These examples point to the fact that exploration should not be constrained by geologic didactics that exclude certain targets because of seemingly inappropriate lithology, tectonic setting, or epoch. Some of the great discoveries of the past have been made by thinking intuitively and "out of the box." Section II presents a variety of techniques that expand the armory of exploration tools available to the geoscientist: from microscopic and laboratory techniques involving mineral cathodoluminescence and isotope vectoring, to big data approaches aimed at geophysically imaging the Earth's crust. Although this book covers but a small fraction of the advances currently being made in mineral exploration science, it is timely because these innovations will catalyze the implementation of resource utilization policies that will, in the future, be more sustainable and environmentally responsive than at any time in the past.

Sophie Decrée
Royal Belgium Institute of Natural Sciences
and
Geological Survey of Belgium

Laurence Robb
University of Oxford
and
CIMERA – University of the Witwatersrand/
University of Johannesburg

Section I
Characteristics of Atypical Mineral Deposit Styles

1

Origin and Exploration of the Kola PGE-bearing Province: New Constraints from Geochronology

Felix P. Mitrofanov[1], Tamara B. Bayanova[1], John N. Ludden[2], Alexey U. Korchagin[1,3], Victor V. Chashchin[1], Lyudmila I. Nerovich[1], Pavel A. Serov[1], Alexander F. Mitrofanov[4], and Dmitry V. Zhirov[1]

ABSTRACT

The NE Fennoscandian Shield comprises the Northern (Kola) Belt in Finland and the Southern Belt in Karelia. The belts host mafic-ultramafic layered Cu-Ni-Cr and Pt-Pd-bearing intrusions. They were studied using precise isotope analyses with U-Pb on zircon and baddeleyite and Sm-Nd on rock-forming silicates and sulfides. The analyses indicate the 130 Ma magmatic evolution with major events at 2.53, 2.50, 2.45, and 2.40 Ga. It is considered to be governed by the long-lived mantle plume activity. Barren phases were dated at 2.53 Ga for orthopyroxenites and olivine gabbro in the Fedorovo-Pansky massif. Main PGE-bearing phases of gabbronorite (Mt. Generalskaya), norite (Monchepluton), and gabbronorites (Fedorovo-Pansky and Monchetundra massifs) yielded ages of 2.50 Ga. Anorthosites of Mt. Generalskaya, the Fedorovo-Pansky and Monchetundra massifs occurred at the 2.45 Ga PGE-bearing phase. According to regional geochronological correlations, this widespread event emplaced layered PGE-bearing intrusions of Finland (Penikat, Kemi, Koitelainen) and mafic intrusions in Karelia. Dikes of the final mafic magmatic pulse at 2.40 Ga are present in the Imandra lopolith. Slightly negative εNd values and ISr values of 0.703–0.704 suggest the layered intrusions to originate from an enriched EM-1-like mantle reservoir.

1.1. INTRODUCTION

Magmatic sulfide Ni-Cu-PGE and low-sulfide Pd-Pt deposits are best-valued commercial types of the Pd-Pt mineralization. In Russia, there is a well-known Ni-Cu-PGE deposit in Norilsk and a low-sulfide Pd-Pt deposit at the Monchegorsk and Fedorovo-Pansky massifs (Sluzhenikin et al., 1994). These deposits differ by their PGE mineralization. In the sulfide type, PGEs are accompanying components, and ferrous metals play the lead role, whereas in the low-sulfide type, Pd, Pt, and Rh are major, while nonferrous metals are secondary. Dividing PGE ores into the sulfide and low-sulfide types (groups) provides a basis for the classification proposed in Naldrett (2003), Dodin et al. (2001), and Likhachyov (2006).

In the 21st century, up to 90% of the platinum-group metals (PGM) production was related to processing of the Norilsk high-grade Ni-Cu-PGE ore. PGE were by-products, though in 2000–2001, their contribution to the price structure in the world's market was about 50%. According to Russian and American specialists (Dodin et al., 2001), the PGE production in Russia will be mainly related to mining of low-sulfide ores. Its resources in the Norilsk district are estimated at thousands of tons (Starostin & Sorokhtin, 2010). In contrast, PGE resources of the Kola region are estimated at hundreds of tons as of 2010.

Though the Kola low-sulfide PGE ores are a minor source of PGE in the global scope, they are widespread in the Kola region and require a detailed study (Mitrofanov

[1] *Geological Institute, Kola Science Centre, Russian Academy of Sciences (GI KSC RAS), Apatity, Russia*
[2] *British Geological Survey, Keyworth, Nottingham, UK*
[3] *JSC "Pana", Apatity, Russia*
[4] *SRK Consulting, Toronto, Canada*

Ore Deposits: Origin, Exploration, and Exploitation, Geophysical Monograph 242, First Edition.
Edited by Sophie Decrée and Laurence Robb.

et al., 2013). This paper provides a comprehensive study of the age distribution in the layered complexes. As some of them are barren, the geochronology may be used as a guide to explore or at least to understand their mineralization and magmatic settings.

1.2. LIPS AND LOW-SULFIDE DEPOSITS: GEOLOGICAL SETTING

Large Igneous Provinces (LIPs) are considered as derivatives of deep mantle plumes (Campbell and Griffits, 1990). In addition to alkaline and komatiite LIPs, a special group of LIPs comprises mafic intraplate continental provinces (Bleeker & Ernst, 2006) and consists of rift-related thick sedimentary and volcanic sequences, dike swarms, and intrusions of mafic and ultramafic rocks.

Some researchers provide geological, geophysical, and geochemical evidence of links between LIPs and deep mantle plumes (Grachev, 2003; Pirajno, 2007; Bogatikov et al., 2010). The plumes are considered to be active in the Precambrian regions, although many of ancient geological and geophysical features of terrestrial structures cannot be detected. Nevertheless, several indicators of an ancient intracontinental mafic LIP can be proposed (Mitrofanov et al., 2013; Robb, 2008; Rundkvist et al., 2006; Smol'kin et al., 2009):

- widespread areas of rocks associated with deep gravity anomalies that were caused by a granulite-mafic layer at the base of the crust;
- a rift-related (anorogenic) assembly, discordant with older basement structures. It occurs as multiphase extensional faulting that controls the arrangement of grabens, volcanic belts, extended dike swarms, and radial intrusive bodies;
- long-term, multistage, and pulsatory tectonics and magmatism;
- breaks in sedimentation and related erosion;
- early manifestations of tholeiitic basaltic (trap), high-magnesian (boninite-like) and alkaline magmatism in domains with the continental crust; formation of leucogabbro-anorthosite complexes;
- sills, lopoliths, sheetlike intrusions, large dikes, and dike swarms;
- multiphase and layered intrusions that differ from spreading and subduction-related rocks by geochemistry (Bleeker & Ernst, 2006). They have fine-scale fractionation (layering) and minor intermediate and felsic rocks, often with final leucogabbro and anorthosite and abundant pegmatoid mafic varieties;
- characteristic undepleted mantle geochemistry of rocks and ores with anomalously high contents of siderophile-chalcophile elements and LILE marked by $^{143}Nd/^{144}Nd$, $^{87}Sr/^{86}Sr$, $^{187}Os/^{188}Os$, and $^{3}He/^{4}He$ isotope ratios;
- large orthomagmatic Cr, Ni, Cu, Co, PGE ± Au, Ti, and V deposits.

The eastern Baltic (Fennoscandian) Shield hosts the vast Palaeoproterozoic East Scandinavian mafic LIP. Its current remnants cover about 1 mln km^2. The shield basement formed as a mature Archaean granulite and gneiss-migmatite crust 2550 Ma ago. It is exposed in the Kola-Lapland-Karelia Craton. Main structural features of the East Scandinavian mafic LIP and its Pd-Pt and Ni-Cu-PGE deposits are described in Mitrofanov et al. (2013). According to geophysical data, the lower crust in the eastern part of the shield is composed of a transitional crust-mantle layer ($Vp = 7.1–7.7$ km/s). Deep xenoliths of granulites and garnet anorthosite are dated ~2460 Ma. They were taken out from this layer by the Kandalaksha explosion pipes. Compositionally, these rocks are close to the bodies exposed at the surface (Verba et al., 2005). It implies that masses of deep magma did not only ascend as volcanic rocks, dikes, and intrusions, but also underplated the crust (Mitrofanov, 2005). The exposed part of the shield extends beneath the sedimentary cover toward the northern Russian Platform as a vast Palaeoproterozoic Baltic-Mid-Russia wide arc-intracontinental orogen (Mints, 2011).

The geological map of the Fennoscandian Shield (2005) clearly shows the anorogenic pattern of grabens, dike swarms, and belts (trends) of intrusive bodies independent of the Archaean gneiss-migmatite framework. These intrusions, related deposits, and occurrences make up extended belts in the northern part of the province: the NW-trending Kola Belt and the NE-trending Karelian Belt with a concentration of intrusions in the Monchegorsk ore cluster (Fig. 1.1) (Bayanova et al., 2009).

The long Early Palaeoproterozoic (2530–2400 Ma) geological history of the East Scandinavian Mafic LIP (ESMLIP) comprises several stages. They are separated by breaks in sedimentation and magmatic activity often marked by uplift erosion and deposition of conglomerates. The Sumian stage (2550–2400 Ma) is crucial for the metallogeny of Pd-Pt ores. It can be related to the emplacement of high-Mg and high-Si boninite-like and anorthosite magmas (Mitrofanov, 2005; Sharkov, 2006). The ore-bearing intrusions were emplaced in the Kola Belt (Fedorovo-Pansky and other intrusions, 2530–2450 Ma) and in the Fenno-Karelian Belt (2450–2400 Ma) (Bayanova et al., 2009).

Recently, the Baltic Shield has been defined as the PGE-bearing ESMLIP of plume nature (Bayanova et al., 2009), or the Baltic LIP with igneous rocks rich in Mg and Si (Bogatikov et al., 2010), or the Kola-Lapland-Karelian plume province (Smol'kin et al., 2009). These Early Palaeoproterozoic geological settings fill a substantial gap in understanding of geological events and Pd-Pt and Ni-Cu metallogeny of the Late Neoarchaean-Early Palaeoproterozoic transitional period in the Earth's evolution (2.7–2.2 Ga ago). In classic metallogenic summaries (Naldrett, 2003; Groves et al., 2005), this period is characterized by the Stillwater, Great Dike of Zimbabwe, Bushveld, and Sudbury ore-bearing complexes. However,

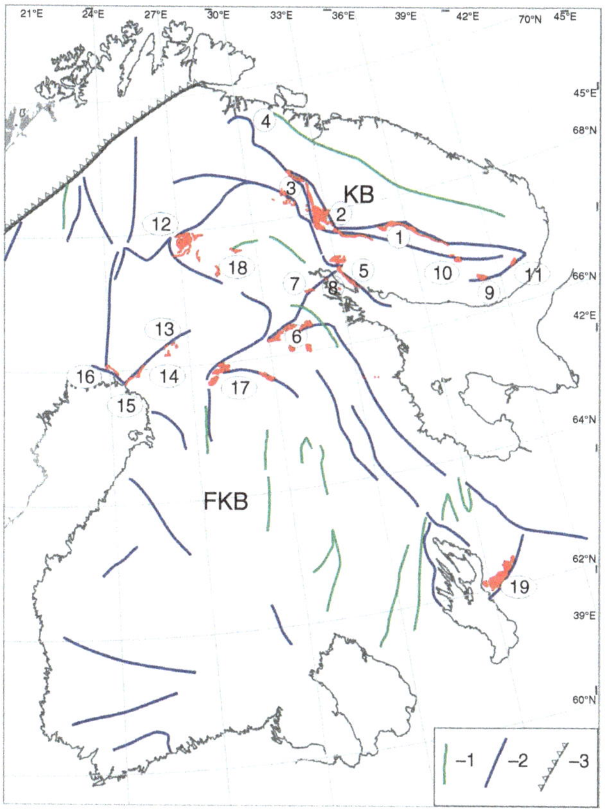

Figure 1.1 Rift belts and known Paleoproterozoic mafic complexes in northern ESMLIP. KB, Kola Belt; FKB, Fenno-Karelian Belt. 1 - Archean belts; 2 - Paleoproterozoic belts; 3 - Thrust caledonides. Main layered complexes (numerals in figure): 1, Fedorov Pana; 2, Monchepluton; 3, Monchetundra, Volchetundra, gabbro of the Main Range; 4, Mt. General'skaya; 5, Kandalaksha and Kolvitsa intrusions; 6, Lukkulaisvaara; 7, Kondozero massif; 8, Tolstik; 9, Ondomozero; 10, Pesochny; 11, Pyalochny; 12, Keivitsa; 13, Portimo Complex (Kontijarvi, Siikakama, Ahmavaara); 14, Penikat; 15, Kemi; 16, Tornio; 17, Koillismaa Complex; 18, Akanvaara; 19, Birakov-Aganozero massif. Hundreds of intrusive bodies are out of scale (Mitrofanov et al., 2013).

their geological setting cannot be coordinated in space and time with regional geological frameworks.

The Neoarchaean and Palaeoproterozoic deposits (2.7–2.5 and 2.0–1.9 Ga) host the world's main resources of Pd-Pt ores in layered intrusions (~60 kt). Neoarchaean komatiites, Mesoproterozoic, and Late Paleozoic deposits contain Ni ores (Groves et al., 2005). These epochs coincide with the existence of the thick (250–150 km) continental lithosphere, completion of collision, and ascent of superplumes that developed over more than 200 Ma (Condie, 2004). The structures that host low-sulfide Pd-Pt deposits were typically within-plate (Groves et al., 2005).

Thus, recent studies of global geodynamics and metallogeny emphasize the importance of the period in the Earth's evolution 2.7–2.2 Ga ago, when the Neoarchaean to Palaeoproterozoic plume tectonics gave way to plate tectonics. It is particularly evident for the Kaapvaal and East European cratons (Glikson, 2014).

1.3. ANALYTICAL PROCEDURES, ISOTOPE U-PB METHOD

1.3.1. U-Pb (TIMS) Method with $^{208}Pb/^{235}U$ Tracer

The method proposed by Krogh (Krogh, 1973) was used to dissolve samples in strong (48%) hydrofluoric acid at a temperature of 205–210 °C over 1–10 days. In order to dissolve fluorides, the samples were reacted with 3.1 N HCl at a temperature of 130 °C for 8–10 hours. To determine the isotope composition of lead and concentrations of Pb and U, a sample was divided into two aliquots in 3.1 N HCl, then a mixed $^{208}Pb/^{235}U$ tracer was added. Pb and U were separated on an AG 1 × 8, 200–400 mesh anion exchanger in Teflon columns. A laboratory blank for the whole analysis was < 0.1–0.08 ng for Pb and 0.01–0.04 ng for U. All isotope determinations for zircon and baddeleyite were made on Finnigan MAT-262 and MI 1201-T mass spectrometers. The Pb isotope composition was analyzed on a secondary-ion multiplier on a Finnigan MAT-262 in an ion-counting mode. Measurements of the Pb isotope composition are accurate to 0.025% (Finnigan MAT-262) and 0.15% (MI 1201-T) when calibrated against NBS SRM-981 and SRM-982 standards, respectively. U and Pb concentrations were measured in a single filament mode with H_3PO_4 and silica gel added. The method described in Scharer and Gower (1988) and Scharer et al. (1996) was used. Pb and U concentrations were measured in temperature ranges of 1350–1450°C and 1450–1550°C, respectively. Isotope ratios were corrected for mass discrimination during static processing of replicate analyses of the SRM 981 and SRM-982 standards (0.12 ± 0.04% for the Finnigan MAT-262 and 0.17 ± 0.05% per a.m.u.). Errors in the U-Pb ratios were calculated during the statistical treatment of replicate analyses of the IGFM-87 standard. They were assumed equal to 0.5% for Finnigan MAT-262 and 0.7% for MI 1201-T. Isochrons and sample points were calculated using the Squid and Isoplot programs (Ludwig, 1991, 1999). Age values were calculated with the conventional decay constants for U (Steiger & Jager, 1977). All errors are reported for a 2-sigma level. Corrections for common Pb were made according to Stacey and Kramers (1975). Corrections were also made for the composition of Pb separated from syngenetic plagioclase or microcline, if the admixture of common Pb was > 10% of the overall Pb concentration and the $^{206}Pb/^{204}Pb$ ratios were < 1000.

1.3.2. $^{205}Pb/^{235}U$ Tracer for Single Grains

The U-Pb (TIMS) method was based on the U-Pb method for single grain accessory minerals using ion-exchange chromatography (Corfu et al., 2011). Handpicked crystals are first treated in ultrasonic bath for cleaning in spirit or in acetone, and then in 7 N nitric acid. Then they are heated for about 15 minutes on a warm rangette and are finally flushed three times with recurrent purification water. Chemical mineral decomposition is performed in Teflon bombs, adding 3 to 5 mcl of mixed $^{205}Pb/^{235}U$ tracer in concentrated nitric acid during 5 to 7 days at a temperature of 210°C. The technique is provided in Krogh (1973). After the complete decomposition, the column effluent is evaporated on a warm rangette. Then 10 drops of 6.2 N chlorohydric acid are added. The sample is placed to the thermostat for 8 to 10 hours at a temperature of 140–150°C for homogenization. Pb and U are separated using ion-exchange chromatography in columns with Dowex IX8 200–400 mesh resin. Pb is eluted with 10 drops of 6.2 N chlorohydric acid, when one drop of 0.1 N phosphoric acid is added and the solution is evaporated on a rangette down to 3 mcl. U is eluted separately from Pb, when 20 drops of water with one drop of 0.1 N phosphoric acid are added. It is evaporated on a rangette down to 3 mcl. All chemical procedures are performed in an ultraclean block with blank Pb and U contamination of ca. 1–3 pg and ca. 10–15 pg, respectively. Pb and U isotope composition and concentrations are measured on Re bands at seven-channel mass-spectrometer Finnigan-MAT 262 (RPG), on collectors, with ^{204}Pb and ^{205}Pb measured at a temperature of 1350–1450°C in an ion-counting mode using a multiplier or quadrupole RPG accessory. Silica gel is used as an emitter. U concentrations are detected at a temperature of 1450–1550°C using a collector and a multiplier in a mixed statically dynamic mode. When U concentrations are negligible, a multiplier or quadrupole RPQ accessory is applied in a dynamic mode. All the measured isotope ratios are adjusted for the obtained mass-discrimination, when parallel analyses of SRM-981 are studied and SRM-982 standards are 0.12 ± 0.04%. Coordinates of points and isochrone parameters are calculated according to Ludwig (1991, 1999). Ages are calculated in accordance with the accepted values of U decay constants (Steiger & Jager, 1977), with errors indicated on a 2b level. The Stacey and Kramers model is used to adjust numbers for the admixture of common Pb (Stacey & Kramers, 1975).

1.3.3. Isotope Sm/Nd Method

In order to define concentrations of Sm and Nd, a sample was mixed with a compound tracer $^{149}Sm/^{150}Nd$ prior to dissolution. It was then diluted with a mixture of $HF + HNO_3$ (or + $HClO_4$) in Teflon sample bottles at a temperature of 100°C until complete dissolution. Sm and Nd were extracted by standard procedures with a two-stage ion-exchange and an extraction-chromatographic separation. An ion-exchange tar Dowex 50 × 8 in chromatographic columns employing 2.3 N and 4.5 N HCl was used as an eluent. The separated Sm and Nd fractions were transferred into nitrate form, whereupon the samples (preparations) were ready for mass-spectrometric analysis. Nd-isotope composition and Sm and Nd concentrations were measured by isotope dilution. A multicollector mass-spectrometer in a Finnigan MAT 262 (RPQ) was used in a static mode with Re + Re and Ta + Re filament. The reproducibility measured for ten parallel analyses of the Nd-isotope composition (standard La Jolla = 0.511833 ± 6) was <0.0024% (2σ). The same reproducibility was obtained from 11 parallel analyses of the Japanese standard: Ji Nd1 = 0.512078 ± 5. The error in $^{147}Sm/^{144}Nd$ ratios of 0.2% (2σ), the average of seven measures, was accepted for statistic calculations of the Sm and Nd concentrations using the BCR standard. Blanks for laboratory contamination for Nd and Sm were 0.3 and 0.06 ng, respectively. Isochron parameters were developed from programs made by Ludwig (Ludwig, 1991 and 1999). The reproducibility of measurements was ± 0.2% (2σ) for Sm/Nd ratios and ± 0.003% (2σ) for Nd-isotope analyses. All $^{147}Sm/^{144}Nd$ and $^{143}Nd/^{144}Nd$ ratios were normalized to $^{146}Nd/^{144}Nd = 0.7219$ and adjusted to $^{143}Nd/^{144}Nd = 0.511860$ using the La Jolla Nd standard. The εNd (T) values and model TDM ages were calculated based on the currently accepted parameters of CHUR (Jacobsen & Wasserburg, 1984): $^{143}Nd/^{144}Nd = 0.512638$ and $^{147}Sm/^{144}Nd = 0.1967$ and DM (Goldstein and Jacobsen, 1988): $^{143}Nd/^{144}Nd = 0.513151$ and $^{147}Sm/^{144}Nd = 0.2136$.

Sulfide minerals (pyrite, pentlandite, chalcopyrite, etc.) were chemically and analytically treated for the Sm-Nd study following a modified (Ekimova et al., 2011) conventional technique (Zhuravlyov et al., 1987). To decompose sulfides, a mineral weight (20 to 50 mg) is mixed with a $^{149}Sm/^{150}Nd$ tracer solution, treated with aqua regia ($HCl + HNO_3$) until complete decomposition and evaporated dry. Afterwards, it is converted to chlorides through evaporating the sample in 4.5–6 N HCl. After the fractional acid decomposition, the dry residue is dissolved in ~1ml 2.3 N HCl. Then REEs are separated from the solution via cation-exchange chromatography. A stepwise elution method is applied to 2.3 and 4.5 N HCl in a chromatographic column with cation-exchange resin Dowex 50 × 8 (200–400 mesh). The separated REE fraction is evaporated dry, dissolved in 0.1 N HCl and loaded to the second column with KEL-F solid ion-exchange resin HDEHP. The resin is used to separate Sm and Nd. The selected Sm and Nd fractions are evaporated to get prepared for further mass-spectrometric analysis. The Nd isotope composition and Sm and Nd concentrations were measured by an isotope dilution technique. A 7-channel solid-phase mass-spectrometer Finnigan-MAT 262 (RPQ) was used in a static double-band mode in collectors with Ta + Re filaments. Re filaments were used as ionizers. A sample was applied to a Ta filament with a diluted H_3PO_4 microdrop. The reproducibility error for 11 determinations of the Nd isotope composition of La Jolla = 0.511833 ± 6 (2σ, N = 11) was up to 0.0024% (2σ).

The same error was obtained for 44 parallel analyses of a new Japanese standard, JNdi1 = 0.512072 ± 2 (2σ, N = 44). The error in $^{147}Sm/^{144}Nd$ ratios is accepted for static calculations of the Sm and Nd concentrations in BCR-1 to be 0.2% (2σ). It is an average of 7 measurements. The blank intralaboratory contamination in Nd and Sm is 0.3 ng and 0.06 ng, respectively. The measured Nd isotope rations were normalized per $^{148}Nd/^{144}Nd = 0.241570$ and recalculated for $^{143}Nd/^{144}Nd$ in La Jolla = 0.511860 afterward. The isochron parameters were computed using programs of K. Ludwig (Ludwig, 1991, 1999). Decompositions constants are as per Steiger & Jager (1977). εNd parameters were calculated according to DePaolo (1981) for a one-stage model and according to Liew and Hofmann (1988) for a two-stage model.

1.3.4. Isotope Rb/Sr Method

The samples and minerals were all treated with double distilled acids (HCl, HF, and HNO_3) and H_2O distillate. A sample of 20–100 mg (depending on Rb and Sr contents) was dissolved with 4 ml of mixed HF and HNO_3 (5:1) in corked Teflon sample bottles and left at a temperature of about 200°C for one day. The solution was then divided into three aliquots in order to determine Rb and Sr isotope compositions and concentrations. These were measured by isotope dilution using separate ^{85}Rb and ^{84}Sr tracers. Rb and Sr extraction was performed by eluent chromatography with Dowex tar 50 × 8 (200–400 mesh). 1.5 N and 2.3 N HCl served as an eluent. Tar volumes in the columns were c. 7 and c. 4 cm^3. The separated Rb and Sr fractions were evaporated until dryness, followed by treatment with a few drops of HNO_3. Sr isotope compositions and Rb and Sr contents were measured by a MI-1201-T (Ukraine) mass spectrometer in the two-ribbon mode with Re filaments. The samples were deposited on the ribbons in the form of nitrate. Sr isotope composition in the measured samples was normalized to a value of 0.710235 recommended by NBS SRM-987. Errors on Sr isotope analysis (confidence interval of 95%) are up to 0.04%, and those of Rb-Sr ratio determination are 1.5%. Blank laboratory contamination for Rb is 2.5 ng and for Sr 1.2 ng. The adopted Rb decay constant of Steiger & Jager (1977) was used for age calculations.

1.3.5. Isotope Re/Os Method

The isotope analysis of sulfides has been provided in VSEGEI (Saint-Petersburg). The method of Re and Os chemical extraction described in Birck et al. (1997) has been applied. Samples of minerals with the weight of 50–200 mg were dissociated in a mixture of reagents (1 ml Br2 + 2 ml 7 N HNO3 + 0.5 ml 40% CrO3 in 7 N HNO3) in 5-ml Teflon Savillex vials under the temperature of 90 °C for 48 hours. After that Os was extracted using the microdistillation method. Re was extracted by liquid extraction with the isoamyl alcohol. The isotope dilution method with the mixed tracer $^{185}Re/^{190}Os$ was used to define the Re and Os concentrations and $^{187}Re/^{188}Os$ ratio. The tracer was added until samples were dissociated. Os as bromides was applied on a Pt filament with 0.2 ml of emitter $Ba(OH)_2 + NaOH$. The Os isotope composition was measured with the Triton (Thermo Scientific) solid-phase multicollector mass-spectrometer using ion source in a dynamic mode in negative ions. The inner standard of $^{187}Os/^{188}Os$ is 0.11997 ± 0.00001. The Element-2 (Thermo Scientific) mass-spectrometer with the inductively coupled plasma was used to measure the Re isotope composition. Re was measured from the solution of 3% HNO_3 using a multiplier in a dynamic mode. A quartz nebulizer, Ni cones, and a peristaltic pump were used. The measurements were carried with an average resolution. The Re standard of 10 mg/t was measured in the beginning and in the end of a session. The obtained value was averaged and the correcting factor of the mass deviation was estimated. The analysis accuracy is 0.5%. The measured standard $^{185}Re/^{187}Re$ ratios are within the range of 0.585–0.591, with the table standard of 0.5974 (Gramlich et al., 1973).

1.3.6. Study of Sulfide Mineral Texture and REE

Sulfides were studied using back-scattered electrons with a high-performance LEO 1450 scanning electron microscope. Analyses were carried to study possible inclusions in sulfides with considerable concentrations of REE that might have distorted results of Sm-Nd dating (Elizarova et al., 2009).

To define REE in samples with no preliminary separation and concentration, reference values of REE concentrations in the GSO 2463 standard (apatite), sulfide from the Talnakh deposit, and international standard samples of the Centre of Petrographic and Geochemical Research (Nancy, France) were reproduced using the ELAN 9000 DRC-e (Perkin Elmer, USA) quadrupole mass-spectrometer in ICTREMRM KSC RAS, Apatity. The samples were separated under the conditions provided in Elizarova and Bayanova (2012).

1.3.7. LA-ICP-MS of PGE in Sulfides

To analyze concentrations of Cr, Co, As, Se, Ru, Rh, Pd, Ag, Cd, Sb, Re, Os, Ir, Pt, Au, Tl, Pb, and Bi in sulfides, laser ablation (UP-213 laser) was used on a high-resolution Element-XR mass spectrometer with ionization in an inductively coupled plasma LA-ICP-MS. Measurement parameters were as follows: 40 μm crater diameter, 4 Hz impulse frequency of laser radiation. Samples were analyzed by blocks. They were prepared using the Element XR software with measurement of standard samples in the beginning and in the end of each block. Internal laboratory sulfide standards were used for analysis. The deviation defined from calibration standards is + −10 – 20%. Fe abundance was applied as an internal standard, since (a) its concentrations are high in relation to background values and (b) it occurs in all studied samples being the most homogeneously distributed in phases. The data were processed using the Glitter software (Jackson, 2001).

1.4. FEDOROVO-PANSKY COMPLEX: GEOLOGICAL SETTING

The Fedorovo-Pansky Layered Complex (Fig. 1.2) exposes over an area of >400 km^2. It strikes northwestward for >60 km and dips southwestward at an angle of 30°–35°. The total rock sequence is about 3–4 km thick. Tectonic faults divide the complex into several blocks. The major blocks from west to east (Fig. 1.2) are known as the Fedorov, Lastjavr, Western, and Eastern Pansky (Mitrofanov, 2005). The Fedorovo-Pansky Complex is bordered by the Archaean Keivy terrain and the Palaeoproterozoic Imandra-Varzuga rift. The rocks of the complex crop out close to the Archaean gneisses only in the NW extremities, but their contacts cannot be defined because of their poor exposure. In the north, the complex borders alkaline granites of the White Tundra intrusion. The alkaline granites were proven to be Archaean with a U-Pb zircon age of 2654 ± 15 Ma (Bayanova, 2004; Zozulya et al., 2005). The contact of the Western Pansky Block with the Imandra-Varzuga volcano-sedimentary sequence is mostly covered by Quaternary deposits. However, drilling and excavations to the south of Mt. Kamennik reveal a strongly sheared and metamorphosed contact between the intrusion and overlying Palaeoproterozoic volcano-sedimentary rocks that we consider to be tectonic.

The Fedorovo-Pansky Complex mostly comprises gabbronorites with varying proportions of mafic minerals and different structural features (Fig. 1.3). From the bottom up, the layered sequence is as follows:

• Marginal Zone (50–100 m) of plagioclase-amphibole schists with relicts of massive finegrained norite and gabbronorite, which are referred to as chilled margin rocks;

• Taxitic Zone (30–300 m), which contains an ore-bearing gabbronoritic matrix (2485 Ma) and early xenoliths of plagioclase-bearing pyroxenite and norite (2526–2516 Ma). Syngenetic and magmatic ores are represented by Cu and Ni sulfides with Pt, Pd, and Au,

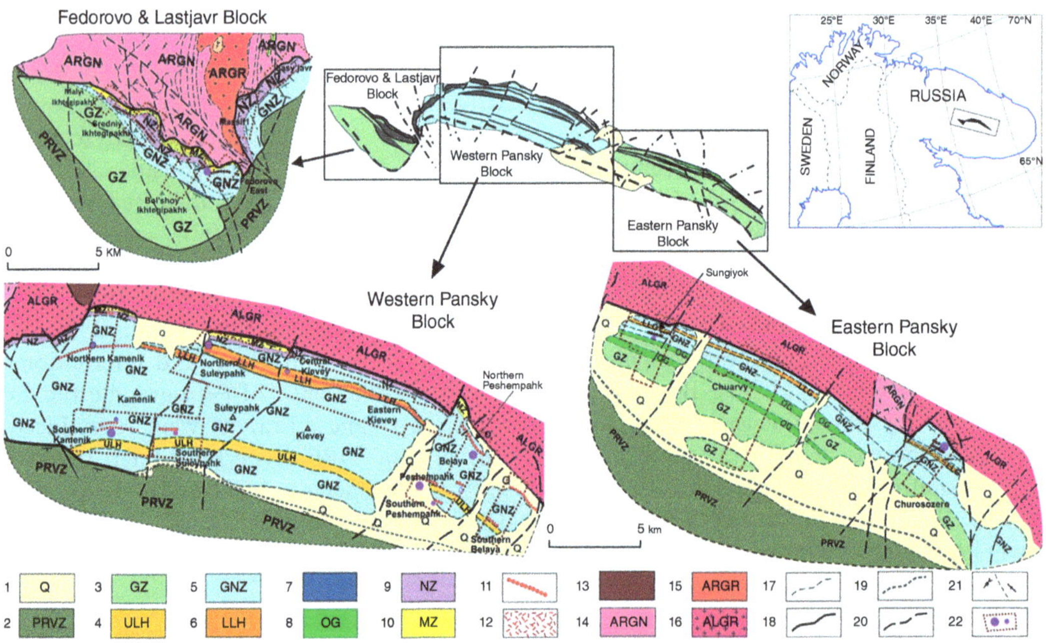

Figure 1.2 General geological map of the Fedorovo-Pansky Layered Complex. (1) Quaternary deposits; (2) The Imandra-Varzuga Proterozoic volcano-sedimentary complex; (3) Gabbro Zone; (4) Upper Layered Horizon; (5) Gabbronorite Zone; (6) Lower Layered Horizon; (7) Alternating gabbro, gabbrnorite and troctolite; (8) Olivine gabbro and gabbronorite horizons; (9) Norite Zone; (10) Marginal Zone – Mafic schists; (11) PGE reef-type mineralization; (12) PGE contact-type mineralization; (13) Early Proterozoic Tsaga gabbro-anorthosite massif; (14) Archaean Kola gneiss; (15) Archaean plagiomicrocline granite; (16) Archaean Keivy alkaline granite; (17) Boundaries between geological units; (18) Reliable boundaries between rock complexes with different ages; (19) Assumed boundaries between rock complexes with different ages; (20) Tectonic dislocations; (21) Schistosity, gneissic banding; (22) Boundaries of license areas with titles of deposits (large circles) or prospects (small circles) (Mitrofanov et al., 2005).

as well as Pt and Pd sulfides, bismuth-tellurides, and arsenides;

• Norite Zone (50–200 m) with cumulus interlayers of harzburgite and plagioclase-bearing pyroxenite that includes an intergranular injection Cu-Ni-PGE mineralization in the lower part. The rocks of the zone are enriched in chromium (up to 1000 ppm) and contain chromite. It is also typical of the rocks of the Penikat and Kemi intrusions (Finland) derived from the earliest magma portion (Iljina & Hanski, 2005). Basal Cu-Ni-PGE deposits of the Fedorov Block were explored and prepared for licensing (Schlissel et al., 2002; Mitrofanov et al., 2005).

• Main Gabbronorite Zone (c. 1000 m) is a thickly layered "stratified" rock series (Fig. 1.3) with a 40–80 m thinly layered lower horizon (LLH) at the upper part.

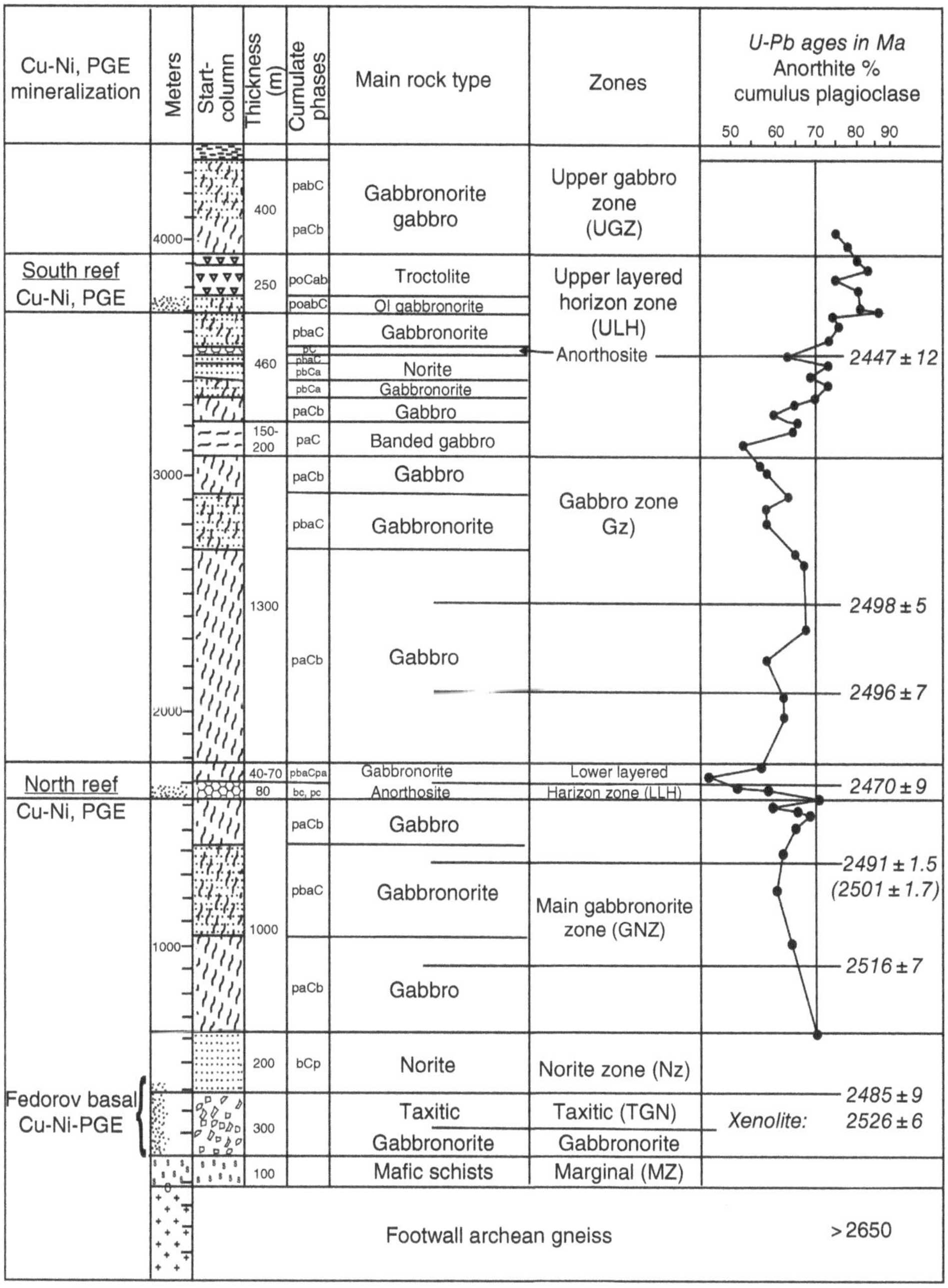

Figure 1.3 Composite "stratigraphic" section of the Fedorovo-Pansky Complex with Cu-Ni and PGE mineralization (modified after Schissel et al., 2002). The cumulate mineral terminology used in this paper is that of cumulate phase minerals in small letters, in order of volume percent, preceding the capital C for cumulate, with postcumulate mineral phases following. Major mineral abbreviations are a = augite; b = bronzite; c = chromite; o = olivine; p = plagioclase (see Table 1.5 for references). Modified after Schissel et al., 2002.

LLH consists of contrasting alteration of gabbronorite, norite, pyroxenite, and interlayers of leucocratic gabbro and anorthosite. LLH contains a reef-type PGE deposit poor in base-metal sulfides. According to field investigations (Latypov & Chistyakova 2000), LLH anorthositic layers intruded later, as shown by cutting injection contacts. It is confirmed by a zircon U-Pb age for the anorthosite of 2470 ± 9 Ma.

• Upper layered horizon (ULH) between the Lower and Upper Gabbro Zones. ULH consists of olivine-bearing troctolite, norite, gabbronorite, and anorthosite (Fig. 1.3). It comprises several layers of rich PGE (Pd S > Pt) ore poor in base-metal sulfides (Mitrofanov et al., 2005). The U-Pb age of the ULH rocks on zircon and baddeleyite is 2447 ± 12 Ma (see below). It is the youngest among those obtained for the Fedorovo-Pansky Complex rocks (Bayanova, 2006; Bayanova et al., 2017).

1.5. MONCHEPLUTON ORE COMPLEX: GEOLOGICAL SETTING

The NE Fennoscandian Shield hosts two large Palaeoproterozoic layered intrusions in its central part, that is, the Monchegorsk (Fig. 1.4) mafic-ultramafic pluton (Monchepluton, 55 km^2) and the substantially mafic Monchetundra massif (120 km^2). They compose the Monchegorsk Complex of layered intrusions (Sharkov, 2006; Sharkov & Chistyakov, 2014; Konnikov & Orsoev 1991; Grokhovskaya et al., 2012) and are incorporated into the Monchegorsk Cr-PGE-Cu-Ni ore district (Korovkin et al., 2003).

The ore potential of the Monchegorsk area is mainly provided by deposits of the Monchepluton. It is one of the most productive plutons among numerous Palaeoproterozoic layered intrusions of the Fennoscandian Shield. There is a series of ore deposits and occurrences related to the pluton in space and origin. Initially, the study of the Monchepluton was focused on complex Ni-Cu-PGE syn- and epigenetic ores. They have been an ore source for the Severonikel Plant for a long time. In the late twentieth century, the large Sopcheozero chrome deposit was discovered by Grokhovskaya et al. (2000) and explored by Chashchin et al. (1999). At the same time, Grokhovskaya et al. (2000), and Ivanchenko (2008, 2009) studied the Monchepluton (Vurechuayvench and Horizon 330 of the Sopcha massif), as well as its southern framing (South Sopcha). The study resulted in the discovery of low-sulfide Pt-Pd ores. It allows us to consider the Monchegorsk area as a large-scale source of Cr, Ni-Cu-PGE, and Pt-Pd ores.

The geochronological data on the Monchetundra massif are clustered into two groups. The first group comprises isotope results on medium-grained mesocratic gabbronorite from the middle part of the massif (2505–2501 Ma) (Mitrofanov & Smol'kin, 2004; Bayanova, 2004). The second group provides isotope results on coarse-grained leucogabbro and leucogabbronorite from the upper part of the massif (2476–2453 Ma) (Mitrofanov et al., 1993; Nerovich et al., 2009; Bayanova et al., 2010). The significantly different ages show that either the massif consists of several intrusive phases, or it was formed over a long time.

Despite numerous geochronological studies of the Monchepluton and Monchetundra, there is still a number of questions to be answered. The most important of these are (a) the age of Pt-Pd reefs and basal ores and (b) the source of the ore matter. The current paper provides a comprehensive study of these issues. Their solution is approached via direct timing of the PGE ore formation using Sm-Nd isotope analysis of sulfide minerals that compose Pt-Pd ores (Serov et al., 2014). We present new results of isotope geochronological U-Pb and Sm-Nd analyses of the low-sulfide PGE mineralization and the Monchepluton host rocks. The study focuses on the critical horizon at the Nyud-II deposit, Horizon 330 of Mt. Sopcha, Vurechuaivench deposit, and massifs from the southern part of the Monchetundra (South Sopcha deposit) and Lake Moroshkovoye.

1.6. MONCHEPLUTON AND ITS SOUTHERN FRAMEWORK

The Monchepluton is located in the central Kola Peninsula at the NW edge of the Palaeoproterozoic Imandra-Varzuga volcanic-sedimentary rift structure. Currently, the pluton is arc shaped and consists of two branches (chambers). The NW branch is more than 7 km in length and comprises the Nittis-Kumuzhya-Travyanaya (NKT) deposit. The nearly latitudinal branch is about 11 km in length and consists of the Sopcha-Nyud-Poaz and Vurechuayvench massifs (Fig. 1.4).

The pluton is differentiated in the vertical and horizontal directions, that is, the rocks become less basic from the bottom up and from west to east. Dunite, harzburgite, orthopyroxenites (NKT), orthopyroxenites (Sopcha), norites (Nyud), and gabbronorites (Poaz, Vurechuayvench) make up a common syngenetic series of rocks (Kozlov, 1973). In the upper part, a continuous orthopyroxenite body of the Sopcha massif is disturbed by Horizon 330. It is a sheetlike body (low-angle syncline), as thick as 1.2 to 14.8 m (3.5 m, on average), 3300 m in extent, and 1200 m wide (Fig. 1.4). Horizon 330 is considered to originate as an injection of an additional magma batch. It is more basic and has higher temperature than the initial melt in the magma chamber (Konnikov & Orsoev 1991; Mitrofanov & Smol'kin (eds) 2004; Sharkov & Chistyakov, 2014). The horizon is characterized by a rhythmic sequence of thin (10–130 cm) layers composed of dunites, harzburgites, olivine orthopyroxenites and feldspatic

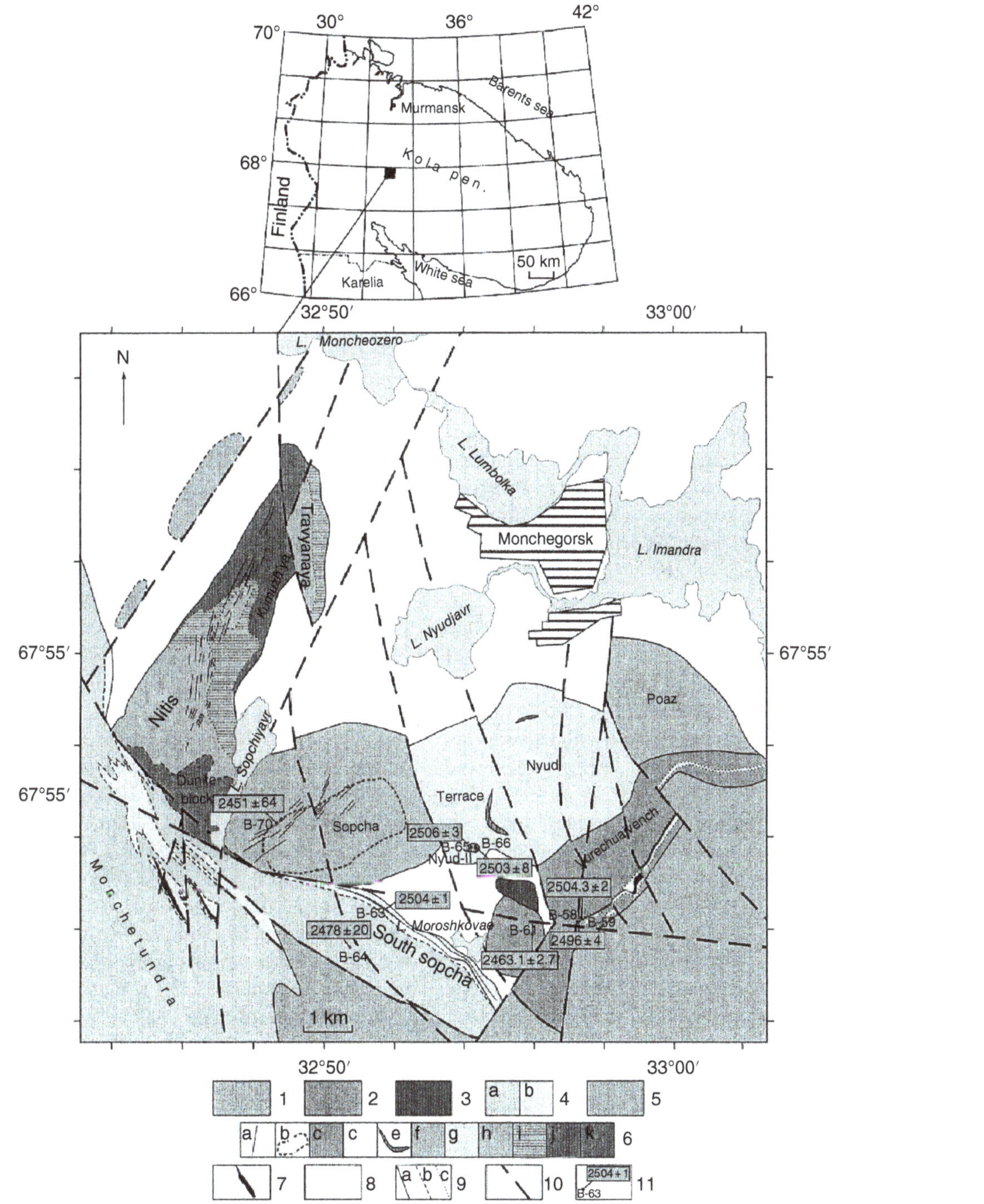

Figure 1.4 Schematic geological map of Nyud massif and section along line I-I. (1) orthopyroxenites from Sopcha massif; (2–5) Nyud massif: (2) irregular alternation of microgabbronorites, micronorites, meso- and melanocratic norites, and plagiopyroxenites (critical horizon); (3) leuco- and mesocratic norites; (4) melanocratic olivine norites; (5) melanocratic norites with locally occurring plagiopyroxenites; (6) mesocratic gabbronorites of Poaz massif; (7) meso- and leucocratic metagabbronorites of Vurechuaivench massif; (8) quartz metagabbro of 10 anomaly massif; (9) Archean quartz diorites and gneissic diorites; (10) faults; (11) geological boundaries: (a) reliable, (b) inferred, and (c) facies; (12) location of geochronological samples and their numbers (after Chaschin et al., 2016).

orthopyroxenites (Kozlov, 1973; Konnikov & Orsoev 1991; Mitrofanov & Smol'kin, 2004). The layering is disturbed by bends and folds formed as a result of melt flow.

The critical horizon occurs in the middle of the Nyud massif section and consists of two parts called Terrace and Nyud-II. The Terrace critical horizon is up to 50 m thick (Fig. 1.4). It is composed of irregularly alternating meso- and melanocratic norites, plagioclase-bearing orthopyroxenites, gabbronorites, harzburgites, microgabbro, and microgabbronorites. The Nyud-II critical horizon is a stocklike body as big as 160 × 70 m. It has a convex bottom and a vertical thickness of about 50 m (Fig. 1.4). Melanocratic poikilitic norite is dominant in this critical horizon, along with mesocratic norites interlayers in the upper part and plagioharzburgites, olivine norites, plagioclase orthopyroxenites interlayers, minor bodies of pegmatoid leucocratic norites in the lower part. There are also isometric bodies of heterogeneous composition and structure. They are composed of fine-grained norites, gabbronorites, and hornfels among melanocratic and olivine norites (Bartenev & Dokuchaeva, 1975).

There are two concepts of the critical horizon origin: (a) it marks a roof of the earlier magma chamber overlain by a later chamber filled with norite-gabbronorites (Kozlov, 1973; Mitrofanov & Smol'kin, 2004), and (b) the critical horizon is an additional intrusive phase of the Monchepluton (Sharkov, 1982; Sharkov & Chistyakov, 2014).

Metagabbronorites and anorthosites of the Vurechuayvench massif occur in the SE Monchepluton at its contact with volcanic rocks of the Imandra-Varzuga riftogenic structure (Fig. 1.4). The massif occurs to the northeast of the Nyud-Poaz massifs and composes their section. Thus, it is the uppermost part of the whole Monchegorsk pluton section (Mitrofanov & Smol'kin, 2004). The Vurechuayvench massif is 1.5–2.0 km wide and 600–700 km thick. It stretches northeastward for 8 km. The massif is not inscribed into the general synformal structure of the Monchepluton, but dips to the southeast at angles of 5°–10° to 20°–30° beneath volcanic rocks of the Imandra-Varzuga structure. They overlie the massif with a 10 m-thick basal bed of residual conglobreccia (Gorbunov,1982). The section of the Vurechuayvench massif is represented by the following rock varieties (from the bottom up): bottom gabbronorites, 5–10 m thick, foliated and brecciated in the contact zone; continuous melano- and mesocratic norites (400–650 m); mesocratic metagabbronorites (300 m) with several metaplagioclasites horizons. The thickness of the latter varies from 10–15 to 40–50 m. This is a light grey rock with large spots containing up to 90–95 vol% of saussuritized and pelitized plagioclase with insignificant amounts of quartz and amphibole. The low-sulfide Pt-Pd ore deposit is spatially and genetically related to a metaplagioclasites horizon (Grokhovskaya et al., 2000).

Metagabbro intrusions of Anomaly 10, Lake Moroshkovoye and South Sopcha are situated in the southern extremity of the Monchepluton (Fig. 1.4). Anomaly 10 is an oval-shaped sheetlike metagabbro massif as big as 300 × 700 m in plan (Fig. 1.4). It is elongated in the latitudinal direction and sheetlike in section. The sheet dips to the northeast at an angle of 45° (Kozlov, 1973) and is composed of amphibolized fine- to coarse-grained leuco- and melanocratic quartz gabbro. Low-sulfide PGE and the associated oxide-sulfide mineralization occur at high levels of the massif near the contact with country diorite. At the bottom, there is a 500 m-long and 2 m-thick stratiform body.

The Lake Moroshkovoye massif occurs to the south of the Nyud massif and adjoins the SW flank of the Vurechuayvench massif. In the north, it contacts with metagabbro of the Anomaly 10 massif. In the west, SW and NW, the Lake Moroshkovoye massif cuts Archaean gneissic diorites (Fig. 1.4). The massif mainly consists of leuco- to mesocratic metanorites that give way to melanocratic metanorite and metagabbronorite in the marginal zone. The SW tectonic contact separates the massif from country gneissic diorites. The border is marked by a zone of shearing and foliation as thick as 5–10 to 35 m, gently dipping to the NE. It is represented by actinolite and actinolite-chlorite schists developed after gabbronorites. The quartz-chlorite schists after diorites contain sulfide mineralization and PGM.

The South Sopcha massif is about 5 km long and up to 1.5 km wide. It is oriented in the NW direction (Fig. 1.4). In the north, the massif borders a wide near-latitudinal tectonic zone of the Sopcha orthopyroxenites. In the NE, it contacts with Archaean gneissic diorites. In the south and SW, the South Sopcha massif is overlain by felsic metavolcanics of the Arvarench Formation in the Imandra-Varzuga structure along the tectonized intrusive contact. It is represented by fine-grained gabbro-amphibolites, up to 200 m thick. The age of the Arvarench Formation is 2429 ± 6.6 Ma (Vrevsky, 2011). In the NW, the South Sopcha massif passes a fault zone to the Monchetundra massif (Fig. 1.4). The South Sopcha massif has a monoclinal structure in section, dipping to the SW at angles of 5°–20° to 45°. It is affected by a large tectonic zone of the Monchetundra Fault, that is, the rocks are intensely foliated and altered.

The internal structure of the South Sopcha massif consists of a lower norite-orthopyroxenite zone and an upper gabbroic zone. The lower zone is 250–300 m thick. It is represented by an irregular alternation of metanorites and metapyroxenites interlayers, 1–20 m in thickness, with schlieren and bodies of pegmatoid rock varieties, irregular in shape, with a subordinate amount of metaperidotites.

Fresh rocks are extremely rare. As a rule, they are intensely amphibolized and saussuritized. The lower zone rocks host sulfide disseminations and pocketlike segregations with the low-sulfide PGE mineralization.

The upper zone of the South Sopcha massif is composed of leuco- to mesocratic coarse-grained mottled metagabbro and metagabbronorites. They are characterized by the constant occurrence of accessory titanomagnetite. These metagabbroic rocks are chemically close to those in the upper zone of Monchetundra massif (Grokhovskaya, 2012), but differ in their high-grade metamorphism and intense foliation, probably due to their thinning and localization in the tectonically active zone. The contact between the rocks of the lower and upper zones is mostly foliated and tectonized. Chlorite-actinolite schist interlayers occur in the lower zone. At the same time, there are sporadic bodies of magmatic breccia with fragments of metanorites and metapyroxenites from the lower zone and cement comparable to the metagabbroic rocks of the upper zone (Rundkvist et al., 2011). These relationships indicate that rocks various in composition are probably related to separate intrusive phases.

1.7. LOW-SULFIDE PGE DEPOSITS AND OCCURRENCES IN THE MONCHEGORSK ORE AREA

The low-sulfide Pt-Pd deposits and occurrences have been recently discovered throughout the Monchegorsk ore area (Chashchin et al., 2016). They are new for the Kola region and divided into two structural types: (1) stratiform reefs conformable to layering in massifs and (2) basal type bodies localized in marginal zones of intrusions. The first type is represented by the Vurechuayvench deposit, Horizon 330, and probably the critical horizon at the Nyud deposit. The second type is represented by the South Sopcha and the Lake Moroshkovoye deposits.

1.7.1. Vurechuayvench Deposit

The Vurechuayvench deposit is a low-sulfide Pt-Pd deposit of the reef type (Grokhovskaya et al., 2000). It is clearly stratiform and related to the anorthosite horizon. The ~2 km-long ore zone consists of several sheetlike and lenticular ore bodies up to 3 m thick and up to 300–500 m long. They are conformable to the massif layering and gently dip to the SE at angles of 2°–5° to 10°–15° (Grokhovskaya et al., 2000). The ore bodies have no distinct borders. Their boundaries are established only by sampling results. The PGE mineralization is closely associated with sulfide disseminations. They develop nonuniformly from 1–2 mm-big sporadic segregations with sulfide contents of about 1 vol% to 1–5 mm-big pockets (2–3 vol%) and sulfide schlieren (5–10 vol%). Sulfides are mainly represented by chalcopyrite (40–90 vol%) and millerite (10–50 vol%) with subordinate amounts of covellite, chalcocite, pentlandite, pyrrhotite, and pyrite. There are nickel sulfoarsenides (gersdorffite) and cobalt sulfoarsenides (cobaltite) as well. PGM are represented by bismuthotellurides (kotulskite, merenskyite, michenerite), arsenides (sperrylite, guanglinite, majakite, etc.) and sulforasenides (hollingworthite, irarsite, platarsite) with dominating Pd minerals. The metal grade in the ore is 1–7 ppm total PGE at Pd/Pt = 3–5; 0.1–0.4 wt% Ni and 0.1–0.5 wt% Cu (Grokhovskaya et al., 2000).

1.7.2. Horizon 330 of Sopcha

The low-sulfide PGE mineralization of Horizon 330 is traced over its entire extent and occurs as separate interlayers. These are 10 cm to 1.5 m thick and closely related to sulfide disseminations. Fine sulfide disseminations (2–3 vol%) occur in the zone of intercalating harzburgites and orthopyroxenites. Their amount is up to 10 vol% in the orthopyroxenite zone. The disseminations are syngenetic and have no reaction relationships with primary silicates. At the same time, there is a distinct resorption of sulfides with late minerals (serpentine, chlorite, carbonate, and pyrite) (Neradovsky et al., 2002). Sulfide mineralization in harzburgites consists of pyrite, millerite, chalcopyrite, and pentlandite. In olivine pyroxenites and orthopyroxenites, it is represented by pyrrhotite, pentlandite, and chalcopyrite. Merenskyite, Pd-Pb, and Pd-Rh-Cu compounds are identified among PGM (Neradovsky et al., 2002; Mitrofanov & Smol'kin, 2004). In addition, Pd occurs as an admixture in pyrrhotite and chalcocite and Ir in pentlandite. The metal grades in the ore are as follows: 0.10–0.77 wt% Ni, 0.02–0.35 wt% Cu, up to 0.25 ppm Pt, and 1.6 ppm Pd at Pd/Pt = 4. The high Rh content (up to 0.1 ppm) is noted (Mitrofanov & Smol'kin, 2004).

1.7.3. Critical Horizon of Nyud

There are two horizons disseminated mineralization. The upper horizon is 5–30 m (up to 65 m) thick. It occupies an area of 700 × 300 m in hanging wall of olivine norites under the critical horizon represented by disseminated and less frequent stringer-disseminated mineralization and pockets of Cu and Ni sulfides. Pyrrhotite, pentlandite, and chalcopyrite dominate in the ore. Magnetite and ilmenite also occur. A segregation of the massive sulfide ore is mined out. It had a shape of a flattened cake, 6.75 m long, 3.5 m wide and 2 m thick. This ore body was composed of pyrrhotite (60–80 vol%), pentlandite (5–20 vol%), chalcopyrite (3–10%), and a great amount of fused silicate xenoliths. The highest Ni and Cu contents were 3.24 and 0.56 wt%, respectively.

The lower horizon occurs in the footwall of olivine norites at the contact with poikilitic norites. It is smaller and its thickness reaches 18.7 m. Fine- and stringer-disseminated ores with small pockets contain 0.2–0.3 wt% of Ni.

The Nyud-II deposit occurred 0.6 km to the SW of the Terrace deposit, hosted in melanocratic norites of the critical horizon. It was mined out in the early 1970s (Fig. 1.4). The sulfide Ni-Cu mineralization has a complex internal structure and comprises veinlet-schlieren, veinlet-disseminated and disseminated types. The veinlet-schlieren mineralization is economically best-valued. The schlieren are sulfide segregations, isometric in shape, and varying in size from a few decimeters to 5–7 m across. They occur at contacts of melanocratic and olivine norites with fine-grained norites and gabbronorites. The schlieren boundaries are both sharp and gradual due to surrounding microveinlets and disseminations. They frequently contain fused fragments of host norite and gabbronorite. The veinlet-disseminated type of mineralization is minor and mainly occurs at margins of schlieren. The disseminated mineralization is widespread as irregularly shaped ore bodies. They are tens of meters across and occur in various rocks (Bartenev & Dokuchaeva, 1975).

Sulfides are represented by pyrrhotite (40–50 vol%), chalcopyrite (20–30 vol%), pentlandite (10–15 vol%), and pyrite (5–10 vol%). There is magnetite as well (10–30 vol%). Mean PGE concentrations are 0.25 ppm Pt and 0.70 ppm Al; Pd/Pt = 2.8.

1.7.4. South Sopcha Deposit

The PGE mineralization is localized in various rocks from the lower marginal norite-pyroxenite zone of the South Sopcha deposit with fine (1–3 vol%) sulfide disseminations (Fig. 1.4). Structures of different ore zones within the deposit are markedly distinct. In the NW part, the ore zone consists of twenty 1–20 m-thick lenticular-stratal ore bodies. They occur throughout the lower zone section and become as thick as 50–60 m in total. In the SE part, the ore bodies are confined to the upper and middle parts of the lower zone and their number is reduced to 10. Their total thickness increases to 55–85 m, while the thickness of separate ore bodies varies from 1 to 65 m in bulges.

Three ore mineral assemblages are distinguished in the mineralized bodies: those with predominance of (1) pyrrhotite, (2) chalcopyrite and Ni-sulfides (violarite, polydymite, millerite, and pentlandite), and (3) sulfide disseminations spatially associated with titanomagnetite. The proportions of the sulfide amount vary widely. Pyrrhotite and pentlandite are frequently replaced with low-temperature marcasite, melnikovite, violarite, and pyrite, whereas chalcopyrite is replaced with chalcocite and covellite. Chalcopyrite and bornite lamellae are typical. Sulfides occur as disseminations and segregations of millerite-bornite-chalcopyrite and pentlandite-chalcopyrite-pyrrhotite assemblages. Their high contents (up to 5–10 vol%) are noted in pegmatoid norites and pyroxenites only. Here, the ore has high PGE contents (up to 0.5–0.9 ppm Pt + Pd). Minerals of the cobaltite-gersdorffite series with PGE admixtures frequently occur at the contact between the lower and upper zones of this massif (Grokhovskaya et al., 2012).

The PGE mineralization is represented by more than 20 mineral species. Palladium bismuthotellurides and arsenides are predominant. Merenskyite is the most abundant. Sperrylite occurs frequently. Sulfides of the braggite-cooperite-vysotskite series and other minerals are less abundant. The PGE grade of ores does not exceed 1–2 ppm with Pd/Pt = 3–4 (Grokhovskaya et al., 2012).

1.7.5. Lake Moroshkovoye Ore Occurrence

The ore body of this occurrence relates to the NW-trending thick tectonic zone in the western part of the massif at the contact of metagabbronorite with Archaean country diorites. The ore body is about 250 m long and up to 6 m thick. It is conformable to the foliation of tectonites, strikes in the NW direction, and dips to the NE at angles of 30°–70°. It is a combination of a veinlet, lenticular, and disseminated mineralization. Thin veinlets and lenses of massive sulfides consist of pyrrhotite-pyrite-chalcopyrite-pentlandite intergrowths. They are oriented conformably to foliation and occasionally contain host schist fragments. The disseminated mineralization is similar in composition and mostly clustered near lenses and veinlets of massive sulfides with sharp boundaries. It is also conformable to schistosity and emphasizes banded structure of the ore. Mean grades of the ore are 2.0 wt% Ni and 0.6 wt% Cu. The total PGE content reaches 1.85 ppm.

1.8. PETROGRAPHY OF SAMPLES

Eight samples have been taken for isotope analyses from the Nyud, Sopcha, Vurechuayvench, South Sopcha, and Lake Moroshkovoye massifs (Fig. 1.4). Two samples have been taken from of the Nyud-II critical horizon (Fig. 1.3). Sample B-65, weighing 68 kg, is composed of fine- to medium-grained olivine orthopyroxenites consisting of orthopyroxene (85–90 vol%), olivine (5 vol%), and plagioclase (1–2 vol%). Secondary minerals are represented by colorless amphibole (5 vol%), which replaces orthopyroxene; phlogopite and sulfides occur as sporadic grains. Sample B-66, weighing 62 kg, has been taken from mineralized medium- to fine-grained meso- to leucocratic taxitic norites (10–40 vol% orthopyroxene, 60–80 vol%

plagioclase, 1–2 vol% quartz). Secondary minerals are represented by colorless and pale green amphibole (2–3 vol%). It develops after orthopyroxene in combination with sulfides (1–3 vol%) and rare grains of accessory apatite.

Sample B-70, weighing 64 kg, has been taken from medium- to fine-grained harzburgites of Horizon 330 in the Sopcha massif (Fig. 1.4). It consists of olivine (65–70 wt%), orthopyroxene (20 vol%), secondary serpentine (5 vol%) replacing olivine, and colorless amphibole (5 vol%) after orthopyroxene and less frequent olivine, magnetite (up to 1 vol%), and sulfides (2–3 vol%).

Two geochronological samples have been taken from the Vurechuayvench massif. Sample B-58, weighing 67 kg, has been taken from fine-grained metaplagioclasite of the PGE-bearing reef (Fig. 1.1). The rock consists of intensely saussuritized (up to 60–70% clinozoisite and chlorite) and pelitized plagioclase (25–30 vol%) and quartz in interstices between plagioclase grains (up to 5 vol%). Amphibole, apatite, scapolite, and muscovite grains are rare. Ore minerals are represented by sulfides (up to 2 vol%). Sample B-59, weighing 62 kg, has been taken from medium-grained leucocratic metagabbronorites underlying PGE-bearing reef (Fig. 1.4). The sample contains (vol%): plagioclase (55–60), colorless amphibole (30), quartz (1–2), chlorite (10) after amphibole, and plagioclase and clinozoisite (2–3) after plagioclase.

Two samples have been taken from the South Sopcha massif. Sample B-63, weighing 44 kg, has been taken from fine-grained leucocratic metanorites of the lower PGE-bearing zone of the massif (Fig. 1.4). The sample contains (vol%): plagioclase (60–65), pale green amphibole (25–30), quartz (1–2), biotite (2–3), and chlorite (2–3) after amphibole and sulfides (2–3). Sample B-4, weighing 60 kg, has been taken from medium-grained mesocratic epidotized quartz-bearing metagabbro (Fig. 1.4). The sample contains (vol%): blue-green amphibole (50–60%), plagioclase (20%), epidote (15%), and quartz (5–10%). Ore minerals are represented by magnetite (2–3) and sporadic sulfide grains.

Sample B-61, weighing 65 kg, has been taken from medium-grained meso- to leucocratic metanorites of the Lake Moroshkovoye massif (Fig. 1.4). The sample contains (vol%): plagioclase (55–60%), orthopyroxene completely replaced with talc (30–40%), pale green amphibole (5%), and quartz (1–2%). Plagioclase is replaced with clinozoisite (2–3%) and amphibole with chlorite (1–2%). Ore minerals are represented by sporadic sulfide grains.

1.9. MONCHEGORSK ORE AREA: ISOTOPE U-PB DATA (ON SINGLE ZIRCON-BADDELEYITE)

The results are provided in Tables 1.1 and 1.2 and Fig. 1.5. Ten mg of zircon grains reflecting three morphotypes have been separated from olivine-bearing orthopyroxenite of the critical horizon in the Nyud-II deposit (sample B-65) (Table 1.2). The first variety is represented by crystal fragments with corroded surface 175 × 175 μm in size. The transparent grains are colored brown. No intraphase heterogeneity has been revealed in BSE images. The procedure of two-stage dissolution with separation of two portions has been applied to these zircons. The second zircon variety is characterized by isometric crystal fragments with a corroded surface 245 × 245 μm in size. The transparent grains are light lilac in color with slightly expressed zoning in BSE images. The near-concordant U-Pb age of these zircons is 2506 ± 3 Ma (Table 1.1). It is interpreted as the time of the orthopyroxenite crystallization in the critical horizon. The lower intersection of discordia with concordia is at the origin. Since the U-Pb system in zircon is not disturbed, this intersection can be considered to mark contemporary loss of Pb. The third zircon variety is crystal fragments with a corroded surface 175 × 175 μm in size. Transparent grains are light yellow in color, with poorly expressed zoning in BSE images. Their concordant age, corresponding to 2670 ± 4 Ma (Table 1.1), characterizes the xenocrystic origin

Table 1.1 U-Pb Zircon (Zr) and Baddeleyite (Bd) Ages of Rocks from Monchegorsk Pluton.

Massif	Rock	Age, Ma	Mineral	Source
NKT	Quartz norite	2507 ± 9	Zr	Mitrofanov & Smol'kin (2004); Bayanova (2004)
Nyud	Gabbro pegmatite	2504.4 ± 1.5	Zr	Amelin et al. (1995)
	Gabbro pegmatite	2500 ± 5	Zr, bad	Mitrofanov & Smol'kin (2004)
	Norite	2493 ± 7	Zr	Balashov et al. (1993)
Nyud-II	Orthopiroxenite	2506 ± 3	Zr	Chashchin et al. (2016)
	Ore norite	2503 ± 8	Zr	
Vurechuaivench	Metagabbronorite	2497 ± 21	Zr, bad	Mitrofanov & Smol'kin (2004)
	Metagabbronorite	2498.2 ± 6.7	Bad	Rundkvist et al. (2014)
	Metagabbronorite	2504.2 ± 8.4	Zr	
	Metaplagioclasite	2507.9 ± 6.6	Zr	
	Metagabbronorite	2504.3 ± 2.2	Zr	Chashchin et al. (2016)
	Ore plagioclasite	2494 ± 4	Zr	Chashchin et al. (2016)

Table 1.2 Isotopic U-Pb Data on Single Zircon Grains from Rocks of Monchegorsk Pluton and Massifs in its Southern Framing.

No.	Weight, mg	Concentration, ppm		Isotope ratios*				Isotope ratios and age, Ma**			Dis., %
		Pb	U	$^{206}Pb/^{204}Pb$	$^{206}Pb/^{238}U \pm 2\sigma$	$^{207}Pb/^{235}U \pm 2\sigma$	$^{207}Pb/^{206}Pb \pm 2\sigma$	$^{206}Pb/^{238}U \pm 2\sigma$	$^{207}Pb/^{235}U \pm 2\sigma$	$^{207}Pb/^{206}Pb \pm 2\sigma$	
Metagabbronorite from Vurechuaivench massif (sample B-59)											
1	0.0200	175.01	240.26	426.9	0.461 ± 0.003	10.465 ± 0.060	0.1649 ± 0.0002	2443 ± 14	2477 ± 14	2504 ± 3	2.4
2	0.0875	62.09	105.09	538.6	0.386 ± 0.001	8.773 ± 0.030	0.1527 ± 0.0003	2106 ± 6	2315 ± 8	2504 ± 4	15.9
3	0.0720	160.42	184.05	339.7	0.340 ± 0.002	7.738 ± 0.083	0.1418 ± 0.0013	1888 ± 12	2201 ± 24	2507 ± 20	24.7
4	0.0880	961.27	754.65	136.1	0.299 ± 0.002	4.258 ± 0.080	0.1009 ± 0.0017	1685 ± 11	1635 ± 31	1612 ± 27	−4.5
Mineralized metanorite from South Sopcha massif (sample B-63)											
1	0.0043	29.36	17.71	541.2	0.477 ± 0.063	10.848 ± 1.508	0.1695 ± 0.0062	2504 ± 331	2505 ± 348	2508 ± 92	0.2
2	0.0114	133.33	308.70	578.4	0.392 ± 0.002	7.149 ± 0.046	0.1323 ± 0.0002	2132 ± 13	2130 ± 14	2129 ± 4	−0.1
Metagabbro from South Sopcha massif (sample B-64)											
1	0.0984	40.78	40.28	60.9	0.420 ± 0.002	9.196 ± 0.184	0.1545 ± 0.0026	2066 ± 21	2237 ± 24	2358 ± 41	15.9
2	0.0700	129.13	201.22	207.5	0.378 ± 0.004	8.149 ± 0.088	0.1526 ± 0.0004	1984 ± 11	2174 ± 4	2396 ± 7	13.8
3	0.2000	67.72	134.61	489.2	0.336 ± 0.005	7.078 ± 0.105	0.1511 ± 0.0006	1869 ± 26	2121 ± 31	2376 ± 10	21.3
Metanorite from massif of Lake Moroshkovoe (sample B-61)											
1	0.0800	67.60	70.85	1352.7	0.436 ± 0.003	9.663 ± 0.063	0.1646 ± 0.0003	2287 ± 14	2403 ± 16	2503 ± 4	8.6
2	0.0212	60.48	71.14	325.5	0.380 ± 0.003	8.408 ± 0.069	0.1626 ± 0.0005	2074 ± 16	2287 ± 19	2483 ± 7	16.5
3	0.0200	2.38	12.06	144.8	0.060 ± 0.003	7.361 ± 0.066	0.1612 ± 0.0038	377 ± 16	872 ± 43	2497 ± 57	84.9

*All ratios are corrected to blank contamination (0.08 ng Pb, 0.04 ng U) and to mass discrimination 0.12 ± 0.04%.

**Correction to common lead has been determined by age according to model of Stacey and Kramers (1975).

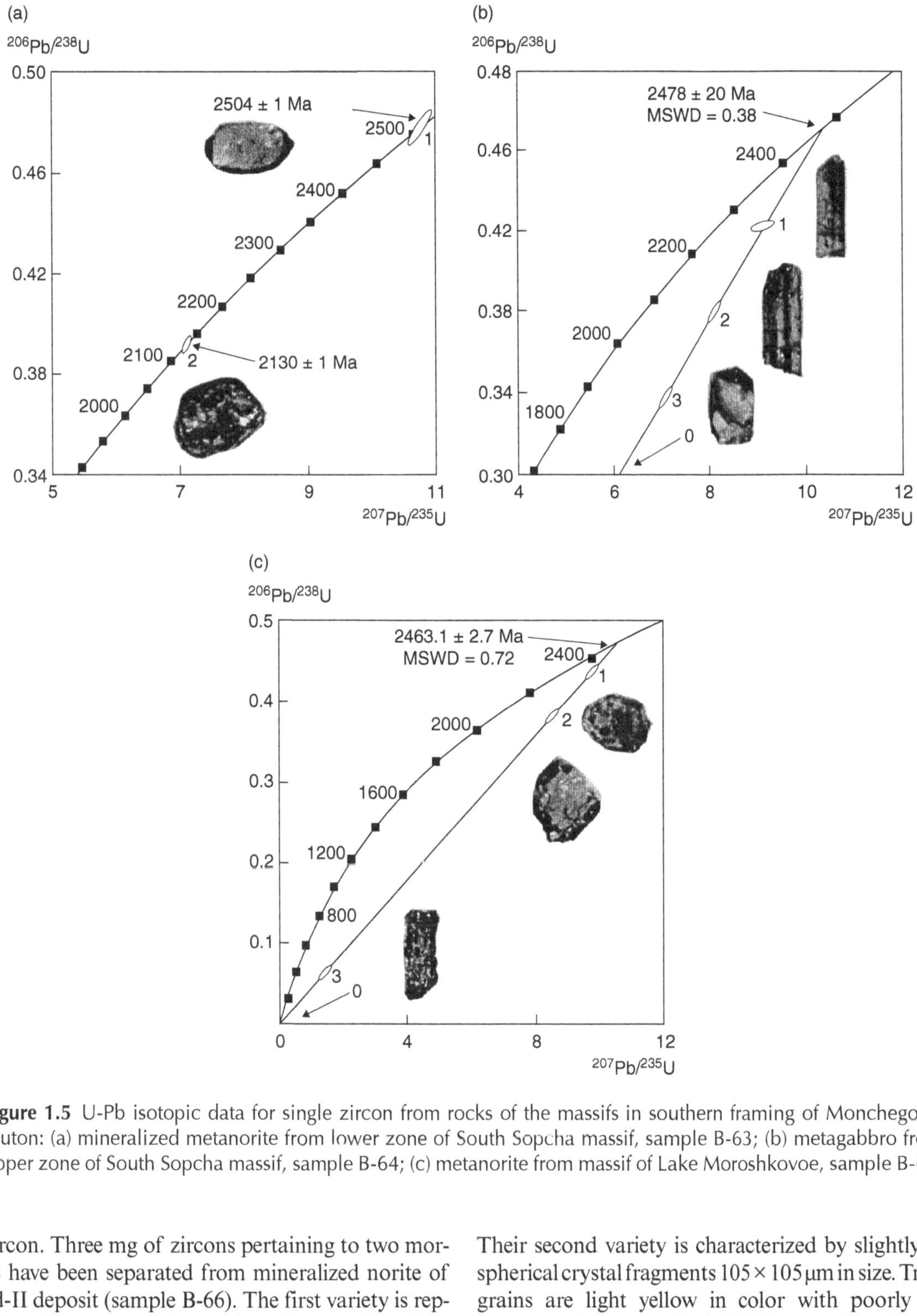

Figure 1.5 U-Pb isotopic data for single zircon from rocks of the massifs in southern framing of Monchegorsk pluton: (a) mineralized metanorite from lower zone of South Sopcha massif, sample B-63; (b) metagabbro from upper zone of South Sopcha massif, sample B-64; (c) metanorite from massif of Lake Moroshkovoe, sample B-61.

of this zircon. Three mg of zircons pertaining to two morphotypes have been separated from mineralized norite of the Nyud-II deposit (sample B-66). The first variety is represented by fragments of long-prismatic crystals with a slightly corroded surface 245 × 105 μm in size; the elongation coefficient is 2.3. The transparent grains are brown in color with distinct intraphase zoning expressed in BSE images. The technique of two-stage dissolution with separation of two portions has been applied to these zircons. Their second variety is characterized by slightly corroded spherical crystal fragments 105 × 105 μm in size. Transparent grains are light yellow in color with poorly expressed intraphase heterogeneity in BSE images. The discordia constructed on the basis of three data points intersects concordia at 2503 ± 8 Ma (Table 1.1). This upper intersection corresponds to the time of the ore-bearing norite crystallization. The lower intersection at zero reflects a contemporary loss of Pb.

From ore-bearing metaplagioclasites (PGE-bearing reef at the Vurechuayvench deposit, sample B-58), 10 mg of zircons reflecting the three morphotypes (Table 1.2) have been separated. The first variety is represented by round crystal fragments with a slightly corroded surface 175 × 140 μm in size; their elongation coefficient is 1.25. The translucent grains are milky in color. The second variety is fragments of long-prismatic crystals with a corroded surface 350 × 140 μm in size; the elongation coefficient is 2.5. The translucent grains are milky in color. The third variety is characterized by round crystal fragments 140 × 140 μm in size. The translucent grains are light yellow in color.

The discordia constructed on the basis of three data points intersects the concordia at 2496 ± 4 Ma (Table 1.1). This age is interpreted as the formation time of PGE-bearing metaplagioclasites from the Vurechuayvench deposit. The lower intersection of discordia with concordia at 486 ± 10 Ma is close in age to the initial stage of the Palaeozoic tectonomagmatic reactivation. It is marked by kimberlite pipes localized at the Tersky Coast (Bayanova et al., 2014).

Five mg of zircons of four morphotypes were separated from metagabbronorites hosted by this deposit (sample B-59) (Table 1.2). The first variety comprises fragments of long-prismatic crystals with slightly corroded surfaces 350 × 175 μm in size; the elongation coefficient is 2.0. The translucent grains are dirty yellow in color. The second variety has fragments of long-prismatic crystals with slightly corroded surfaces 245 × 210 μm in size; elongation coefficient is 1.16. The translucent grains are light yellow in color. The third variety consists of long-prismatic crystal fragments 245 × 210 μm in size; elongation coefficient is 1.16. The translucent grains are dark brown in color. The discordia constructed on the basis of three data points intersects the concordia at 2504.3 ± 2.2 Ma (Table 1.1). The upper intersection corresponds to the crystallization time of gabbronorites from the Vurechuayvench massif. The lower intersection at zero reflects a contemporary loss of Pb. The concordant age of 1689 ± 10 Ma was obtained for the fourth variety. This age corresponds with the time of the Svecofennian Orogeny completion. It is expressed in local tectonic zones of cataclasis and blasomylonitization in rocks of this massif (Sharkov et al., 2006).

Single zircon grains of two morphotypes have been separated from metanorites of the lower zone at the low-sulfide South Sopcha PGE deposit (sample B-63). The first variety is represented by transparent prismatic crystals with smoothed facets and corroded surfaces 105 × 50 μm in size; the elongation coefficient is 2.1. The transparent grains are light brown in color. The concordant age of 2504 ± 1 Ma (Fig. 1.5a) reflects the crystallization time of ore-bearing norites at this deposit. The second variety comprises deeply corroded crystal fragments 122 × 90 μm in size; the elongation coefficient is 1.3. The translucent grains are dark brown in color. The concordant age of this zircon estimated at 2130 ± 1 Ma (Fig. 1.5a) apparently corresponds with the time of tectonic rearrangement in the fault zone that separates the Monchepluton and Monchetundra massifs (Sharkov et al., 2006). Four mg of zircon grains pertaining to three morphotypes were separated from mesocratic medium-grained metagabbro of the upper zone in the South Sopcha massif (sample B-64). The first variety is represented by long-prismatic crystal fragments milky yellow in color with deeply corroded surfaces 350 × 140 μm in size; the elongation coefficient is 2.5. The second variety comprises long-prismatic crystal fragments with deeply corroded surfaces 350 × 140 μm in size; elongation coefficient is 2.5. The translucent grains are dark orange in color. The third variety is characterized by fragments of slightly corroded prismatic crystals 175 × 120 μm in size; the elongation coefficient is 1.5. The transparent grains are spotty in color, from water transparent to orange. The discordia constructed on the basis of three data points intersects the concordia at 2478 ± 20 (Fig. 1.5b). This age apparently characterizes the crystallization time of rocks from the upper zone of the South Sopcha massif. The lower intersection with the concordia is at zero and reflects the closure of the U-Pb isotope system and a contemporary loss of Pb.

One mg of zircon pertaining to three morphotypes was separated from metanorites in the Lake Moroshkovoye massif (sample B-61). The first variety is represented by crystal fragments 120 × 105 μm in size; the elongation coefficient is 1.2. The translucent grains are dark brown in color. The second variety comprises fragments of transparent crystals 120 × 120 μm in size and light brown in color. The third variety is represented by fragments of deeply corroded long-prismatic crystals 157 × 70 μm in size; the elongation coefficient is 2.2. The grains are opaque, dark orange in color. The discordia constructed on the basis of three data points intersects concordia at 2463.1 ± 2.7 Ma (Fig. 1.5c). This age characterizes the crystallization time of norites in the Lake Moroshkovoye massif. The lower intersection with concordia equals zero and reflects contemporary loss of Pb.

1.9.1. Isotope Sm-Nd Ages of Sulfides

Table 1.3 and Fig. 1.6 provide results of the study. Sm-Nd mineral isochron ages have been obtained for four samples: ortopyroxenites from the Nyud-II deposit (2497 ± 36 Ma), harzburgites from Horizon 330 (2451 ± 64 Ma), mineralized metaplagioclasites from the Vurechuayvench massif (2410 ± 58 Ma), and mineralized norites from the Nyud-II deposit (1940 ± 32 Ma).

Table 1.3 Isotopic Geochemical Sm-Nd Data on Rocks and Minerals from Monchegorsk Pluton and Massifs in Its Southern Framing.

Rocks and minerals	Concentration, ppm		Isotope ratios		T_{DM}, Ma	$\varepsilon_{Nd}(T)$
	Sm	Nd	$^{147}Sm/^{144}Nd$	$^{143}Nd/^{144}Nd \pm 2\sigma$		
Nyud-II, sample B-65						
Orthopyroxenite	0.456	2.06	0.1333	0.511530±14		−1.1
Sulf	3.39	19.63	0.1043	0.511059±11		
Opx-1	0.039	0.176	0.1355	0.511599±42		
Opx-2	0.318	1.226	0.1569	0.511961±34		
Pl	0.466	5.33	0.0528	0.510218±15		
Mineralized norite (Nyud-II, sample B-66)						
Py	0.029	0.168	0.1058	0.511086±13		−7.8
Ccp	0.0822	0.556	0.0895	0.510842±72		
Opx	1.66	5.69	0.1763	0.511975±16		
Pl	0.27222	2.25	0.0731	0.510656±14		
Ap	282	772	0.1148	0.511176±7		
330 horizon, Sopcha, sample B-70						
Harzburgite	0.0431	0.149	0.1656	0.511813±25		−6.0
Ol	0.028	0.144	0.1119	0.510982±43		
Sulf	0.034	0.188	0.1106	0.510934±36		
Opx	0.055	0.160	0.2064	0.512499±33		
Vurechuaivench, sample B-58						
Mineralized metaplagioclasite	0.971	4.62	0.1271	0.511408±7	3051	−2.4
Pn	0.109	0.350	0.1884	0.512382±18		
Sulf	0.031	0.116	0.1603	0.511880±87		
Vurechuaivench, sample B-59						
Metagabbronorite	0.639	2.98	0.1298	0.511391±18	3177	−2.82
Lake Moroshkovoe, sample B-61						
Metanorite	0.6111	2.95	0.1251	0.511340±13	3097	−2.68
South Sopcha, sample B-63						
Mineralized metanorite	1.269	5.47	0.1402	0.511596±17		−2.19
South Sopcha, sample B-64						
Metagabbro	2.82	12.25	0.1389	0.511619±17		−1.46

Note: See caption to Figure 1.6 for notation of minerals. Opx-1, orthopyroxene with density of 3.32 g/cm3; Opx-2, orthopyroxene with density of 3.25 g/cm³.

The rocks of the Monchegorsk pluton and the southern boundary massifs have negative εNd(T) and a range of model ages. Moderately radiogenic negative εNd(T) values have been determined for orthopyroxenites from the Nyud-II deposit and metagabbro of the South Sopcha massif (−1.46). Lower εNd(T) values are characteristic of mineralized metanorites from the South Sopcha massif (−2.19), mineralized metaplagioclasites from the Vurechuayvench massif (−2.4), metanorites from the massif of Lake Moroshkovoye (−2.68), and metagabbronorites from the Vurechuayvench massif (−2.82). Anomalously low εNd(T) values were determined for harzburgites from Horizon 330 (−6.0) and mineralized norites from the Nyud-II deposit (−7.8) (Table 1.3, Fig. 1.6). Model ages of protoliths (depleted mantle (T_{DM}) are estimated at 3.05–3.18 Ga for the Vurechuayvench massif and 3.10 Ga for the Lake Moroshkovoye massif (Table 1.3).

Table 1.4 and Figs. 1.7–1.8 provide new isotope-geochemical data on different rocks of the Monchegorsk ore area.

Table 1.5 and Figs. 1.9–1.10 provide new LA-ICP-MS data on sulfide minerals from the Fedorovo-Pansky massif.

Figure 1.11 shows isotope Re-Os data on the Kemi PGE intrusion (Finland) and Monchepluton.

Table 1.6 and Figure 1.12 represent isotope Sm-Nd data on PGE deposits of the Fedorovo-Pansky massif.

1.10. DISCUSSION

The Monchegorsk pluton, along with the Fedorovo-Pansky complex (Amelin et al., 1995; Bayanova, 2004; Bayanova, 2006; Bayanova et al., 2017; Nitkina, 2006; Rundkvist et al., 2006; Elizarova et al., 2009; Smol'kin et al., 2009; Starostin & Sorokhtin 2010; Serov et al., 2014;

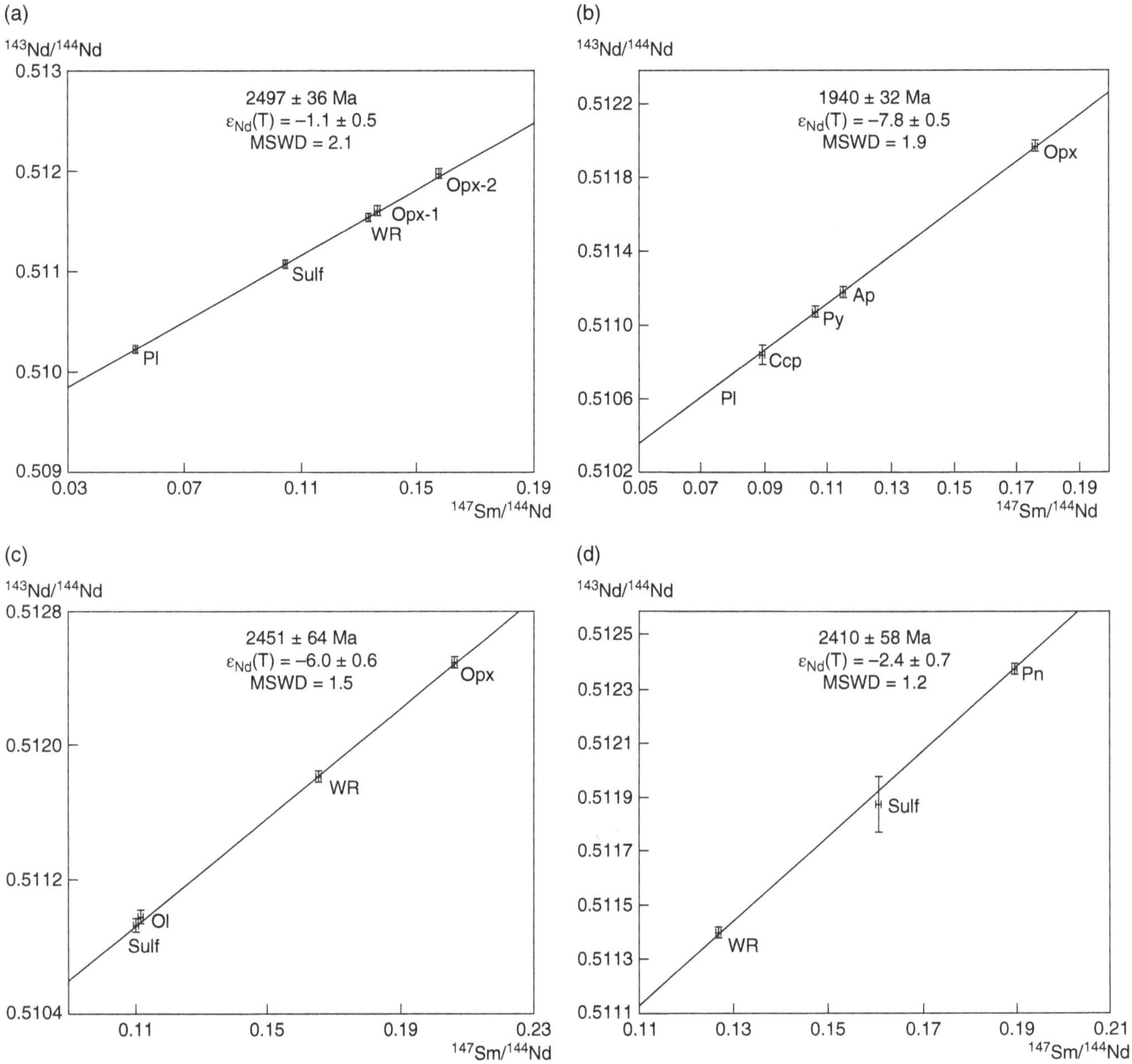

Figure 1.6 Sm-Nd sulfides mineral isochrones for rocks from Monchegorsk pluton: (a) olivine-bearing orthopyroxenite from Nyud-II deposit, sample B-65; (b) mineralized norite from Nyud-II deposit, sample B-66; (c) harzburgite from Horizon 330 of Sopcha deposit, sample B-70; (d) mineralized metaplagioclasite from Vurechuaivench deposit, sample B-58. Symbols of minerals: Ap, apatite; Ccp, chalcopyrite; Ol, olivine; Opx, orthopyroxene; Pl, plagioclase; Pn, pentlandite; Py, pyrite; Sulf, sulfides as a whole; WR, whole-rock sample.

Chaschin et al., 2016; Bayanova et al., 2017), the Ulitaozero massif (Mitrofanov & Smol'kin, 2004), and Mt. Generalskaya (Amelin et al., 1995; Bayanova et al., 1999) are the oldest (~2.5 Ga) layered intrusions of the Fennoscandian Shield. They are related to the initial stage of continental rifting along the northern wall of the Pechenga-Imandra-Varzuga volcanic-sedimentary riftogenic structure. They have anomalously low values of initial εNd, varying from −0.5 to −2.3, the Archaean Sm-Nd model age (2.80–3.15 Ga), and moderate enrichment in LREE (Bayanova, 2004; Bayanova et al., 2009); and negative Ta, Nb, and Ti anomalies in combination with positive Sr anomalies in chondrite-normalized spidergrams (Krivolutskaya et al., 2010). At the same time, a long-term evolution of magmatic system involving a two-phase mechanism has been established for some layered intrusions of the initial stage, that is, the Fedorov Tundra and West Pana. Thus, ore-bearing gabbronorites in the Fedorov Tundra massif formed 2485 ± 9 Ma ago (Groshev et al., 2009). Low-sulfide

Table 1.4 Sm-Nd and Rb-Sr Isotopic Data on Rocks from Dikes and Veins Hosted in the Moncha Tundra Massif.

		Concentration, ppm						
Sample	Rock	Sm	Nd	$^{147}Sm/^{144}Nd$	$^{143}Nd/^{144}Nd$	Age, Ga	$\varepsilon_{Nd}(T)$	T_{DM}
		High-Ti ferrodolerite						
39/205	Dolerite (Ti-dol$_2$)	9.5	47.2	0.1216	0.511666 ± 22	2.47	+4.9	2448
12/106	Plagioamphibolite (Ti-dol$_2$)	9.4	41.8	0.1363	0.511688 ± 34	2.47	+0.63	2867
24/206a	Amphibolite (Ti-dol$_1$)	9.4	43.0	0.1318	0.511894 ± 9	2.47	+6.12	2430
24/206	Amphibolite (Ti-dol$_1$)	9.3	42.7	0.1308	0.511794 ± 9	2.47	+4.47	2486
3/505	Amphibolite (Ti-dol$_1$)	7.4	32.4	0.1387	0.511816 ± 6	2.47	+2.38	2700
		Ferrodolerite						
49/206	Dolerite	2.6	9.7	0.1607	0.512016 ± 27	2.5	–0.61	3249
18/106a	Amphibolite	2.8	10.8	0.1565	0.512234 ± 3	2.5	+5.06	2434
18/106	Amphibolite	2.7	10.3	0.1569	0.512201 ± 25	2.5	+4.27	2541
17/306	Dolerite	3.8	12.9	0.1794	0.512586 ± 5	2.5	+4.55	2505
4/206a	Dolerite	2.6	10.0	0.1577	0.512252 ± 14	2.5	+5.0	2441
4/206	Dolerite	2.4	9.3	0.1590	0.512259 ± 6	2.5	+4.73	2477
		Gabbro-dolerite						
34/106	Dike margin	0.6	2.2	0.1715	0.512368 ± 8	2.5	+2.83	2816
34/206	Dike center	0.8	3.1	0.1626	0.512064 ± 21	2.5	–0.24	3227
		Gabbro-pegmatite and aplite						
36/105	Gabbro-pegmatite	2.7	11.6	0.1405	0.511927 ± 11	2.445	+3.79	2539
36/105a	Gabbro-pegmatite	2.8	12.5	0.1364	0.511799 ± 6	2.445	+2.58	2654
44/205	Gabbro-pegmatite	4.6	20.9	0.1342	0.511524 ± 23	2.445	–2.12	3102
3/306	Aplite	3.8	18.8	0.1238	0.511449 ± 7	1.9	–5.45	2869

		Concentration, ppm		Isotopic ratios			
Sample	Rock	Rb	Sr	$^{87}Rb/^{86}Sr$	$^{87}Sr/^{86}Sr$	Age, Ga	$I_{sr}(T)$
39/205	Dolerite	22.9	649.7	0.1020	0.70644 ± 8	2.47	0.7028
12/106	Plagioamphibolite	28.1	231.9	0.3513	0.71535 ± 11	2.47	0.7028
18/106	Amphibolite	10.5	124.8	0.2439	0.71002 ± 8	2.5	0.7012
34/206	Gabbro-dolerite	2.7	230.1	0.0341	0.70392 ± 6	2.5	0.7027

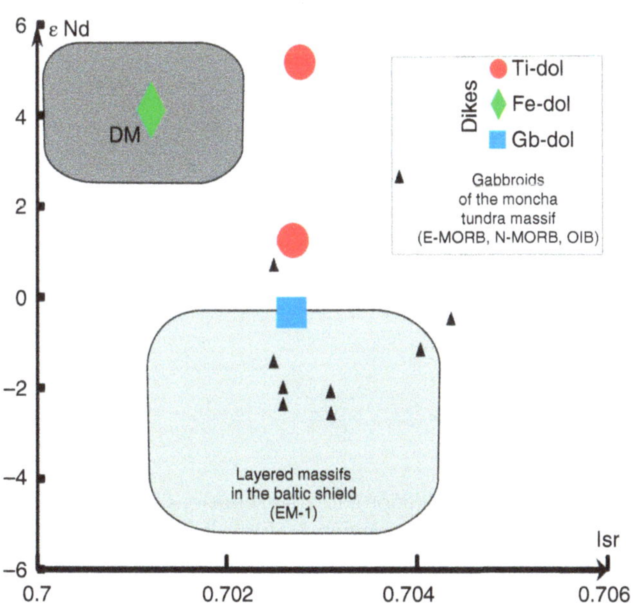

Figure 1.7 εNd-ISr diagram for dolerites and rocks of the layered complex of the Moncha Tundra Massif. DM is the depleted mantle (Hofmann 1997). The composition of Paleoproterozoic layered massifs in the Baltic Shield is shown according to Bayanova et al. (2009) and Nerovich et al. (2014).

PGM reefs in the West Pana massif are related to the younger anorthosite veins (~2470 and ~2450 Ma) hosted in the older (2495 ± 5 Ma) roughly layered gabbronorites (Bayanova 2004; Nitkina, 2006).

In contrast, results of our U-Pb isotope geochronological study combined with available data (Balashov et al., 1993; Amelin et al., 1995; Bayanova 2004; Mitrofanov & Smol'kin 2004; Rundkvist et al., 2014) show that ages of all rocks in the Monchepluton from quartz norites of the NKT bottom zone to plagioclasites that terminate the section in the Vurechuayvench refer to a narrow time interval of 2507 to 2493 Ma. It gives evidence of their similar age within limits of uncertainty, including rocks from separate

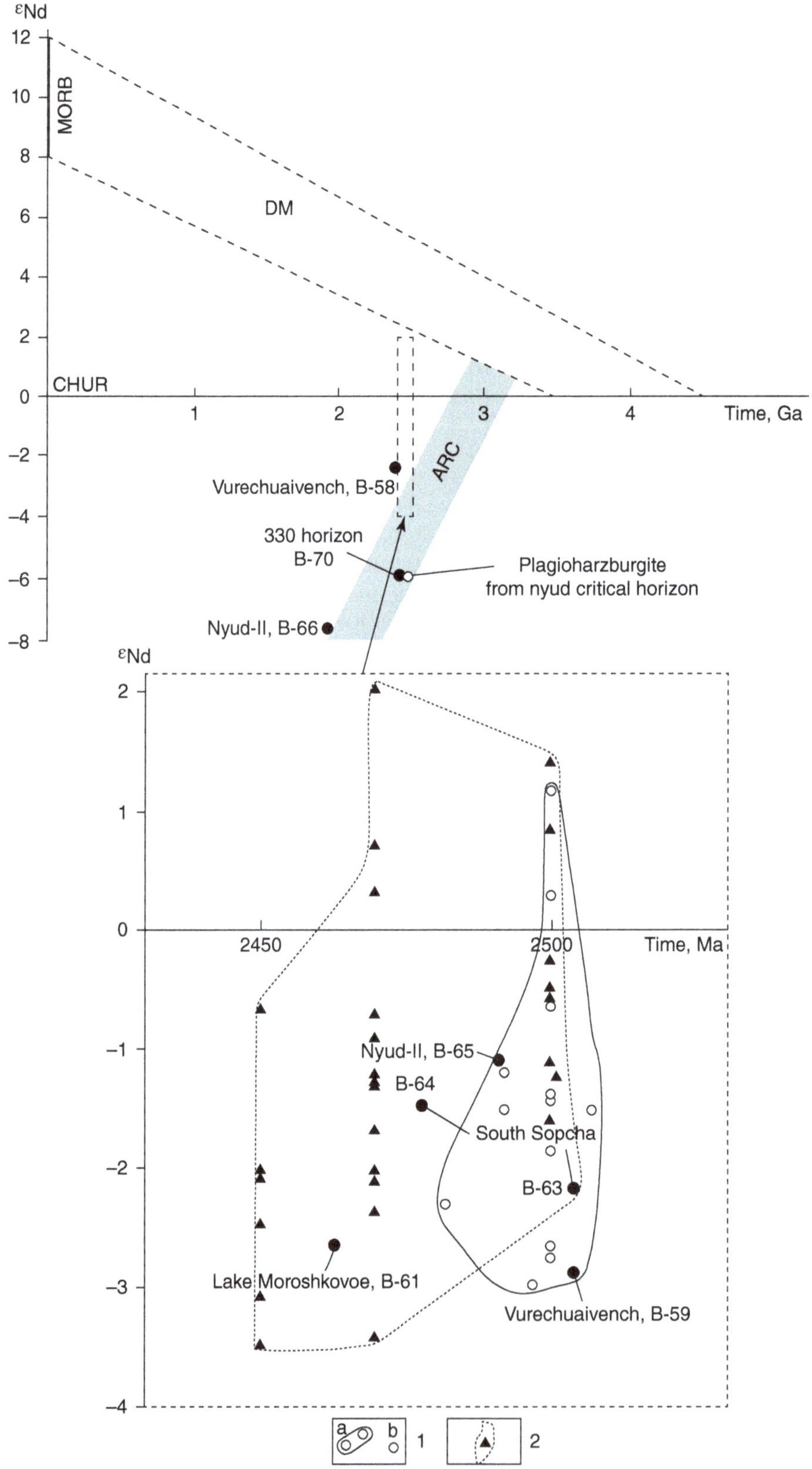

Figure 1.8 εNd versus time (according to Chaschin et al., 2016) for Early Paleoproterozoic intrusions of Kola region. (1) Monchegorsk pluton (a) and quartz norites of its bottom zone (b) (Mitrofanov & Smol'kin, 2004); (2) Monchetundra massif (Mitrofanov & Smol'kin, 2004; Nerovich et al., 2009); CHUR, chondrite uniform reservoir; DM, depleted mantle; MORB, midocean ridge basalt, according to Smith and Ludden (1989); ARC, Archean crust (Patchett & Kouvo, 1986).

Table 1.5 Concentrations (ppm) of Different Elements in Sulfide Parageneses of the Fedorova Tundra Deposit.

Sample	Mineral	Number of analyses	Cr	Co	As	Se	Ru	Rh	Pd	Ag	Cd	Sb	Re	Os	Ir	Pt	Au	Tl	Pb	Bi
237/ 132.0	Pentlandite	8	2.80–16.49	6811.43–10681.10	0.31–0.65	92.40–182.80	0.49–1.02	0.06–6.91	258.30–1221.08	0.39–5.78	–	0.06–0.10	–	0.06–0.12	0/01–0.28	0.01–1.19	0.01	0.05–0.79	0.18–8.12	0.02–0.17
	Chalcopyrite	6	2.79	0.27–141.49	0.33	178.0	0.18	0.12–9.07	18.8	0.17–5.28	14.02	0.08–0.17	0.02	0.01–0.14	0.26	0.01	0.029	–	4.48–7.98	–
487/ 50.5	Pentlandite	10	0.92–5.87	8395	0.25–0.43	98.7	0.82	1.47–9.05	442.1	0.03–0.74	–	0.10	0.05	0.15–0.26	0.582	0.03–0.17	–	0.04–0.43	0.09–2.29	0.02
	Pyrrhotite	8	0.58–452.91	58.90–15612.25	0.22–2.46	102.31–222.40	0.30–1.50	1.21–2.87	0.08–636.40	0.10–1.60	–	0.07	0.03–0.09	0.15–0.56	0.58–1.08	0.181	0.01	0.12–1.38	0.10–5.80	0.01–0.04
	Chalcopyrite	5	–	–	0.54–0.72	98.3	0.17	8.80	17.7	0.35	4.60	0.14	–	–	0.003	–	–	–	1.48	–
495/ 76.5	Pentlandite	10	2.08	5498.70–12079.02	0.42	117.8	0.60–1.23	0.08–0.61	246.54–1487.41	0.30–1.05	–	0.10	–	0.03–0.24	0.01–0.37	0.18–1.95	0.01	0.07–0.87	0.11–1.46	0.03–0.07
	Pyrrhotite	9	0.70–15.18	65.76–118.73	–	131.7	0.20	0.05–0.15	0.06–4.14	0.38–2.13	–	0.11	0.03	0.113	0.19–0.34	0.04–0.30	0.01	0.02–0.05	0.05–6.26	0.02–0.60
	Chalcopyrite	8	2.64	0.53–487.25	0.66	141.8	0.12–0.22	11.10	22.72–125.22	0.61–1133.19	2.90–9.14	–	0.042	0.02	0–0.02	0.02–0.11	0.02	0.04–0.73	0.55–4.10	0.04–0.16

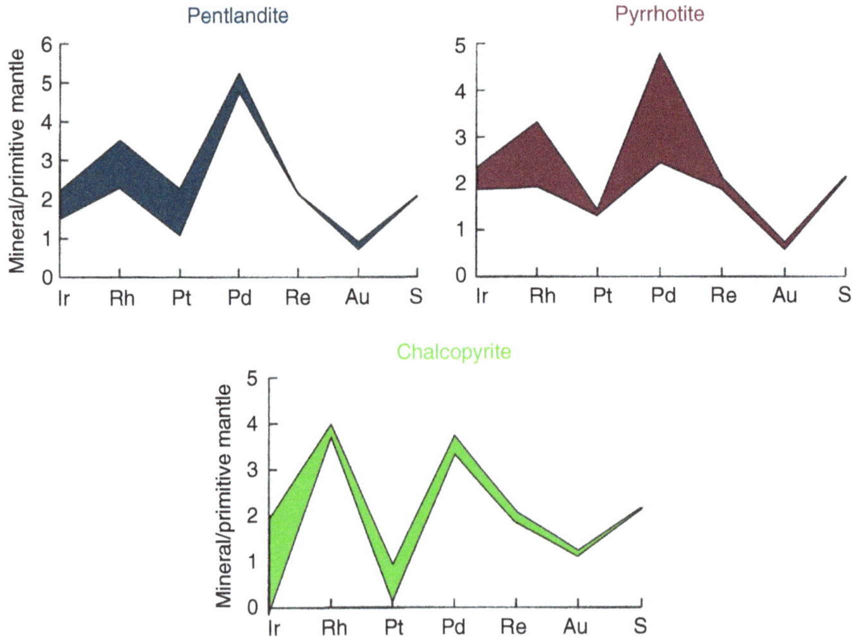

Figure 1.9 Spider diagram (after Mitrofanov et al., 2013) of distribution of PGEs, Au, and Re (composition of the primitive mantle is taken from Fischer-Godde et al., 2010).

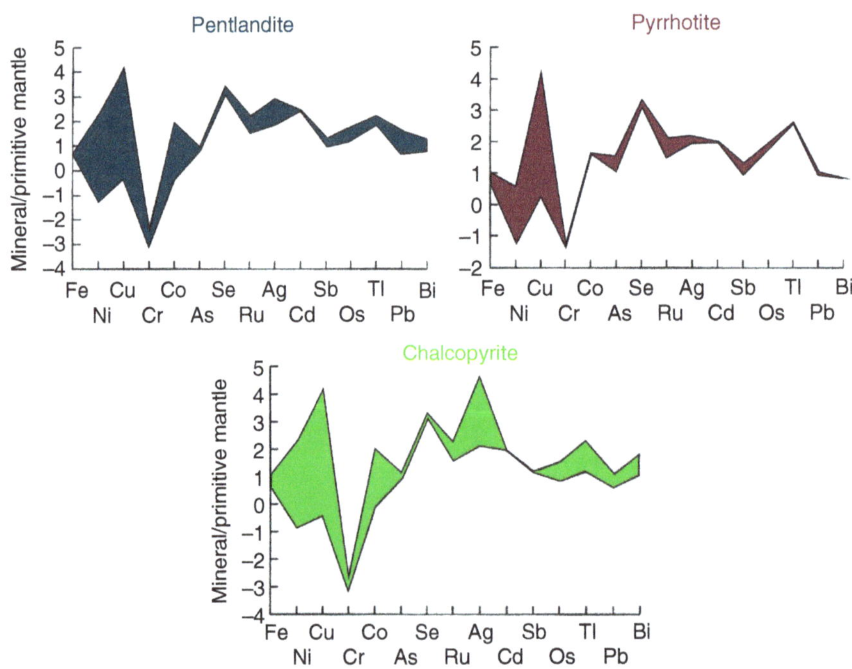

Figure 1.10 Spider diagram (after Mitrofanov et al., 2013) of distribution of other analyzed elements (composition of the primitive mantle) is taken from Lyubetskaya and Korenaga (2007a), Lyubetskaya and Korenaga (2007b).

PGE-bearing reefs, for example, the Nyud critical horizon and the Vurechuayvench metaplagioclasites. These data corroborate a conclusion drawn by Kozlov (1973) that all above-mentioned rocks were formed in the same process of intrachamber melt differentiation.

Ore-bearing metanorites of the lower zone in the South Sopcha massif are close in age (2504 ± 1 Ma) to plagiopyroxenites of the Monchetundra massif near the Pentlandite Gorge (2502.3 ± 5.9 Ma) (Bayanova et al., 2014) and to main volume rocks of the Monchegorsk

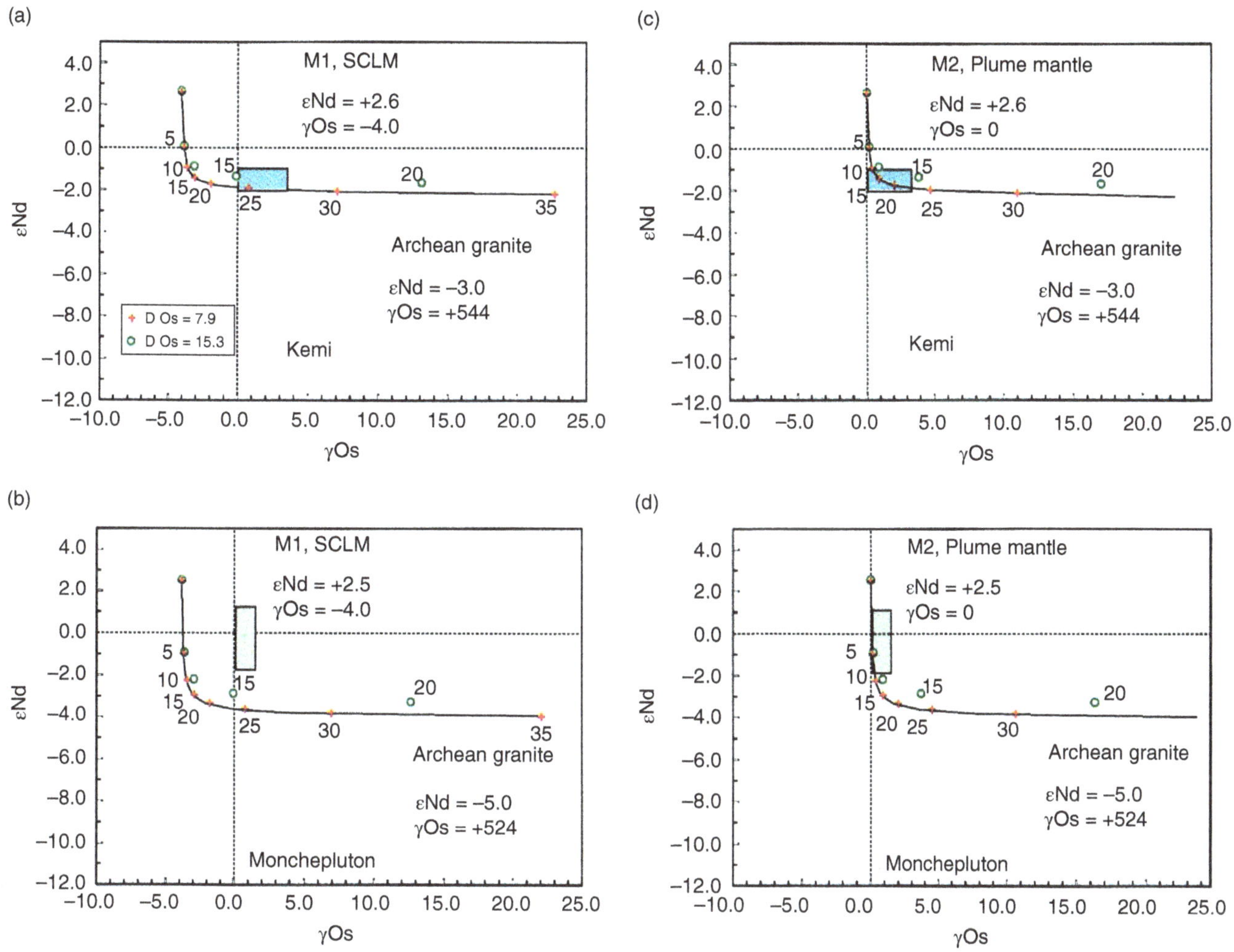

Figure 1.11 Model calculations showing the effects of AFC on εNd and γOs of the Kemi and Monchepluton magmas (Yang et al., 2016). M1 represents a nonmetasomatized SCLM source with γOs of −4.0 and εNd of +2.6 at 2.44 Ga (a, b), and γOs of −3.8 and εNd of +2.5 at 2.50 Ga (estimated from Peltonen & Brugmann, 2006; Nagler & Kramers, 1998); M2 represents a plume mantle source with γOs of 0 and εNd of +2.6 at 2.44 Ga and +2.5 at 2.50 Ga (c, d) (estimated from Puchtel et al., 1997, 2001; Nagler & Kramers, 1998). Crustal contaminant is estimated to have εNd of −3.0 and −5.0 for Kemi and Monchepluton, with Nd abundance of 54 ppm, and γOs of +542 at 2.44 Ga and +524 at 2.50 Ga with Os abundance of 0.0075 ppb (Hanski et al., 2001). The partition coefficient of Nd is assumed to be 0.001 for olivine (McKenzie & O'Nions, 1991). The bulk partition coefficient of Os is assumed to be 7.9 (Puchtel & Humayun, 2001), and 15.3 (Yang et al., 2013). The ratio r = mc (mass of crystallization) to ma (mass of assimilation) is assumed to be 0.4 in the AFC modeling (Ersoy & Helvaci, 2010). The numbers 5 to 35 represent the percentage of fractionation.

pluton. Metagabbro in the upper zone of the South Sopcha massif is younger (2478 ± 20 Ma). It is evidence of intrusive relationship between rocks of the lower and upper zones, as supported by field observations (Rundkvist et al., 2011; Grokhovskaya et al., 2012). Moreover, the obtained age of metagabbro is similar to that of medium- to coarse-grained leucocratic gabbronorites in the SE Monchetundra massif: 2471 ± 9 and 2476 ± 17 Ma (Bayanova et al., 2010). The obtained age estimates, together with the similar petrography and chemistry of gabbroic rocks in the South Sopcha and Monchetundra massifs (Grokhovskaya et al., 2012), show that the South Sopcha massif belongs to the Monchetundra group of intrusions.

A still younger age (2463.1 ± 2.7 Ma) was determined for rocks in the massif of Lake Moroshkovoye. It suggests that they were formed during late phases of the Sumian magmatism in the Monchegorsk ore district. In particular, these data are close to the age of leuconorites from the marginal zone (2463 ± 2.4 Ma) and leucogabbro from the main zone (2467 ± 8 Ma) of gabbroanorthosites in the Volchetundra massif (Chashchin et al., 2012).

Table 1.6 Sm-Nd Isotope Data on Sulfides for PGE Fedorovo-Pansky and Ahmavaara Deposits.

Sample	Concentration, mkg/g		Isotopic ratios		T_{DM}, Ma	£M(T)
	Sm	Nd	$^{147}Sm/^{144}Nd$	$^{143}Nd/^{144}Nd$		
	Densely disseminated massive ore of the Ahmavaara deposit (sample F-28)					
WR	1.132	6.01	0.1136	0.511195±20	2964	−2.1
Pn	0.151	0.842	0.1089	0.511129±27		
Po	0.073	0.294	0.1358	0.511549±26		
Ccp	0.761	5.14	0.0893	0.510804±11		
	Redeposited ore of the Ahmavaara deposit (sample F-27)					
WR	2.49	8.41	0.1791	0.512302±11	2912	−1.4
Po	0.263	1.617	0.0982	0.511272±10		
Py	0.157	0.934	0.1057	0.511372±49		
Pn	0.192	4.99	0.0433	0.510605±6		
Ccp	0.183	3.04	0.0636	0.510843±26		
	Gabbro-anorthosite of the Kievey deposit (sample MP-1)					
WR	1.038	4.99	0.1263	0.511441±10	2967	−1.3
Po	0.033	0.147	0.1144	0.511217±69		
Pn	0.011	0.041	0.1160	0.511259±53		
Pl	0.332	2.30	0.0853	0.510738±24		
Cpx+Opx-1	4.75	16.44	0.1747	0.512209±7		
Cpx+Opx-2	2.54	9.34	0.1641	0.512033±9		
Ccp+Pn	0.022	0.124	0.1106	0.511143±27		
	Ore gabbronorite of the Kievey deposit (sample FPM-1)					
WR	0.563	3.12	0.1096	0.511125±14	2949	−1.7
Po	0.028	0.176	0.1050	0.511044±26		
Pn+Py+Ccp	0.424	1.663	0.1521	0.511821±23		
Ccp	0.049	0.248	0.1086	0.511132±60		
	Metagabbro of the Fedorov Tundra (sample BGF-616)					
WR	1.313	5.77	0.1377	0.511727±18	2841	−1.2
Py	0.082	0.452	0.1089	0.511251±50		
Pl-1	1.351	7.34	0.1108	0.511283±17		
Pl-2	1.042	8.31	0.0757	0.510707±14		
Ccp	0.104	0.597	0.1046	0.511165±29		
Py+Pn	0.153	0.912	0.1008	0.511130±48		

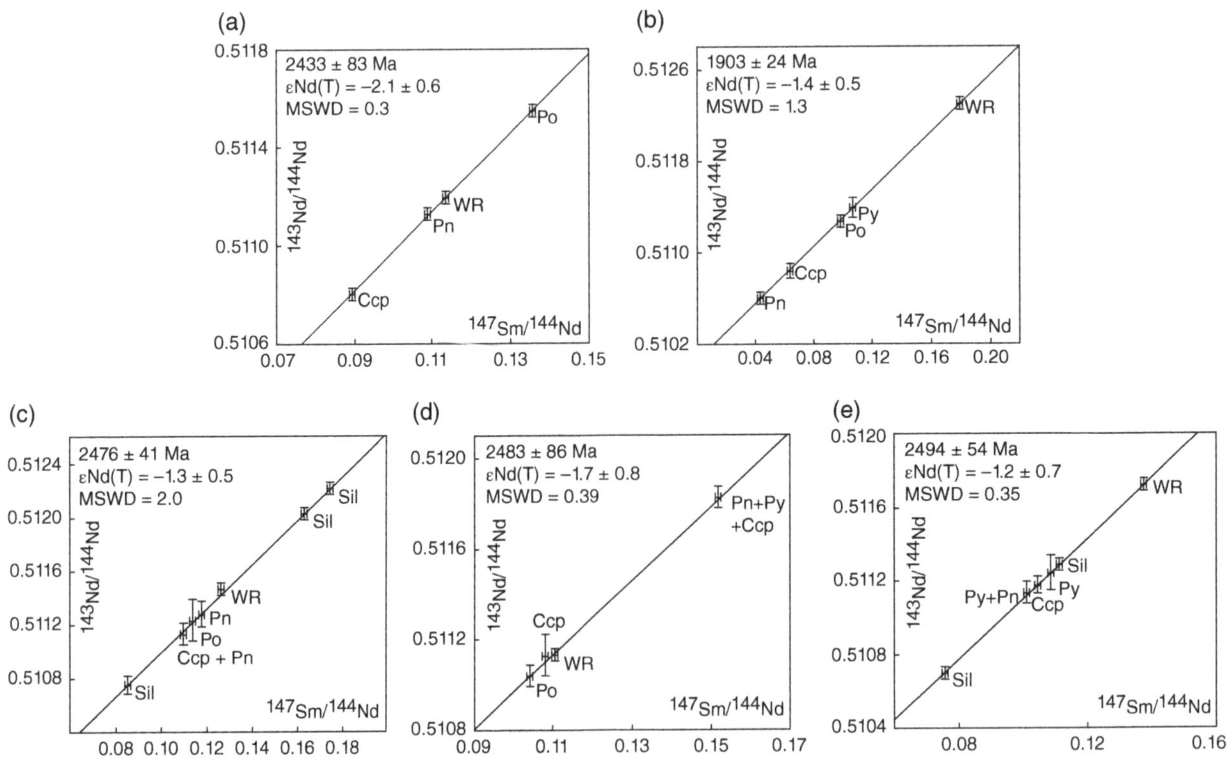

Figure 1.12 Isotope Sm-Nd data for PGE Ahmavaara, Finland (a, sample F-28; b, sample F-28) and Fedorovo-Pansky deposits (c, sample MP-1; d, sample FPM-1; e, sample BGF-616).

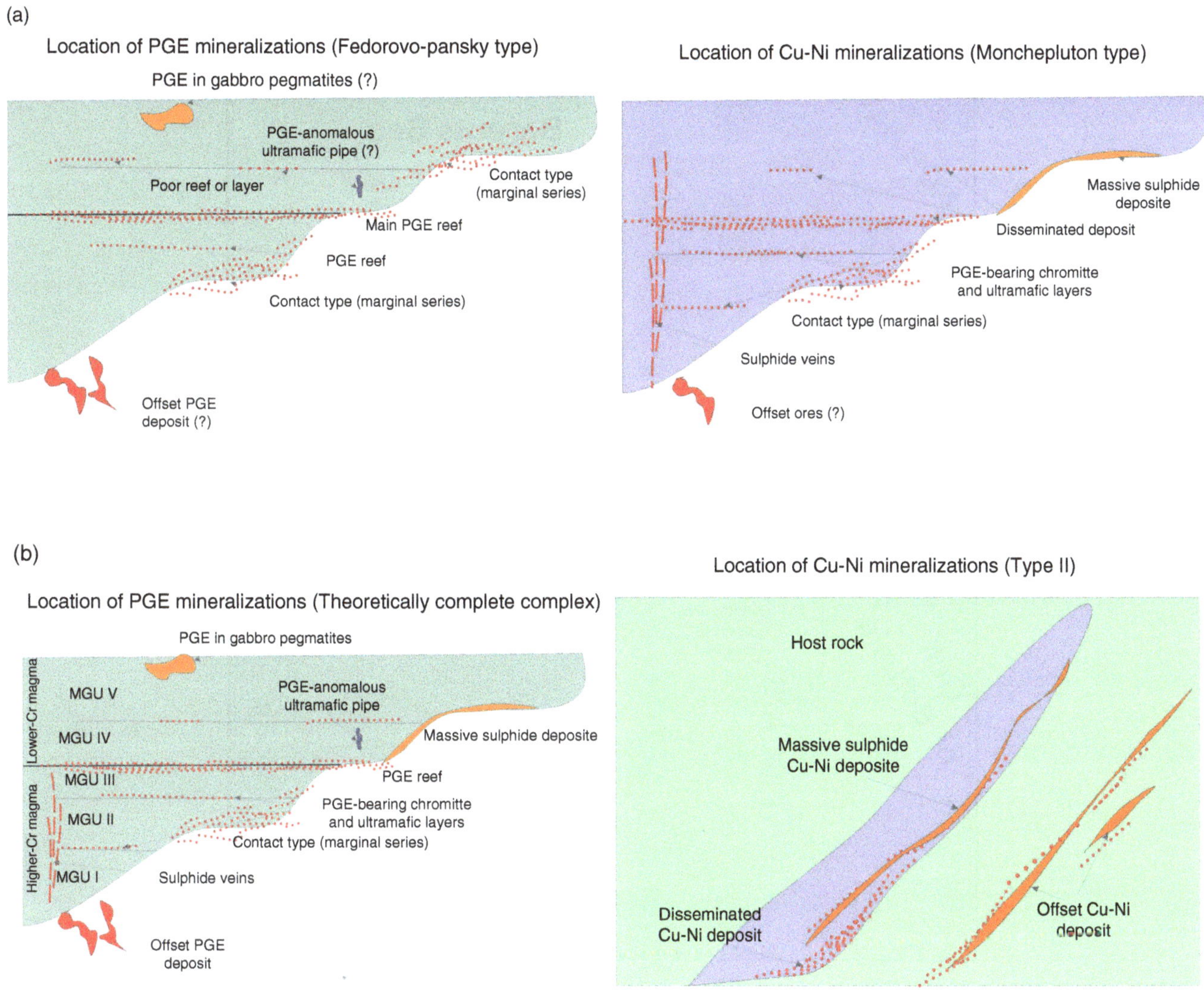

Figure 1.13 (a) Comparison of the ore mineralization controlling factors in the PGE-bearing layered intrusions of the Fedorovo-Pansky type (I) (left) and Moncheplutan type (II) (right). (b) Comparison of the ore mineralization controlling factors in the PGE-bearing layered intrusions with the theoretically complete location (I) (left) and Cu-Ni mineralization in the mafic-ultramafic rocks of the Pechenga / Noril'sk-type intrusions (II) (right).

Low Nd and Sm concentrations in rocks and minerals of the studied areas do not allow us to achieve good certainty for age determinations with the Sm-Nd method, which commonly yields big errors (1.4–2.6%). Nonetheless, results of the Sm-Nd isotope geochronological study of orthopyroxenites from the Nyud-II deposit (2497 ± 6 Ma) are close, within uncertainty limits, to the data of the U-Pb geochronology (2506 Ma). It testifies to the validity of both results.

A younger Sm-Nd age relative to the Moncheplutan main volume rocks has been obtained for harzburgites of Horizon 330 in the Sopcha massif (2451 ± 64). These data are consistent with geological reasoning. It assumes formation of this horizon to be a result of postdated injection of high-temperature magma, which is coeval with younger layered intrusions of the Imandra Complex (Amelin et al., 1995; Bayanova, 2004). The Sm-Nd age of mineralized metaplagioclasites from the Vurechuayvench deposit (2410 ± 58 Ma) markedly differs from the U-Pb determinations. These data are close to U-Pb ages of hydrothermal metasomatic alteration of anorthosites from the Volchetundra massif dated at 2407 ± 3 Ma (Chashchin et al., 2012) and to the early-stage metamorphism of the Monchetundra massif dated at 2406 ± 3 (Mitrofanov et al., 1993). Thus it cannot be ruled out that the obtained Sm-Nd isochron age of minerals from ore-bearing metaplagioclasites of the Vurechuayvench deposit correspond to the time of metamorphic and hydrothermal metasomatic alteration of the massif and its associated PGE mineralization, which postdates the crystallization

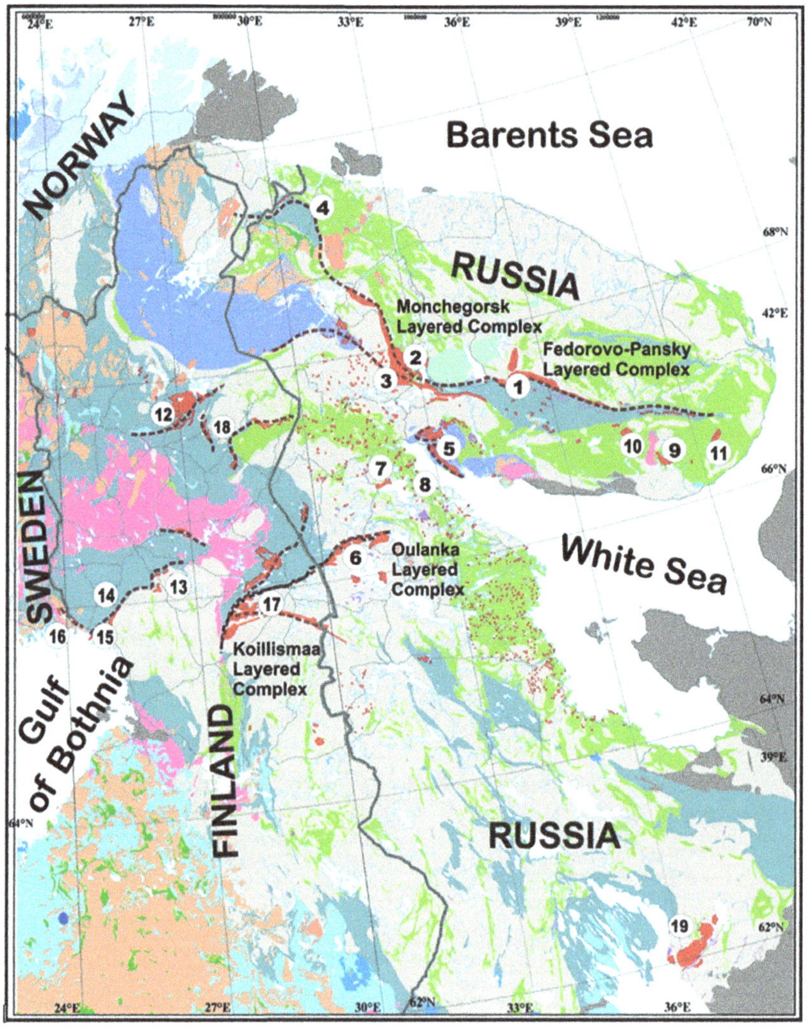

Figure 1.14 Generalized geological map of the northeastern part of the Baltic Shield and the location of Paleoproterozoic mafic layered intrusions (Mitrofanov et al., 2005).
In terms of geological interpretation this map is based on the Geological map of the Fennoscandian Shield 1:2 000 000 (GTK, NGU, GUS, MPR, 2009).
– Main Palaeoproterozoic layered intrusions with PGE (rich and poor) mineralization: 1, Fedorovo-Pansky; 2, Monchepluton; 3, Monchetundta, Volch'ya Tundra massifs, Main Ridge gabbros; 4, General'skaya Mt.; 5, Kandalaksha and Kolvitsa massifs; 6, Lukkulaisvaara; 7, Kovdor massif; 8, Tolstik; 9, Ondomozero; 10, Pesochny; 11, Pyalochny; 12, Keivitsa; 13, Portimo Complex (Kontijarvi, Siika-Kama; Ahmavaara); 14, Penikat; 15, Kemi; 16, Tornino; 17, Koilismaa Complex; 18, Akanvaara (Ahanvaara); 19, Birakov-Aganozero massif.

of igneous rocks and ore horizons. It seems especially plausible considering that sulfide-pentlandite mixture has been used for Sm-Nd dating. A much younger Sm-Nd age (1940 ± 32 Ma) has been obtained for mineralized norites at the Nyud-II deposit. It fits the time of the Svecofennian metamorphism (Chaschin et al., 2016) that partly affected the Monchepluton ore-magmatic system and led to the rearrangement of the Sm-Nd system and its incomplete closure.

All obtained initial εNd values are negative and characterized by a significant dispersion. The lowest εNd values were determined for orthopyroxenites from the Nyud-II deposit (−1.1), metagabbro from the upper zone of South Sopcha massif (−1.46), metaplagioclasites from Vurechuayvench deposit (−2.4), metanorites from the Lake Moroshkovoye massif (−2.68), and metagabbronorites from the Vurechuayvench massif (−2.82). The Nd isotope composition of these rocks is evidence of an enriched mantle source close to the present-day reservoir EM-1 (Zindler & Hart, 1986). The subcontinental lithospheric mantle (SCLM) (Farmer, 2003) of the Fennoscandian Shield may have been such a source. SCLM is considered to be a derivative of the Palaeoproterozoic plume-related magmatism (Bayanova et al., 2009). In εNd value, orthopyroxenites from the Nyud-II deposit are close to olivine pyroxenites of the

Nyud massif (Mitrofanov & Smol'kin, 2004). It testifies to the identity of their magmatic sources. The εNd value for metagabbronorites from the Vurechuayvench massif (-2.82) turned out to be similar with that previously determined for this rock (Mitrofanov & Smol'kin, 2004). It is close to this parameter for melanocratic norites of the Nyud critical horizon and for norites from the Poaz massif (Mitrofanov & Smol'kin, 2004), that is, for the rocks pertaining to the uppermost part of the composite Monchepluton section. This allows us to assume a moderate contamination of the Monchepluton magmatic chamber roof with crustal material.

The lowermost εNd values were determined in harzburgites from Horizon 330 of the Sopcha (-6.0) and in mineralized norites from the Nyud II deposit (-7.8). They are evidence of substantial crustal contamination of the magma. Notably, εNd value of harzburgites from Horizon 330 is almost identical to that of plagioharzburgites from the Nyud critical horizon (Mitrofanov & Smol'kin, 2004). Thus, both rocks from Horizon 330 of the Sopcha and Nyud critical horizon were significantly contaminated with the crustal material (Chashchin et al., 2016).

Hence, the isotope geochronological study of ore-magmatic systems with low-sulfide PGE mineralization show PGE-bearing reefs of the Monchepluton to result from the intrachamber fractionation of the initial magmatic melt (PGE-bearing reef at the Vurechuayvench deposit, critical horizon of the Nyud-II deposit) and from an injection of additional magma batch (Horizon 330 of the Sopcha). The basal-type low-sulfide South Sopcha PGE deposit formed at the initial stage of the massif development.

It has to be noted that the main rocks of the Monchegorsk pluton (orthopyroxenites and mineralized norites of the Nyud-11 deposit), plagioclasites of the PGE reef, and gabbronorites of the Vurechuaivench massif were geochronologically investigated with high precision by the U-Pb method on single zircon and baddeleyite grains at the Kola Science Center (RAS, Apatity); local SHRIMP-11 techniques (VSEGEI, St. Petersburg) were used to verify and reproduce the results of geochronological data for the most essential PGE deposits of the world. This allows us to believe that the reliable ages were determined to be comparable with the isotope results obtained in Toronto (Mungall et al., 2016) for the reefs and main rocks of the Bushveld Complex.

Application of various isotope systematics (Sm-Nd, Rb-Sr, Os, He, etc.) for the PGE deposits of the Monchegorsk, Fedorovo-Pansky, and other ore complexes located in the Baltic part of the Fennoscandian Shield provides advantages as compared with other studies based on the single-isotope systematics, which is easier to explain, understand, and accept as evident. In contrast, different isotope systematics allow correlating the measured geochronological data on igneous (early?) rock-forming, ore and accessory (post-magmatic zircon or early-magmatic baddeleyite inside zircon or orthopyroxene?) minerals.

1.10.1. Timing, Pulsation, and Total Duration of Magmatic Activity

The comprehensive geochronological study has been given to the largest and richest ore deposits of the Fedorovo-Pansky Complex (Fig. 1.13). The Fedorov Block of the Fedorovo-Pansky Complex is an independent magma chamber, whose rocks and ores significantly differ from those of the Western Pansky Block (Schissel et al., 2002). The 2 km-thick rock sequence from the Marginal Zone to the Lower Gabbro Zone is a layered or differentiated syngenetic series of relatively melanocratic pyroxenite-norite-gabbronorite-gabbro dated at 2526 ± 6 and 2516 ± 7 Ma. The Taxitic Zone is penetrated by concordant and cutting Cu-Ni-PGE-bearing gabbronorite (Fedorov deposit) of the younger second-pulse magmatic injection (2485 ± 9 Ma).

The Western Pansky Block in the Main Gabbronorite Zone, without LLH and probably without the upper part (above 3000 m), also can be considered as a single syngenetic series of relatively leucocratic, mainly olivine-free gabbronorite-gabbro crystallized within the interval of 2526–2485 Ma. There are the Norite and Marginal Zones in the lower part of the Block. The Marginal Zone contains the poor-disseminated Cu-Ni-PGE mineralization. This rock series can be correlated with certain parts of the Fedorov Block. The 40–80 m-thick LLH is prominent because of its contrasting structure with predominant leucocratic anorthositic rocks. The exposed part of the horizon strikes for almost 15 km and can be traced in boreholes down to a depth of 500 m (Mitrofanov et al., 2005). By its morphology, the horizon seems to be part of a single layered series. Nevertheless, there are anorthositic bodies that expose cross-cutting contacts and apophyses (Latypov & Chistyakova, 2000). The cumulus plagioclase compositions in the horizon rocks differ from those in the surrounding rocks. The age of PGE-bearing leucogabbro-pegmatite is precisely defined by concordant and near-concordant U-Pb data on zircon as 2470 ± 9 Ma. It is slightly younger than surrounding rocks (e.g., 2491 ± 1.5 Ma and 2500 ± 3 Ma). LLH rocks, especially anorthosite and the PGE mineralization, are likely to represent an independent magmatic pulse.

The upper part and the olivine-bearing rocks of the Western Pansky Block and anorthosite of ULH with the Southern PGE Reef have been poorly explored. They differ from the main layered units of the block in the rock, mineral, and PGE mineralization composition (Mitrofanov et al., 2005). Up to this date, only one reliable U-Pb age (2447 ± 12 Ma) has been obtained for PGE-bearing anorthosite of the block. It may represent another PGE-bearing magmatic pulse. Sm-Nd age is 2467 ± 39 Ma, which complies with the U-Pb data.

The early magmatic activity of about 2.5 Ga reflected in gabbronorite of the Monchetundra (2505 ± 6 and 2501 ± 8 Ma) and Mt. Generalskaya (2496 ± 10 Ma). The magmatic activity (~2470 and ~2450 Ma) produced anorthosite. It also contributed to layered series of the Chunatundra (2467 ± 7 Ma) and Mt. Generalskaya (2446 ± 10 Ma), Monchetundra gabbro (2453 ± 4 Ma) (Bayanova et al., 2009), and pegmatoid gabbronorite of the Ostrovsky intrusion (2445 ± 11 Ma).

The Imandra lopolith is the youngest large layered intrusion within the Kola Belt. It differs from other intrusions both in its emplacement age and its metallogeny. There are five U-Pb zircon and baddeleyite ages for the rocks of the main magmatic pulse represented by norite, gabbronorite, leucogabbro-anorthosite, gabbrodiorite, and granophyre; all formed within the interval of 2445–2434 Ma.

Thus, several eruptive pulses of magmatic activity have been established in the complex intrusions of the Kola Belt. There were at least four pulses (phases) in the Fedorovo-Pansky Complex: a 2526–2516 Ma barren pulse and three ore-bearing pulses of 2505–2485 Ma, 2470 and 2450 Ma.

The multiphase magmatic duration of the Fenno-Karelian Belt intrusions was short-term and took place about ~2.44 Ga years ago. However, there are only few U-Pb age estimations for the Fenno-Karelian Belt intrusions (Iljina & Hanski, 2005). The Kola results show that layering of the intrusions with thinly-differentiated horizons and PGE reefs was not syngenetic with each intrusion, defining its own metallogenic trends in time and space.

1.10.2. Metallogenic Characteristics

The distribution of rare and precious metals in sulfide parageneses has been first studied in detail using LA-ICP-MS. It allowed us to estimate the distribution of key elements with a high degree of accuracy. The results clearly show that pentlandite in sulfide parageneses of the Fedorov Tundra deposit is the main concentrator of the PGE mineralization and best-valued economically.

The Palaeoproterozoic magmatic activity in the eastern Fennoscandian Shield is associated with the formation of widespread ore deposits (Fig. 1.14): Cu-Ni (± PGE), Pt-Pd (Rh, ± Cu, Ni, Au), Cr, Ti-V (Mitrofanov & Golubev, 2008; Richardson & Shirey, 2008). The basal ores of the Fedorov deposit are best-valued for platinum-group elements (Pt, Pd, Rh), but nickel, copper, and gold are also economically important (Schissel et al., 2002). Ore-forming magmatic and postmagmatic processes are closely related to the Taxitic Zone gabbronorite of the 2485 ± 9 Ma magmatic pulse. Reef-type deposits (Pt-Pd [± Cu, Ni, Rh, Au]) and ore occurrences of the Western Pansky Block (Fedorovo-Pansky Complex) seem to be genetically associated with pegmatoid leucogabbro and anorthosite rich in late-stage fluids. Portions of this magma produce additional injections of c. 2500 Ma, c. 2470 Ma (the Lower, Northern PGE reef), and ca. 2450 Ma (the Upper, Southern PGE reef of the Western Pansky Block and PGE-bearing mineralization of the Mt. Generalskaya intrusion). These different magma injections are quite similar in terms of composition, prevalence of Pd over Pt, ore mineral composition (Mitrofanov et al., 2005), and isotope geochemistry of the Sm-Nd and Rb-Sr systems. εNd values for these rocks vary from −2.1 to −2.3. It probably indicates a single long-lived enriched magmatic source.

High Cr concentrations (>1000 ppm) are typical of lower mafic-ultramafic rocks of layered intrusions in the Baltic Shield (Alapieti, 1982; Iljina & Hanski, 2005). The chromite mineralization is known in basal series of the Monchepluton, Fedorovo-Pansky Complex, Imandra lopolith (Russia), Penikat and Narkaus intrusions (Finland), chromite deposits of the Kemi intrusion (Finland), and Dunite Block (Monchepluton, Russia). In contrast, the Fe-Ti-V mineralization of the Mustavaara intrusion (Finland) tends to occur in the most leucocratic parts of layered series, as well as in leucogabbro-anorthosite and gabbro-diorite of the Imandra lopolith (Russia) and Koillismaa Complex (Finland).

Thus, PGE-bearing deposits of the region are represented by the basal and reeflike types. According to modern economic evaluation, the basal type is preferable for mining, even if the PGE concentration (1–3 ppm) is lower compared to the reef-type deposits (>5 ppm). Basal deposits are thicker and contain more platinum, copper, and, especially, nickel. These deposits are accessible to open pit mining.

ESMLIP complies with all characteristics of modern LIP (Campbell & Griffiths, 1990; Ernst & Buchan, 2003; Bleeker & Ernst, 2006; Ernst, 2014; Yi-Gang Xu, 2007) as derivatives of deep mantle plume or asthenospheric upwelling processes. These are intimately LIPs associated with thick riftogenic sedimentary and volcanic rocks cogenetic with dike swarms and mafic-ultramafic intrusions (Klimentyev, 1995; Grachyov, 2003; Dedyukhin, 2005; Ivanchenko, 2006; Pirajno, 2007; Ivanchenko, 2009; Bayanova 2010; Bogatikov et al., 2010; Smol'kin et al., 2010; Chashchin et al., 2012; Grokhovskaya, 2012; Ernst, 2014; Chashchin et al., 2016; Bayanova, 2017). We confirm that ESMLIP has the following indicators proposed to be typical of intraplate mafic LIPs (Table 1.7):

• presence of gravity anomalies caused by a crust-mantle layer at the bottom of the crust;

• riftogenic (anorogenic) structural ensembles with manifestations of multipath extensional fault tectonics identified by the distribution of grabens and volcanic belts, elongated dike swarms, and radial intrusive belts;

Table 1.7 Prediction and Search Indicators for the Origination Conditions Complex Industrial Mineralization.

Low-sulphide Pt-Pd (with accompanying Ni, Cu, Au, Co, Rh)	Sulphide Cu-Ni (with accompanying Co, S, PGM, Se, Te, etc.)
Geophysics	
Presence of the granulite-mafic (anortosite) layer with the crust-mantle characteristics (Vp = 7.7-7.1 km/s) formed as a result of plume underplating (composition of the layer is defined on the basis of deep crustal xenoliths in the volcanic pipes) detected by the deep geophysical methods in the foot of the crust.	Presence of local gravity anomalies concentrated in narrow linear zones in accordance with the geophysical data. The ascent of the Moho discontinuity from the level of 40–42 km in the framing up to 39–38 km in the ore-controlling series.
Structure	
Regional: distribution of a discordant ensemble of rift-related volcano-sedimentary flexures, dikes, and polyphase layered mafic intrusions over a vast area of Archaean basement domains. Local: ore bodies occur at basal (lower) contacts, extended reef beds, in the deposits of pegmatoid mafic rocks, in veined and offset settings.	Regional: narrow extensive belts in the whole composite ensemble of the Palaeoproterozoic orogens within the crystalline shields (e.g., Pechenga structure). Ore-bearing intrusive bodies are injected in the upper part of the Early Palaeoproterozoic volcano-sedimentary cross-section. Local: ore locates in the basal intrusive contacts in the redeposited veined bodies, including offset setting.
Geodynamic setting	
Large-scale, long-term, and pulsating style of deep plume or asthenosphere-related upwelling processes causing the formation of the vast non-subduction-type igneous mafic intraplate continental province (LIP's). Change of geodynamic Archaean orogenic regime with intracontinental rifting (with origination of variously oriented ensialic belts). Ore-controlling mafic-ultramafic intrusions form at an initial (pre-rift) stage of continental rifting.	Ore genesis processes and magmatism tend in time and space during the period of the geodynamic regime interchange from the intracontinental rifting (ensialic) to the Red Sea-type (ensimathic) early spreading. Ore-controlling mafic-ultramafic intrusions are generated at a final stage of the continental rifting.
Composition	
Siliceous high-Mg (boninite-like) and anorthositic magmas. Cyclic (regular poly-stage style) structure of the layered intrusions and abrupt variability of the cumulus association stratigraphy and geochemical melt profile. There are two to five and more megacycles in the majority of the Palaeoproterozoic layered intrusions. The megacycles represent regularly layered series from ultramafic varieties to gabbroids. The ore is confined to the most contrasting series of alternating thin rock layers differing in composition from leuco- and mesocratic gabbro to norite, anorthosite, plagiopyroxenite, inequigranular and inhomogeneous textures (e.g., varitextured gabbro), leucocratic varieties (leucogabbro, anorthosite, spotted gabbro), inequigranular, coarse-grained and pegmatoid rocks with eruptive magmatic relationships. All known stratiform reef-type deposits are confined to the borders of the megacycles, which mainly reflect the interchange of the high-Cr magma with the low-Cr one. Intense manifestation of deep reducing fluids enriched with the compounds of C, F, Cl, H, etc. is typical in the rock associations. Mineralogical factors: PGMs associate with the disseminated sulphide mineralization, anomalously high concentration of PGEs in sulphides, platinum metal distribution coefficient between liquating silicate and sulphide melts of >100000.	Initial magma is depleted and similar to the Mid-Ocean Ridge Basalt (MORB) in terms of rare earths distribution. Ferropicritic Fe-Ti enriched magma derivatives generate single volcano-plutonic rock series. For intrusive ore bodies, gabbro-wehrlite composition, subvolcanic and hypabyssal crystallization setting, wide rock differentiation with the formation of syngenetic wehrlite-clinopyroxenite-gabbro- orthoclase gabbro sequence are typical.

(Continued)

Table 1.7 (Continued)

Low-sulphide Pt-Pd (with accompanying Ni, Cu, Au, Co, Rh)	Sulphide Cu-Ni (with accompanying Co, S, PGM, Se, Te, etc.)
Isotope geochemistry	
Deep mantle magma source initially is enriched with ore components (fertile source) and lithophile elements. It is reflected in such isotope indicators as εNd(T) = -1 to -3, $I_{Sr} = {}^{87}Sr/{}^{86}Sr = 0.702–0.705$, ${}^{3}He/{}^{4}He = n\bullet(10^{-5}–10^{-6})$ where n denotes a natural number of 1 to 9. Magma and ore source differs from that of Mid-Ocean Ridges and subduction zones.	Upper mantle source of the depleted magma with isotope indicators: εNd(T) = +0.5 to +4, $I_{Sr} = {}^{87}Sr/{}^{86}Sr = 0.703–0.704$, ${}^{187}Os/{}^{188}Os = 0.935 \pm 0.03$ (single measurement).
Geochronology	
Intraplate mafic extensive igneous provinces with low-sulphide platinum-palladium deposits (East Scandinavian Province on the Fennoscandian (or Baltic) shield, East Sayany Province at the prominence of the Siberian Platform basement, Huronian Province on the Canadian shield) are generated at the very beginning of the supercontinent break-up epochs, mostly at the Archaean – Palaeoproterozoic geochronological border, or 2600-2400 million years ago. For the East-Scandinavian province, it is the Sumi – Early Sariola epoch, or 2530-2400 million years ago. Ore-magmatic complexes evolve for a long time and in a pulsating manner (2490 ± 10 Ma; 2470 ± 10 Ma; 2450 ± 10 Ma phases) with the interchange of the boninitic magmas with the anorthositic ones, and Cr and Cu + Ni ore profile with Pt + Pd and Ti + V one.	Spreading mafic magmatism in the crystalline shields occurred at a late stage of the intracontinental rifting, finishing the Transitional period and starting the typical Lithospheric Plate Tectonic epoch (2200-1980 Ma). In the Fennoscandian Shield, this is the Svecofennian paleoocean origination stage.
Metamorphism	
Known commercial deposits occur in the regionally non-metamorphosed rocks. Only Pt-Pd ore prospects are found in the regionally metamorphosed mafic complexes. There are data demonstrating that the exceeded PT parameters of the mid-temperature amphibolites facies result in the impoverishment of the ore.	Collision metamorphism results in the formation of redeposited (remobilized) ore bodies both inside ore-bearing bodies and offset settings.

• long duration, polyphase and pulsating nature of tectonics and magmatism, continental discontinuities and erosion with early stages of tholeiite-basalt (trapps), boninite-like and subalkaline magmatism in the continental crust, and possible closing stages of incipient rift (Red Sea-type) spreading magmatism;

• intrusive sills, lopoliths, sheetlike bodies, large dikes, and dike swarms. The intrusions are often layered, being of a different nature than rocks formed in subduction and spreading zones (Bleeker & Ernst, 2006), with trends of thin differentiation layering and limited development of intermediate and felsic rocks, often with leucogabbro and anorthosite end members and abundant pegmatoid mafic varieties;

• typical mantle geochemistry of rocks and ores, as registered by isotope mantle tracers: ${}^{143}Nd/{}^{144}Nd$, ${}^{87}Sr/{}^{86}Sr$, ${}^{187}Os/{}^{188}Os$, ${}^{3}He/{}^{4}He$;

• mafic intracontinental LIPs accommodate large orthomagmatic Cr, Ni, Cu, Co, PGE (± Au), Ti, V deposits.

1.11. CONCLUSIONS

New U-Pb data on rocks of the Monchegorsk pluton show that the formation of orthopyroxenites and mineralized norites at the Nyud-II deposit, as well as plagioclasites of a PGE-bearing reef and gabbronorites at the Vurechuayvench deposit, fall into the same time interval of 2496–2506 Ma. It corresponds with the known age determinations of the Monchepluton.

The age of harzburgites from a PGE-bearing reef called Horizon 330 is determined by the Sm-Nd mineral isochron based on rock-forming minerals and sulfides. It is 2451 ± 64 Ma at initial εNd = -6.0. This estimate is consistent with the geological data, indicating that this reef resulted from an additional injection of high-temperature ultramafic magma, which experienced significant crustal contamination.

The low-sulfide South Sopcha PGE deposit formed in the lower marginal zone of the intrusion 2504 ± 1 Ma ago. It occurred synchronously with the initial stage of the mantle-derived magma crystallization. Thus, it is coeval with the Monchepluton.

The Sm-Nd isotope data on PGE-bearing metaplagioclasites at the Vurechuayvench deposit and norites at the Nyud-II deposit may indicate that the hydrothermal metasomatic changes significantly affected the PGE ore formation.

Detailed study of the rare and precious metals distribution in sulfide parageneses has been first performed using the laser ablation method (LA-ICP-MS). It allowed us to estimate quantitative patterns of the distribution of all the above-mentioned elements with a high degree of accuracy.

The obtained results clearly show that pentlandite in sulfide parageneses of the Fedorov Tundra deposit is the main concentrator of PGE mineralization and the most economically significant mineral.

According to petrological and geodynamic interpretations, EMSLIP is a product of a large long-lived plume. The evidence includes the homogenous and enriched isotope characteristics of the magmas, as well as their large volume and widespread distribution. It is quite possible and fully consistent with our observations, that the geochemical signatures of the LIP magmas may have been partly inherited from the subcontinental lithosphere, as described recently for Os isotope characteristics for the Bushveld magmas (Richardson & Shirey, 2008).

Currently, ESMLIP occupies an area of ca. 1,000,000 km^2 in the NE part of the Fennoscandian Shield. Its basement is represented by the mature Archaean granulite and gneiss-migmatite crust formed >2550 Ma ago. The province had several stages of magmatism and sedimentation separated by breaks (conglomerates). The Sumi (2550–2400 Ma) stage was crucial for the production of Pt-Pd ores related to the intrusive siliceous, high Mg boninite-like and anorthositic magmatism (Mitrofanov, 2005; Sharkov, 2006). The ore-bearing intrusions formed in the Kola Belt (Fedorovo-Pansky and other intrusions) earlier. They covered a surprisingly long 80-Ma period of multiphase magmatic activity (2530–2450 Ma). The main magmatism occurred in the Fenno-Karelian Belt later, between 2450 and 2400 Ma (Iljina & Hanski, 2005; Kullerud et al., 2006; Bayanova et al., 2009; Ekimova et al., 2011; Mitrofanov et al., 2013).

ACKNOWLEDGMENTS

This paper is dedicated to the memory of the outstanding researchers E. B. Bibikova (1934–2016) and J. Wasserburg (1927–2016).

Many thanks to the late G. Wasserburg for providing ^{205}Pb artificial spike; J. Ludden for 91500 and Temora standards; F. Corfu, W. Todt, and U. Poller for assistance in establishing of the U-Pb method for single zircon and baddeleyite grains at the Kola Science Centre. The authors express their gratitude to the following colleagues: L. Koval for baddeleyite, zircon, rock-forming, and sulfides minerals separation from rock samples; L. Lyalina for baddeleyite and zircon analyses using a Cameca MS-46 and for taking images of baddeleyite crystals; N. Levkovich for the chromatographic separation of U and Pb for analyses by mass spectrometry in the Geological Institute, Kola Science Centre, Russian Academy of Sciences.

We express our special gratitude to unknown reviewers for their contribution to the improvement of the paper.

The investigation is supported by RFBR No. 18-05-70082 and Program of the Presidium of the Russian Academy of Sciences No. 1.48.

REFERENCES

Alapieti, T. (1982). The Koillismaa layered igneous complex, Finland – its structure, mineralogy and geochemistry, witli emphasis on the distribution of chromium. *Geological Survey of Finland, Bulletin*, *319*, 116.

Amelin, Y.V., Heaman, L. M., & Semenov, V. S. (1995). U-Pb geochronology of layered mafic intrusions in the eastern Baltic Shield: Implications for the timing and duration of Palaeoproterozoic continental rifting. *Precambrian Res.*, *75*, 31–46.

Balashov, Y. A., Bayanova, T. B., & Mitrofanov, F. P. (1993). Isotope data on the age and genesis of layered basic-ultrabasic intrusions in the Kola Peninsula and northern Karelia, northeastern Baltic Shield. *Precambrian Research*, *64*, (14), 197–205.

Bartenev, I. S., & Dokuchaeva, V.S. (1975).Geological-structural features and conditions of formation of the Nyud-II Deposit. In G. I. Gorbunov (Ed.), *Mafic and Ultramafic Rocks of the Kola Peninsula and their Metallogeny*. Apatity: KolFAN SSSR.

Bayanova, T. B. (2004). Age of reference geological complexes of the Kola region and the duration of igneous processes. *Saint-Petersb., Nauka*, *174*.

Bayanova, T. B. (2006). Baddeleyite: A promising geochronometer for alkaline and basic magmatism. *Petrology*, *14* (2), 187–20.

Bayanova, T. B, Ludden, J., & Mitrofanov, F. P. (2009).Timing and duration of Palaeoproterozoic events producing ore-bearing layered intrusions of the Baltic Shield: metallogenic, petrological and geodynamic implications. In S.M. Reddy, R. Mazumder, D.A.D. Evans, & A.S. Collins (ed.), *Palaeoproterozoic Supercontinents and Global Evolution. Geological Society*, London, Special Publications, *323*, 165–198.

Bayanova, T. B., et al. (1999). U-Pb age of rocks of the Mt. General'skaya layered intrusion, Kola Peninsula. *Geochem. Int.*, *37* (1), 1–10.

Bayanova, T. B., Mitrofanov, F. P., Serov, P. A., Nitkina, E. A., & Kamensky, I. L. (2014). Layered PGE Palaeoproterozoic (LIP) intrusions in the NE part of the Fennoscandian Shield: Isotope Nd-Sr and 3He/4He data, summarizing U-Pb ages (on baddeleyite and zircon), Sm-Nd data (on rock-forming and sulfide minerals), duration and mineralization.

Geochronology: Methods and Case Studies, Morner, N.A. (eds), INTECH, 143–193; doi: 10.5772/58835.

Bayanova, T. B., Nerovich, L. I., Mitrofanov, F. P., Zhavkov, V.A., & Serov, P. A. (2010).The Monchetundra basic massif of the Kola region: New geological and isotope geochronological data. *Doklady Earth Sciences*, *431* (1), 288–293.

Bayanova, T. B., Rundkvist, T. V., Serov, P. A., Korchagin, A.U., & Karpov, S. M. (2017). The Palaeoproterozoic Fedorovo-Pansky layered PGE Complex of the northeastern Baltic Shield, Arctic Region: New U-Pb (Baddeleyite) and Sm-Nd (Sulfide) data. *Doklady Earth Sciences*, *472*, Pt. 1, 1–5.

Birck, J. L., Barman, M. R., & Campas, F. (1997). Re-Os isotope measurements at the femtomole level in natural samples. *Geostandards Letter*, *20* (1), 19–27.

Bleeker, W., & Ernst, R. (2006).Short-lived mantle generated magmatic events and their dike swarms: the key unlocking Earth's Palaeogeographic record back to 2.6 Ga, in dike swarms-time marker of crustal evolution. *Rotterdam: Balkema Publ.*, 1–20.

Bogatikov, O. A., Kovalenko, V. I., & Sharkov, E.V. (2010). Magmatism, tectonics, and geodynamics of the Earth. *Moscow, Nauka*, *606*.

Campbell, L. H., & Griffiths, R.W. (1990). Implications of mantle plume structure for the evolution of flood basalts. *Earth and Planetary Science Letters*, *99*, 79–93.

Chashchin, V. V., Bayanova, T. B., Mitrofanov, F. P., & Serov, P. A. (2016). Low-sulfide PGE ores in Palaeoproterozoic Monchegorsk Pluton and massifs of its southern framing, Kola Peninsula, Russia: geological characteristic and isotope geochronological evidence of polychronous ore-magmatic systems. *Geology of Ore Deposits*, *58* (1), 37–57.

Chashchin, V. V., Bayanova, T. B., Serov, P. A., & Yelizarova, I. R. (2012). The Volchetundrovsky massif of the autonomous anorthosite complex of the Main Range, the Kola Peninsula: geological, petrogeochemical, and isotope-geochronological studies. *Petrology*, *20* (5), 467–490.

Chashchin, V. V., Galkin, A. S., Ozeryansky, V. V., & Dedyukhin, A. N. (1999). The Sopcheozero chromite deposit and its PGE content, Monchepluton (Kola Peninsula). *Geology of ore deposits*, *41* (6), 507–526.

Condie, K. (2004). Supercontinents and superplume events: distinguishing signals in the geology record, *Phys. Earth and Planet. Inter.*, *146*, 319–332.

Corfu, F., Bayanova, T.B., Shchiptsov, V.V., & Frantz, N. (2011). U-Pb ID-TIMS age of the Tiksheozero carbonatite: expression of 2.0 ga alkaline magmatism in Karelia, Russia. *Cent. Eur. J. Geosci.*, *3* (3), 302–308.

DePaolo, D. J. (1981). Neodymium isotopes in the Colorado Front Range and crust-mantle evolution in the Proterozoic. *Nature*, *291*, 21, 193–196.

Dodin, D. A., Chernyshov, N. M., & Cherednikova, O. I. (2001). Russian platinum-group element mineralogy. *Moscow, Geoinformmark*, *302*.

Ekimova, N. A., Serov, P. A., Bayanova, T. B., Yelizarova, I. R., & Mitrofanov, F. P. (2011). The REE distribution in sulfide minerals and Sm-Nd age determination for the ore-forming processes in mafic layered intrusions, *Doklady RAS*, *436* (1), 75–78.

Elizarova, I. R., & Bayanova, T. B. (2012). Mass-spectrometric REE analysis in sulfide minerals. *Journal of Biology and Earth Sciences*, *2* (1), E45–E49.

Elizarova, I. R., Bayanova, T. B., Mitrofanov, F. P., & Kalinnikov, V.T. (2009). Methodology of defining REE in standard geological samples using mass-spectrometer ELAN 9000. *Mass Spectrometry and Its Applied Issues: Proceedings of III All-Russ. Conf. with Int. Part. M.: NU-8. 110* (in Russian).

Ernst, R. E. (2014). Large Igneous Provinces. Cambridge University Press, Cambridge.

Ernst, R.E., & Buchan, K. L. (2003). Recognizing mantle plumes in the geological record. Annu. Rev. *Earth Planet. Sci.*, *31*, 469–523.

Ersoy, Y., & Helvaci, C. (2010). FC-AFC-FCA and mixing modeler: A Microsoft Excel spreadsheet program for modeling geochemical differentiation of magma by crystal fractionation, crustal assimilation and mixing. *Comput. Geosci*, *36*, 383–390.

Farmer, G. L. (2003). Continental basaltic rocks. *Treatise on Geochemistry*, edited by R. L. Rudnick, *3*, 85–121.

Fischer-Godde, M., Becker, H., & Wombacher, F. (2010). Rhodium, gold and other highly siderophile element abundances in chondritic meteorites, *Geochim. Cosmochim. Acta*, *74*, 356–379.

Glikson, A.Y. (2014). The Archaean: Geological and geochemical windows into the Early Earth, *Springer*, *14*, 223.

Goldstein, S. J., & Jacobsen, S. B. (1988). Nd and Sr isotope systematics of river water suspended material implications for crystal evolution, *Earth Plan. Sci. Letters*, *87*, 249–265.

Gorbunov, G. I., ed (1982). *Imandra-Varzuga Zone of Karelides (Geology, Geochemistry and Evolution)*. Leningrad: Nauka.

Grachev, A. F. (2003). Identification of mantle plumes based on studying the composition of volcanic rocks and their isotope-geochemical characteristics, *Petrology*, *11*, (6), 562–596.

Gramlich, J. W., Murphy, T. J., Garner, E. L., & Shields, W. R. (1973). Absolute isotope abundance ratio and atomic weight of a reference sample of Rhenium, *J. Research National Bureau Standards: A. Physics and Chemistry*, *77A*, (6), 691–698.

Grochovskaya, T. L., Bakaed, G. F., Sholohnev, A. B., Lapina, M. I., Muravizkaya, G. H., & Voitechovich, V. C. (2003). Ore platinometal mineralization in layered Monchegorsk magmatic complex (Kola Peninsula, Russia). *Geol. Ore Deposits*, *45* (4), 329–352.

Grokhovskaya, T. L., et al. (2000). PGE mineralization in the Vurechuaivench gabbronorite massif, Monchegorsk pluton (Kola Peninsula, Russia). *Geol. Ore Deposits*, *42* (2), 133–146.

Grokhovskaya, T. L., et al. (2012). Geology, mineralogy, and genesis of PGE mineralization in the South Sopcha Massif, Monchegorsk Complex, Russia. *Geol. Ore Deposits*, *54* (5), 347–369.

Groshev, N.Y., Nitkina, E. A., & Mitrofanov, F. P. (2009). Two-phase mechanism of the formation of platinum-metal basites of the Fedorov Tundra Intrusion on the Kola Peninsula: New data on geology and isotope geochronology, *Dokl. Earth Sci.*, *427A* (6), 1012–1016.

Groves, D. I., Vielreicher, R. M., & Goldfarb, R. J. (2005). Controls on the heterogeneous distribution of mineral deposits through time. *Miner. Deposits and Earth Evolution, Geol. Soc. Sp. Publ.*, *248*, 71–101.

Hanski, E., Walker, R. J., Huhma, H., & Suominen, I. (2001). The Os and Nd isotope systematics of the c. 2.44 Ga Akanvaara and Koitelainen mafic layered intrusions in northern Finland. *Precam. Res.*, *109*, 73–102.

Hofmann, A. W. (1997). Mantle geochemistry: the message from oceanic volcanism. *Nature*, *385*, 219–229.

Iljina, M. (1994). The Portimo Layered Igneous Complex: with emphasis on diverse sulphide and platinum-group element deposits. Acta Universitatis Ouluensis Series A, ScientiaeRerum Naturalium, *258*, 158.

Iljina, M., & Hanski, E. (2005). Layered mafic intrusions of the Tornio-Näränkävaara belt. In Lehtinen, M., Nurmi, P. A. & Rämo, O.T. (Eds.), *Precambrian Geology of Finland: Key to the Evolution of the Fennoscandian Shield*, Elsevier B.V., Amsterdam, 101–138.

Ivanchenko, V. N., & Davydov, P. S. (2009). Petrological characteristics of ultramafic (komatiitic) formations in the Savukoski area, NE Finland. An Interreg-Tacis Project: Strategic Mineral Resources of Lapland – Base for the Sustainable Development of the North. Project publication, volume II. – Apatity: KSC RAS, 65–70.

Ivanchenko, V. N., Davydov, P. S., Dedeev, V. A., & Knauf, V. V. (2008). Major features of the Vuruchuaivench (Vurechuaivench) deposit geological structure. The neighborhood cooperation and experience exchange of geological prospecting and survey of PGE deposit in the Northern Fennoscandia. Kola Science Centre RAS publication, Apatity, 82–87.

Jacobsen, S. B., & Wasserburg, G. J. (1984). Sm-Nd Isotope Evolution of Chondrites and Achondrites. *II. Earth Planet. Sci. Lett.*, *67*, 137–150.

Jackson S.E. (2001). The application of Nd:YAG lasers in LA-ICP-MS. Principles and applications of laser ablation mass-spectrometry in the Earth. *Mineralogical Association of Canada Short Course Series*, *29*, 29–46.

Konnikov, E. G., & Orsoev, D. A. (1991). Nature of rhyrthmic layered horizon of the Sopcha Massif, Monchegorsk Pluton. *Dokl. Akad. Nauk SSSR*, *320*, (3), 696–699.

Korovkin, V. A., et al. (2003). *Interior of the Northwestern Russian Federation*. St. Petersb.: VSEGEI.

Kozlov, E. K. (1973). *Natural series of nickel-bearing rocks and their metallogeny*. Leningrad: Nauka.

Krivolutskaya, N. A., et al. (2010). Geochemical features of the drusite massifs, the central part of the Belomorian Mobile Belt: I. Distribution of major and trace elements in the rocks. *Geochem. Int.*, *48* (3), 465–491.

Krogh, T. E. (1973). A low-contamination method for hydro-thermal dissolution of zircon and extraction of U and Pb for isotope age determinations, *Geochim. Cosmochim. Acta*, *37*, 485–494.

Kullerud, K., Skjerlie, K. P., Corfu, F., & de la Rosa, J. D. (2006). The 2.40 Ga Ringvassøy mafic dykes, West Troms Basement Complex, Norway: The concluding act of early Palaeoproterozoic continental breakup. *Precambrian Research*, *150*, 183–200.

Latypov, R. M., & Chistyakova, S.Y. (2000). Mechanism for differentiation of the Western-Pana layered intrusion. *Apatity: Publ. of KSC RAS*, 315.

Liew, I. C., Hofmann, A.W. (1988). Precambrian crustal components, plutonic associations, plate environment of the Hercinian Fold Belt of central Europe: indications from a Nd and Sr isotope study. *Contribs mineral and petrol.*, *98* (2), 129–138.

Likhachyov, A. P. (2006). Platinum-copper-nickel and platinum deposits. *M., Nauka. 496*

Ludwig, K. R. (1991). PBDAT: A Computer Program for Processing Pb-U-Th isotope Data, Version 1.22, Open-file report 88-542, *US Geol. Surv.*, *38*.

Ludwig, K. R. (1999). ISOPLOT/Ex: A geochronological toolkit for Microsoft Excel, Version 2.05. *Berkeley Geochronology Center Special Publication*, *1a*, 49.

Lyubetskaya, T., & Korenaga, J. (2007a). Chemical composition of Earth's primitive mantle and its variance: 2. Implications for global geodynamics. *J. Geophys. Res.*, *112*, B03212; doi:10.1029/2005JB004224

Lyubetskaya, T., & Korenaga, J. (2007b). Chemical composition of Earth's primitive mantle and its variance: 1. Method and results,*J. Geophys. Res.*,*112*,B03211;doi:10.1029/2005JB004223

McKenzie, D., & O'Nions, R. K. (1991). Partial melt distributions from inversion of rare earth element concentrations. *J Petrol*, *32*, 1021–1091.

Mints, M. V. (2011). 3D model of deep structure of the Early Precambrian crust in the East European Craton and paleogeodynamic implications. *Geotectonics*, *45*, (4), 267–290.

Mitrofanov, A. F., Kogarko, L. N., Anosova, M. O., & Kostitsyn, Y. A. (2013a). Features of the distribution of precious metals in sulfide parageneses of the Fedorova Tundra Deposit (Kola Peninsula). *Doklady Earth Sciences*, *451*, Pt 2, 875–878.

Mitrofanov, F. P. (2005). New types of mineral raw materials in the Kola Province: discoveries and perspectives. *Smirnovsky Collection of works. Moscow: VINITI RAS*. 39–53 (in Russian).

Mitrofanov, F. P., & Smol'kin, V. F. (eds.) (2004). *Layered intrusions of the Monchegorsk ore district: Petrology, mineralization, isotopy and deep structure*, Apatity: KSC RAS.

Mitrofanov, F. P., Balagansky, V.V., Balashov, Y. A., Gannibal, L.F., Dokuchaeva, V.S., Nerovich, L. I., Radchenko, M. K., & Ryungenen, G. I. (1993). U-Pb age for gabbroanorthosite of the Kola Peninsula. *Doklady RAN*, *331* (1), 95–98.

Mitrofanov, F. P., Bayanova T. B., Korchagin A. U., Groshev N.Y., Malitch K. N., Zhirov D.V., & Mitrofanov A. F. (2013). East Scandinavian and Noril'sk Plume Mafic Large Igneous Provinces of Pd–Pt Ores: Geological and Metallogenic Comparison. *Geology of Ore Deposits*, *55* (5), 305–319.

Mitrofanov, F. P., Korchagin, A.U., Dudkin, K. O.,& Rundkvist, T. V. (2005). Fedorovo-Pansky layered mafic intrusion (Kola Peninsula, Russia): Approaches, methods and criteria for prospecting PGEs. Exploration for platinum-group elements deposits. *Short Course delivered on behalf of the Mineralogical Association of Canada in Oulu, Finland*, *35*, 343–358.

Mungall, J. E., Kamo S. L., & McQuade, S. (2016). U-Pb geochronology documents out-of-sequence emplacement of ultramafic layers in the Bushveld Igneous Complex of South Africa. *Nature Communicaitons*. doi: 10/1038/ncomms13385.

Nagler, T. F., & Kramers, J. D. (1998). Nd isotope evolution of the upper mantle during the Precambrian: Models, data and the uncertainty of both. *Precam. Res.*, *91*, 233–252.

Naldrett, A. J. (2003). Igneous sulfide deposits of copper-nickel and PGE ores. *S.-Pb., SpbGU*, *487*.

Neradovsky, Y. N., et al. (2002). On problem of the platinum potential of Mt. Sopcha ore bed-330 and its economic use (Monchegorsk Pluton). *Bulletin of the Murmansk State Technical University*, *5*, 85–90.

Nerovich, L. I., Bayanova, T. B., Savchenko, E. E., Serov, P. A., & Ekimova, N. A. (2009). New data on geology, petrography, isotope geochemistry and PGE mineralization of the Monchetundra pluton. *Bulletin of the Murmansk State Technical University*, *12*, (3), 461–477.

Nerovich, L. I., Bayanova, T. B., Serov, P. A., & Elizarov, D. V. (2014). Magmatic sources of dikes and veins in the Monchetundra mssif, Baltic Shield: isotope-geochronologic and geochemical evidence. *Geochemistry International*, *52* (7), 548–566.

Nitkina, E. A. (2006). U-Pb Zircon Dating of Rocks of the Platiniferous Fedorovo-Pansky Layered Massif, Kola Peninsula. *Doklady Earth Sciences*, *408* (4), 551–554.

Patchett, P. J., & Kuovo, O. (1986). Origin of continental crust of 1.9–1.7 Ga age: Nd isotope and U-Pb zircon ages in the Svecokarelian terrain of south Finland. *Contrib. Mineral. Petrol.*, *92*, 1–12.

Peltonen, P., & Brugmann, G. (2006). Origin of layered continental mantle (Karelian craton, Finland): geochemical and Re-Os isotope constraints. *Lithos*, *89*, 405–423.

Pirajno, F. (2007). Mantle plumes, associated intraplate tectono-magmatic processes and ore systems. *Episodes*, *30* (1), 6–19.

Puchtel, I. S., & Humayun, M. (2001). Platinum-group element fractionation in a komatiitic basalt lava lake. *Geochim Cosmochim Acta*, *65*, 2979–2993.

Puchtel, I. S., Brugmann, G. E., Hofmann, A. W., Kulikov, V. S, & Kulikova, V. V. (2001). Os isotope systematics of komatiitic basalts from the Vetreny Belt, Baltic Shield: Evidence for a chondritic source of the 2.45 Ga plume. *Contrib Mineral Petrol.*, *140*, 588–599.

Puchtel, I. S., Haase, K. M., Hofmann, A. W., Chauvel, C., Kulikov, V. S., Garbe-Schonberg, C. D., & Nemchin, A. A. (1997). Petrology and geochemistry of crustally contaminated komatiitic basalts from the Vetreny Belt, southeastern Baltic Shield: Evidence for an early Proterozoic mantle plume beneath rifted Archaean continental lithosphere. *Geochim. Cosmochim. Acta*, *61*, 1205–1222.

Richardson, S. H., & Shirey, S. B. (2008). Continental mantle signature of Bushveld magmas and coeval diamonds. *Nature*, *453*, 910–913.

Robb, L. (2008). *Introduction to Ore-forming processes*. Oxford: Blackwell Publ., 373.

Rundkvist, D. V., et al. (2006). *Large and superlarge ore deposits, Vol. 1: Global trends of localization*. Moscow: IGEM RAN.

Rundkvist, T. V., et al. (2011). Geological structure and features of localization of PGE mineralization in the eastern South Sopcha mafic-ultramafic massif, Kola Peninsula. *Rudy Met.* (5), 58–68.

Rundkvist, T. V., et al. (2014). The Palaeoproterozoic Vurechuayvench layered Pt-bearing pluton, Kola Peninsula: new results of the U-Pb (ID-TIMS, SHRIMP) dating of baddeleytte and zircon. *Dokl. Earth Sci.*, *454*, (1), 1–6.

Scharer, U., & Gower, C. F. (1988). Crustal evolution in Eastern Labrador: Constraints from precise U-Pb ages. *Precambrian Research*, *38*, 405–421.

Scharer, U., Wilmart, E., & Duchesne, J.-C. (1996). The short duration and anorogenic character of anorthosite magmatism: U-Pb dating of the Rogaland Complex, Norway. *Earth Planet. Sci. Lett.*, *139*, 335–350.

Schissel, D., Tsvetkov, A. A., Mitrofanov, F. P., & Korchagin, A.U. (2002). Basal platinum-group element mineralization in the Fedorov Pansky Layered Mafic Intrusion, Kola Peninsula Russia. *Econ. Geol.*, *97*, 1657–1677.

Serov, P. A., Ekimova, N. A., Bayanova, T. B., & Mitrofanov, F. P. (2014). Sulfide minerals as new geochronometers during Sm-Nd dating of the ore genesis for layered mafic and ultramafic intrusions. *Lithosphere*, (*4*), 11–21.

Sharkov, E. V. (1982). Critical horizon of the Monchegorsk Pluton: Additional intrusive phase. *Zap. Vseross. Mineral. Ova*, *111* (6), 656–663.

Sharkov, E. V. (2006). *Formation of Layered Intrusions and Related Mineralization*. Moscow: Scientific World, 364.

Sharkov, E. V., & Chistyakov, A. V. (2014). Geological and petrological aspects of Ni-Cu-PGE mineralization in the Early Palaeoproterozoic Monchegorsk Layered Mafic-Ultramafic Complex, Kola Peninsula. *Geol. Ore Deposits*, *56* (3), 147–168.

Sharkov, E. V., et al. (2006). Age of the Monchetundra Fault, Kola Peninsula: Evidence from the Sm-Nd and Rb-Sr isotope systematics of metamorphic assemblages. *Geochem. Int.*, *44*, (4), 355–364.

Sluzhenikin, S. F., Distler, V. V., Dyuzhikov, O. A., Kravzov, V. F., Kunilov, V. T., Laputina, N. P., & Turovzev, D. M. (1994). Low-sulfide PGE mineralization in the Norilsk differentiated intrusions. *Geology of Ore Deposits*, *36* (3), 195–217.

Smith, A. D., & Ludden, J. N. (1989). Nd-isotope evolution of the Precambrian mantle. *Earth Planet. Sci. Lett.*, *93*, 14–22.

Smol'kin, V. F., Kremenetsky, A. A., & Vetrin, V. R. (2009). Geological and geochemical model of the Palaeoproterozoic ore-magmatic systems, *Baltic Shield, National Geology*, *3*, 54–62.

Stacey, J. S., & Kramers, J. D. (1975). Approximation of terrestrial lead isotope evolution by a two-stage model. *Earth Planet. Sci. Lett.*, *26*, 207–221.

Starostin, V. I., & Sorokhtin, O. G. (2010). Recycling origin model for sulfide copper-nickel deposits with PGEs of Norilsk type, *Proceedings of the Earth Sciences Division, RANS*, *19*, 5–10.

Steiger, R. H., & Jager, E. (1977). Subcommission on geochronology: convention on the use of decay constants in geo- and cosmochronology, *Earth Planet. Sci. Lett.*, *36*, 359–362.

Verba, M. L., Vinogradov, A. N., & Mitrofanov, F. P. (2005). Evolution of the Earth crust and mineragenic potential of the Barents region. *Structure of the Lithosphere in the Russian Part of the Barents Region, Petrozavodsk, Karelian Science Centre RAS*, 256–311.

Vrevsky, A. B. (2011). Petrology, age, and polychronous sources of the initial magmatism of the Imandra-Varzuga paleorift, Fennoscandian Shield. *Petrology*, *19* (5), 521–548.

Yang, A.Y., Zhao, T. P., Zhou, M. F., Deng, X. G., Wang, G. Q., & Li, J. (2013b). Os isotope compositions of MORBs from the ultra-slow spreading Southwest Indian Ridge: constraints on the assimilation and fractional crystallization (AFC) processes. *Lithos*, *179*, 28–35.

Yang, S.-H., Hanski, E., L. C., Maier, W., Huhma, H., Mokrushin, A.V., Latypov, R., Lahaye, Y., O'Brien, H., & Qu, W.-J. (2016). Mantle source of the 2.44–2.50-Ga mantle plume-related magmatism in the Fennoscandian Shield: evidence from Os, Nd, and Sr isotope compositions of the Monchepluton and Kemi intrusions. *Miner. Deposita*. Doi: 10.1007/s00126-016-0673-9.

Yi-Gang, X. (2007). Mantle plume, large igneous provinces, and lithospheric evolution, *Episodes*, *30*, (1), p. 5.

Zhuravlyov, A. Z., Zhuravlyov, D. Z., Kostitsyn, Y. A., & Chernyshov, I.V. (1987). Determination of the Sm-Nd ration for geochronological purposes. *Geochemistry*, *8*, 1115– 1129.

Zindler, A., & Hart, S. (1986). Chemical geodynamics. *Ann. Rev. Earth Planet. Sci.*, *14*,493–571.

Zozulya, D. R., Bayanova, T. B., & Nelson, E. G. (2005). Geology and age of the late Archaean Keivy Alkaline Province, Northeastern Baltic Shield. *Geology*, *113*, 601–608.

2

Geochemical, Microtextural, and Mineralogical Studies of the Samba Deposit in the Zambian Copperbelt Basement: A Metamorphosed Paleoproterozoic Porphyry Cu Deposit

S. Master[1,2] and N. M. Ndhlovu[2]

ABSTRACT

Ever since the ca. 50 Mt, 0.5% Cu Samba prospect (12.717°S, 27.833°E), hosted by porphyry-associated quartz-sericite-biotite-schists in Zambia, was described in 1978 as a porphyry-type Cu deposit, there has been controversy about its origin and significance. This is because it is situated in the basement to the world's largest stratabound sediment-hosted copper province, the Neoproterozoic Central African Copperbelt. The Samba mineralization consists of chalcopyrite and bornite, occurring as disseminations, stringers, and veinlets, found in a zone >1 km along strike, in steeply dipping lenses up to 10 m thick and >150 m deep. Our new geochemical data show that the host rocks are mainly calc-alkaline metadacites. Cu is correlated with Ag (Cu/Ag ~10,000:1) with no Au or Mo. We confirm that there is alteration similar to that accompanying classical porphyry Cu mineralization, including potassic (biotite-sericite-quartz), propylitic (clinozoisite-chlorite-saussuritized plagioclase), phyllic (sericite-quartz-pyrite-hydromuscovite/illite), and argillic (kaolinite-chlorite-dolomite) alteration. All the rocks show penetrative deformational textures and fabrics, and mineralization and alteration predated deformation and metamorphism. Samba, together with the Mkushi deposits, is part of a long-lived (>100 Ma) Palaeoproterozoic porphyry-Cu province in the Zambian Copperbelt basement, and ore genetic theories for the Copperbelt mineralization must now seriously take this into account

2.1. INTRODUCTION

Wakefield (1978) described a porphyry-type metamorphosed Cu prospect, the Samba deposit, in the basement of the Zambian Copperbelt. Although few other studies have been made of this deposit, there has been controversy about its origin and significance, because it is situated in the basement to the world's largest stratabound sediment-hosted copper province, the Central African Copperbelt (Hitzman ct al., 2012), of which the Zambian Copperbelt forms a part. In the Central African Copperbelt, there are several dozen stratabound sediment-hosted copper (-cobalt) deposits, as well as several vein-hosted copper deposits, situated in the Neoproterozoic Katanga Supergroup (Taylor et al., 2013; Zientek et al., 2014), and some important deposits, such as Lumwana, which are hosted along major shear zones that also involve pre-Katangan basement rocks (Cosi et al., 1992; Bernau, 2007; Bernau et al., 2012; Sillitoe et al., 2015; Turlin et al., 2016).

The presence of mineralization in the pre-Katangan basement has long been noted (e.g., Truter, 1935; Brummer, 1948; Askew & Schmitz, 1960; Pienaar, 1961; Freeman, 1988), and it has played a prominent role in ore genetic models for the stratabound mineralization in the Katanga Supergroup. For example, a number of authors have suggested that copper mineralization in the pre-Katangan basement may have been recycled into the

[1] *Economic Geology Research Institute, School of Geosciences, University of the Witwatersrand, Johannesburg, South Africa*

[2] *School of Geosciences, University of the Witwatersrand, Johannesburg, South Africa*

Ore Deposits: Origin, Exploration, and Exploitation, Geophysical Monograph 242, First Edition.
Edited by Sophie Decrée and Laurence Robb.

Katangan Basin through sedimentological processes of erosion and redeposition (e.g., Binda, 1975, 1993, 1994; Fleischer et al., 1976; Routhier, 1980, 1983; Sweeney, 1985; Vaughan & Binda, 1991; Sweeney & Binda, 1994; El Desouky et al., 2010). Others have suggested that the circulation of fluids through copper-rich basement may have leached out the metals that are found concentrated in the Katangan orebodies (Dejonghe & Ngoyi, 1995; Koziy et al., 2009; Van Wilderode et al., 2014; Sillitoe et al., 2015).

2.1.1. Geological Setting

Although the Central African Copperbelt has a strike length of 700 km in the 900-km-long Neoproterozoic Lufilian arc (Porada & Berhorst 2000), the pre-Katangan basement is mainly exposed in Zambia and immediately adjacent areas of the DRC. In the Zambian Province, the basement rocks are exposed in a structural high known as the Kafue Anticline (Garlick, 1961a) (Fig. 2.1). Samba lies on the western edge of the Kafue Anticline (Fig. 2.1), close to the Katangan unconformity, in the basement of the Zambian Copperbelt (Wakefield, 1978).

In the Central African Copperbelt, the oldest pre-Katangan basement consists of a Paleoproterozoic magmatic arc sequence defined by Rainaud et al. (2005a) as the Lufubu Metamorphic Complex. The Lufubu Metamorphic Complex comprises the Lufubu schists (Mendelsohn, 1961b) and intrusive granitoids, dated at between 1994 and 1873 Ma (Rainaud et al., 2005a). These Paleoproterozoic basement rocks are overlain unconformably by quartzitic and metapelitic metasedimentary rocks of the pre-Katangan Muva Supergroup (Garlick, 1961a). This was then intruded by the 883 ± 10 Ma Nchanga Granite (Armstrong et al., 2005), and unconformably overlain by the Katanga Supergroup, which hosts the stratiform sediment-hosted Cu ± Co orebodies of the Central African Copperbelt. The Katanga Supergroup and its basement were deformed and metamorphosed to greenschist facies in Zambia northeast of the Copperbelt, and to upper greenschist and amphibolite facies in the Solwezi area southwest of the Copperbelt (Mendelsohn, 1961c; McGowan, 2006), during the Lufilian orogenic event at ca. 600–500 Ma (John et al., 2004; Rainaud et al., 2005b). The uppermost Katangan sediments are unmetamorphosed, and were

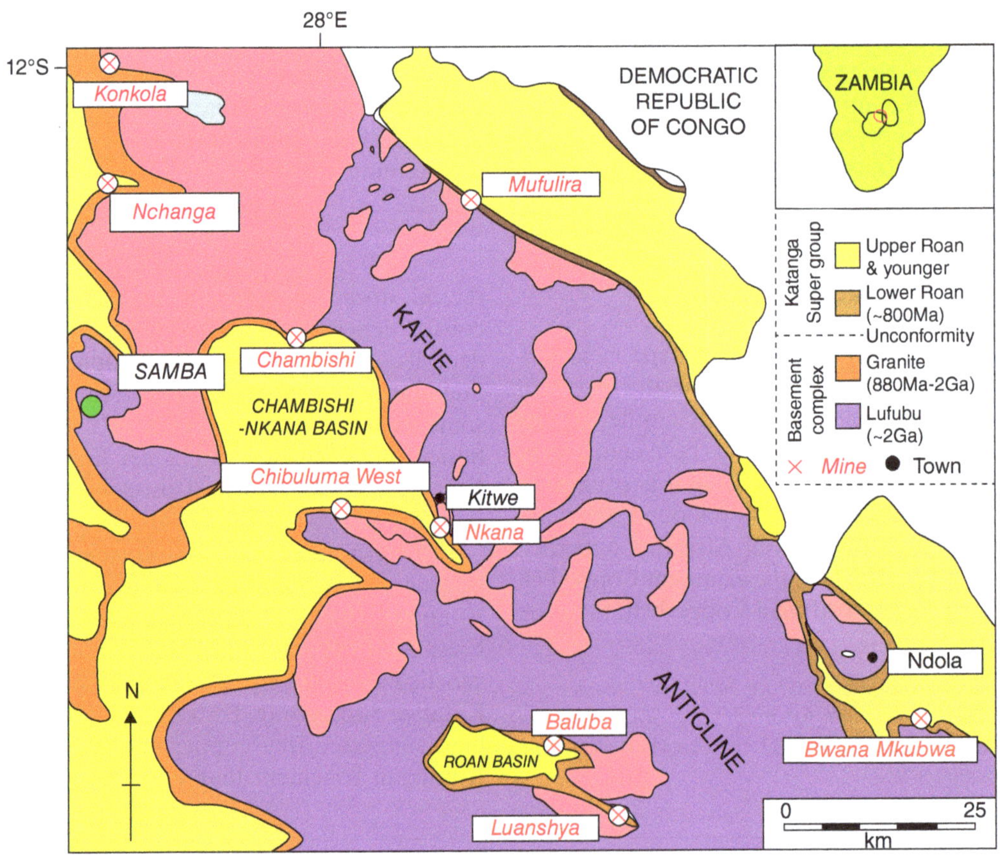

Figure 2.1 Geological map of the Zambian Copperbelt showing the location of Samba deposit, in relation to other major stratiform sediment-hosted copper-cobalt deposits (modified after Fleischer et al., 1976, and Mendelsohn, 1961a).

deposited in syntectonic foreland basins after 573 Ma (Master et al., 2005).

The Samba mineralization is hosted by a sequence of metamorphic rocks, including quartz-mica schists and porphyries. The region around the Samba prospect was mapped by Garrard (1965, 1996). He regarded the host rocks as belonging to the Muva Supergroup. However, U-Pb zircon dating of a felsic metavolcanic quartz-K-feldspar-biotite-muscovite schist from Samba has yielded an age of 1964 ± 12 Ma (Rainaud et al., 2005a), while detrital zircons from Muva metaquartzites from near Mufulira have yielded a maximum age of sedimentation of 1941 ± 40 Ma (Rainaud et al., 2003), hence the metavolcanic schists from Samba are older than the Muva, and belong to the Lufubu Metamorphic Complex (Rainaud et al., 2005a). The quartzites and associated metapelites in the area around Samba (Garrard, 1965, 1996) may belong to the Muva Supergroup, but they must unconformably overlie the schists and deformed granitoid porphyries that host the Samba, deposit. Geochemical analyses of Lufubu schists and porphyritic intrusions (Wakefield, 1978; Rainaud et al., 2005a) are consistent with the Lufubu schists being mainly calc-alkaline metavolcanic rocks, regarded as having formed in a subduction-related magmatic arc (Rainaud et al., 1999).

2.2. PREVIOUS WORK

2.2.1. Discovery, Prospecting and Exploration

The earliest recorded work carried out in the Samba area was in 1949 by Roan Selection Trust (RST) when a regional self-potential survey was carried out and a slight anomaly was detected in the center of the Samba area. Geophysical surveys carried out at Samba include an induced polarization (IP) survey, which identified a long, broad, east-west trending zone of conductivity approximately 1000 m long, and an airborne electromagnetic (EM) survey. In 1950, geochemical surveys, including soil sampling, augering, and pitting, were performed on the soils near the headwaters of the Mwambashi River, and yielded anomalous copper values in weathered schists over a strike length of 1.9 km (Watts et al., 1991). On the east end of the anomaly a small area containing over 0.5% copper was delineated (Fig. 2.2). The auger sludge was also analyzed for cobalt, nickel, zinc, and lead; but no anomalous values were detected (Hayward, 1996).

The results of the auger-drilling program led to an immediate further investigation by means of diamond drilling. Twenty-four boreholes with a maximum depth of 250 m were drilled between 1967 and 1970 (Kihn & Wheeler, 1973). The core was sampled for copper,

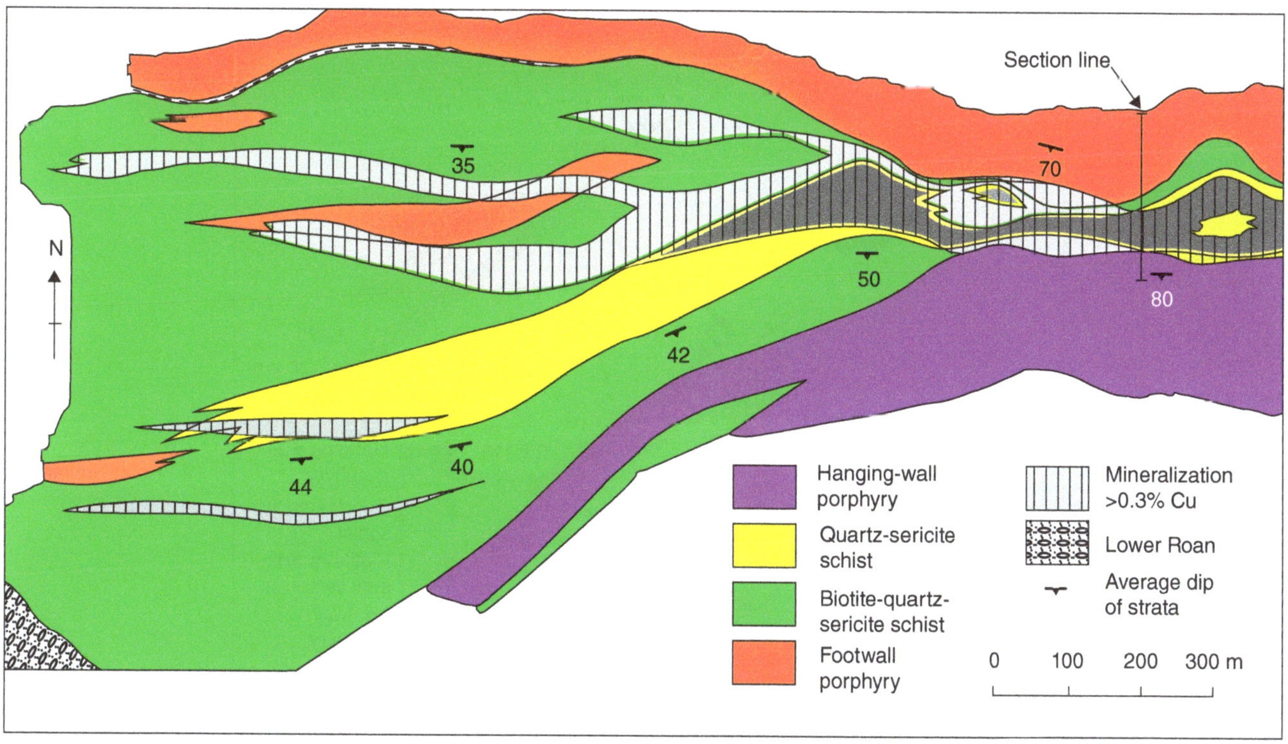

Figure 2.2 Geological plan of the Samba copper deposit at level 100 m level (after Wakefield, 1978). The section line for Figure 2.3 is indicated with an arrow.

cobalt, nickel, and zinc; again only copper anomalies were observed. The copper sulfides occurred in the form of disseminations, stringers, and veinlets of pyrite, chalcopyrite, and bornite, with traces of chalcocite and, at shallow depths, minor malachite and chrysocolla (Wakefield, 1978; Hayward, 1996).

Thin zones of relatively high-grade (±2% Cu) mineralization were observed. These zones are separated by zones of little or no mineralization, and can seldom be correlated between adjacent drill-holes (Figs. 2.2 and 2.3). The conclusion was that mineralization occurs as several discontinuous bands or that, less likely, there is a tight isoclinal fold in which two or more mineralized zones are repeated (Hayward, 1996). There is, however, from the available drilling information, no evidence of isoclinal folding, and no indication that any mineralization is controlled by, or concentrated in, fold structures. The drilling by RST (Kihn & Wheeler, 1973), and further economic evaluation by Mindeco-Noranda (1974), led to the delineation of a copper resource of ca. 50 Mt at 0.5% Cu, to a depth of 175 m below surface (Wakefield, 1978). In 1996, Avmin Zambia reevaluated the Samba deposit, and reassayed some of the Mindeco-Noranda borehole core for Cu, Ag, Au, Co, and Mo (Hayward, 1996). They also carried out geophysical surveys on the strike extension of the Samba deposit.

2.2.2. Mineralization and Origin

The geology of the Samba deposit was described for the first time by Wakefield (1978), who used Garrard's (1965) mapping and stratigraphic correlations as the basis for his interpretation of the geology of the Samba deposit. His work was derived exclusively from drill-holes, since there was no surface outcrop. He defined four main rock units that occur east of the Samba deposit (see Fig. 2.3), as follows:

1. Unit 1, Hangingwall porphyry: Feldspar-porphyritic quartz monzonite grading to sericite-quartz-biotite schist
2. Unit 2, Quartz-sericite schist with minor biotite
3. Unit 3, Biotite-porphyroblastic quartz-sericite schist grading to sericitized feldspar-biotite porphyry
4. Unit 4, Footwall porphyry: feldspar-porphyritic grey microgranodiorite, grading to sericite-quartz-biotite schist

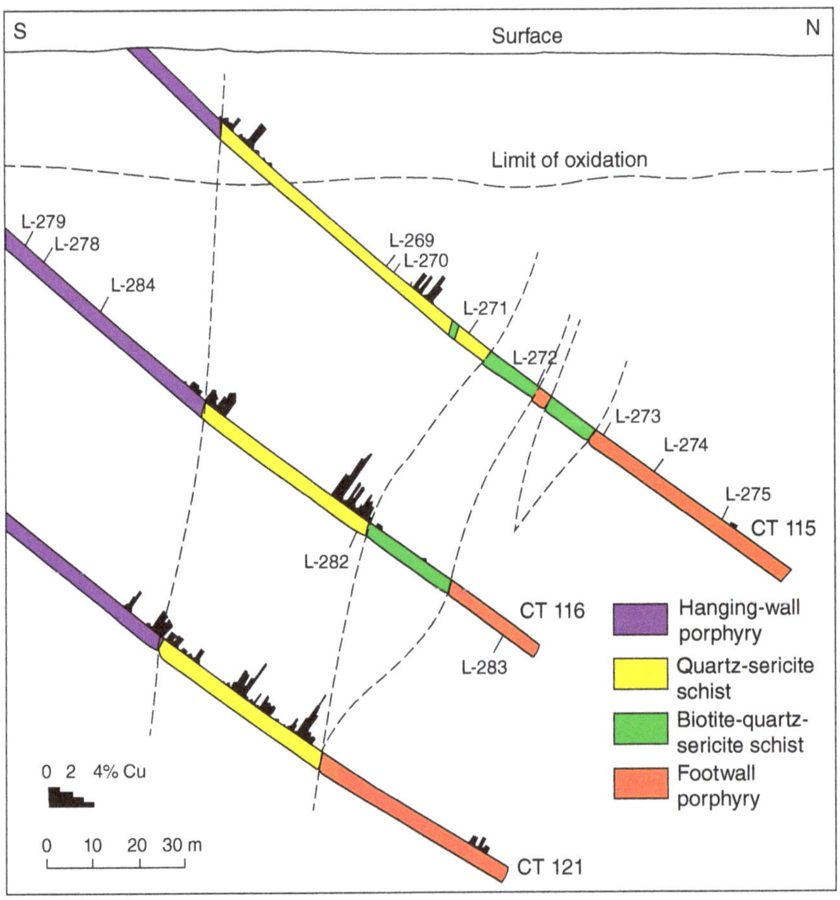

Figure 2.3 A S-N geological cross section through part of the Samba deposit, along drill section line CT 115-116-121, shown on Figure 2.2 (after Wakefield, 1978).

The Samba mineralization is hosted by a sequence of metamorphic rocks, including quartz-sericite porphyries, quartz-sericite-biotite schists, biotite-chlorite schists. The orebodies contain concentrations of the copper sulfides chalcopyrite and bornite, occurring as disseminations, stringers, and veinlets, found in a zone extending for over 1 km along strike, in two elongate lenses up to 10 m thick and extending to a depth of over 150 m. Wakefield (1978) distinguished four main ore types in the Samba deposit as follows:

1. Stringer sulfide: fine stringers of chalcopyrite and bornite in highly deformed quartz-sericite schist
2. Bleb sulfide: subequidimensional chalcopyrite aggregates disseminated in poorly foliated rocks, commonly associated with chlorite-bearing schist of unit 3
3. Vein sulfide: blebs of chalcopyrite with minor bornite in pre-S1 quartz sericite veinlets, most common in the unit 3 schists
4. Late quartz-vein sulfide: occasional, very coarse segregation of chalcopyrite and minor bornite in postdeformational quartz veins

A number of features were found by Wakefield (1978), including association with porphyritic intrusions, and mineralization and alteration styles similar those associated with porphyry Cu deposits, which suggested that the Samba deposit was a metamorphosed porphyry-type deposit. Since that time, the Samba deposit has been reevaluated, and a number of authors have pronounced that from their examination of the drill-cores, it does not resemble typical porphyry-style mineralization. Lobo-Guerrero (2005), who only examined four samples petrographically, argued that because of its deformation, few aspects of a typical porphyry copper system were readily evident in the Samba deposit. For example, he noted that he did not recognize a porphyritic intrusion, and did not see potassic alteration, epidote, or argillic alteration, or typical crackle breccias, and suggested that Samba could be related to a high sulfidation system, but without providing any supporting evidence. Bernau (2007) and Bernau et al. (2012) likened Samba to the style of mineralization in the Lumwana deposit, associated with shears along the contact of Katangan metasedimentary rocks and pre-Katangan basement rocks. They noted the widespread alteration of feldspars to sericite in both deposits, but also noted that the alteration at Samba predated deformation. Hitzman et al. (2012) suggested that it is a late-stage impregnation of copper mineralization into the basement, and was presumably formed by metamorphic fluids derived from the Katangan Basin during regional deformation and metamorphism. This inference was prompted by a ca. 480 Ma ^{40}Ar-^{39}Ar age from the Samba deposit (Hitzman et al., 2012). Hitzman et al. (2012) stated that in their opinion, the Samba mineralization was the latest copper deposit to have formed in the Central African Copperbelt. However, if the Samba deposit really is so young, then it must have postdated the Lufilian deformation and metamorphic events, which are dated at ca. 560–510 Ma (Rainaud et al., 2005b), and the deposit ought to be undeformed and unmetamorphosed.

2.3. THIS STUDY

In our study, we have examined the microtextures and mineralogy of the Samba ores, in relation to alteration and deformation; looked at the geochemistry and classification of the host rocks; and the trace metal contents of the ores; and compared the tonnage-grade relationships with other deposits worldwide. We have set out to test the rival hypotheses for the origin of the Samba deposit: it either represents a deformed and metamorphosed porphyry-style copper deposit related to porphyritic intrusions in a magmatic arc setting (Wakefield, 1978), or it represents a deposit formed in a late-stage (post-tectonic) shear zone, with ore-bearing fluids derived by metamorphic dewatering of the Katangan basin (Hitzman et al., 2012). In Table 2.1 we set out the relevant features of each style of deposit, which would be consistent with each hypothesis.

2.3.1. Analytical Methods

Two borehole cores from the Samba deposit, CT-115 and CT-116, were logged in detail at the Kalulushi core storage facility in Zambia, and samples were collected for petrographic and geochemical studies.

Table 2.1 Features That Would Be Expected to Be Present for the Two Models to Be Applicable.

Porphyry Cu Model	Late Impregnation Cu Model
• Must be related to calc-alkaline magmatic arc • Related to porphyritic intrusion • Mineralization style in stringers, veins, breccias • Timing of mineralization: Early, pretectonic • Alteration types: potassic, propylitic, phyllic, argillic • Metals: Cu (Mo, Au, Ag) • Tonnage and grade must conform to porphyry Cu model.	• Rock types unimportant, except Eh-pH, fO_2 • Not related to intrusions • Mineralization style–veins, impregnations • Timing of mineralization: Late, post-tectonic • Alteration types: only one, if wall-rock reactions involved • Metals: Cu, Co, for fluids derived from Katangan basin • Tonnages and grades–any

Note: Based on the characteristics of porphyry copper deposits, after Lowell and Guilbert (1970), and Titley (1982); and shear-zone hosted mineralization models, after Craw (1990) and Ettner et al. (1993).

Table 2.2 Major Element Analyses of Samba Samples (in Weight %).

Sample #; Borehole, depth,	Major element oxides (weight %)											
Lithology	SiO_2	TiO_2	Al_2O_3	Fe_2O_3	MnO	MgO	CaO	Na_2O	K_2O	P_2O_5	LOI	TOTAL
L-268; CT115, c.70 m Qz-sericite schist	69.51	0.69	16.52	4.32	0.01	0.35	0.13	0.54	4.53	0.17	3.62	100.39
L-269; CT115, 73.5 m, Qz-sericite schist	75.84	0.31	13.87	3.02	0.00	0.00	0.20	0.31	3.67	0.21	3.03	100.46
L-270; CT115, 76.9 m, Qz-sericite schist	77.73	0.30	11.64	1.70	0.01	0.22	0.12	0.34	3.25	0.10	1.99	97.40
L-271; CT115, 95.3 m, Qz-sericite schist	69.55	0.77	14.95	5.01	0.01	0.84	0.30	0.40	4.72	0.20	4.07	100.82
L-272; CT115, 113.4 m, Feldspar-qz-biotite porphyry	66.82	0.60	15.17	4.37	0.12	2.58	2.22	3.53	3.25	0.16	1.95	100.77
L-273; CT115, 130.5 m, Feldspar-qz-biotite porphyry	65.92	0.52	14.92	4.14	0.12	1.78	3.54	2.94	3.12	0.18	2.98	100.16
L-275; CT115, 160.3 m, Biotite p'blastic qz-bt schist	92.49	0.10	1.04	1.47	0.02	0.60	0.24	0.00	0.62	0.09	0.83	97.50
L-278; CT116, 63.8 m, Feldspar-qz-biotite porphyry	67.86	0.70	14.29	4.00	0.09	1.49	3.31	3.43	3.44	0.18	1.30	100.09
L-279; CT116, 59.5 m, Feldspar-qz-biotite porphyry	67.94	0.63	14.39	4.41	0.08	1.31	3.25	3.31	3.08	0.22	1.29	99.91
L-282; CT116, 156.2 m, Biotite schist	54.00	2.12	15.94	9.49	0.16	6.47	2.20	0.15	6.18	0.47	2.49	99.67
L-283; CT116, 194.3 m, Dol. p'blastic chlorite schist	41.86	0.63	9.03	9.44	0.40	13.43	8.18	0.00	0.04	0.30	15.49	98.80
L-284; CT116, 81.2 m, Feldspar-qz-biotite porphyry	67.61	0.68	13.78	4.47	0.07	2.38	1.93	2.40	3.92	0.17	2.08	99.49
CT127/370 Feldspar-qz-biotite porphyry	65.41	0.57	14.38	4.27	0.09	1.89	3.48	3.93	2.89	0.16	2.65	99.72
CT169/1256, Feldspar-qz-biot.-ms schist	65.99	0.56	14.08	4.29	0.12	2.77	2.71	2.65	3.83	0.15	2.42	99.57

Table 2.3 Trace Element Analyses of Samba Samples (in Parts per Million).

	Trace Elements (parts per million)											
Sample #	Rb	Sr	Y	Zr	Nb	Co	Ni	Cu	Zn	V	Cr	Ba
L-268	178	37	48	275	22	10	16	942	14	72	187	1116
L-269	74	20	16	179	19	6	10	6	11	27	185	291
L-270	88	30	15	174	14	17	1279	10569	49	17	263	319
L-271	174	26	45	317	14	14	1032	37	14	75	209	1191
L-272	162	181	38	191	11	7	726	24	56	74	160	862
L-273	128	110	31	173	14	18	8	49	38	77	206	966
L-275	17	5	2	47	0	38	2637	17543	36	4	579	55
L-278	131	208	45	295	13	8	811	12	43	60	173	1032
L-279	123	204	39	285	19	7	9	6	51	64	302	810
L-282	347	60	32	202	16	76	571	1052	79	232	125	1013
L-283	12	49	15	126	5	48	701	10	225	132	1153	5
L-284	175	118	47	294	14	27	876	613	60	70	192	866
CT127-370	125	156	34	186	11	10	645	15	40	66	138	894
CT169-1256	144	57	28	184	7	14	786	3	57	72	172	1049

Major and trace elements (Tables 2.2 and 2.3) were analyzed using X-Ray Fluorescence Spectrometry (XRF), on Philips PW 1400 or PW 2404 spectrometers in the EarthLab at the School of Geosciences, University of the Witwatersrand, Johannesburg, South Africa. Major element geochemistry was done using the fusion technique of Norrish and Hutton (1969). Trace element geochemistry was done on pressed powder pellets by XRF using the method of Feather and Willis (1976). X-Ray Diffraction (XRD) identification of minerals was done on powders using a Philips PW 1380 diffractometer, in the School of Chemistry, University of the Witwatersrand.

2.3.2. Geochemistry of Host Rocks

In terms of their compositions, the Samba samples from this study, together with those from Wakefield (1978), predominantly plot in the acid subalkaline dacite field in the total alkalis-silica (TAS) diagram (Fig. 2.4), and in the field of high-K calc-alkaline dacites in the K_2O vs. SiO_2 diagram (Le Maitre et al. (1989) (Fig. 2.5). In the AFM diagram of Irvine and Barager (1971), the samples are depleted in MgO and enriched in alkalis, and plot in the calc-alkaline field (Fig. 2.6). Since these rocks have been altered and metamorphosed, they may have suffered some alkali element mobility. Using the immobile, incompatible trace element Zr/TiO_2 vs. Nb/Y classification diagram of Winchester and Floyd (1977), the Samba metavolcanic rocks plot mainly in the field of rhyodacite/dacite, with a few samples plotting in the fields of andesite, alkali basalt, and trachyandesite (Fig. 2.7). In addition to classification based on major oxides, major cation compositions were plotted on the R1-R2 (De la Roche et al., 1980) and Al-(Fe_{tot}+Ti)-Mg (Jensen, 1976) diagrams for subalkaline volcanic rocks (Ndhlovu, 2007). The cation plots confirmed that the rocks are indeed mainly calc-alkaline granodiorites/dacites. In the Sr/Y vs. Y discrimination diagram of Martin (1999), the samples plot in the field of normal arc volcanics, and not in the field of adakites (Ndhlovu, 2007). The calc-alkaline composition of the host rocks for the Samba deposit is significant, because it is one of the defining characteristics of porphyry copper deposits that they are hosted by calc-alkaline intrusions in magmatic arcs related to subduction zones (e.g., Sillitoe & Perelló, 2005; Berger et al., 2008).

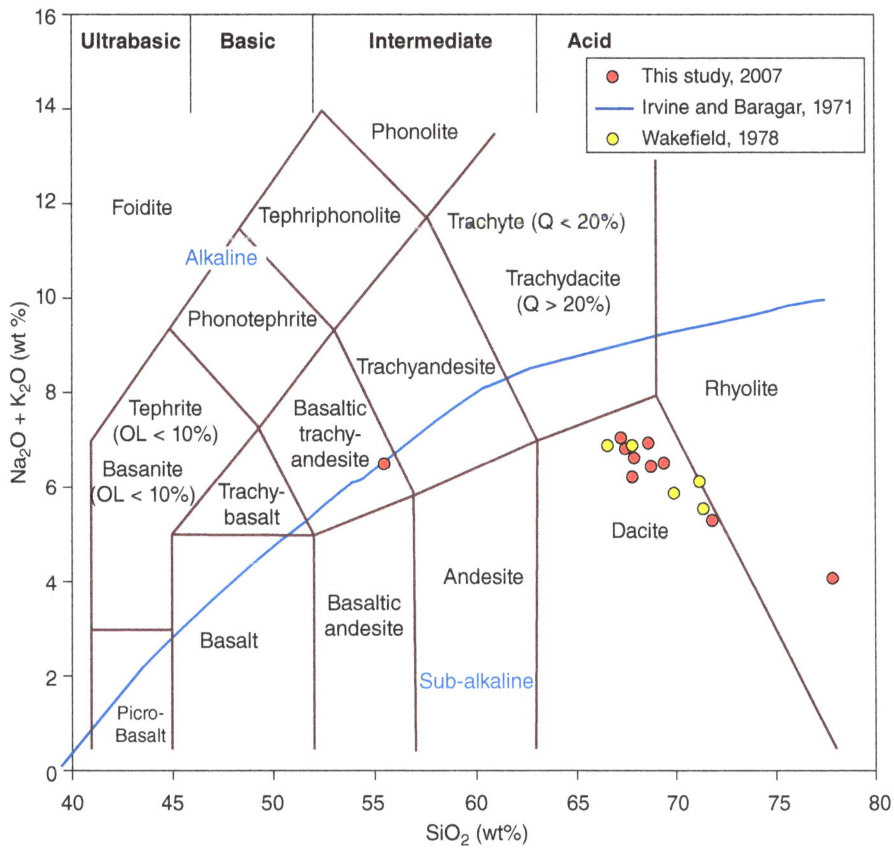

Figure 2.4 The total alkali-silica (TAS) diagram (after Le Maitre et al., 1989). Analyses from this study are plotted together with those from Wakefield (1978).

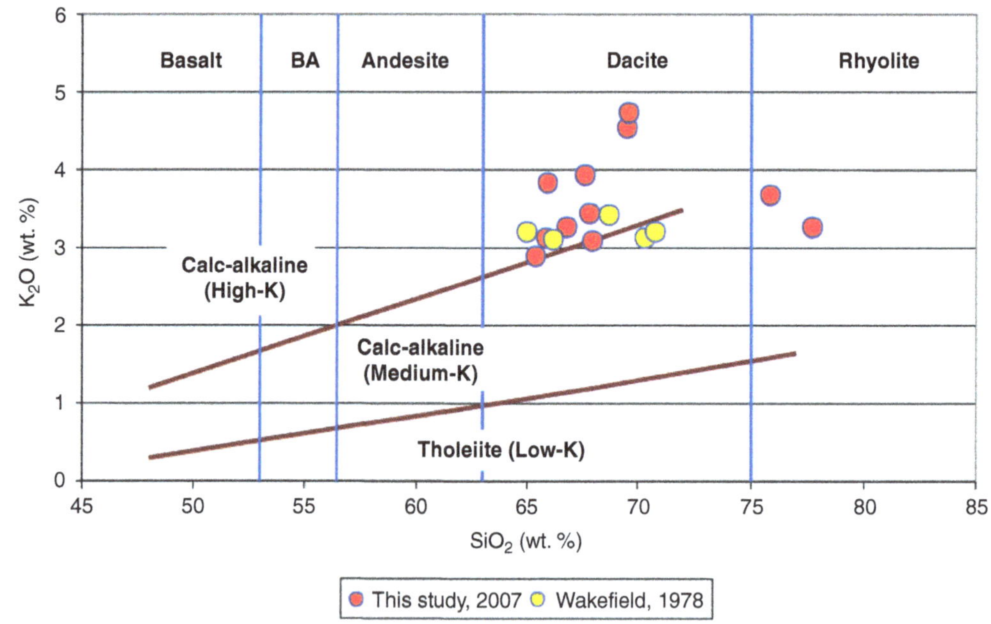

Figure 2.5 K_2O vs. SiO_2 diagram (after Le Maitre et al., 1989). Analyses from this study are plotted together with those from Wakefield (1978). BA = Basaltic andesite.

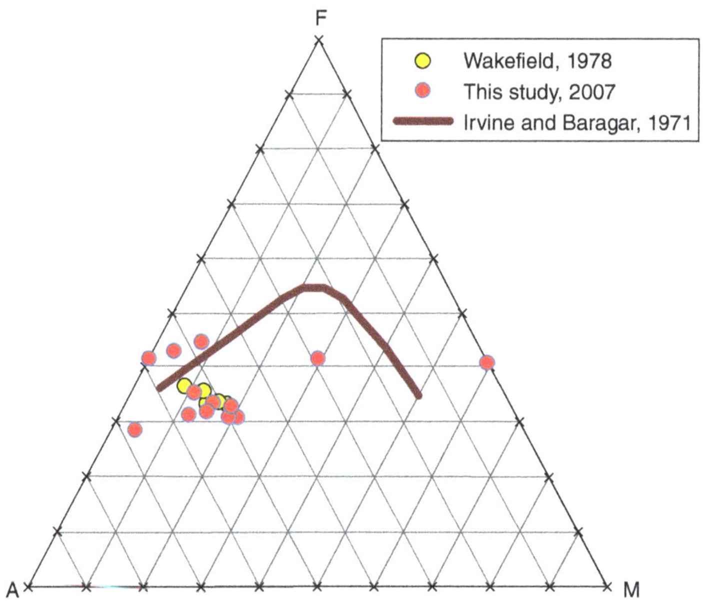

Figure 2.6 AFM diagram (Irvine and Baragar, 1971). A = $Na_2O + K_2O$; F = FeO + $0.8998Fe_2O_3$; M = MgO. Analyses from this study are plotted together with those from Wakefield (1978).

2.3.3. Mineralogy and Textures

2.3.3.1. Mineralization: Metal Endowment, Tonnages, and Grades

The Samba mineralization consists mainly of chalcopyrite-bornite veins and disseminated pyrite. When plotted on metal grade and tonnage distribution plots based on a worldwide database of 422 porphyry copper deposits (Singer et al., 2008), the Samba deposit plots well within the cluster, toward the low copper tonnage side (Fig. 2.8). The average grade of Samba (0.5 % Cu) is close to the median grade of 0.44% for the global porphyry copper dataset (Singer et al., 2008) (Fig. 2.9). The estimated tonnage of the Samba deposit (50 Mt) is at the low end of the global tonnage distribution (the lower ten percentile of the tonnage is 33 Mt, Singer et al., 2008) (Fig. 2.10).

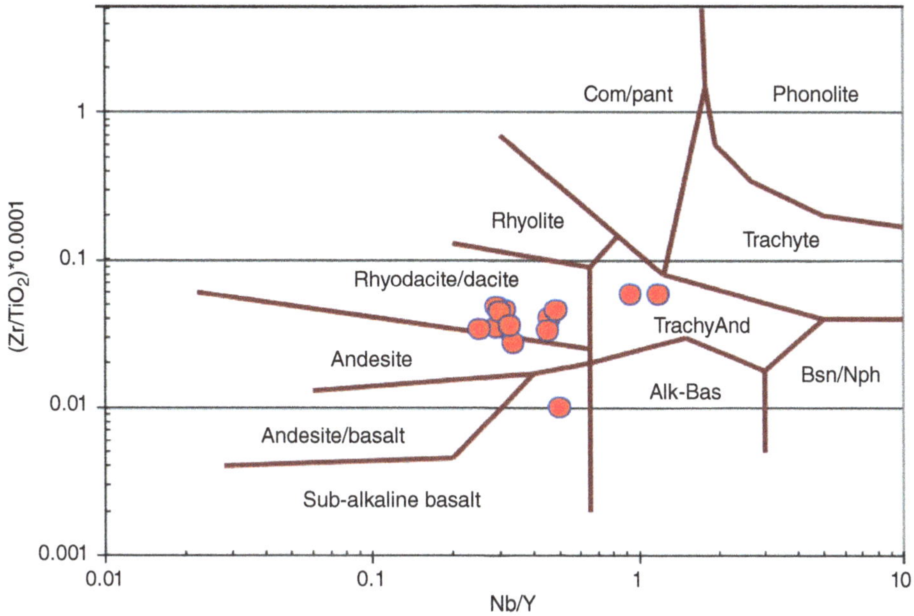

Figure 2.7 Zr/TiO$_2$ vs. Nb/Y geochemical classification diagram (Winchester & Floyd, 1977). Abbreviations: com/pant = comendite/pantellerite; TrachyAnd = trachyandesite; Bsn/Nph = basanite/nephelinite; Alk-Bas = alkaline basalt.

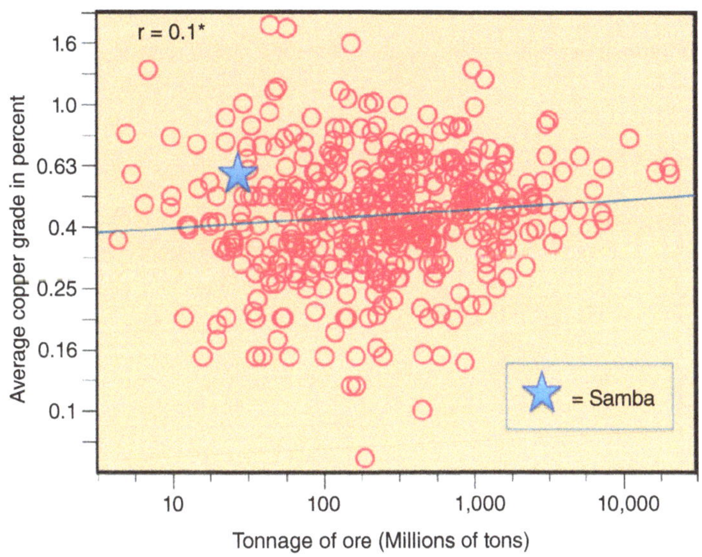

Figure 2.8 Grade versus tonnage plot for a global dataset of 422 porphyry copper deposits, after Singer et al. (2008). Samba is plotted as a blue star.

The copper is associated with silver (with no molybdenum or gold). In the mineralized samples analyzed by Hayward (1996), the Ag values ranged from <1 to 7.3 g/t, with a positive correlation between Cu and Ag (Fig. 2.11). For copper grades of around 1%, the Ag grade is around 2 ppm, while for copper grades of 4%, the Ag grade is around 5 ppm, giving a Cu:Ag ratio of ~10,000:1. The average Ag grade (n = 30) is 0.5 ppm (Hayward, 1996). This ore mineral association together with the grade and tonnage has been observed in most calc-alkaline hosted porphyry copper deposits in Western Canada (Gilmour, 1982). Analyses done by Avmin Zambia show that the Samba ores are poor in gold, the average gold value being <0.08 ppm. They also show that in most samples, there is little or no Mo, with the highest value being 0.03% Mo (Hayward, 1996). Twenty-three analyses of various Samba ore samples with copper values ranging between 1.02% and 4.03% all had a cobalt

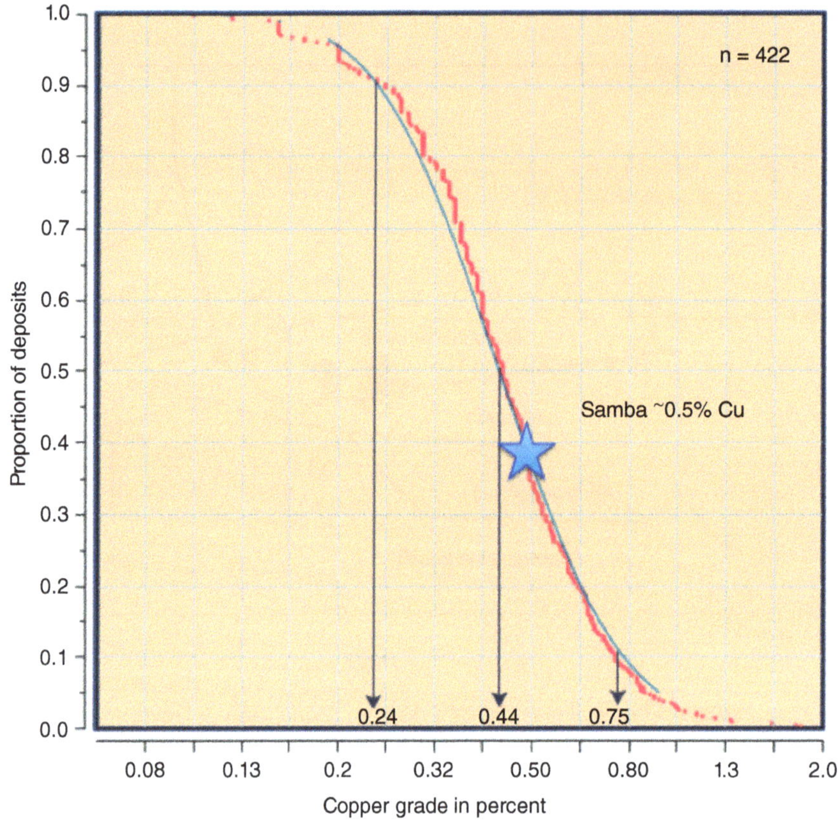

Figure 2.9 Plot of proportion of deposits versus copper grade for a global dataset of 422 porphyry copper deposits (after Singer et al., 2008).

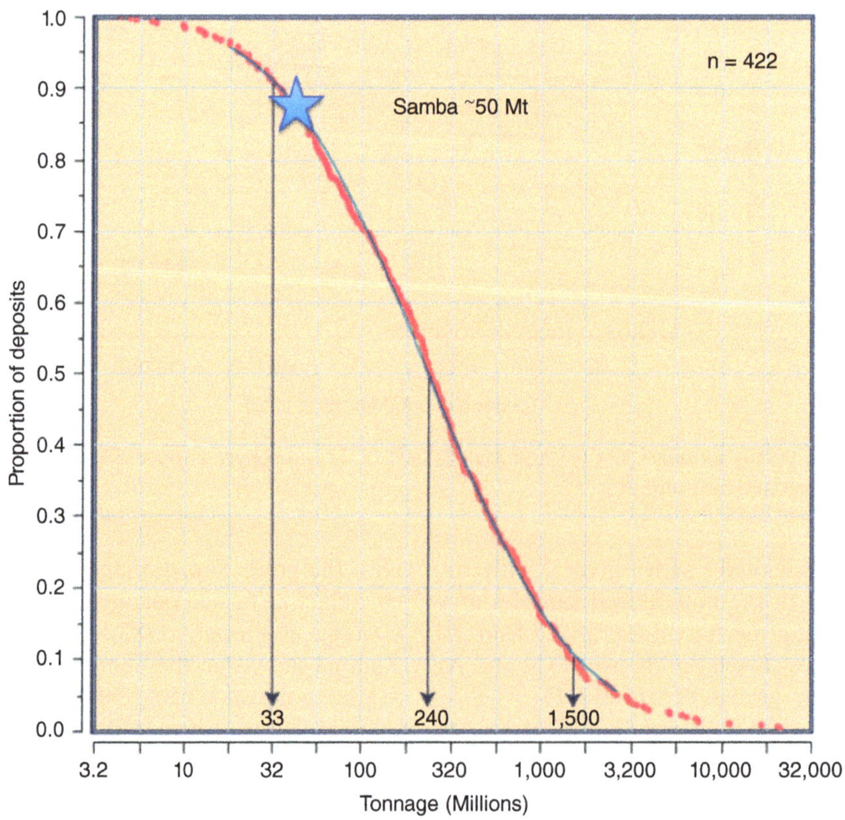

Figure 2.10 Plot of proportion of deposits versus tonnage for a global dataset of 422 porphyry copper deposits (after Singer et al., 2008).

content less than the detection limit of 0.01% (Hayward, 1996), while in 14 other samples analyzed, the highest Co value was 76 ppm (Ndhlovu, 2007).

2.3.3.2. Mineralization: Wallrock Alteration and Deformation

2.3.3.2.1. Potassic (K-silicate) Zone Secondary (hydrothermal) euhedral brown biotite and quartz-filled fractures have been observed in the quartz-feldspar-biotite porphyry (L-272, L-273) (Fig. 2.12a). This brecciated rock contains polycrystalline quartz and feldspar phenocrysts (<5 mm) in moderately foliated sericite-quartz-biotite matrix with clinozoisite. The plagioclase is replaced by sericite. The secondary biotite is thought to be recrystallized original igneous biotite (rather than alteration of hornblende) because of its brown color, which is a result of high Fe^{2+} and low Ti contents compared to the igneous biotite, which is more greenish (Tröger, 1956). The Ti from the original igneous biotite was possibly lost during alteration to sphene, which is also present in the feldspar-quartz-biotite porphyries. The crackle breccia containing biotite is considered to have formed early in the deformation history, since the biotite has been partly replaced by later retrograde chlorite (Fig. 2.12b,c). Such breccias were also recorded by Wakefield (1978). The rock also contains early euhedral pyrite, with quartz-rich pressure shadows, showing the predeformational origin of the mineralization. This assemblage is interpreted in porphyry copper deposits as potassic (Lowell & Guilbert, 1970) or K-silicate alteration (Creasey, 1966; Meyer & Hemley, 1967; Titley, 1982).

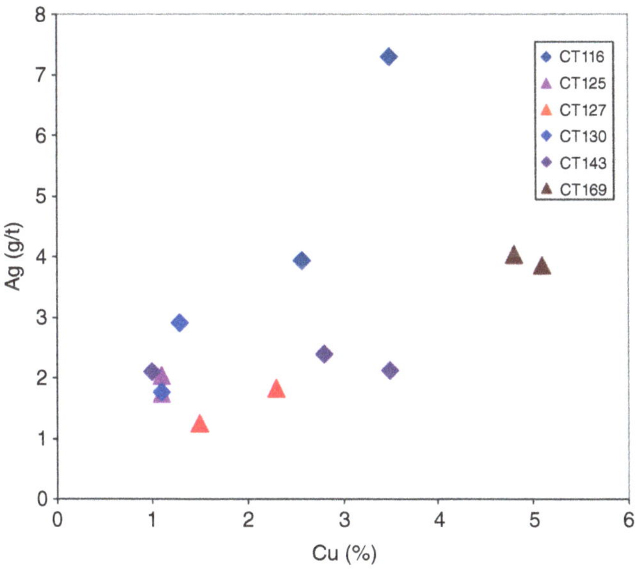

Figure 2.11 Ag vs. Cu plot for quartz-sericite schists from different boreholes drilled at Samba.

2.3.3.2.2. Phyllic (Sericitic) Zone Pervasive sericitization of most minerals, associated with relatively large volumes (>5%) of pyrite and chalcopyrite, has been observed in the quartz-blastoporphyritic sericite schists (formed from altered quartz porphyries) (Figs. 2.13a and 2.14). The sericite defines a strong S1 cleavage, which has been crenulated by S2 (Fig. 2.14b,c). Early pyrite

Figure 2.12 (a) Crackle brecciated quartz-feldspar-biotite porphyry L-272, with biotite along fractures; (b) plane polarized photomicrograph showing biotite replaced by chlorite; (c) cross-polarized view of (b). Bt = biotite; chl = chlorite; qz = quartz.

Figure 2.13 Photomicrographs of quartz blastoporphyritic quartz-sericite schist L-271 (altered quartz porphyry): (a) Relict quartz phenocryst with fine-grained matrix of sericite and quartz; (b) cross-polarized transmitted light view of fractured euhedral pyrite crystal, replaced by sericite, with a quartz pressure shadow at the bottom; (c) reflected light view of fractured pyrite crystal of (b), showing replacement along fractures by hematite and sericite. Qz = quartz; py = pyrite; hm = hematite; ser = sericite.

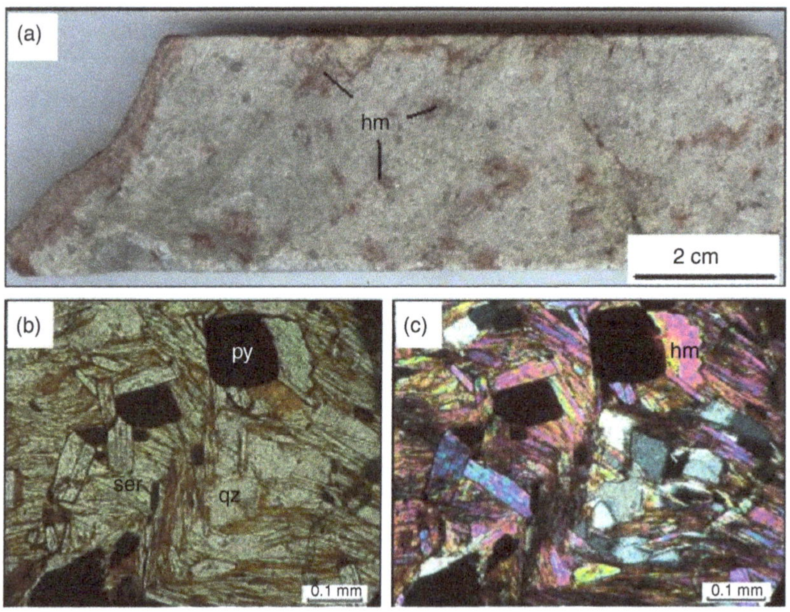

Figure 2.14 Quartz blastoporphyritic crenulated quartz-muscovite schist L-268, with hematite pseudomorphs after pyrite: (a) cut core specimen showing reddish hematite clusters after pyrite; (b) plane polarized photomicrograph showing pyrite (black) partly replaced by hematite, surrounded by crenulated intergrowth of sericite and quartz; (c) cross-polarized view of (b) revealing pressure shadows around pyrite grains, and S1/S2 fabrics wrapping around early pyrite grains.

has been fractured and replaced by sericite, and later pseudomorphed by hematite (Fig. 2.13b,c). The S1 and S2 fabrics wrap around early formed pyrite, which has quartz-filled pressure shadows (Fig. 2.14b,c). XRD results have also identified the presence of illite (hydromuscovite) in one of the samples of this rock type. This type of alteration assemblage has been identified by Lowell and Guilbert (1970) as phyllic alteration, and by Meyer and Hemley (1967) as sericitic alteration. Other authors defined this sericite alteration as the assemblage of sericite, quartz, and pyrite (Creasey, 1966; Rose, 1970). Phyllic alteration is the most dominant alteration of the four types observed at Samba.

2.3.3.2.3. Argillic Zone XRD results have identified the presence of kaolinite associated with chlorite in the dolomite-blastoporphyritic chlorite schist (L-283; CT116, 194.3m). The kaolinite is intergrown with chlorite, and is found as inclusions within later dolomite porphyroblasts, showing its early, premetamorphic origin. Kaolinite is the alteration product of alkali feldspar under acidic conditions, and is observed to be weakly developed compared with the other alteration assemblages. The kaolinite and chlorite characterize the argillic alteration assemblage (Creasey, 1966; Meyer & Hemley, 1967; Rose, 1970). The argillic zone is often poorly developed or absent in porphyry copper systems (Lowell & Guilbert, 1970).

2.3.3.2.4. Propylitic Zone In quartz-feldspar-biotite porphyry L-273, chlorite and clinozoisite is present in the same sample (Fig. 2.15a). Chlorite replaces biotite from the grain boundary inward, while clinozoisite replaces plagioclase feldspar, which is also saussuritized (Fig. 2.15b). Early pyrite has quartz-filled pressure shadows (Fig. 2.15b,c). This alteration assemblage has been interpreted in most well-studied porphyry copper deposits as the propylitic alteration (Lowell & Guilbert, 1970; Beane, 1982; Titley, 1982).

All four typical porphyry copper alteration types (Lowell & Guilbert, 1970) are present in the Samba suite, however the characteristic mineral assemblages of both argillic and propylitic alteration have been observed in the same sample. This observation is not really problematic with respect to alteration zonation, and may be explained by temporal overprinting or by the presence of kaolinite in the propylitic alteration zone, as has been recorded in many porphyry copper deposits (Lowell & Guilbert, 1970). Wakefield (1978) suggested that this mineral assemblage might represent a metamorphosed propylitic assemblage.

2.3.4. Deformation

The Samba host rocks are mineralized, altered, and metamorphosed, and the relationship between the mineralization and alteration has been obscured by metamorphism. There are two tectonic deformation events identified at

Figure 2.15 Photomicrographs of quartz-feldspar-biotite porphyry L-273: (a) Chlorite (green) replacing biotite (brown); (b) saussuritized plagioclase; (c) pyrite grain with pressure shadow (reflected light); (d) pyrite grain with quartz pressure shadow (transmitted light, cross polarized). chl = chlorite; pl = plagioclase; py = pyrite; qz = quartz.

Samba. The first deformation (D1) is defined by parallel alignment of the phyllosilicate minerals sericite, muscovite, biotite, and chlorite, defining a well-developed cleavage (S1). This alignment of micas is often accompanied by compositional segregation defined by quartz-rich and mica-rich microlithons. The S1 cleavage was later folded during a second deformation event (D2), producing crenulation cleavage (S2), which is associated with quartz-rich hinges. The S2 cleavage is restricted to the schists only, while the S1 cleavage is present throughout the whole suite of samples.

The following evidence indicates that mineralization and alteration predates the earliest deformation: (a) the S1 foliation is present in alteration minerals such as sericite, secondary biotite, and chlorite; (b) pressure shadows occupied by metamorphic minerals are present around early euhedral pyrites (Figs. 2.13–2.15); (c) sulfide aggregates aligned parallel to S1 foliation, giving the host-rock a banded texture, (d) the S1 and S2 foliations wrap around fragmented sulfides (Fig. 2.14); and (d) mineralized quartz veins containing copper sulfides are folded.

2.3.5. Fractures

Limonite-filled fractures are found in weathered quartz-sericite schists. The limonite represents altered chalcopyrite veins. The relationship between the limonite and the fractures suggest structurally controlled primary copper mineralization, prior to weathering. Other fracture sets are found in the feldspar-quartz-biotite porphyry units. The fractures are filled with recrystallized biotite (partly replaced by chlorite) associated with very fine-grained disseminated pyrite (Fig. 2.12). These are similar to crackle breccias formed during mineralization in porphyry copper systems (e.g., Titley, 1982). Because of the partial replacement of the biotite by retrogressive chlorite, the crackle breccias are constrained to have formed early, prior to greenschist facies metamorphism.

2.4. DISCUSSION

Hitzman et al. (2012) obtained $^{40}Ar/^{39}Ar$ ages of 490–460 Ma on mica from Cu mineralized rocks of the Samba deposit. They claimed that this gave the age of the mineralization, and that Samba was the youngest copper deposit to have formed in the Zambian Copperbelt, (formed by late stage impregnation of Cu-bearing fluids from the Katangan basin into the pre-Katangan basement). However, ^{40}Ar–^{39}Ar ages in the range of 490 to 450 Ma are common in the entire Copperbelt (Rainaud et al., 2005b), and confirm the original findings of Cahen and Snelling (1971), based on K-Ar dating, of a widespread regional post-tectonic thermal event at this time, related to uplift and cooling of the Lufilian orogen, and closure of the K-Ar system in the metamorphic rocks.

An important question is the origin of the metals. If Samba was a late impregnation, then the implication is that the metals were derived from the Katanga Basin and "remobilized" into the basement. Examples of veins derived from such remobilization are common in the mines of the Zambian Copperbelt, for example, at Baluba, where quartz-calcite-anhydrite-sulfide veins derived from the overlying mineralized Katangan rocks are found in the Lufubu schists (Pienaar, 1961). However, nowhere in the Copperbelt does one find large-scale deposits formed by this kind of remobilization. The size of the Samba orebody (50 Mt @ 0.5% Cu; 250 kt Cu) is comparable to that of one of the nearby Katangan stratabound sediment-hosted copper deposits, such as Fitula (4.5 Mt @ 5% Cu; 225 kt Cu) and Mimbula (46.9 Mt @ 1.2 % Cu) (Zientek et al., 2014), in terms of contained metal. If it were formed by remobilization, it would imply the complete remobilization of a deposit the size of Fitula, in terms of contained Cu metal. Presumably, any late fluids coming out of the Katangan basin would have to be of metamorphic origin, such as the cupriferous fluids implicated in the origin of syntectonic mineralization at Nchanga Mine (McGowan et al., 2003, 2006) and at Lumwana (Bernau et al., 2012; Turlin et al., 2016). However, such cupriferous fluids are invariably also enriched in cobalt, which is not the case at Samba.

Most primary copper mineralization hosted by pre-Katangan basement rocks, along contacts, and in shear zones, as well as secondary copper mineralization in veins emplaced from the overlying metal-rich Katangan rocks, is also enriched in cobalt (Pienaar, 1961; Daly et al., 1984; Vaughan & Binda, 1991; Bernau et al., 2012). Soil sampling by RST over the Mimbula and Fitula deposits, the nearest copper deposits to Samba that are hosted in Katangan rocks, showed that there is a good correlation of Cu and Co values, with Co in soil ranging between 10 and 100 ppm, going up to over 200 ppm, and exceptionally even higher than 1000 ppm (Garlick, 1961b; Cornwall, 1970). This is explained by the solubility of both copper and cobalt as chloride complexes in saline solutions (Pan & Susak, 1989). By contrast, the Samba deposit contains no cobalt, implying a source other than from remobilization from Katangan orebodies.

In terms of tonnage and grade, the Samba deposit falls squarely within the range of 422 known porphyry copper deposits worldwide (Singer et al., 2008). The mean grade of Samba (0.5 % Cu) is just slightly higher than the median grade (0.44%) of the global porphyry copper dataset, while the mean tonnage of Samba, 50 Mt, is above the lower quartile, 33 Mt, of the global range (Singer et al., 2008). In terms of associated metal content- the Samba deposit is associated with Ag, but with no Au or Mo. Such metal associations are also found in a number of porphyry copper systems worldwide (Kirkham & Sinclair, 1996; Singer et al., 2008).

In this study, we have found all of the features that are to be expected from the porphyry copper model (Table 2.1), which is thus strongly supported. We find a number of features, however, that contradict the expectations of the late-stage impregnation model (early, pretectonic, rather than late, post-tectonic, introduction of metals; four types of alteration, rather than one; strong spatial association with porphyritic intrusion; metal association different from those in Katanga basin). Thus, the late impregnation model (Table 2.1) for the origin of the Samba deposit is refuted.

Precambrian porphyry type Cu ± Mo ± Au deposits, associated with calc-alkaline subduction zone volcanics, or I-type granitoids, are present in many parts of the world, and range in age from 3.3 Ga to 561 Ma (Sikka & Nehru, 1997). Paleoproterozoic porphyry-type deposits include the 2.49 Ga Malanjkhand deposit in the Singhbum craton, India (Stein et al., 2004), the 2.16 Ga Gaoua deposits of Burkina Faso (Le Mignot et al., 2017), the 2.12 Ga Yongkuanyu deposit in the South China Craton (Jiang et al., 2014); the 1.9 Ga Aitik and Tallberg deposits in the Skellefte District of Sweden (Weihed & Schöberg, 1991; Weihed, 1992; Wanhainen et al., 2012); and the ca. 1.9 Ga Haib and Lorelei prospects in southern Namibia (Minnitt, 1986; Viljoen et al., 1986).

To this list can now be added the fertile Paleoproterozoic basement of the Zambian Copperbelt and surrounding areas, which contains porphyry-style mineralization at Samba and also in the historic Mkushi District of central Zambia, east of the Copperbelt, where copper mineralization in veins and aplites related to intrusive porphyritic granites has been mined in the past (Stillman, 1965; Legg, 1976). Whereas the Mkushi mineralization was thought to have been related to intrusive granites of the late Mesoproterozoic Irumide Belt (Rainaud et al., 2005a), the Mkushi granite has now been dated at 2044 ± 68 Ma, and the mineralized aplites have been redated at 2052 ± 18 Ma (Master et al., 2015). Thus, the Zambian Copperbelt basement records long-lived magmatic arc systems, which contained porphyry-style copper deposits of Paleoproterozoic age, as well as younger Mesoproterozoic mineralization (Sillitoe et al., 2015). As already hinted at by Routhier (1980, 1983), these rich metal endowments in the basement may have been recycled into the overlying Neoproterozoic Katanga Basin, and may explain the unique size and metal enrichments found in it.

2.5. CONCLUSIONS

In our study of the Samba deposit, we have been able to find all four classic styles of mineralization-related alteration associated with porphyry Cu deposits, including potassic, propylitic, phyllic, and argillic alteration. The presence of these different alteration assemblages, with implied changes in ore fluid pH and cationic composition, supports the porphyry copper model for Samba (Wakefield, 1978), and refutes a late introduction model, for which a single alteration type would be implied.

The textures indicating predeformational sulfide mineralization and alteration, overprinted by two phases of deformation, are completely incompatible with any theory, whether explicit or implied, such as that suggested by Hitzman et al. (2012), which calls for a late introduction of mineralization into the host rocks. Once again, the porphyry copper model is supported by this evidence, and the late introduction theory is refuted.

Samba, together with the Mkushi deposits of central Zambia, is part of a long-lived (>100 Ma) Paleoproterozoic porphyry-Cu province in the Zambian Copperbelt basement, and ore genetic theories attempting to explain the uniquely huge size of the Copperbelt mineralization must now seriously take this into account.

While our results strongly support the deformed porphyry copper deposit model proposed by Wakefield (1978), our interpretations differ from his in some important ways. Whereas Wakefield regarded the Samba host rocks as belonging to the "Muva System" (after Garrard, 1965), we now know that the host rocks are Paleoproterozoic in age, and are part of the Lufubu Metamorphic Complex. This age difference also explains why Wakefield (1978), like us, found two cleavages in the Samba host rocks, but only one in outcropping rocks of the Muva System. Wakefield (1978) suggested that areas of basement mineralization may be spatially associated with enhanced mineralization in overlying Katangan rocks, containing detrital sulfides (Binda, 1975), derived from the underlying basement mineralization, for example, at Chingola (Voet & Freeman, 1972). However, these early reports of detrital sulfides have been refuted (Annels, 1979), and the enrichment in copper of the Nchanga Granite at Chingola (Voet & Freeman, 1972) is due to infiltration along fractures from the overlying mineralized Katangan rocks, since the Nchanga Granite intrusion lacks primary copper sulfide mineralization (Garlick, 1973). We regard the presence of porphyry copper deposits in the basement of the Central African Copperbelt as an indication of a copper-enriched provenance for the Katanga basin, which is derived from denudation of this basement (Master et al., 2005). This could have provided metal-enriched detrital components into the basin, such as copper-rich magnetites derived from arc rocks (Jenner et al., 2010), which would have been a good intrastratal source of copper and other metals (Ilton & Eugster, 1989) for basinal brines, which were responsible for the stratabound sediment-hosted copper mineralization in the Katangan rocks.

ACKNOWLEDGMENTS

We are grateful to Avmin Zambia for access to the cores, and to unpublished company reports. We acknowledge the logistical help received for this project from Laurence Robb, Tim Williams, and the late Dave Armstrong. S. Master thanks the School of Geosciences, University of the Witwatersrand, and the South African NRF for funding to attend the 2015 EGU Meeting in Vienna, where this paper was first presented.

REFERENCES

Annels, A. E. (1979). Mufulira greywackes and their associated sulphides. *Trans. Inst. Min. Metall., Sect. B, 88*, 15–23.

Armstrong, R. A., Master, S., & Robb, L. J. (2005). Geochronology of the Nchanga Granite, and constraints on the maximum age of the Katanga Supergroup, Zambian Copperbelt. *Journal of African Earth Sciences, 42* (1–5), 32–40.

Askew, J. F. R., & Schmitz, J. G. (1960). An occurrence of mineralized Basement, 6 Limb. Unpubl. Rept., Roan Antelope Copper Mines Limited, Zambia.

Beane, R. E. (1982). Hydrothermal alteration in silicate rocks. In S. R. Titley (Ed.), *Advances in Geology of the Porphyry Copper Deposits: Southwestern North America*. The University of Arizona Press, Tucson, 118–123.

Berger, B. R., Ayuso, R. A., Wynn, J. C., & Seal, R. R. (2008). Preliminary model of porphyry copper deposits. U.S. Geological Survey Open-File Report 2008-1321.

Bernau, R. (2007). The geology and geochemistry of the Lumwana basement hosted copper-cobalt (uranium) deposits, NW Zambia. D. Phil. Thesis, School of Ocean and Earth Science, University of Southampton, England, UK.

Bernau, R., Roberts, S., Richards, M., Nisbet, B., Boyce, A., & Nowecki, J. (2012). The geology and geochemistry of the Lumwana Cu (± Co ± U) deposits, NW Zambia. *Mineralium Deposita*; doi: 10.1007/s00126-012-0424-5.

Binda, P. L. (1975). Detrital bornite grains in the Late Precambrian B Greywacke of Mufulira, Zambia. *Mineralium Deposita, 10*, 101–107.

Binda, P. L. (1993). In search of modern analogues of the Zambian copper orebodies: The Cenozoic exotic deposits of northern Chile. Abstracts, IGCP Project 302: The structure and metallogenesis of Central African Late Proterozoic Belts, Copperbelt Field Conference, Kalulushi, Zambia, 23–31 July 1993.

Binda, P. L. (1994). Metallogenic models and Cenozoic analogues for the Katanga copper deposits. Résumé des communications, Colloque International de Cornet, 5–9 septembre 1994, Faculté Polytechnique de Mons, Belgique, 11.

Brummer, J. J. (1948). Pyrrhotite–bearing quartz veins in the Basement Complex; Small mineralized fault in the Basement pegmatites. Unpubl. Rept., Roan Antelope Copper Mines Limited, Zambia.

Cahen, L., & Snelling, N. J. (1971). Données radiométriques nouvelles par la méthode K-Ar. Existence d'une importante élévation de température post-tectonique dans les couches katangiennes du sud du Katanga et de la Zambia. *Annales de la Société Géologique de Belgique, 94*, 199–209.

Cornwall, F. W. D. (1970). Discovery and exploration of the Fitula copper deposit, Nchanga area, Zambia. In M. J. Jones (Ed.), *Mining and Petroleum Geology*. Proc. 9th Commonw. Min. Metall. Congr., London, 1969, vol. *2*, Inst. Min. Metall., London, 535–560.

Cosi, M., De Bonis, A., Gosso, G., Hunziker, J., Martinotti, G., Moratto, S., Robert, J. P., & Ruhlman, F. (1992). Late Proterozoic thrust tectonics, high pressure metamorphism and uranium mineralization in the Domes area, Lufilian Arc, northwestern Zambia. *Precambrian Research, 58*, 215–240.

Craw, D. (1990). Regional fluid and metal mobility in the Dalradian metamorphic belt, Southern Grampian Highlands, Scotland. *Miner. Deposita, 25*, 281–288.

Creasey, S. C. (1966). Hydrothermal alteration. In S. R. Titley and C. L. Hicks (Eds.). *Geology of the Porphyry Copper Deposits Southwestern North America*. The University of Arizona Press, Tucson, 51–75.

Daly, M. C., Chakroborty, S. K., Kasolo, P., Musiwa, M., Mumba, P., Naidu, B., Namateba, C., Ngambi, O., & Coward, M. P. (1984). The Lufilian Arc and Irumide Belt of Zambia: Results of a geotraverse across their intersection. *Journal of African Earth Sciences, 2*, 311–318.

Dejonghe, L., & Ngoyi, K. (1995). Le gisement de Kinsenda (sud-est du Shaba, Zaïre): une concentration cuprifère stratoïde dans les formations détritiques du Roan (Protérozoïque supérieur). Chronique de recherches minière, Orléans, No. 521, 19–37.

De la Roche, H., Leterrier J., & Grande Claude, P. (1980). A classification of volcanic and plutonic rocks using R1–R2 diagrams and major element analyses: its relationship and current nomenclature. *Chemical Geology, 29*, 183–210.

El Desouky, H., Muchez, P., Boyce, A., Schneider, J., Cailteux, J., Dewaele, S., & von Quadt, A. (2010). Genesis of sediment-hosted stratiform copper-cobalt mineralization at Luiswishi and Kamoto, Katanga Copperbelt (Democratic Republic of Congo). *Mineralium Deposita, 45*(8), 735–763.

Ettner, D. C., Bjorlykke, A., & Andersen, T. (1993). Fluid evolution and Cu-Au genesis along a shear zone: A regional and fluid inclusion study of shear zone-hosted gold and copper mineralization and alteration in the Kautokeino greenstone belt, Finnmark, Norway. *Journal of Geochemical Exploration, 49*, 233–267.

Feather, C. E. & Willis, J. P. (1976). A simple method for background and matrix correction of spectral peaks in trace element determination by X-ray fluorescence spectrometry. *X-Ray Spectrom., 5*, 41–48.

Fleischer, V. D., Garlick, W. G., & Haldane, R. (1976). Geology of the Zambian Copper Belt. In K. H. Wolf (Ed.), *Handbook of Strata-bound and Stratiform Ore Deposits*, Vol. *6*. Elsevier, New York, 223–352.

Freeman, P. V. (1988). Description of some mineral deposits discovered or re-investigated in the post-World War II period. Unpubl. Rept., Zambia Consolidated Copper Mines Limited, Zambia.

Garlick, W. G. (1961a). Basement complex: Muva system. In F. Mendelsohn (Ed.), *The Geology of the Northern Rhodesian Copperbelt*. Macdonald, London, 21–30.

Garlick, W. G. (1961b). Geochemistry: (2) Rhodesian Selection Trust. In F. Mendelsohn (Ed.), *The Geology of the Northern Rhodesian Copperbelt*. Macdonald, London, 208–212.

Garlick, W. G. (1973). The Nchanga granite. In L. A. Lister (Ed.), *Symposium on Granites, Gneisses and Related Rocks*, Geol. Soc. S. Afr. Spec. Publ., *3*, 455–474.

Garrard, P. (1965). Chingola, Kalulushi and Ndola (Rural) Districts, Chingola Sheet. Annual Report for 1964, *Geological Survey of Zambia*, 2–3.

Garrard, P. (1996). The geology of the Chingola area: Explanation of Degree Sheet 1227, SE Quarter. Rep. *Geological. Survey of Zambia,* No. 66. (Written 1965, published 1996).

Gilmour, P. (1982). Grades and tonnages of porphyry copper deposits. In S. R. Titley (Ed.), *Advances in Geology of the Porphyry Copper Deposits: Southwestern North America*. The University of Arizona Press, Tucson, 18–20.

Hayward, G. A. (1996). Geological evaluation of the Samba copper deposit, Zambia. Unpublished Report, Avmin Zambia, Kitwe.

Hitzman, M. W., Broughton, D., Selley, D., Woodhead, J., Wood, D. & Bull, S. (2012). *The Central African Copperbelt: Diverse Stratigraphic, Structural, and Temporal Settings in the World's Largest Sedimentary Copper District. Sillitoe Volume*, Society of Economic Geologists, Special Publication, 16, 487–514.

Ilton, E. S., & Eugster, H. P. (1989). Base metal exchange between magnetite and a chloride-rich hydrothermal fluid. *Geochimica et Cosmochimica Acta*, *53*, 291–301.

Irvine, T. N., & Baragar, N. B. W. (1971). A guide to the chemical classification of the common volcanic rocks. *Canadian Journal of Earth Sciences*, *8*, 523–548.

Jenner, F. E., O'Neill, H. St. C., Arculus, R. J., & Mavrogenes, J. A. (2010). The magnetite crisis in the evolution of arc-related magmas and the initial concentration of Au, Ag and Cu *Journal of Petrology*, *51*(12), 2445–2464.

Jensen, L. S. (1976). A new cation plot for classifying subalkalic volcanic rocks. Ontario Div. Mines. Misc. Pap. 66.

Jiang, Y., Niu, H., Bao, Z., Li, N., Shan, Q., & Yang, W. (2014). Fluid evolution of the Tongkuanyu porphyry copper deposit in the Zhongtiaoshan region: Evidence from fluid inclusions. *Ore Geology Reviews*, *63*, 498–509.

John, T., Schenk, V., Mezger, K., & Tembo, F. (2004). Timing and PT evolution of whiteschist metamorphism in the Lufilian Arc- Zambezi Belt Orogen (Zambia); implications for the assembly of Gondwana. *Journal of Geology*, *112*(1), 71–90.

Kihn, L. M., & Wheeler, J. C. (1973). Final Report on Samba Exploration, Area EL15, Chisangwa Mines Limited. Unpubl. Rept., Roan Consolidated Copper Mines Limited, RCM Geological Services, Kalulushi, Zambia.

Kirkham, R. V., & Sinclair, W. D. (1996). Porphyry copper, gold, molybdenum, tungsten, tin, silver. In O. R. Eckstrand, W. D. Sinclair, & R. I. Thorpe (Eds), *Geology of Canadian Mineral Deposit Types, Geology of Canada*, No. 8: Geological Survey of Canada, 421–446.

Koziy, L., Bull, S., Large, R., & Selley, D. (2009). Salt as a fluid driver, and basement as a metal source, for stratiform sediment-hosted copper deposits. *Geology*, *37*, 1107–1110.

Legg, C. (1976). The geology and mineralization of the Mkushi copper deposits. Ministry of Mines and Industry, Geological Survey Department, Zambia, Lusaka, Report No. 38.

Le Maitre, R. W., Baterman, P., Dudek, A., Séller, J., Lameyere, Le Bas, M. J., Sabine, P. A., et al. (1989). *A Classification of Igneous Rocks and Glossary of Terms*. Blackwell Scientific Publications, Oxford.

Le Mignot, E., Siebenaller, L., Béziat, D., André-Mayer, A.-S., Reisberg, L., Salvi, S., Velasquez, G., et al. (2017). The Paleoproterozoic copper-gold deposits of the Gaoua District, Burkina Faso: Superposition of orogenic gold on a porphyry copper occurrence? *Economic Geology*, *112*, 99–122.

Lobo-Guerrero, A. (2005). Geochemistry, Geochronology and Metallogeny of Pre-Katangan and Post-Katangan granitoids in the greater Lufilian Arc, Zambia and Namibia. Ph.D. thesis, School of Geosciences, University of the Witwatersrand, Johannesburg, South Africa.

Lowell, J. D., & Guilbert, J. M. (1970). Lateral and vertical alteration-mineralization zoning in porphyry ore deposits. *Economic Geology*, *65*, 373–406.

Martin, H. (1999). Adakite magma: Modern analogues of Archean granitoids. *Lithos.*, *46*, 411–429.

Master, S., Frei, D., & Milani, L. (2015). A revised Palaeoproterozoic age for the granite-hosted (porphyry-style?) copper mineralization at Mkushi, Irumide Belt, Zambia, and regional implications. Abstract, 7th Igneous and Metamorphic Studies Group Conference, University of Pretoria, Pretoria, South Africa, 12–14 January 2015.

Master, S., Rainaud, C., Armstrong, R. A., Phillips, D., & Robb, L. J. (2005). Provenance ages of the Neoproterozoic Katanga supergroup (Central African Copperbelt), with implications for basin evolution. *Journal of African Earth Sciences*, *42*, 41–60.

McGowan, R. R., Roberts, S., & Boyce, A. J. (2006). Origin of the Nchanga copper-cobalt deposits of the Zambian Copperbelt. *Mineralium Deposita*, *40*, 617–638.

McGowan, R. R., Roberts, S., Foster, R. P, Boyce, A. J., & Coller, D. (2003). Origin of the copper-cobalt deposits of the Zambian Copperbelt: An epigenetic view from Nchanga. *Geology*, *31*, 497–500.

Mendelsohn, F. (Ed.) (1961a). *The Geology of the Northern Rhodesian Copperbelt*. Macdonald, London.

Mendelsohn, F. (1961b). Basement complex: Lufubu system. In F. Mendelsohn (Ed.), *The Geology of the Northern Rhodesian Copperbelt*. Macdonald, London, 18–21.

Mendelsohn, F. (1961c). Metamorphism. In F. Mendelsohn (Ed.), *The Geology of the Northern Rhodesian Copperbelt*, Macdonald, London, 106–116.

Meyer, C., & Hemley, J. J. (1967). Wall rock alteration. In H. L. Barnes (Ed.), *Geochemistry of Hydrothermal Ore Deposits*, Holt, Rinehart and Winston, New York, 166–235.

Mindeco-Noranda (1974). Samba ore reserve estimation. Unpublished Report, Mindeco Zambia.

Minnitt, R. C. A. (1986). Porphyry copper-molybdenum mineralization at Haib River, S.W. Africa. In C. R. Anhaeusser & S. Maske (Eds.), *Mineral Deposits of Southern Africa*. Geol. Soc. S. Afr., Johannesburg, Vol. II, 1567–1585.

Ndhlovu, N. M. (2007). Mineralization and origin of the Samba copper deposit, in the basement of the Zambian Copperbelt.

Geology Honours Dissertation, School of Geosciences, University of the Witwatersrand, Johannesburg, South Africa.
Norrish, K., & Hutton, J. T. (1969). An accurate X-ray spectrographic method for the analysis of a wide range of geological samples. *Geochim. Cosmochim. Acta*, *33* (4), 431–453.
Pan, P., & Susak, N. J. (1989). Co(II)-chloride and -bromide complexes in aqueous solutions up to 5 m NaX and 90 °C: Spectrophotometric study and geological implications. *Geochim. Cosmochim. Acta*, *53*, 327–341.
Pienaar, P. J. (1961). Basement complex: Mineralization in the basement. In F. Mendelsohn (Ed.), *The Geology of the Northern Rhodesian Copperbelt*. Macdonald, London, 30–41.
Porada, H., & Berhorst, V. (2000). Toward a new understanding of the Neoproterozoic–Early Palaeozoic Lufilian and northern Zambezi belts in Zambia and the Democratic Republic of Congo. *Journal of African Earth Sciences*, *30*, 727–771.
Rainaud, C., Armstrong, R. A., Master, S., & Robb, L. J. (1999). A fertile Palaeoproterozoic magmatic arc beneath the Central African Copperbelt. In C. J. Stanley et al. (Eds.), Mineral deposits: Processes to processing. Proc. 5th Biennial SGA Meeting & 10th Quadrennial IAGOD symposium, London, 22–25 August 1999,1427–1430.
Rainaud, C., Master, S., Armstrong, R. A., & Robb. L. J. (2003). A cryptic Mesoarchaean terrane in the basement to the Central African Copperbelt. *Journal of the Geological Society, London*, *160*, 11–14.
Rainaud, C., Master, S., Armstrong, R. A., & Robb, L. J. (2005a). Geochronology and nature of the Palaeoproterozoic basement in the Central African Copperbelt, with regional implications. *Journal of African Earth Sciences*, *42*(1–5), 1–31.
Rainaud, C., Master, S., Armstrong, R. A., Phillips, D., & Robb, L. J. (2005b). Monazite U-Pb dating and 40Ar/39Ar thermochronology of metamorphic events in the Central African Copperbelt during the Pan-African Lufilian orogeny. *Journal of African Earth Sciences*, *42*(1–5), 183–199.
Rose, A. W. (1970). Zonal relations of wallrock alteration and sulfide distribution at porphyry copper deposits. *Economic Geology*, *65*, 920–936.
Routhier, P. (1980). Ou sont les métaux pour l'avenir? Les provinces métalliques. Essai de métallogénie globale. Mém. du BRGM No. 105. Éditions du BRGM, Orléans.
Routhier, P. (1983). Where are the metals for the future? The Metal Provinces. An essay on global metallogeny. Éditions du BRGM, Orléans (Contains a section on metallogenic inheritance in the Central African Copperbelt, 215–220).
Sikka, D., & Nehru, C. E. (1997). Review of Precambrian porphyry Cu ± Mo ± Au deposits with special reference to Malanjkhand porphyry copper deposit, Madhya Pradesh, India. *Journal of the Geological Society of India*, *49*(3).
Sillitoe, R. H., & Perelló, J. (2005). Andean copper province: Tectonomagmatic settings, deposit types, metallogeny, exploration, and discovery. Society of Economic Geologists, Economic Geology 100th Anniversary Volume, 845–890.
Sillitoe, R. H., Perelló, J., Creaser, R. A., Wilton, J., & Dawborn, T. (2015). Two ages of copper mineralization in the Mwombezhi Dome, Northwestern Zambia: Metallogenic implications for the Central African Copperbelt. *Economic Geology*, *110*(8), 1917–1923.
Singer, D. A., Berger, V. I., & Moring, B. C. (2008). Porphyry copper deposits of the world: Database, and grade and tonnage models (2008). USGS Open File Report 2008–1155.
Stein, H. J., Hannah, J. L., Zimmerman, A., Markey, R. J., Sarkar, S. C., & Pal, A. B. (2004). A 2.5 Ga porphyry Cu-Mo-Au deposit at Malanjkhand, central India: Implications for Late Archean continental assembly. *Precambrian Research*, *134*, 189–226.
Stillman, C. J. (1965). The geology of the Musofu River and the Mkushi areas. Explanation of Degree Sheet 1329, part of NW Quarter and SW Quarter. Rep. Geol. Surv. Zambia, 12.
Sweeney, M., & Binda, P. L. (1994). Some constraints on the formation of the Zambian Copperbelt deposits. *Journal of African Earth Sciences*, *19*(4), 303–313.
Sweeney, M. A. (1985). Diagenetic Processes in Ore Formation with Special Reference to the Zambian Copperbelt and Permian Marl Slate. Ph.D. dissertation, University of Aston, Birmingham, UK.
Taylor, C. D., Causey, J. D., Denning, P. D., Hammarstrom, J. M., Hayes, T. S., Horton, J. D., Kirschbaum, M. J., et al. (2013). Descriptive models, grade-tonnage relations, and databases for the assessment of sediment-hosted copper deposits, with emphasis on deposits in the Central African Copperbelt, Democratic Republic of Congo and Zambia. *U.S. Geol. Surv. Sci. Investig. Rep*. 2010-5090-J.
Titley, S. R. (1982). The styles and progress of mineralization and alteration in porphyry copper systems. In S. R. Titley (Ed.), *Advances in Geology of the Porphyry Copper Deposits: Southwestern North America*. The University of Arizona Press, Tucson, 93–114.
Tröger, W. E. (1956). Optische Bestimmung der gesteinsbildenden Minerale. E. Schweizerbart'sche Verlagsbuchhandel, Stuttgart.
Truter, F. C. (1935). The geology and petrography of the Mkushi-Lunsemfwa area, Northern Rhodesia. Unpublished thesis, University of Cape Town.
Turlin, F., Eglinger, A., Vanderhaeghe, O., André-Mayer, A.-S., Poujol, M., Mercadier, J., & Bartlett, R. (2016). Synmetamorphic Cu remobilization during the Pan-African orogeny: Microstructural, petrological and geochronological data on the kyanite-micaschists hosting the Cu(–U) Lumwana deposit in the Western Zambian Copperbelt of the Lufilian belt. *Ore Geology Reviews*, *75*, 52–75.
Van Wilderode, J., El Desouky, H. A., Elburg, M. A., Vanhaecke, F., & Muchez, P. (2014). Metal sources for the Katanga Copperbelt deposits (DRC): Insights from Sr and Nd isotope ratios. *Geologica Belgica*, *17*, 137–147.
Vaughan, D. J., & Binda, P. (1991). The geochemistry of the basement complex of the Zambian Copperbelt- implications for mineralization. In M. Pagel & J. L. Leroy (Eds.), *Source, Transport and Deposition of Metals*. Balkema, Rotterdam, 359–362.
Viljoen, R. P., Minnit, R. C. A., & Viljoen, M. J. (1986). Porphyry copper-molybdenum mineralization at the Lorelei, S.W.A./Namibia. In C. R. Anhaeusser & S. Maske (Eds.), *Mineral Deposits of Southern Africa. Geol. Soc. S. Afr., Johannesburg*, Vol. *II*, 1559–1566.
Voet, H. W., & Freeman, P. V. (1972). Copper orebodies in the basal Lower Roan meta-sediments of the Chingola open pit area, Zambian Copperbelt. *Geologie en Mijnbouw*, *51*(3), 299–308.

Wakefield, J. (1978). Samba: A deformed porphyry-type copper deposit in the basement of the Zambian Copperbelt. *Transitions Institution of Mining and Metallurgy, section B*, *87*, B43–52.

Wanhainen, C., Broman, C., Martinsson, O., & Magnor, B. (2012). Modification of a Palaeoproterozoic porphyry-like system: Integration of structural, geochemical, petrographic and fluid inclusion data from the Aitik Cu-Au-Ag deposit, northern Sweden. *Ore Geology Reviews*, *48*, 306–331.

Watts, Griffis, & McOuat Limited (1991). Assessment of Mineral Exploration Opportunities in Zambia-Appendices. Unpublished Report for Geological Survey of Zambia; Watts, Griffis and McOuat Limited, Toronto, Canada, 13 August 1991.

Weihed, P. (1992). Lithogeochemistry, metal and alteration zoning in the Proterozoic Tallberg porphyry-type deposit, northern Sweden. *Journal of Geochemical Exploration*, *42*(2–3), 301–325.

Weihed, P., & Schöberg, H. (1991). Age of porphyry-type deposits in the Skellefte District, northern Sweden. *Geologiska Föreningen i Stockholm Förhandlingar*, *113*(4), 289–294.

Winchester, J. A., & Floyd, P. A. (1977). Geochemical discrimination of different magma series and their differentiation products using immobile trace elements. *Chemical Geology*, *20*, 325–343.

Zientek, M. L., Bliss, J. D., Broughton, D. W., Christie, M., Denning, P. D., Hayes, T. S., Hitzman, M. W., et al. (2014). Sediment-hosted stratabound copper assessment of the Neoproterozoic Roan Group, Central African Copperbelt, Katanga Basin, Democratic Republic of the Congo and Zambia. U. S. Geol. Surv., Global Min. Resource Assessment, *Sci. Investig. Rep.*, 2010-590-T.

3

The Geology of the Mufulira Deposit: Implications for the Metallogenesis of Arenite-Hosted Ore Deposits in the Central African Copperbelt

Philippe Muchez[1], Maarten Minnen[1], Stijn Dewaele[2,3], and Niels Hulsbosch[1]

ABSTRACT

The Neoproterozoic rocks of the Zambian Copperbelt host world-class sediment-hosted stratiform and strata-bound Cu-Co deposits in both fine- and coarse-grained metamorphosed siliciclastic rocks. Ore deposits in the coarse-grained siliciclastics occur below (footwall deposits) and above (hanging-wall deposits) the fine-grained Ore Shale Formation and are termed *arenite-hosted deposits*. The arenite-hosted deposit at Mufulira is studied and compared with similar deposits in the Central African Copperbelt to propose an integrated ore-forming model. The arenites are characterized by several alteration, dissolution, and cementation phases. Authigenic phases include quartz, albite, and K-feldspar overgrowths and calcite and dolomite cements filling pores. Carbonaceous matter or pyrobitumen postdates compaction and coat detrital grains. The pyrobitumen formed as a residue of oil cracking and/or as a byproduct of sulfate reduction and the oxidation of hydrocarbons. Chalcopyrite, digenite, chalcocite, covellite, carrolite, pyrrhotite, malachite, and iron oxides occur as finely disseminated ores or as replacive blebs in sericitic quartzites. The ore minerals replace the compacted rock, and may enclose all diagenetic phases and occur together with quartz and calcite in veins, indicating they formed late in the paragenesis. Ore minerals precipitated as the result of the mixing of a highly saline, metal-bearing brine with an H_2S-rich hydrocarbon reservoir.

3.1. INTRODUCTION

The Central African Copperbelt straddles the border between Zambia and the Democratic Republic of Congo (DRCongo) and is one of the largest and richest metallogenic provinces in the world. It contains 200 Mt base metals (Cu, Co) and forms the richest source of cobalt in the world (Selley et al., 2005). Since the discovery of these deposits in the 1900s, they have continuously been the object of numerous geological studies with the origin of the metals and the nature and origin of the mineralizing fluids the subject of intense discussion. Several metallogenic models have been proposed including detrital synsedimentary, early to late diagenetic, and synorogenic to postorogenic origin. A historical overview of these genetic theories is given by Cailteux et al. (2005) and Selley et al. (2005). The current consensus suggests a multiphase origin of high-grade Cu-(Co) deposits in the Central African Copperbelt (Cailteux et al., 2005; Dewaele et al., 2006; El Desouky et al., 2009; Selley et al., 2005; El Desouky et al., 2010) with several hypogene phases ranging from early diagenetic up to synorogenic. An important, later supergene event is also recognized (De Putter et al., 2010; Decrée et al., 2010). Observation suggests that diagenesis led

[1] *KU Leuven, Geodynamics and Geofluids Research Group, Department of Earth and Environmental Sciences, Leuven, Belgium*

[2] *Department of Geology and Mineralogy, Royal Museum for Central Africa, Tervuren, Belgium*

[3] *Department of Geology and Soil Science, Ghent University, Ghent, Belgium*

Ore Deposits: Origin, Exploration, and Exploitation, Geophysical Monograph 242, First Edition.
Edited by Sophie Decrée and Laurence Robb.

to the creation of the main ore phase in the Katangan part of the Copperbelt (Dewaele et al., 2006; El Desouky et al., 2010). However, the recognition of a diagenetic main Cu-mineralization is, however, problematic in the Zambian Copperbelt (Sillitoe et al., 2010). Clear petrographic evidence is for the most part lacking in this region, generating a discussion between diagenetic and synorogenic models (Muchez et al., 2010).

The Central African Copperbelt consists of two major ore districts, the Zambian and the Congolese (Katanga) Copperbelt. There are important geological and mineralogical differences between these two mineralized zones.

Old stratigraphy		Min	Lithology	Thickness	New stratigraphy
Intermediate	Dolomite		Dolomite		Kirilabombwe subgroup
Hanging wall formation	Glassy quartzite	Py, Cp Bn	Quartzite and dolomite	0–12 m	
	Upper argillaceous quartzite		Argillite and argillaceous quartzite	0–20 m	
	Marker grit		Quartzite, conglomerate, shale and dolomite	0–6 m	
	Lower argillaceous quartzite		Argillite with dolomite banc Dolomite Argillite and argillaceous quartzite	0–27 m	Kitwe subgroup
Ore formation	A orebody	Cc, Cp, Bn	Grey quartzite and graywacke Pink quartzite	0–10 m	
	Inter A/B horizon		Pink quartzite	0–7 m	
			Banded shale and quartzite, dolomite	0–5 m	
			Lower dolomite with conglomerate layer	0–6 m	
	B orebody	Bn, Py, Cc	Grey quartzite and graywacke Banded silty quartzite	0–12 m	
	Inter B/C horizon	Bn, Cp, Py	White quartzite and dolomite Banded silty quartzite	0–9 m	
	C orebody	Cp, Bn, Cc, Py	Grey quartzite and graywacke Well-bedded quartzite and conglomerate	0–15 m	
Footwall formation	Upper		White quartzite and conglomerate		Mindola subgroup
	Middle		Well-banded quartzite and conglomerate	0–150 m	
	Lower		Dark grey finely bedde quartzite		
	Basal		Basal conglomerate	0–15m	
LUFUBU			Quartz-mica-chlorite schistes, intruded by granite		

Figure 3.1 Stratigraphy of the Neoproterozoic Katanga Supergroup at Mufulira, Zambian Copperbelt (after Cailteux et al., 2007; Brandt et al., 1961).

The Congolese Copperbelt contains allochthonous dolomitic rock-hosted orebodies disrupted into irregular fragments (Cailteux et al., 2005). Due to their disruption and allochthonous nature, the stratigraphic position of the deposits relative to the basement is unknown (Cailteux et al., 2005; Jackson et al., 2003). The Zambian deposits are stratiform to stratabound and are both argillite (70% of ore) and arenite hosted (30%) (Selley et al., 2005). The Zambian deposits are variably folded, but tectonically intact (Selley et al., 2005). The majority of these deposits are situated in the lower part of the Roan Group at the base of the Neoproterozoic Katanga Supergroup and locally in direct contact with basement rocks (Fig. 3.1). The giant arenite-hosted deposits are Mufulira and partly Nchanga-Chingola (Fig. 3.2a). The Zambian Copperbelt contains 46% of the 200 Mt base metals (Cu, Co) in the Central African Copperbelt (Selley et al., 2005). Recent exploration demonstrated the presence of similar sandstone-hosted ore deposits higher in the stratigraphic sequence in the Copperbelt and in other parts of Central Africa. Examples of such mineralization include Lufukwe in the foreland of the Copperbelt (El Desouky et al., 2008a) and the Ghanzi project in Botswana (Gorman et al., 2013).

The aim of this paper is to discuss the geology of the Mufulira (Zambia) deposit and to compare these data with the geological setting and ore-forming process of other arenite-hosted deposits in the Zambian part of the Copperbelt. These deposits are further compared with

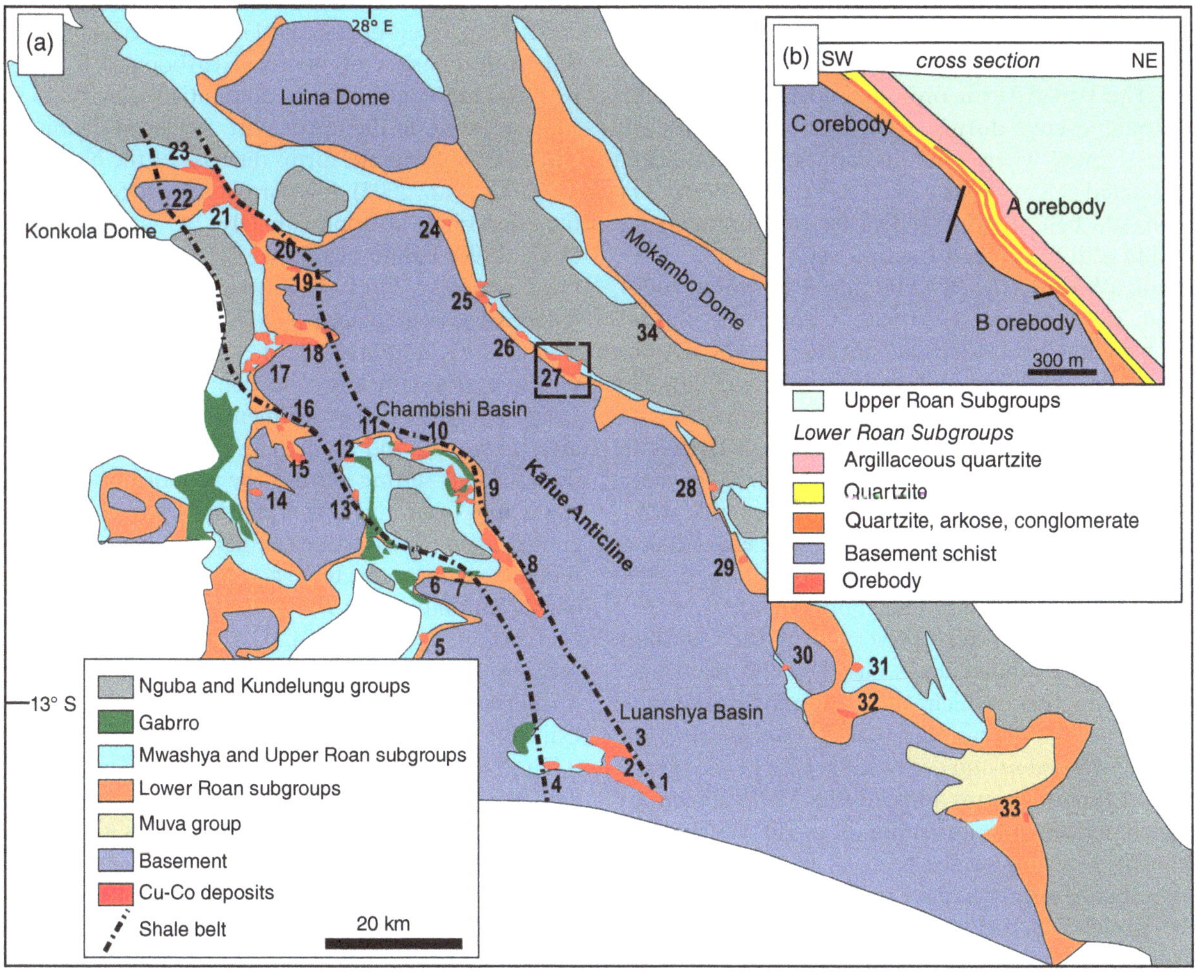

Figure 3.2 (a) Geological map showing the most important Cu-Co deposits in the Zambian Copperbelt. Deposits: 1 = Luanshya, 2 = Roan Extension, 3 = Baluba, 4 = Lufubu South, 5 = Chibuluma South, 6 = Chibuluma West, 7 = Chibuluma, 8 = Nkana-Mindola, 9 = Chambishi South East, 10 = Chambishi, 11 = Pitanda, 12 = Mwambashi A, 13 = Mwambashi B, 14 = Samba, 15 = Fitula, 16 = Mimbula, 17 = Chingola A-F, 18 = Nchanga, 19 = Fitwaola, 20 = Konkola, 21 = Konkola Deep, 22 = Konkola North, 23 = Musoshi, 24 = Lubembe, 25 = Luansobe, 26 = Kasaria, 27 = Mufulira, 28 = Frontier (Lufua), 29 = Mwekera, 30 = Ndola West, 31 = Itawa, 32 = Bwana Mkubwa, 33 = Lonshi, 34 = Mokambo. (b) Schematic cross section through the Mufulira deposit with the stratigraphic position of the three ore bodies A, B, and C, from top to bottom (after Hitzman et al., 2012).

siltstone- to sandstone-hosted deposits in other parts of the Copperbelt, which were discovered during the last decade, and in the world. This study forms the basis for further exploration in the Neoproterozoic siliciclastics in the broader Central African region and world wide.

3.2. GEOLOGICAL SETTING OF ZAMBIAN COPPERBELT

The Zambian Copperbelt forms part of the arc-shaped Lufilian belt that extends in northwest-southeast direction, with the Kafue Anticline as a prominent structural feature. The Lufilian belt is 700 km long and 150 km wide between the Congo and Kalahari Cratons (Selley et al., 2005). This belt formed due to Neoproterozoic extensional tectonics during the breakup of the Rodinia supercontinent followed by the late Neoproterozoic to early Phanerozoic collision, deformation, and metamorphism. The Pan-African orogeny encompasses a series of tectonic events during Neoproterozoic and early Paleozoic times when the supercontinent Gondwana amalgamated (Kampunzu & Cailteux, 1999). The Lufilian fold belt consists of five tectonic domains, all exhibiting a different deformation style (Porada, 1989; Kampunzu & Cailteux, 1999; Porada & Berhorst, 2000), with the classic Central African Copperbelt and all major known Cu-Co deposits located in the Domes region and external fold and thrust belt (Fig. 3.2a). The Zambian Copperbelt is located in the eastern part of the Domes region. The Domes region is characterized by Neoproterozoic metasedimentary rocks, intruded by mafic (0.73–0.76 Ga) and felsic (0.56–0.53 Ga) magmatic rocks, overlying a 2.05–1.8 Ga basement (De Waele et al., 2006).

The Neoproterozoic Katanga Supergroup is subdivided into the Roan, Nguba, and Kundelungu Groups (Fig. 3.1). The latter is only locally preserved in the Zambian Copperbelt. The subdivision is based on the occurrence of two diamictites, which are regional stratigraphic markers. They are termed the *Grand Conglomérat* and the *Petit Conglomérat* and occur at the base of the Nguba and Kundelungu Groups respectively. The sedimentary succession reflects the formation of a rift system with subsequent widening of the basin. The Roan rocks are further subdivided into a lower part containing the dominantly siliciclastic rocks of the Mindola and Kitwe Subgroups and an upper part consisting of dolomites, dolomitic shales, shales, and evaporites of the Kirilabombwe Subgroup and the Mwashia shales at the top. The Mindola Subgroup consists of texturally immature conglomerates and subarkosic sandstones that were deposited in fluvial, alluvial fan, eolian, and fan-delta environments and is characterized by significant lateral and vertical facies variations (Jordaan, 1961). The sediments were deposited in narrow troughs or subbasins bounded by basement highs (Selley et al., 2005). The Kitwe Subgroup reflects a transgression and marine sedimentation conditions. Sedimentation no longer occurred in troughs, but in larger, but still fault-bounded, basins. The Kitwe Subgroup includes a succession of arenites, silty to sandy arenites, and shales northeast of the Kafue Anticline and dolomitic shales and dolomites southwest of this anticline. The Ore Shale Formation, which hosts 70% of the ore in the Zambian Copperbelt, is situated in the lower part of the Kitwe Subgroup and is characterized by organic-rich fine-grained siliciclastics. This formation is overlain by arkoses, argillaceous sandstones, dolomitic siltstones, and dolomites (Cailteux et al., 2007). The overlying Kirilabombwe Subgroup contains dolomites and arenitic dolomites interbedded with dolomitic shales. Interbedded gabbros, having a mafic to intermediate composition intruded at 760 ± 5 Ma (Key et al., 2001). The overlying Mwashia Subgroup consists of platform carbonates in the lower part and more open marine dolomitic shales, black shales, and quartzites in the upper part. A glacial diamictite, the Grand Conglomerate, forms the base of the Nguba Group (Cailteux et al., 2005). It is correlated with Sturtian diamictites deposited at approximately 740 Ma (e.g., Hoffman et al., 1998; Fanning & Link, 2004). The Nguba assemblage is composed of siliciclastics, carbonates, and mafic igneous rocks emplaced in a proto-ocean similar to the Red Sea (Kampunzu et al., 1991). The Kundelungu Group starts with a second diamictite, that is, the Petit Conglomerate. Further sedimentation consists of carbonates, siltstones, and mudstones (Selley et al., 2005). The depositional setting during the Kundelungu marks the transition from this open marine environment toward compression related to the Lufilian orogeny. These deposits represent synorogenic to postorogenic continental molasse, which continued into the early Paleozoic (Kampunzu & Cailteux, 1999; Cailteux et al., 2005). During the Lufilian orogeny, the Katanga basin closed and the sediments underwent deformation and metamorphism (Kampunzu & Cailteux, 1999; Porada & Berhorst, 2000).

The Neoproterozoic rocks were metamorphosed and deformed during the Lufilian orogeny, which is closely linked to the collision of the Congo and Kalahari cratons at ~530 Ma (John et al., 2004; Porada & Berhorst, 2000; Rainaud et al., 2005b). The metamorphic grade in the Copperbelt decreases north-westward from upper greenschist at Nkana to anchimetamorphism at Kamoto (Eglinger et al., 2013). However, the highest metamorphic grades of the Neoproterozoic rocks were reached in the Domes region. Recent dating of metamorphic minerals shows that the Lufilian orogeny spans a period of about 60 my. The oldest metamorphic ages are given by U-Pb monazite (592 ± 22 Ma) and ^{40}Ar-^{39}Ar biotite (585.8 ± 0.8 Ma) data from greenschist facies rocks in the Zambian Copperbelt (Rainaud et al., 2005b). The main phase of the orogeny took place between 560 and 530 Ma. This is based on U-Pb zircon dating of syntectonic to posttectonic granites and rhyolites in the Katangan high (Hanson et al., 1993). These data are supported by a peak metamorphic age of ~530 Ma from a U-Pb monazite dating of whiteschist facies rocks in the central and western part of the Domes region (John et al., 2004). ^{40}Ar-^{39}Ar biotite and Rb-Sr muscovite and biotite ages of 510–465 Ma probably represent postorogenic cooling (John et al., 2004; Rainaud et al., 2005a). Deformation is characterized by high-amplitude folds, well exposed in open pits and underground mines.

3.3. METHODOLOGY

Microthermometric investigations have been performed on 300 μm-thick doubly polished sections from carbonate cements associated with ore minerals at the Mufulira deposit. Fluid inclusion analyses were carried out on a Linkam THMSG 600 stage mounted on an Olympus BX60 microscope. Reproducibility was within 0.2 °C for the melting temperature of ice and clathrate and ≤ 5 °C for the total homogenization temperature. Fluid salinities of fluids whose composition is broadly H_2O-NaCl-CO_2 were calculated with the program Q2 of Bakker (1997), using the equation of state (EOS) of Duan et al. (1996). Salinities of fluid inclusions, which did not show clathrate dissociation have been calculated with the equation of Bodnar (1993) for the H_2O-NaCl system.

The carbon and oxygen isotopic composition of the carbonate cements in arenites and of dolomitic rocks at Mufulira has been determined at the University of Erlangen-Nürnberg (Germany). Sample powders were drilled at the KU Leuven. They reacted with 100% phosphoric acid. Oxygen isotopic compositions of dolomite were corrected for fractionation effects resulting from the reaction with the hot acid, using the fractionation factors given by Rosenbaum and Sheppard (1986). The analyses were performed on a Kiel III carbonate preparation line connected online to a ThermoFinnigan 252 mass spectrometer. All values are reported in per mil relative to the belemnite standard of the Cretaceous Peedee Formation (‰ V-PDB). Reproducibility was checked by replicate analysis of laboratory standards and is better than ± 0.03 ‰ for $\delta^{13}C$ and ± 0.05 for $\delta^{18}O$ at the 1σ level.

3.4. ARENITE-SANDSTONE HOSTED MINERALIZATION

3.4.1. Stratigraphy, Lithology, Diagenesis, and Alteration

Arenite-hosted deposits occur both below (footwall deposits) and above (hanging-wall deposits) of the Ore Shale Formation, respectively in the Mindola Subgroup and in the middle part of the Kitwe Formation (Selley et al., 2005). Footwall deposits directly underlie the Ore Shale Formation or are separated from it by barren sandstones and conglomerates. Hanging-wall ore bodies are stratigraphically separated. They are less abundant than footwall ore bodies and often form part of giant ore deposits consisting of several vertically stacked ore bodies that occur in both hanging-wall and footwall positions (Selley et al., 2005). Mufulira, Kinsenda, and Nchanga-Chingola form such giant deposits. The footwall deposits are stratabound and mostly lenticular. The geometry of the deposits is related to dimensions of the troughs in which the arenites were deposited. In contrast, the hanging-wall deposits are laterally extensive (e.g., Nchanga Upper Orebody and ore bodies A and B at Mufulira), likely reflecting the continuity of the Kitwe Subgroup above the Ore Shale Formation.

The Mufulira deposit can be subdivided into three ore bodies (Figs. 3.2b and 3.3 A, B, C, from top to bottom) and two barren interore bodies (Maree, 1963). It reflects a cyclic sedimentation with a relatively slow transgression followed by a fast regression (Van Eden, 1974). One cycle shows a reduction in grain size from conglomeratic sandstone to argillaceous sandstone in the orebodies, followed by dolomite and argillite. The mineralized rocks are massive to poorly bedded quartzite units (Brandt et al., 1961). They are overlain by dolomitic siltstones, massive dolomites, or poorly bedded argillites. So characteristic is the cover of the mineralized horizons by impermeable rocks. The quartzites consist of quartz, K-feldspar (orthoclase and microcline), and plagioclase. The feldspars are intensely altered to sericite (Fig. 3.4a). Some sericitized grains show pressure solution with a neighboring particle, indicating that sericitization started before the end of intense compaction. The latter is reflected by the suture contacts between the quartz grains. Rare tourmaline and epidote are observed. The rocks show several alteration and diagenetic phases. Authigenic quartz overgrows detrital quartz grains and predates and postdates compaction. The latter is indicated by authigenic quartz overgrowths both almost completely surrounding detrital quartz as well as enveloping point contacts. Less extensive is authigenic albite, but later K-feldspar alteration may be important (Fig. 3.4b). Calcite and dolomites are the diagenetic carbonates present (Fig. 3.4c). Calcite is fine grained and replaces feldspars and fills pores. Dolomite forms large patches and replaces detrital grains and their overgrowths calcite and fills pores. Especially important is the presence of carbonaceous matter coating detrital grains (Fig. 3.4d). The carbonaceous matter does not completely surround the grains and especially occurs in pores between intensely compacted (sutured) quartz

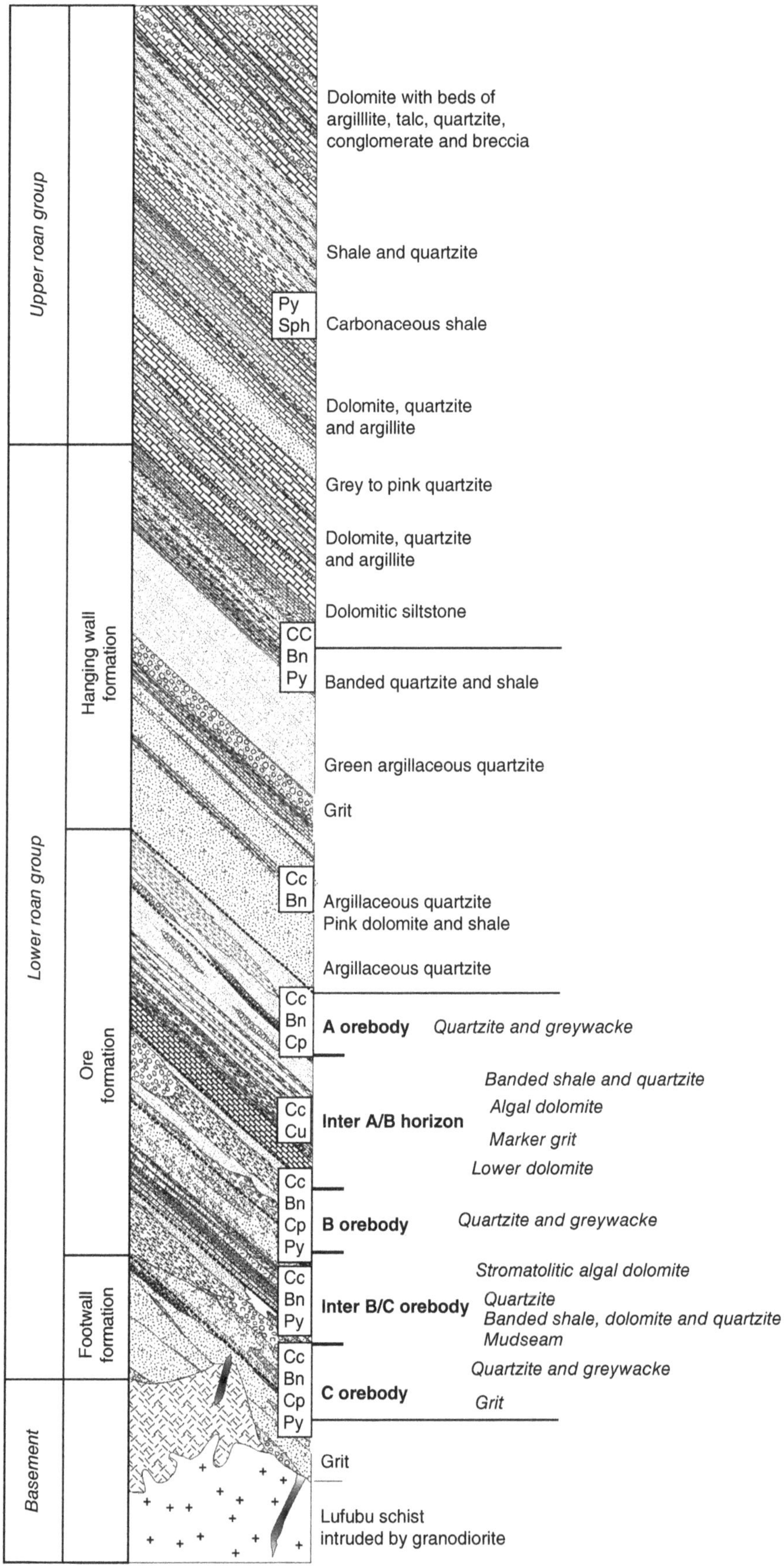

Figure 3.3 Schematic geological profile through the Roan Group at the Mufulira deposit. The stratigraphic position of the three ore bodies, A, B, and C, and the interhorizons are indicated (after Greyling, 2009).

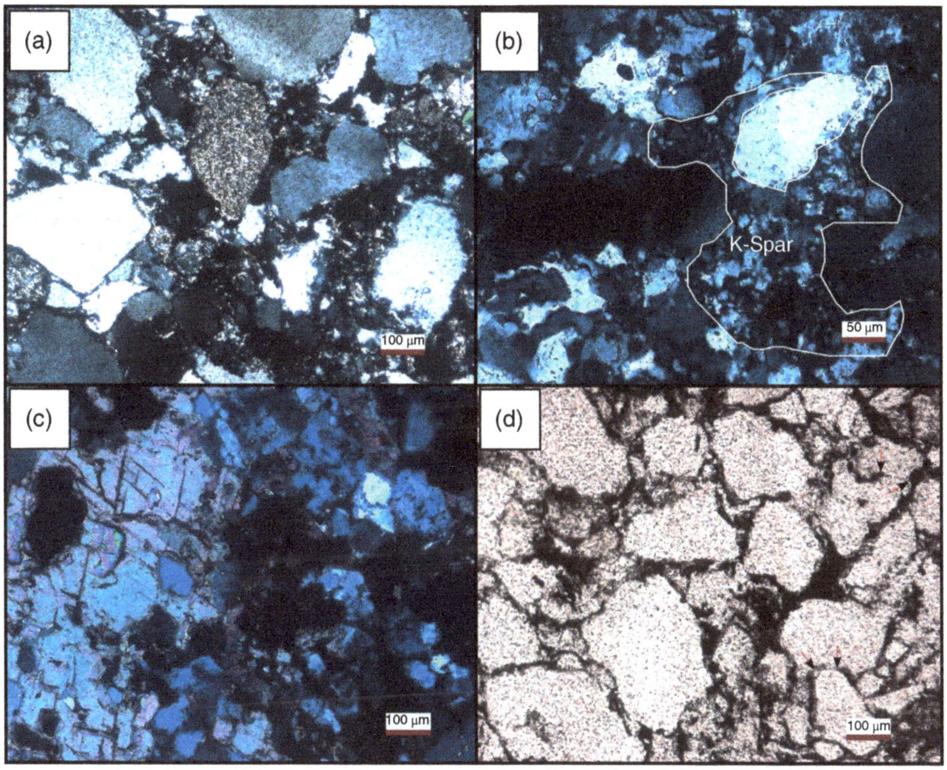

Figure 3.4 Alteration and mineral precipitation at the Mufulira deposit: (a) intense alteration of feldspar to sericite; (b) K-felspar alteration of the host rock; (c) coarse-grained dolomite replacing the siliciclastic host-rock; (d) carbonaceous matter coating detrital grains.

grains, indicating it postdates compaction. The dolomites are made up of anhedral to euhedral dolomite and of quartz crystals. Crystals form a mosaic texture with a slightly preferred orientation. Muscovite and chlorite are accessory minerals. The argillites and siltstones consist of fine-grained quartz, K-feldspar, plagioclase, biotite (phlogopite), and muscovite and are embedded in a groundmass of fine-grained phyllosilicates. The latter can be described as sericitic. Anhydrite is absent in the orebodies at Mufulira but abundantly present at their peripheries (Annels, 1979). Veins are abundant and are filled by quartz, carbonate, and sulfide minerals were the rock is mineralized.

The Kinsenda deposit is situated northwest of the Mufulira deposit less than 10 km across the border with the DRC. Both deposits occur in the same SW-NW oriented zone of wacke-arenite orebodies as defined by Binda and Mulgrew (1974). The copper deposit is present in arkoses from the Mutonda Formation, belonging to the Mindola Subgroup (Ngoyi & Dejonghe, 1995). Main copper minerals are chalcopyrite, chalcocite, digenite, and covellite (Dejonghe, 1997). Bornite is less abundant. Cobalt is present in cattierite, cobaltite, and carrollite. Mineralization fills pores after authigenic overgrowths of quartz and feldspar (albite) on detrital grains and thus certainly after early diagenesis (Ngoyi & Dejonghe, 1995). Both framboidal and euhedral pyrite are present and replaced by copper sulfides. The framboidal pyrites reflect bacterial sulfate reduction during early diagenesis (Ngoyi & Dejonghe, 1995). Intense sericitization is often associated with mineralization. Also comparable with the Mufulira deposit is the occurrence of organic matter (Fleischer et al., 1976).

The arenite-hosted deposits at Chibuluma at the southern flank of the Chambishi Nkana basin show a similar paragenesis (Winfield & Robinson, 1963). The dominant ore mineral chalcopyrite is present in a feldspathic quartzite (Garlick, 1962). Chalcopyrite is later than pyrite and cobalt sulfides. Sericite or muscovite replaced feldspar and microcline was replaced peripherally or completely by albite. Sulfides are interstitial and replace authigenic quartz and carbonates.

3.4.2. Tectonic Setting

The Mufulira deposit is located at the NW limb of the Kafue anticline (Fig. 3.2). The bedding of the sedimentary strata dips 45° to the northeast. Sedimentation started in a trough bounded by WNW- and NNW-trending segmented faults with detrital products originating from the bordering basement blocks (Selley et al., 2003). The Lufilian orogeny deformed the sediments into their NW-SE-oriented and NW-dipping fold pattern. It also reactivated the rift-related fault pattern (Broughton, 2003), and therefore the geometry of the structures generated during Lufilian folding exhibit inheritance from the earlier rift architecture. Fold patterns mimic the shape and distribution of basin compartments (Selley et al., 2003). The SW-limb of the Mufulira syncline exhibits several pod folds and there is a relationship between flexures in the basement rock and the pod folds. These folds could have resulted from the draping of sediments over basement hills as a result of compaction, and this geometry became incorporated into the large-scale Lufilian fold pattern (Annels, 1979).

At Chibuluma West, the orebody is fairly close to an anticline axis and is intersected by two tight drag folds (Winfield & Robinson, 1963). The ore shoot parallels the basement quartzites that are exposed in an anticline. A Lower Orebody and an Upper Orebody occur toward the top of arkosic units and within the base of overlying shales at Nchanga (Voet & Freeman, 1972; McGowan et al., 2006). The ore zones in the Nchanga Upper Orebody are located in a series of folds that developed above a shear zone (McGowan et al., 2003). Faulting and folding are interpreted as largely synchronous since

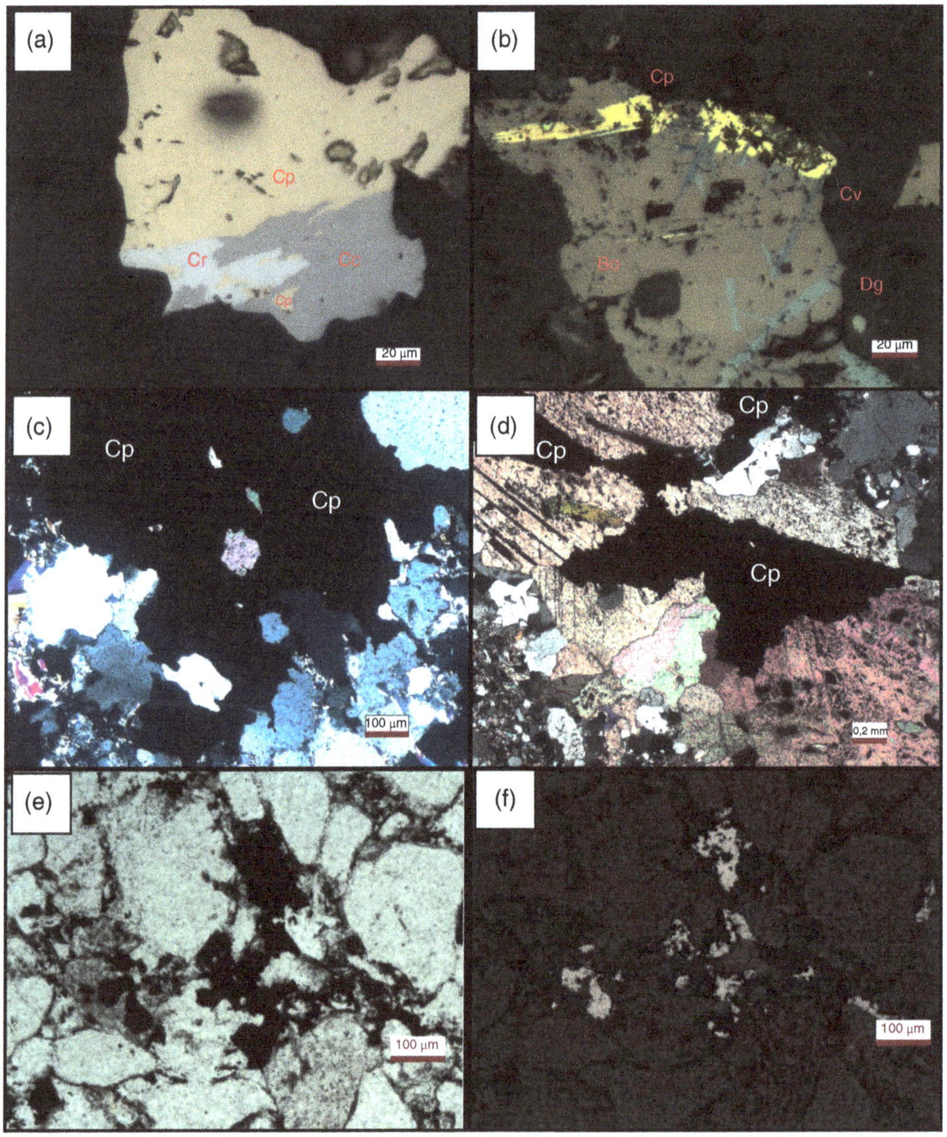

Figure 3.5 Mineralization at the Mufulira ore deposit: (a) chalcocite and carrollite replacing chalcopyrite; (b) digenite and covellite filling cracks in bornite; (c) chalcopyrite replacing the intense compacted host-rock at Mufulira; (d) chalcopyrite occurring with quartz and calcite in a vein cross-cutting the siliciclastic host-rock; (e and f) close spatial association of bornite and organic matter.

folding predominantly took place at the tips of the propagating thrust faults (McGowan et al., 2006). Ore formation has been related to the presence of hydrocarbons trapped in these structural sites. The arenite-hosted Mwambashi B deposit shows a half-graben geometry with mineralization at the top of the Mindola Subgroup immediately below a carbonaceous shale (Selley et al., 2005). The Mindola Sugroup is absent or largely reduced in thickness above the graben shoulders and devoid of ore-grade mineralization. Microfractures have been observed in all arenite-hosted deposits and exert a control on the sulfide distribution (McGowan et al., 2003).

3.4.3. Paragenesis

At Mufulira, ore minerals are mainly bornite and chalcopyrite. They both occur finely disseminated or as replacive blebs, mainly in sericitic quartzites. Also chalcocite, carrollite, covellite, digenite, malachite, pyrrhotite, and iron oxides are present (Fig. 3.5a, b). Ore minerals replace the intense compacted host rock (Fig. 3.5c) and may enclose all diagenetic phases (authigenic quartz, feldspar, and carbonates) present in the sericitic quartzites, indicating they formed very late in the paragenesis. Sulfides occur associated with quartz and calcite in veins (Fig. 3.5d). However, sulfides are most abundant where authigenic carbonates are present and carbonates just predate the sulfides in the paragenetic sequence. Also at the Mwambashi B deposit, calcite is in close spatial association with the copper mineralization. However, no quantitative relationship between carbonate and copper grade has been observed at Mufulira and Mwambashi B (Selley & Scott, 2003).

Pyrite, chalcopyrite, bornite, chalcocite, and carrollite are the sulfides at the Nchanga deposit (McGowan et al., 2006). Pyrite replaces nodular anhydrite within the Lower Shale and copper and cobalt sulfides replace pyrite resulting in an inverse relationship between both minerals. Malachite, chrysocola, and copper oxides represent supergene alteration, abundantly present in the Lower Orebody. The predominance of chalcocite as sulfide mineral in the Lower Orebody together with the abundance of supergene copper oxides, carbonates, and silicates favors a supergene origin of the chalcocite (cf. Konkola deposit; Torremans et al., 2013). At Kambove West (DRC), abundant supergene chalcocite occurs associated with iron and other (hydro)oxides and cobaltian dolomite (Van Langendonck et al., 2013). Hydrothermal alteration is characterized by the occurrence of dolomite, phlogopite, sericite, quartz, tourmaline, and quartz. Phlogopite is the most important alteration mineral and can constitute 40% of high-grade copper- and cobalt-rich ore in the Lower Orebody at Nchanga (McGowan et al., 2006). Sericite is also widespread and is often accompanied by silicification. In the Upper Orebody, dolomitization is the dominant alteration (5%–30%), but also phlogopite and sericite can make up 10% of the altered rock sequence.

3.5. MINERAL GEOCHEMISTRY AND FLUID CHARACTERISTICS

The carbon and oxygen isotopic composition of carbonate cements in mineralized arenites and of dolomite rocks in the same sequence have been analyzed at Mufulira (Fig. 3.6; Table 3.1). The carbon isotopic values of the dolomite cement (n = 6) ranges between −2.40 and −14.64‰ V-PDB and of the dolomite rock (n = 4) between 2.43 and 5.54‰ V-PDB. The oxygen isotopic values of the dolomite cements and of the dolomite rocks range between 17.29‰ and −14.31 V-PDB and one value of −7.58‰ V-PDB, and between −11.33 and −8.23‰ V-PDB, respectively.

The carbon isotopic composition of the dolomite rocks largely falls within the range typical for Neoproterozoic marine carbonates (−4 to +4‰) (Veizer & Hoefs, 1976; Lindsay et al., 2005). The oxygen isotopic values are slightly lower than those characteristic for these marine carbonates (−8 to −4‰ V-PDB) (Veizer & Hoefs, 1976; Lindsay et al., 2005), implying the oxygen isotopic values are modified during subsequent diagenesis. Both the

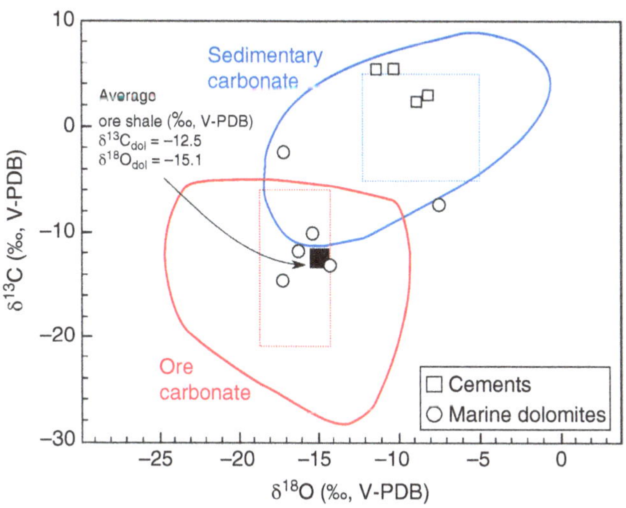

Figure 3.6 Plot showing $\delta^{18}O$ and $\delta^{13}C$ (‰, V-PDB) isotopic signatures of carbonate cements in mineralized arenites and of dolomite rocks in the same sequence at Mufulira in comparison with marine-derived carbonates (in blue) and dolomite associated with ore zones (in red). Full lines: Ndola west, Mufulira, Nkana-Mindola, Mwambashi, Konkola North, Konkola, Chambishi, Chibuluma west data from Bull et al. (2002) and Selley et al. (2005). Dotted lines: Konkola, Nchanga, and Chambishi data from McGowan et al. (2006; and references therein). Average ore shale isotopic data from Bull et al. (2002).

Table 3.1 Carbon and Oxygen Isotopic Data of Carbonate Cements in Mineralized Arenites and of Dolomite Rocks in the Same Sequence at Mufulira.

Sample		Type	$\delta^{13}C$ (‰ V-PDB)	$\delta^{18}O$ (‰ V-PDB)	$\delta^{18}O$ (‰ V-SMOW)
MM10MF12	C-Orebody	cement	−10.3	−15.31	15.13
MM11MF12	C-Orebody	cement	−11.89	−16.32	14.09
MM15MF12	C-Orebody	cement	−14.64	−17.29	13.09
MM17MF12	inter B-C	cement	−2.39	−17.25	13.13
MM20MF12	B-Orebody	cement	−13.17	−14.31	16.16
MM21MF12	inter A-B	marine dolomite	2.92	−8.13	22.53
MM27MF12	A-Orebody	cement	−2.4	−7.58	23.10
MM30MF12	Hanging wall	marine dolomite	2.43	−8.90	21.73
MM32MF12	Hanging wall	marine dolomite	5.54	−11.33	19.23
MM34MF12	Hanging wall	marine dolomite	5.54	−10.29	20.30

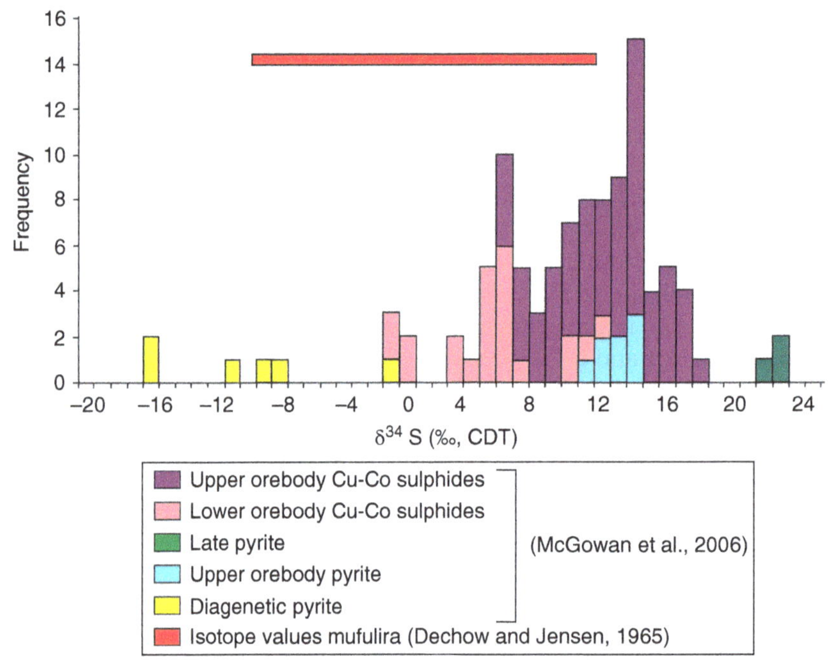

Figure 3.7 Plot of $\delta^{34}S$ showing data of pyrite and copper sulfides at Mufulira (Dechow & Jensen, 1965) and of various settings with the Nchanga deposit (McGowan et al., 2006).

carbon and oxygen isotopic values of the dolomite cements are significantly lower than of the Neoproterozoic carbonates. Similar low carbon isotopic values (below -10‰ V-PDB) have been reported from intensely mineralized strata from the arenite-hosted Mwambashi B deposit (−25 and −17‰ V-PDB) (Selley & Scott, 2003) and argillite-hosted deposits in the Zambian Copperbelt (Selley et al., 2005; Muchez et al., 2010). Such low carbon isotope values have not been reported from the Nchanga deposits, where the lowest carbon isotopic values (−2.9 to −8.3‰ V-PDB) occur in the shear zone and alteration assemblages in the Upper Orebody (McGowan et al., 2006). Low carbon isotopic values have been related to abiotic reactions involving the oxidation of organic material or methane (cf. Irwin et al., 1977). The lower oxygen isotopic values can be related to a precipitation temperature of the carbonate cement significantly higher than of marine carbonates. Carbon and oxygen isotopic values of cements outside the low range (cluster) can be due to mixing with an enriched carbon and/or oxygen isotopic source such as the marine dolomites present in the sequence.

The $\delta^{34}S$ values of pyrite and copper sulfides at Mufulira range between −10.5 and +10.9‰ V-CDT (Fig. 3.7 Dechow & Jensen, 1965). From their data, it is not clear if the negative values correspond only to pyrite or also to copper sulfides. All copper sulfide values shown in their Figure 3a are positive. Sulfur isotopic composition of diagenetic pyrites at Nchanga range between -17 and −1‰V-CDT (Fig. 3.7; McGowan et al., 2003, 2006). These values contrast with the $\delta^{34}S$ values of copper and cobalt sulfides from both the Lower Orebody (-1 to +12‰V-CDT) and the Upper Orebody (+5 to +18‰ V-CDT). The low $\delta^{34}S$ values of the pyrite at Nchanga and Mufulira clearly reflect bacterial sulfate reduction as the process for the generation of the H_2S. The positive sulfur isotope values, however, indicate thermochemical sulfate reduction (TSR). The lower $\delta^{34}S$ values in the Lower Orebody compared with the Upper Orebody are likely due to the incorporation of light sulfur from the pyrite present in this Lower Orebody (McGowan et al., 2006). Such a mixture of BSR- and TSR-derived sulfur has been reported from successive vein generations at Nkana, with an increase in TSR-derived sulfur in the later veins (Muchez et al., 2010). Microthermometric measurements have been performed on fluid inclusions from the carbonate cements to obtain insight into the fluids present in the deeper subsurface before precipitation of the sulfides and therefore constrain minimum temperatures, since the metal-bearing fluids migrated upward to the ore deposit site (Selley et al., 2005).

Clusters of fluid inclusions (FI) that define nice FI assemblages are present in the carbonates from Mufulira (Fig. 3.8; Table 3.2). The fluid inclusions are three phase (aqueous liquid, gaseous liquid, and a vapor phase) or two phase (aqueous liquid and vapor) at room temperature. The fluid has a H_2O-NaCl-$CaCl_2$-$MgCl_2$-CO_2 as indicated by the presence of three phases at room temperature, the melting of liquid CO_2, and clathrate, and the first melting of the solid aqueous phase at and below -52 °C (Table 3.2). Most melting temperatures of CO_2

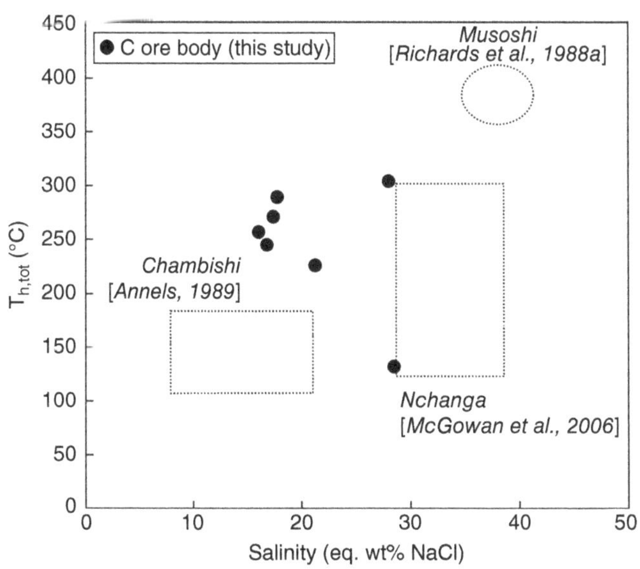

Figure 3.8 Salinity versus temperature of homogenization of fluid inclusion data for Mufulira compared with data ranges for Chambishi (Annels, 1989), Nchanga (McGowan et al., 2006), and Musoshi (Richards et al., 1988a).

($T_m(CO_2)$) are around -56.6 °C, indicating that CO_2 was the only gaseous phase. The $T_m(CO_2)$ values of -57.5 and -58.9 °C are due to the presence of one or more other gaseous phases (e.g., methane or nitrogen). The minimum temperature of the fluid as indicated by the minimum homogenization temperature is 225 °C, with one outlier at 130 °C (Table 3.2). Two values of the fluid salinity based on the clathrate melting temperature are reported (14 and 17 eq. wt% NaCl). Maximum salinities based on the final melting temperature of ice are between 15.1 and 21.1 eq. wt% NaCl, with two outliers around 28 eq. wt% NaCl (Table 3.2) and comparable to the salinities derived from the clathrate melting. At Nchanga, high salinity (31–38 eq. wt% NaCl) H_2O-NaCl-$CaCl_2$-$MgCl_2$(-CO_2) fluids are reported from quartz vein cements that cut the Upper and Lower Orebodies (McGowan et al., 2006). A wide range of homogenization temperatures (105–300 °C) has been measured, possibly indicating the presence of multiple fluid generations or modification of the original fluid inclusions after entrapment. McGowan et al. (2006) proposed a temperature between 220 °C and 290 °C for the ambient fluid. This minimum temperature is comparable with the temperature proposed for the fluid causing the precipitation of the carbonates at Mufulira.

Taking into account a minimum precipitation temperature of 225 °C, the oxygen isotopic composition of the ambient fluids can be calculated from the oxygen isotopic composition of the minerals. Minnen (2013) calculated an oxygen isotopic composition of the fluid between +3.8 and +6.9‰ V-SMOW for the carbonate cements at Mufulira for $\delta^{18}O$ carbonate values ranging between -17.3 and -14.3‰ V-PDB. A similar composition has been calculated for the fluids at Nchanga, which caused the precipitation of the vein quartz within the orebodies. These values fall within the range of both basinal brines and metamorphic fluids (Sheppard, 1986; Hoefs, 2009) and reflect intense water-rock interaction (Clayton et al., 1966; Hitchon & Friedman, 1969; Suchecki & Land, 1983; Muchez et al. 1991).

High salinity (39 wt% NaCl and maximum 15 wt% KCl) H_2O-NaCl-(KCl-$FeCl_3$-CO_2) fluid inclusions with even higher total homogenization temperatures (dissolution temperature of halite of 397 ± 5 °C) are present in quartz-hematite veins in the footwall of the Musoshi deposit (Richards, et al., 1988a). The age of this mineralization is 514 ± 2 Ma determined by U-Pb dating of rutile in a sulfide-bearing vein (Richards et al., 1988a) and is confirmed by U-Pb age of 514 ± 3 Ma from five uraninite samples, which cut the ore shale (Richards et al., 1988b). The homogenization temperature of the fluid inclusions is significantly higher than the regional metamorphic grade, indicating a deep seated fluid source. This mineralization type, however, crosscuts the stratiform copper ore and remobilized sulfides from an earlier mineralization.

Table 3.2 Microthermometric Data from Fluid Inclusions in the Carbonates in the Ore Zone at Mufulira. Tm: melting temperature, Th: homogenization temperature, Td: decrepitation temperature, Ts: dissolution temperature.

Sample		Zone (# FI)	$Tm(CO_2)$ (°C)	Tm (ice) (°C)	Tm (clath) (°C)	Th (CO_2) (°C)	Mode	Th Tot (°C)	Td (°C)	Ts(halite) (°C)	Wt % $NaCl_{equiv}$
MM11MF12	C-Orebody	1(1)		–15.3	–0.8	30.0	L				18.9
MM11MF12	C-Orebody	1(2)	–56.9	–13.0		29.8	L	245			16.9
MM11MF12	C-Orebody	1(3)	–56.9			29.8	L		260		
MM11MF12	C-Orebody	1(4)		–14.0		29.2	L		269		17.8
MM11MF12	C-Orebody	2(1)				30.0	L		265		
MM11MF12	C-Orebody	3(1)		–14.0	1.1	30.0	L				17.8
MM11MF12	C-Orebody	3(2)		–11.1							15.1
MM11MF12	C-Orebody	3(3)		–13.7							17.5
MM11MF12	C-Orebody	4(1)	–56.5	–12.6					265		17.5
MM11MF12	C-Orebody	4(2)							320		16.5
MM11MF12	C-Orebody	4(3)						301		190	
MM11MF12	C-Orebody	4(4)		–18.2					297		21.1
MM11MF12	C-Orebody	4(5)	–56.7	–13.5	–4.7						17.3
MM11MF12	C-Orebody	4(6)		–14.5							18.2
MM11MF12	C-Orebody	5(1)		–13.4							17.3
MM11MF12	C-Orebody	6(1)	–56.7			29.4	L				
MM11MF12	C-Orebody	6(2)		–13.1							17
MM11MF12	C-Orebody	6(3)	–56.6	–13.7		29.3	L	270			17.5
MM11MF12	C-Orebody	6(4)		–14.2	–1.0	29.0	L				18
MM11MF12	C-Orebody	7(1)				30.5	L		311		
MM13MF12	C-Orebody	1(1)						307			
MM13MF12	C-Orebody	1(2)		–29.0				303			28
MM13MF12	C-Orebody	1(3)									
MM13MF12	C-Orebody	1(4)		–29.7				130			28.5
MM13MF12	C-Orebody	1(5)									
MM13MF12	C-Orebody	2(1)	–57.5	–14.1		29.2	L		276		17.9
MM13MF12	C-Orebody	2(2)	–58.9	–18.4				225			21.3
MM13MF12	C-Orebody	2(3)	–59.1	–15.3		29.5	L		262		18.9
MM10MF12	C-Orebody	1(1)		–14.0				289			17.8
MM10MF12	C-Orebody	1(2)		–12.3				255			16.2
MM10MF12	C-Orebody	1(3)						276			

Disseminated carrollite from the Upper Orebody at Nchanga together with chalcopyrite and bornite from veins at Nkana-Mindola and Chibuluma West gave a Re-Os isochron with an age of 576±41 Ma (Selley et al., 2005).

3.6. DISCUSSION

3.6.1. Ore-Forming Processes

Pyrobitumen commonly forms as a residue of oil cracking to gas and as a byproduct of sulfate reduction and the oxidation of hydrocarbons (Landis & Castaño, 1995; Machel et al., 1995). Annels (1979) suggested that the carbonaceous material at Mufulira was introduced as mobile hydrocarbons. In addition, the arenite-hosted ore are typically related to structural positions (traps) in which hydrocarbons can be accumulated (Annels, 1979). At Mufulira the orebodies occupy a common, low amplitude anticline above a semirestricted basin. Mineralization also occurs preferentially in linkage zones between the WNW- and NNW-trending faults bordering the basement highs (Selley et al., 2003). The possible high permeability of these interference zones could have focused both hydrothermal fluids and hydrocarbons (Selley et al., 2003). The late Lufilian age deduced from the structural position of the arenite-hosted deposits has recently been confirmed by absolute age dating (Sillitoe et al., 2017). The latter authors found by Re-Os dating of molybdentite present at Chibuluma South and Mufulira an age of 519.9–500.7 Ma and 490.9 Ma, respectively.

The source of the sulfate that became reduced by the hydrocarbons could have been the diagenetic sulfate present in the arenites or basinal fluids. Taking into account the timing of mineralization, for example, after major compaction or even folding of the strata and related to hydrocarbon generation, sulfate reduction was thermochemical (Machel et al., 1995) . A likely source of the hydrocarbons is the Ore Shale Formation. Metal-bearing fluids could have originated from the underlying basement and Mindola siliciclastics. Mass balance calculations, however, showed that the latter are insufficient to account for the majority of the ore metals (Hitzman, 2000).

Alteration phases include K-feldspar alteration, albitization, and most dominant are sericitization and carbonatization. Carbonatization seems to be spatially and temporally associated with sulfide mineralization. In addition, widespread sericitization of the mineralized quartzites might have provided the necessary permeability after compaction of the rocks during deep burial conditions (cf. Surdam et al., 1984; Muchez et al., 1992). K-alteration (K-felspar, phlogopite, sericite-muscovite) is omnipresent in the Zambian Copperbelt, but cannot be used as a vector toward mineralization (Selley et al., 2005; Hitzman et al., 2012).

The characteristic occurrence of the carbonates is interpreted as the result of thermochemical sulfate reduction. During sulfate reduction, sulfate is reduced to HS^- (Machel, 2001), which later reacts with ions from the metal-bearing fluid to form the sulfides.

$$\text{hydrocarbons} + SO_4^{-2} \rightarrow \text{altered hydrocarbons} + H_2S(HS^-) + HCO_3^-(CO_2) + H_2O$$

$$Cu^{+2} + Fe^{+2} + 2HS^- \leftrightarrow CuFeS_2 + 2H^+$$

A second reactant is HCO_3^- that forms with Ca^{+2} and Mg^{+2} in the basinal fluids dolomite and/or calcite.

$$2HCO_3^- + Ca^{+2} + Mg^{+2} \leftrightarrow CaMg(CO_3)_2 + 2H^+$$

The low carbon isotope composition of the carbonates further supports this interpretation. The fact that a qualitative but no quantitative relationship exists between the amount of calcite and sulfide is explained by the fact that precipitation does not occur by the same reaction but can be explained by successive reactions. Between these reactions, the reactants may migrate laterally and vertically in the permeable horizons. Moreover, filling of the pores by carbonates reduces the available pore space for subsequent sulfide precipitation. Processes are even more complicated since both carbonate and sulfide replace matrix components and earlier authigenic phases.

K-feldspar alteration is characteristic at Mufulira, Chibuluma, Kinsenda, and Nchanga. Sericitization of K-feldspar is due to a lowering of the pH:

$$3KAlSi_3O_8 + 2H^+ \leftrightarrow KAl_2(Si_3AlO_{10})(OH)_2 + 6SiO_2 + 2K^+$$

The necessary acidic conditions may have been generated during chemical reactions involving hydrocarbons. The alteration reaction of K-feldspar explains the association of silicification with sericitization observed at Nchanga.

3.6.2. Comparison with Other Neoproterozoic Deposits in the Lufilian Arc

Sandstone-hosted stratiform copper mineralization is present in the Lufilian foreland, north of the Lufilian arc, at Lufukwe and Mwitapile The stratiform deposits are situated in the Nguba and Kundelungu Groups (El Desouky et al., 2008a). At Lufukwe, the host rock is the Monwezi Sandstone (Nguba Group), a finely laminated to massive fine- to medium-grained feldspar-rich sandstone with variable amounts of clay, lithic fragments, and calcite. The copper minerals present are hypogene chalcopyrite, bornite, and chalcocite and supergene digenite and covellite. The hypogene copper minerals replace pyrite. Mineralization took place after strong compaction, authigenic quartz overgrowths and intense feldspar dissolution. Copper content above 1.25% occurs in coarser-grained (>175μm) siliciclastics, 35% of altered feldspars and little or no fine-grained matrix (El Desouky et al., 2008a). Characteristic for the mineralized intervals is the presence of pyrobitumen, demonstrating the former occurrence of hydrocarbons. Mineralization is interpreted to post-date the Lufilian folding of the rocks.

The host-rock of the mineralization at Mwitapile is the Sonta Sandstone of the Ngule Subgroup (Kundelungu Group). The Sonta Sandstone is an intensely compacted, arkosic to calcareous sandstone (El Desouky et al., 2008b). Copper mineralization occurred after compaction, quartz overgrowths, sericitization, kaolinitization, and dissolution of feldspar and calcite. Pyrobitumen has been observed as intergranular coatings to the detrital grains and as tiny particles filling the primary porosity. Larger pyrobitumen particles are found replacing altered feldspars. The hypogene copper minerals are chalcopyrite, bornite, and chalcocite. Two generations of these hypogene copper minerals have been recognized, which are separated by authigenic quartz cementation (El Desouky et al., 2008b). Copper minerals replacing pyrite have been observed in a few scattered grains. Supergene copper minerals are digenite, covellite, and malachite.

The mineralization at Lufukwe and Mwitapile show many similarities with the arenite-hosted copper mineralization in the Copperbelt. Mineralization occurs in feldspar-rich sandstones characterized by sericitization and/or dissolution of the feldspars. Copper mineralization postdates diagenetic processes such as compaction and authigenic quartz cementation and fill the remaining primary and/or secondary pores between the framework grains. Earlier pyrite forms a sulfur source for the copper minerals and additional reduction of sulfate could have been the result from thermochemical sulfate reduction by hydrocarbons (oil and gas). The geological structure of the ores, i.e. in anticlinal structures allows the ponding of these hydrocarbons or reduced sulfur until the arrival of the metal-bearing fluid. Migration of the metal-bearing fluid is thought to have taken place along faults (cf. Ngoyi & Dejonghe, 1995; Selley et al., 2005; El Desouky et al., 2008a).

Sulfide precipitation due to the interaction of a metal-bearing fluid with a reduced, carbon-rich reservoir has also been proposed for vein-type Zn-Cu and Cu-Ag deposits in the Neoproterozoic of the Lufilian arc. The Kipushi deposit, one of Africa's largest producers of Zn and Cu, is characterized by the presence shungite, which represents metamorphosed bitumen (Heijlen et al., 2008). This bitumen originated from petroleum. In addition, Heijlen et al. (2008) proposed that the reduced reservoir was fed by a high amount of H_2S (sour gas) resulting from thermochemical reduction of seawater-derived sulfate. The ore at Kipushi is located in an anticlinal structure between a large fragment consisting of silty, dolomitic shales (Grand Lambeau) (De Magnée & François, 1988), massive dolomites of the Kakontwe and Kipushi formations and the shaly dolomites of the Lower Mwashya. This structure could have allowed the accumulation of the hydrocarbons and sour gas. The Cu-Ag deposit at Dikulushi in the Lufilian foreland is situated in an anticlinal structure above a thrust fault and crosscut by strike- and oblique-slip faults (Haest et al., 2007). Based on the presence of pyrobitumen in the deposit and the high sulfur isotopic composition of the primary sulfides (11.3 – 13.6 ‰V-CDT), Haest et al. (2009) proposed that sulfates were reduced at the mineralization site by TSR.

3.6.3. Comparison with Other Sandstone-Hosted Cu Deposits

Sandstone-hosted deposits Cu deposits occur worldwide in sediments ranging from the Proterozoic to the Cenozoic. The Spar Lake stratabound Cu-Ag deposit is an example of a mineralization in Mesoproterozoic quartzites (Hayes et al., 2012). As has been observed in sandstone-hosted deposits in the Central African Copperbelt, ore formation took place due to the mixing of an oxidized metal-rich brine and a reservoir of hydrocarbons, in this case a sour H_2S natural gas (Hayes et al., 2012). The calcite cements present in the quartzites have low $\delta^{13}C$ values (i.e., up to -18.4‰ V-PDB). This indicates a major contribution of light carbon derived from the oxidation of the organics. Sulfide originated both from the sour gas and from thermochemical sulfate reduction. The Camaquã copper mine in southern Brazil is an example of a Neoproterozoic to Early Paleozoic sandstone and conglomerate-hosted deposit (Laux et al., 2005). As in the Lufilian foreland deposits, sulfide mineralization occurred after authigenic quartz precipitation and microthermometric measurements reflect an evolution from a low saline fluid (≤2.7 eq wt% in fluid inclusions in quartz) to an intermediate salinity of 14 eq. wt% NaCl in carbonates postdating the sulfide stage (Laux et al., 2005). Historical important stratiform sandstone-hosted copper deposits have been exploited from the Lower Cambrian Timna Formation in the Timna Valley in Israel (Asael et al., 2007). The Bronze Age copper mines have been exploited for over 5000 years, from the Chalcolithic to the Mamluk periods (Weisgerber, 2006). Mesozoic deposits are present in the Ravar Copper Belt in Iran (Azaraien et al., 2017), and in northern Chile (Kojima et al., 2003; Cisternas & Hermosilla, 2006). The arenite host rock in the Ravar Copper Belt formed in a fluviatile to deltaic environment (Azaraien et al., 2017). Mineralization typically occurs in organic-rich rocks and sulfur isotopes suggest that sulfide originates from bacterial sulfate reduction. The high salinity of the fluids, which is regarded as important for the transport of the Cu as chloride complexes, was derived from the dissolution of evaporates in lower sequences. In northern Chile, the Cu-Ag mineralization is hosted by Lower Cretaceous volcanic and volcanoclastic rocks of the Copiao area (Cisternas & Hermosilla, 2006). As in the deposits in the Lufilian foreland, hydrocarbons played an important role in the reduction of the sulfur and the precipitation of the Cu sulfides. The former presence of hydrocarbons is indicated by the current presence of pyrobitumen, which is proposed as an exploration tool for these deposits (Cisternas & Hermosilla, 2006). A similar association between hydrocarbons and copper mineralization is present in the Late Cretaceous red bed sequence in the Neuquén Basin in Argentina (Rainoldi et al., 2015b; Pons et al., 2017), which is also characterized by the presence of numerous hydrocarbon fields (Pons et al., 2015). The red bed sequence was deposited in a fluvial environment. The migration of hydrocarbons and warm reduced formation waters caused the dissolution of early cements (hematite, kaolinite, calcite, and barite) and clasts causing an increase in the porosity and an important change in color from red to yellow-green-grey (Rainoldi et al., 2014; Pons et al., 2017). This decoloration forms a halo, surrounding the copper deposits. The presence of bitumen and organic-rich fluid inclusions form the relics of the former presence of the hydrocarbons in the exposed mineralized rocks (Rainoldi et al., 2015a). Mesozoic and Cenozoic sandstone-hosted copper deposits in South China occur in fault-bounded, intermontane, and piedmont basins. The host rocks were deposited in fluviolacustrine delta and shore facies (Chen, 1988). Interestingly also in this case, an organic carbon anomaly is reported by the latter author.

3.7. CONCLUSION

The Zambian Copperbelt forms part of the arc-shaped Lufilian belt that extends in northwest-southeast direction, with the Kafue Anticline as a prominent structural feature.

Arenite-hosted deposits occur both below (footwall deposits) and above (hanging-wall deposits) the Ore Shale Formation, respectively in the Mindola Subgroup and in the middle part of the Kitwe Formation. The Zambian deposits are stratiform to stratabound and are both argillite and arenite-hosted. The Zambian deposits are variably folded.

The rocks show several alteration and diagenetic phases. Authigenic quartz overgrowths occur on detrital quartz grains and predate and postdate compaction. Other phases are authigenic albite and K-feldspar overgrowths, and calcite and dolomite cements. Especially important is the presence of carbonaceous matter or pyrobitumen coating detrital grains, which has been interpreted to postdate compaction. Ore minerals are chalcopyrite, digenite, chalcocite, covellite, carrollite, pyrrhotite, malachite, and iron oxides that occur as finely disseminated or as replacive blebs, mainly in sericitic quartzites. Ore minerals replace the intense compacted host rock and may enclose all diagenetic phases, indicating they formed very late in the paragenesis. Sulfides occur associated with quartz and calcite in veins. However, sulfides are most abundant where authigenic carbonates are present and carbonates just predate the sulfides in the paragenetic sequence.

Fluid inclusion studies indicate that the mineralizing fluid was a high-saline, high-temperature fluid with an H_2O-NaCl-$CaCl_2$-$MgCl_2$-CO_2 composition. Stable isotope investigation indicates values that the oxygen isotopic values of the ambient fluid fall within the range of both basinal brines and metamorphic fluids and reflect intense water-rock interaction. Pyrobitumen forms as a residue of oil cracking and/or as a byproduct of sulfate reduction and the oxidation of hydrocarbons. The carbonaceous material can be introduced as mobile hydrocarbons. In addition, the arenite-hosted ore in the Zambian Copperbelt is typically related to structural positions (traps) in which hydrocarbons can be accumulated. In addition, mineralization also occurs preferentially in linkage zones between the WNW- and NNW-trending faults bordering the basement highs, which facilitates a high permeability that could have focused both hydrothermal fluids and hydrocarbons. Sulfide mineralization occurred when the highly saline, metal-bearing fluids encountered an H_2S-rich hydrocarbon reservoir during their fault-controlled expulsion.

A comparison with sandstone-hosted deposits in the Lufilian arc and world wide shows similar main features regarding alteration, cementation, and mineralization and most often a characteristic association with hydrocarbons or indications of their former presence. Therefore, the identification of "hydrocarbon" systems is an important tool to prospect for this type of arenite-hosted mineralization. A full basin study approach as performed in the hydrocarbon reservoir studies is thus highly recommended in this but also sediment-hosted type deposits in general.

ACKNOWLEDGMENTS

We thank Dr. Sophie Decrée for editing this paper and the anonymous reviewer for constructive suggestions. We are grateful to Prof. M. Joachimski (University of Erlangen-Nürnberg Germany) for the stable isotope analysis. We also thank Herman Nijs for the careful preparation of the thin and doubly polished sections.

REFERENCES

Annels, A. E. (1979). Mufulira greywackes and their associated sulphides. *Trans. Inst. Mining Metall.*, *B88*, 15–23.

Annels, A. E. (1989). Ore genesis in the Zambian Copperbelt, with particular reference to the northern sector of the Chambishi Basin, in R.W. Boyle, A. C. Brown, C. W. Jefferson, E. C. Jowett, R.V. Kirkham (Ed.). *Sediment-hosted stratiform copper deposits* (pp. 427–452). Geological Association of Canada Special Paper 36.

Asael, D., Matthews, A., Bar-Matthews, M., & Halicz, L. (2007). Copper isotope fractionation in sedimentary copper mineralization (Timna Valley, Israel). *Chem. Geol.*, *243*, 238–254; doi: 10.1016/j.chemgeo.2009.01.015.

Azaraien, H., Shahabpour, J., & Aminzadeh, B. (2017). Metallogenesis of the sediment-hosted stratiform Cu deposit of the Ravar Copper Belt (RCB), Central Iran. *Ore Geol. Rev.*, *81*, 369–395; doi: 10.1016/j.oregeorev.2016.09.035.

Bakker, R. J. (1997). Clathrates: computer programs to calculate fluid inclusion V-X properties using clathrate melting temperatures. *Computers and Geosciences*, *23*, 1–18.

Binda, P. L., & Mulgrew, J. R. (1974). Stratigraphy of copper occurrences in the Zambian Copperbelt, in P. Bartholomé (Ed.), *Gisements Stratiformes et Provinces Cuprifères. Centenaire de la Société Géologique de Belgique* (pp. 215–233). Liège, Belgium.

Bodnar, R. J. (1993). Revised equation and table for determining the freezing point depression of H_2O-NaCl solutions. *Geochim. Cosmochim. Acta*, *57*, 683–684.

Brandt, R. T., Burton, C. C. J., Maree, S. C., & Woakes, M. E. (1961). Mufulira, in F. Mendelsohn (Ed.), *The Geology of the Northern Rhodesian Copperbelt* (pp. 441–461). MacDonald, London, UK.

Broughton, D. (2003). Stratigraphy, regional alteration and mineralization of the Zambian Copperbelt. *RAC/AMIRA P544 Final Report*, 2.1–2.8.

Bull, S., Croaker, M., McGoldrick, P., Pollington, N., Scott, R., Selley, D., & Large, R. (2002). Carbon and oxygen isotopic composition of ore-related and sedimentary carbonates in the Roan: Zambian Copper belt. *AMIRA P544 Proterozoic Sediment Hosted Copper deposits, Progress Report*, *4*.1–4.8.

Cailteux, J. L. H., Kampunzu, A. B., & Lerouge, C. (2007). The Neoproterozoic Mwashya-Kansuki sedimentary rock succession in the Central African Copperbelt, its Cu-Co mineralization, and regional correlations. *Gondwana Res.*, *11*, 414–431; doi: 10.1016/j.gr.2006.04.016.

Cailteux, J. L. H., Kampunzu, A. B., Lerouge, C., Kaputo, A. K., & Milesi, J. P. (2005). Genesis of sediment-hosted stratiform copper-cobalt deposits, central African Copperbelt. *J. Afr. Earth Sci.*, *42*, 134–158; doi: 10.1016/j.jafrearsci.2005.08.001.

Chen, W. (1988). Mesozoic and Cenozoic sandstone-hosted copper deposits in south China. *Min. Deposita*, *23*, 262–267.

Cisternas, M. E., & Hermosilla, J. (2006). The role of bitumen in strata-bound copper deposit formation in the Copiapo area, Northern Chile. *Miner. Deposita*, *41*, 339–355.

Clayton, R. N., Friedman, I., Graf, D., Mayeda, T. K., Meents, W. F., & Shimp, N. F. (1966). The origin of saline formation waters, 1. Isotopic composition. *J. Geoph. Res.*, *71*, 3869–3882.

Dechow, E., & Jensen, M. L. (1965). Sulfur isotopes of some Central African sulphide deposits. *Econ. Geol.*, *60*, 894–941.

Decrée, S., Deloule, E., Ruffet, G., Dewaele, S., Mees, F., Marignac, C., & de Putter, T. (2010). Geodynamic and climate controls in the formation of Mio-Pliocene world-class oxidized cobalt and manganese ores in the Katanga province, DR Congo. *Miner. Deposita*, *45*, 621–629; doi: 10.1016/j.gexplo.2015.10.005.

Dejonghe, L. (1997). Features and origin of the strata-bound siliciclastic-hosted Kinsenda copper deposit contained in the Roan of south-east Shaba, Zaire, in J.-M. Charlet (Ed.), *International Cornet Symposium, Strata-bound Copper Deposits and Associated Mineralizations* (pp. 145–158). Royal Academy of Overseas Sciences.

De Magnée, I., & François, A. (1988). The origin of the Kipushi (Cu, Zn, Pb) deposit in direct relation with a Proterozoic salt diapir, Copperbelt of central Africa, Shaba, Republic of Zaïre, in G. H. Friedrich & P. M. Herzig (Ed.), *Base metal sulphide deposits* (pp. 165–183). Springer-Verlag, Berlin Heidelberg.

De Putter, T., Mees, F., Decrée, S., & Dewaele, S. (2010). Malachite, an indicator of major Pliocene Cu remobilization in a karstic environment (Katanga, Democratic Republic of Congo). *Ore Geol. Rev.*, *38*, 90–100; doi: 10.106/j.oregeorev.2010.01.001.

De Waele, B., Liégeois, J. P., Nemchin, A. A., & Tembo, F. (2006). Isotopic and geochemical evidence of Proterozoic episodic crustal reworking within the Irumide belt of south-central Africa, the southern metacratonic boundary of an Archaean Bangweulu Craton. *Precam. Res.*, *148*, 225–256; doi: 10.1016/precamres.2006.05.006.

Dewaele, S., Muchez, P., Vets, J., Fernandez-Alonzo, M., & Tack, L. (2006). Multiphase origin of the stratiform Cu-Co ore deposits in the western part of the Lufilian fold-and-thrust belt, Katanga (Democratic Republic of Congo). *J. Afr. Earth Sci.*, *46*, 455–469; doi: 10.1016/j.afrearsci.2006.08.002.

Duan, Z., Moller, N., & Weare, J. H. (1996). A general equation of state for supercritical fluid mixtures and molecular dynamics simulation of mixture PVTX properties. *Geochim. Cosmochim. Acta*, *60*, 1209–1216.

Eglinger, A., André-Mayer, A.-S., Vanderhaeghe, O., Mercadier, J., Cuney, M., Decrée, S., Feybesse, J.-L., & Milesi, J.-P. (2013). Geochemical signatures of uranium oxides in the Lufilian belt: From unconformity-related to syn-metamorphic uranium deposits during the Pan-African orogenic cycle. *Ore Geol. Rev.*, *54*, 197–213; doi: 10.1016/j.oregeorev.2013.04.003.

El Desouky, H. A., Muchez, P., Boyce, A. J., Schneider, J., Cailteux, J. L. H., Dewaele, S., & Von Quadt, A. (2010). Genesis of sediment-hosted stratiform copper-cobalt mineralization at Luiswishi and Kamoto, Katanga Copperbelt (Democratic Republic of Congo). *Miner. Deposita*, *45*, 735–763.

El Desouky, H. A., Muchez, P., & Cailteux, J. (2009). Two Cu-Co sulfide phases and contrasting fluid systems in the Katanga Copperbelt, Democratic Republic of Congo. *Ore Geol. Rev.*, *36*, 315–332; doi: 10.1016/j.oregeorev.2009.07.003.

El Desouky, H. A., Muchez, P., Dewaele, S., Boutwood, A., & Tyler, R. (2008a). Post-orogenic origin of the stratiform Cu mineralization at Lufukwe, Lufilian Foreland, Democratic Republic of Congo. *Econ. Geol.*, *103*, 555–582.

El Desouky, H. A., Muchez, P., & Tyler, R. (2008b). The sandstone-hosted stratiform copper mineralization at Mwitapile and its relation to the mineralization at Lufukwe, Lufilian foreland, Democratic republic of Congo. *Ore Geol. Rev.*, *34*, 561–579; doi: 10.1016/j.oregeorev.2008.09.004.

Fanning, C. M., & Link, P. K. (2004). U-Pb SHRIMP ages of Neoproterozoic (Sturtian) glaciogenic Pocatello Formation, Southeastern Idaho. *Geology*, *32*, 881–884.

Fleischer, V. D., Garlick, W. G., & Haldane, R. (1976). Geology of the Zambian Copperbelt, in *Handbook of Strata-Bound and Stratiform ore Deposits*, vol. 6, edited by K. H. Wolf, (pp. 223–352). Elsevier, Amsterdam.

Garlick, W. G. (1962). Geology of the Chibuluma mine, *Stratiform Copper Deposits in Africa. 1st part, Lithology Sedimentology*, edited by J. Lombard & P. Nicolini, (pp. 137–149). Association des Services Géologiques Africains, Paris.

Gorman, A. R., Jenkin, G. R. T., Morgan, K. L., & Catterall, D. (2013). Developing genetic models for copper-silver mineralisation in the Kalahari Copperbelt, Botswana: Mineralogy, geochemistry and structure, in *Mineral deposit research for a high-tech world. Proceedings of the 12th Biennial SGA Meeting* (pp. 624–627). Uppsala, Sweden.

Greyling, L. N. (2009). Fluid evolution and characterization of mineralizing solutions in the Central African Copperbelt. School of Geosciences, University of Witwatersrand. Unpublished PhD thesis.

Haest, M., Muchez, P., Dewaele, S., Boyce, A., von Quadt, A., & Schneider, J. (2009). Petrographic, fluid inclusion and isotopic study of the Dikulushi Cu-Ag deposit, Katanga (D.R.C.): implications for exploration. *Miner. Deposita*, *44*, 505–522; doi: 10.1007/s00126-009-0230-x.

Haest, M., Muchez, P., Dewaele, S., Franey, N., & Tyler, R. (2007). Structural control on the Dikulushi Cu-Ag deposit, Katanga, Democratic Republic of Congo. *Econ. Geol.*, *102*, 1321–1333.

Hanson, R. E., Wardlaw, M. S., Wilson, T. J., & Mwale, G. (1993). U-Pb zircon ages from the Hook granite massif and Mwembeshi dislocation: constraints on Pan-African deformation, plutonism, and transcurrent shearing in Central Zambia. *Precam. Res.*, *63*, 189–209.

Hayes, T. S., Landis, G. P., Whelan, J. F., Rye, R. O., & Moscati, R. J. (2012). The Spar Lake strata-bound Cu-Ag deposit formed across a mixing zone between trapped natural gas and metals-bearing brine. *Econ. Geol.*, *107*, 1223–1249.

Heijlen, W., Banks, D. A., Muchez, P., Stensgard, B. M., & Yardley, B. W. D. (2008). The nature of mineralizing fluids of the Kipushi Zn-Cu deposit, Katanga, Democratic Republic of Congo: Quantitative fluid inclusion analysis using laser ablation ICP-MS and bulk crush-leach methods. *Econ. Geol.*, *103*, 1459–1482.

Hitchon, B., & Friedman, I. (1969). Geochemistry and origin of formation waters in the western Canada sedimentary basin: I. Stable isotopes of hydrogen and oxygen. *Geochim. Cosmochim. Acta*, *33*, 1321–1349.

Hitzman, M. W. (2000). Source basins for sediment-hosted stratiform Cu deposits: Implications for the structure of the Zambian Copperbelt. *J. Afr. Earth Sci.*, *30*, 855–863.

Hitzman, M. W., Broughton, D., Selley, D., Woodhead, J., & Bull, S. (2012). The Central African Copperbelt: Diverse stratigraphic, structural and temporal settings in the world's largest sedimentary copper district. *Society of Economic Geologists, Special Publication*, *16*, 487–514.

Hoefs, J. (2009). *Stable Isotope Geochemistry*, 6th edition. Springer-Verlag, Berlin-Heidelberg.

Hoffmann, P. F., Kaufman, A. J., & G. P. Halverson (1998). A Neoproterozoic snowball earth. *Science*, *281*, 1342–1346.

Irwin, H., Curtis, C., & Coleman, M. (1977). Isotopic evidence for source of diagenetic carbonates formed during burial of organic-rich sediments. *Nature*, *269*, 209–213.

Jackson, M. P. A., Warin, O. N., Woad, G. M., & Hudec, M. R. (2003). Neoproterozoic allochthonous salt tectonics during the Lufulian orogeny in the Katanga Copperbelt, central Africa. *Geol. Soc. Am. Bull.*, *115*, 314–330.

John, T., Schenk, V., Mezger, K., & Tembo F. (2004). Timing and PT evolution of whiteschist metamorphism in the Lufilian Arc-Zambezi Belt Orogen (Zambia): Implications for the assembly of Gondwana. *J. Geol.*, *112*, 71–90.

Jordaan, J. (1961). Nkana, in Mendelsohn, F. (Ed.), *The Geology of the Northern Rhodesian Copperbelt*: (pp.297–328). London, MacDonald.

Kampunzu, A. B., & Cailteux, J. (1999). Tectonic evolution of the Lufilian arc during Neoproterozoic Pan-African orogenesis. *Gondwana Res.*, *2*, 401–421.

Kampunzu, A. B., Kapenda, D. & Manteka, B. (1991). Basic magmatism and geotectonic evolution of the Pan-African Belt in Central Africa – Evidence from the Katangan and West Congolian segments. *Tectonophysics*, *190*, 363–371.

Key, R. M., Liyungu, A. K., Njamu, F. M., Somwe, V., Banda, J., Mosley, P. N., & Armstrong, R. A. (2001). The western arm of the Lufilian Arc in NW Zambia and its potential for copper mineralization. *J. Afr. Earth Sci.*, *33*, 503–528.

Kojima, S., Astudillo, J., Rojo, J., Tristá, D., & Hayashi, K.-i. (2003). Ore mineralogy, fluid inclusion, and stable isotopic characteristics of stratiform copper deposits in the Coastal Cordillera of northern Chile. *Miner. Deposita*, *38*, 208–2016.

Landis, C. R., & Castaño, J. R. (1995). Maturation and bulk chemicals properties of a suite of solid hydrocarbons. *Organic Geochemistry*, *22*, 137–149.

Laux, J. H., Lindenmayer, Z. G., Teixeira, J. B. G., & Neto, A. B. (2005). Ore genesis at the Camaquá copper mine, Neoproterozoic sediment-hosted deposit in Southern Brazil. *Ore Geol. Rev.*, *26*, 71–89; doi: 10.1016/j.oregeorev.2004011.01.

Lindsay, J. F., Kruse, P. D., Green, O. R., Hawkin, E., Brasier, M. D., Cartlidge, J., & Corfield, R. M. (2005). The Neoproterozoic–Cambrian record in Australia: a stable isotope study. *Precam. Res.*, *143*, 113–133.

Machel, H. G. (2001). Bacterial and thermochemical sulfate reduction in diagenetic settings: Old and new insights. *Sediment. Geol.*, *140*, 143–175.

Machel, H. G., Krouse, H. R., & Sassen, R. (1995). Products and distinguishing criteria of bacterial and thermochemical sulphate reduction. *Appl. Geochem.*, *10*, 373–389.

Maree, S. C. (1963). Structural features of the Mufulira area, Northern Rhodesia, in *Stratiform Copper Deposits in Africa. Part 2 Tectonics*, edited by J. Lombard & P. Nicolini, (pp. 158–177). Association des Services Géologiques Africains, Paris.

McGowan, R., Roberts, S., Foster, R. P., Boyce, A. J., & Coller, D. (2003). Origin of the copper-cobalt deposits of the Zambian Copperbelt: An epigenetic view from Nchanga. *Geology*, *31*, 497–500.

McGowan, R. R., Roberts, S., & Boyce, A. J. (2006). Origin of the Nchanga copper-cobalt deposits of the Zambian Copperbelt. *Miner. Deposita*, *40*, 617–638; doi: 10.1007/s00126-005-0032-8.

Minnen, M. (2013). Petrographical, mineralogical and geochemical study of the sandstone hosted Cu-Co mineralization of the Zambian type at Mufulira (Zambia). MSc thesis Geology, Department of Earth and Environmental Sciences, KU Leuven, Belgium.

Muchez, P., Brems, D., Clara, E., De Cleyn, A., Lammens, L., Boyce, A., De Muynck, D., et al. (2010). Evolution of Cu-Co mineralizing fluids at Nkana Mine, Central African Copperbelt, Zambia. *J. Afr. Earth Sci.*, *58*, 457–474; doi: 10.1016/j.afrearsci.2010.05.003.

Muchez, P., Viaene, W., & Dusar, M. (1992). Diagenetic control on secondary porosity in flood plain deposits: an example of the Lower Triassic of northeastern Belgium. *Sediment. Geology*, *78*, 285–298.

Muchez, P., Viaene, W. A., Keppens, E., Marshall, J. D., & Vandenberghe, N. (1991). Vein cements and the geochemical evolution of subsurface fluids in the Visean of the Campine Basin (Poederlee borehole, Belgium). *J. Geol. Soc., London*, *148*, 1005–1017.

Ngoyi, K., & L. Dejonghe (1995). Géologie et genèse du gisement stratoide cuprifère de Kinsenda (SE du Shaba, Zaire). *Bull. Soc. belge Géol.*, *104*, 245–281.

Pons, M. J., Franchini, M., Giusiano, A., Patrier, P., Beaufort, D., Impiccini, A., Rainoldi, A. L., & Meinert, L. (2017). Alteration halos in the Tordillos-hosted copper deposit of the Neuguén Basin, Argentina. *Ore Geol. Rev.*, *80*, 691–715; doi: 10.1016/j.oregeorev.2016.06.011.

Pons, M. J., Rainoldi, A. L., Franchini, M., Giusiano, A., Cesaretti, N., Beaufort, D., Patrier, P., & Impiccini, A. (2015). Mineralogical signature of hydrocarbon circulation in Cretaceous red beds of the Barda Gonzáles area, Neuquén Basin, Argentina. *Am. Assoc. Petrol. Geol., Bulletin*, *99*, 525–554; doi: 10.1306/08131413170.

Porada, H. (1989). Pan-African rifting and orogenesis in southern to equatorial Africa and eastern Brazil. *Precam. Res.*, *44*, 103–136.

Porada, H., & Berhorst, V. (2000). Towards a new understanding of the Neoproterozoic Early Paleozoic Lufilian and northern Zambezi belts in Zambia and Congo/Zaire. *J. Afr. Earth Sci.*, *30*, 727–771.

Rainoldi, A. L., Franchini, M., Beaufort, D., Mozley, P., Giusiano, A., Nora, C., Patrier, P., Impiccini, A., & Pons, J. (2015a). Mineral reactions associated with hydrocarbon paleomigration in the Huincul High, Neuquén Basin, Argentina. *Geol. Soc. Am. Bull.*, *127*, 1711–1729; doi: 10.1130/B31201

Rainoldi, A. L., Franchini, M., Beaufort, D., Patrier, P., Giusiano, A., Impiccini, A., & Pons, J. (2014). Large-scale bleaching of red beds related to upward migration of hydrocarbons: Los Chihuidos High, Neuquén Basin, Argentina. *J. Sediment. Res.*, *84*, 373–393

Rainoldi, A. L., Franchini, M. B., Boyce, A., Pons, M. J., Giusiano, A., & Cesaretti, N. N. (2015b). The role of hydrocarbons in the genesis of the sediment-hosted Stratiform copper deposits, Neuguén Basin Argentina, in A.-S. André-Mayer et al. (Eds.), *Mineral Resources in a Sustainable World* (pp. 1989–1992). Proceedings of the 13th Biennial SGA Meeting, vol. *5*, Nancy, France.

Rainaud, C., Master, S., Armstrong, R. A., & Robb, L. R. (2005a). Geochronology and nature of the Palaeoproterozoic basement in the Central African Copperbelt (Zambia and the Democratic Republic of Congo), with regional implications. *J. Afr. Earth Sci.*, *42*, 1–31.

Rainaud, C., Master, S., Armstrong, R. A., Phillips, D., & Robb, L. R. (2005b). Monazite U-Pb dating and ^{39}Ar-^{40}Ar thermochronology of metamorphic events in the Central African Copperbelt during the Pan-African Lufilian Orogeny. *J. Afr. Earth Sci.*, *42*, 183–199.

Richards, J. P., Cumming, G. L., Krstic, D., Wagner, P. A., & Spooner, E. T. C. (1988a). Pb isotopic constraints on the age of sulphide ore deposition and U-Pb age of late uraninite veining at the Musoshi stratiform copper deposit, Central African Copper Belt, Zaire. *Econ. Geol.*, *83*, 724–741.

Richards, J. P., Krogh, T. E., & Spooner, E. T. C. (1988b). Fluid inclusion characteristics and U-Pb rutile age of late hydrothermal alteration and veining at the Musoshi stratiform copper deposit, Central African Copper Belt, Zaire. *Econ. Geol.*, *83*, 118–139.

Rosenbaum J., & Sheppard S. M. F. (1986). An isotopic study of siderites, dolomites and ankerites at high temperatures. *Geochim. Cosmochim. Acta*, *50*, 1147–1150.

Selley, D., & Scott, R. (2003). Mwambashi B: Keys to arenite-hosted Cu-Co mineralization. *RAC/ZAMIRA P544 Final Report*, pp. 15.1–15.6.

Selley, D., Broughton, D., Scott, R., Hitzman, M., Bull, S., Large, R., McGoldrick, P., Croaker, M., Pollington, N., & Barra, F. (2005). A new look at the geology of the Zambian Copperbelt, in *Economic Geology 100th Anniversary Volume*, pp. 965–1000, Society of Economic Geology, Tulsa.

Selley, D., Bull, S., Scott, R., Croaker, M., Broughton, D., & Pollington, P. (2003). Development of the Lower Roan basin system: Controls on Cu-Co mineralization. *RAC/AMIRA P544 Final Report*, pp. 4.1–4.6.

Sheppard, S. M. F. (1986). Characterization and isotopic variations in natural waters, in J. W. Valley et al. (Eds.), *Stable isotopes in high temperature geological processes* (pp. 165–184). Reviews in Mineralogy 16.

Sillitoe, R. H., Perelló, J., & García, A. (2010). Sulfide-bearing veinlets throughout the stratiform mineralization of the Central African Copperbelt: Temporal and genetic implications. *Econ. Geol.*, *105*, 1361–1368.

Sillitoe, R. H., Perelló, J., Creasar, R. A., Wilton, J., Wilson, A. J., & Dawborn, T. (2017). Age of the Zambian Copperbelt. *Miner. Deposita*, *52*, 1245–1268.

Suchecki, R. K., & Land, L. S. (1983). Isotopic geochemistry of burial-metamorphosed volcanogenic sediments, great Valley sequence, northern California. *Geochim. Cosmochim. Acta*, *47*, 1487–1499.

Surdam, R. C., Boese, S. W., & Crossey, L. J. (1984). The chemistry of secondary porosity, in D. A. McDonald & R. C. Surdam, *Clastic Diagenesis* (pp. 127–151). American Association of Petroleum Geologists Memoir 37.

Torremans, K., Gauquie, J., Boyce, A. J., Barrie, C. D., Dewaele, S., Sikazwe, O., & Muchez, P. (2013). Remobilisation features and structural control on ore grade distribution at the Konkola stratiform Cu-Co ore deposit, Zambia. *J. Afr. Earth Sci.*, *79*, 10–23; doi: 10.1016/j.afrearsci.2012.10.005.

Van Eden, J. G. (1974). Depositional and diagenetic environment related to sulphide mineralization, Mufulira, Zambia. *Econ. Geol.*, *69*, 59–79.

Van Langendonck, S., Muchez, P., Dewaele, S., Kaputo Kalubi, A., & Cailteux, J. (2013). Petrographic and mineralogical study of the sediment-hosted Cu-Co ore deposit at Kambove West in the central part of the Katanga Copperbelt (DRC). *Geol. Belgica*, *16*, 91–104.

Veizer, J., & Hoefs, J. (1976). The nature of $^{18}O/^{16}O$ and $^{13}C/^{12}C$ secular trends in sedimentary carbonate rocks. *Geochim. Cosmochim. Acta*, *40*, 1387–1395.

Voet, H. W., & Freeman, P. V. (1972). Copper orebodies in the basal Lower Roan meta-sediments of the Chingola open pit area Zambian Copperbelt. *Geol. Mijnbouw*, *51*, 299–308.

Weisgerber, G. (2006). The mineral wealth of ancient Arabia and its use I: Copper mining and smelting at Feinan and Timna: Comparison and evaluation of techniques, production, and strategies. *Arab. Archaeol. Epigr.*, *17*, 1–30; doi: 10.1111/j.1600-0471.2006.00253.x.

Winfield, O., & Robinson, I. C. (1963). The structures of Chibuluma mine, in J. Lombard, & P. Nicolini (Eds.), *Stratiform Copper Deposits in Africa. Part 2 Tectonics* (pp. 192–202). Association des Services Géologiques Africains, Paris.

4

Nb-Ta-Sn-W Distribution in Granite-related Ore Systems: Fractionation Mechanisms and Examples from the Karagwe-Ankole Belt of Central Africa

Niels Hulsbosch

ABSTRACT

Tantalum, niobium, tin, and tungsten mineralization typically occurs worldwide in close proximity to fractionated peraluminous leucogranites. These granite-related ore deposits show a diversity of mineralization modes, ranging from plutonic-hosted breccia and hydrothermal vein-stockwork systems to (peri)batholitic greisens, skarns, pegmatites, and hydrothermal veins. In the Mesoproterozoic Karagwe-Ankole orogenic belt (KAB) of Central Africa, numerous early Neoproterozoic Nb-Ta-Sn-W rare-metal deposits formed as primary mineralization in hydrothermal quartz veins, magmatic lithium-cesium-tantalum (LCT) family pegmatites and associated intrapegmatitic greisens. This chapter reviews the petro- and metallogenesis of Nb-Ta-Sn-W granite-related ore deposits, as they occur in the Rwandese part of the KAB, and provides a general overview of the geochemical mechanisms behind the formation of pegmatite- and quartz vein-hosted deposits and the distribution and enrichment of the ore elements Nb, Ta, Sn, and W in the different metallogenic subsystems (i.e., granites, pegmatites, and veins). Moreover, this chapter assesses the observed close spatiotemporal association between leucogranites, pegmatites, and quartz veins in terms of a direct genetic link between the mineralization and the felsic magmatism. Based on this review and element distribution calculations, an integrated orthomagmatic metallogenic model for Nb-Ta-Sn-W mineralization in the KAB is demonstrated. This model can form a major tool in the exploration for granite-related ore deposits in general.

4.1. INTRODUCTION

Granites are able to concentrate a large number of chemical elements, such as Li, Rb, Cs, Be, Cu, Mo, Sn, Zr, Th, U, Nb, Ta, W, Au, and rare earth elements, several order of magnitude above their respective crustal abundances (Černý et al., 2005; Taylor & McLennan, 1985). This chapter will mainly focus on rare-element mineralization associated with peraluminous granites, pegmatites, and related quartz veins, which are typically enriched in Li, Rb, Cs, Ta, Nb, Ta, Sn, and W (Linnen & Cuney, 2005; Raimbault et al., 1995). Rb and Cs are geochemically classified as large-ion lithophile element (LILE) because of their small charge to ionic radius ratios. Gast (1972) also included Li as a LILE since it has a large radius to charge ratio, despite its small radius. Nb, Ta, W, and probably Sn are High Field Strength Elements (HFSE) with high charge-to-ionic-radius ratios. Both LILE and HFSE do not substitute into common rock-forming minerals and thus behave incompatibly during the crystallization of silicate melts (e.g., Breiter et al., 2007; Linnen & Cuney, 2005). Except for Sn, these metals have been classified based on their economic importance and supply risk as critical raw materials for the high-tech industry (EC, 2017).

KU Leuven, Geodynamics and Geofluids Research Group, Department of Earth and Environmental Sciences, Leuven, Belgium

Ore Deposits: Origin, Exploration, and Exploitation, Geophysical Monograph 242, First Edition.
Edited by Sophie Decrée and Laurence Robb.

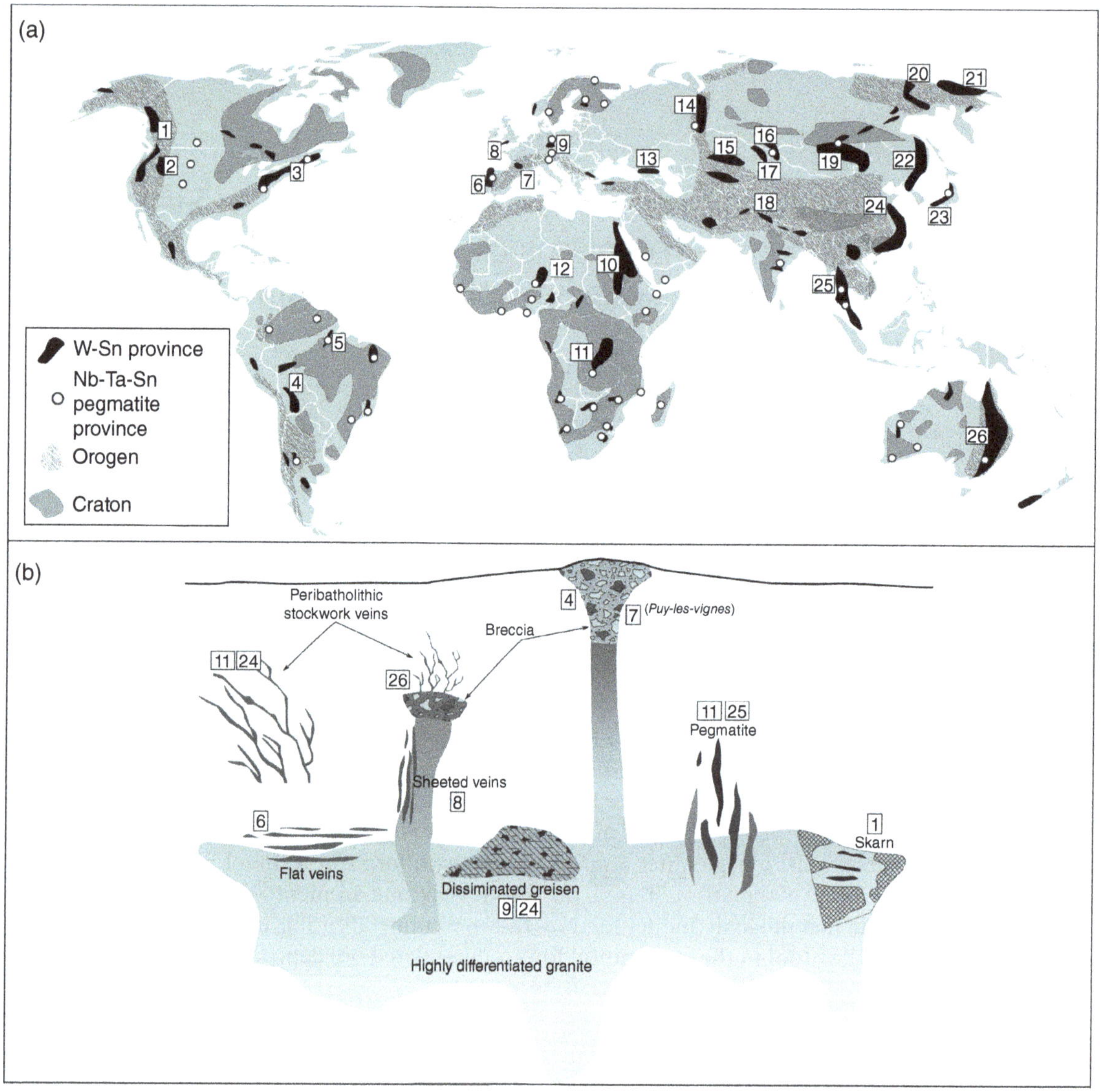

Figure 4.1 (a) Overview of major W-Sn provinces in the world in relationship to Nb-Ta-Sn pegmatite provinces. 1: North American Cordillera; 2: Colorado Plateau; 3: Appalachian; 4: Bolivian; 5: east Brazilian;, 6: Iberian-Panasqueira; 7: Massif Central; 8: Cornwall; 9: Erzgebirge/Krušné Hory; 10: east African; 11: Kibara Metallogenic; 12: Nigerian; 13: north Caucasian; 14: east Uralian; 15: central Kazakhstan; 16: Gorno-Altai; 17: Kalba-Narym; 18: south Tien Shan; 19: Mongolian Transbaikalian; 20: Verkhoyansk-Kolyma; 21: Chukotka; 22: Sikhote-Alyn; 23: Japanese; 24: southeast China; 25: Myanmar-Thailand-Malaysia; and 26: east Australian province (after Taylor, 1979; Beus, 1986; Romer and Kroner, 2016; and Melcher et al., 2017). (b) Schematic diagram illustrating primary granite-related W-Sn deposit styles in combination with their type-localities. Styled partly after Baker et al. (2005).

Three major categories of granite-related ore deposits -sensu stricto- could develop in the vicinity of granite intrusions and are observed worldwide (Fig. 4.1): (1) disseminated magmatic mineralization in granites themselves, (2) mineralization in late-magmatic pegmatites intruding as endo- and exodikes, and (3) mineralization in magmatic-hydrothermal or metasomatic deposits imprinted on the granites, pegmatites, or on the immediate host rocks (i.e., veins, greisens or skarn; Baker et al., 2005; Beus, 1986; Černý et al., 2005; Štemprok, 1987; Taylor, 1979). Peraluminous rare-metal granites with magmatic disseminated mineralization are locally associated with vein- and greisen-type mineralization (Breiter et al., 1999), both requiring water saturation and under hydrostatic regime (Černý et al., 2005). However, significant hydrothermal deposits are generated from granites that lack magmatic disseminated ore bodies or show no direct relationships with a parental granite (Audétat et al., 2000a; Pettke et al., 2005). Reciprocally, most evolved

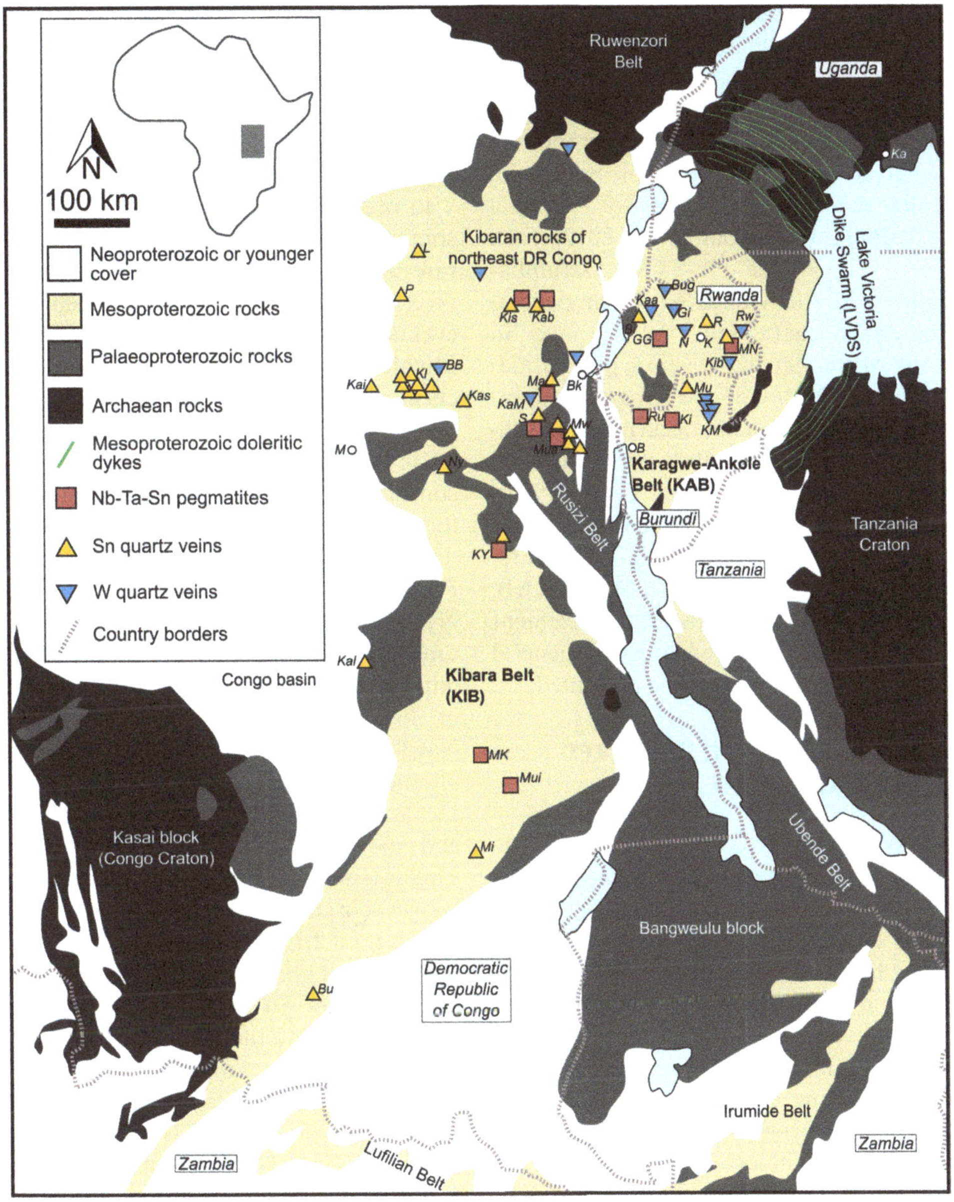

Figure 4.2 Conceptual outline of the geographical and geological setting of the major rare-metal deposits of the Karagwe-Ankole (KAB) and Kibara (KIB) belts in the Kibara Metallogenic Province of the Central African Great Lakes region; B: Bujumbura; BB: Bengo-Biri; Bi: Bisesero; Bk: Bukavu; Bu: Busanga; Bug: Bugarama; GG: Gatumba-Gitarama pegmatite field; Gi: Gifurwe K: Kigali; Ka: Kampala; Kaa: Kabaya; Kab: Kabunga; Kai: Kailo; Kal: Kalanda; Kas: Kasese; Kl: Kalima; Ki: Kivuvu; Kib: Kibungo; KaM: Kamituga-Mobale; KM: Kirundo-Muyinga; KY: Kabambare – Yungwe; L: Lowa; Kis: Kiseke M: Maniema; Ma: Matale; Mi: Mitwaba; MK: Manono-Kitotolo; MN: Musha-Ntunga; Mu: Mulehe; Mua: Muana; Mui: Muika; Mw: Mwenga; N: Nyakabingo; Ny: Nyangulube P: Punia; R: Rutongo; Ru: Ruhembe; Rw: Rwinkwavu; S: Sasa (after Dewaele et al., 2015; Fernandez-Alonso et al., 2012; Koegelenberg et al., 2015; Pohl et al., 2013).

granites are not associated with W-Sn mineralization (Romer & Kroner, 2015; Romer & Kroner, 2016). In addition, pegmatites are often not associated with disseminated magmatic or hydrothermal mineralization (Černý et al., 2005).

The co-occurrence of Nb-Ta-Sn pegmatite-type and Sn-W quartz vein-type mineralization in the Karagwe-Ankole Belt (KAB) region of the Kibara Metallogenic Province in Central Africa (Fig. 4.2) provides a first-class and, moreover, relatively rare opportunity to investigate the geochemical evolution of a global granite-related ore system and to evaluate the element distribution in space and time between pegmatites and quartz veins during the magmatic-hydrothermal transition. Another example of granite-related, rare-element mineralization in cogenetic pegmatites and quartz veins is described for the Neo-Archean Preissac-Lacorne batholith (Superior Province, Canada; Mulja et al., 1995b).

The Rwandan part of the KAB is well known for its richness in granite-related ore deposits spatiotemporally associated to the barren early Neoproterozoic Kibara Sn granites or locally named G4 granites of Rwanda (Figs. 4.2 and 4.3; Pohl, 1994; Pohl et al., 2013; Tack et al., 2010), for example, the Nb-Ta-Sn Gatumba-Gitarama pegmatite field, the Ntunga Nb-Ta-Sn pegmatite, the Nyakabingo W quartz vein, the Rutongo Sn quartz vein, and the Musha Sn quartz vein type localities (Fig. 4.3). An important feature of the Kibara Sn granites (G4) in the KAB is their frequent occurrence in well-defined composite plutons dominantly composed of an older generation (~1375 Ma) of granites (Pohl, 1994). This work will provide a review, discussion, and interpretation of the metallogenesis of Nb-Ta-Sn-W granite-related ore deposits based on the example of the KAB. Main fractionation processes responsible for element enrichment will be discussed. As a conclusion, a general orthomagmatic metallogenic model will be presented.

4.2. ORTHOMAGMATIC OR RECYCLED METAL ORIGIN

Nb, Ta, Sn, and W deposits typically occur worldwide in close proximity to differentiated, late- to post-orogenic granites mainly peraluminous in composition (Baker et al., 2005; Černý et al., 2005; Melcher et al., 2015; Melcher et al., 2017; Romer & Kroner, 2016). One of the major research axes on these deposits is focused around the question of whether a close spatiotemporal association implies a direct genetic link between the mineralization and the granitic magmatism, especially for the case of peribatholitic deposits (Černý et al., 2005; Förster et al., 2009; Harlaux et al., 2017a; Wilkinson, 1990). This debate opposes two end-member models as conceptualized in Figure 4.4: (1) the granitic magma acted as the direct source of the ore metals and the fluids (i.e., the orthomagmatic hypothesis: Audétat et al., 2000b; Moura et al., 2014; Reyf, 1997; Varlamoff, 1972; Webster, 2004), or (2) the intrusion induced thermal convection of metamorphic fluids, which extracted ore metals from surrounding rocks, mainly metasediments, within the circulation cell (i.e., recycling hypothesis: Linnen & Williams-Jones, 1995; Marignac & Cuney, 1999; Marignac & Cathelineau, 2010; Vindel et al., 1995). Mixing and cooling of the magmatic fluid with (preheated) meteoric water, circulating in the upper parts of the granite cupola, has been proposed by Audétat et al. (2000b) at the Mole Granite (Australia) and led toward precipitation of W in intragranitic or perigranitic quartz veins. While Nb-Ta-oxide mineralization in pegmatites is generally accepted to be formed cogenetic with magmatic granite-pegmatite differentiation (Breiter et al., 2007; Černý et al., 2005; Linnen et al., 2014; London, 2008), the contribution of granite-derived fluids as direct W and Sn metal sources is still debated for peribatholitic quartz-wolframite-cassiterite veins (Hulsbosch et al., 2016; Marignac & Cathelineau, 2010).

In order to relate the Kibara Sn granites as potential sources to the pegmatite- and quartz vein hosted rare metal mineralization, the geochemical distribution of elements among the different deposit styles should be evaluated. Previous metallogenic research in the Kibara Metallogenic Province, applying petrographical, geochemical, and fluid inclusion studies, focused mainly on the W and Sn quartz vein deposits (e.g., Dewaele et al., 2010; Pohl, 1994; Pohl & Günther, 1991). This has led to contrasting interpretations concerning the metal source for W and Sn mineralization in the quartz veins (see discussion in De Clercq, 2012; Hulsbosch et al., 2016). For example in the case of the Rutongo Sn and Nyakabingo W quartz veins deposits, the pristine signatures were overprinted by fluid-rock reactions during mineralization at the deposit site, as indicated by studies of fluid inclusions (aqueous-gaseous H_2O-CO_2-CH_4-N_2-NaCl fluids), minerals, and bulk rocks (intense wall-rock interaction), and isotopes ($\delta^{18}O_{quartz}$: between 13.6 and 15.6‰ V-SMOW; highly variable δD_{quartz}: between −47 and −64‰ V-SMOW) (see details in sections 4.3. & 4.4.). This was interpreted as a metamorphic origin for these quartz veins (De Clercq et al., 2008; Dewaele et al., 2010) as opposed to a magmatic-hydrothermal origin (Pohl, 1994; Pohl et al., 2013). Overprinting of the source signature record of Sn-W hydrothermal quartz-vein deposits by fluid-rock interactions is common and also observed elsewhere (see e.g., Chicharro et al., 2016; Heinrich, 1990; Polya et al., 2000; Wood & Samson, 2000). These difficulties imply that genetic metal source studies of Sn-W deposits cannot be firmly based on the investigation of individual quartz-vein systems only.

4.3. ANATECTIC ORIGIN OF PEGMATITES

One of the current discussions in pegmatite research is centered around the question of whether the formation of pegmatites is mainly driven via granitic differentiation or anatectic processes (Simmons & Webber, 2008). Many pegmatite fields worldwide are known to form by differentiation from granitic melts (Černý, 1991c; London, 2008; London, 2014). In many cases, the connection of pegmatites with a parent pluton is inferred from spatial relationships, age determinations, and geochemical characteristics (e.g., London, 2008; Mulja et al., 1995a; Mulja et al., 1995b). However, there is a renewal of interest for the nongranitic or anatectic origin of some pegmatite fields because these pegmatites seem not to be

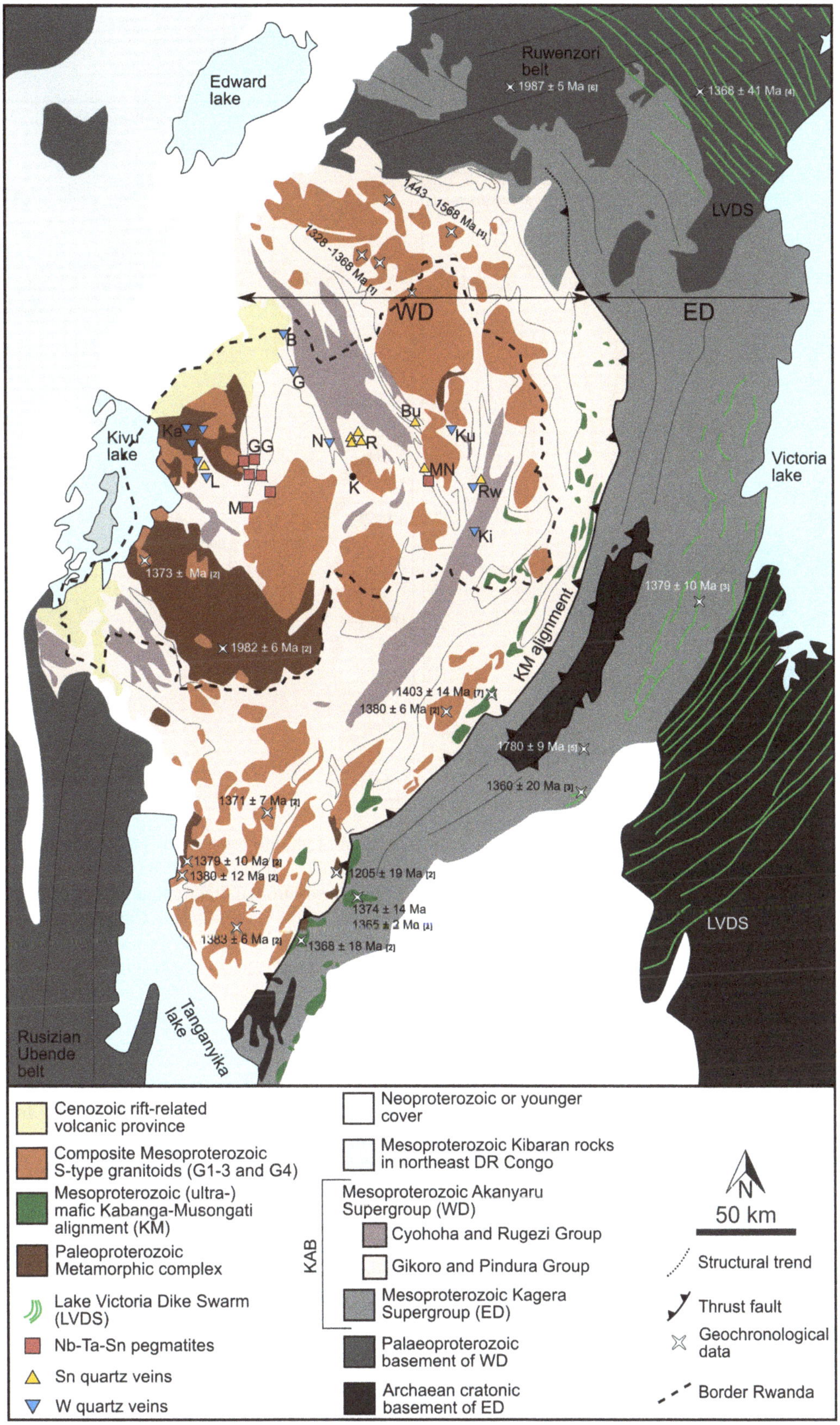

Figure 4.3 Geological map of Rwanda and the Karagwe-Ankole belt (KAB). The most important granite-related ore deposits of Rwanda are indicated. B: Bugarama; Bu: Bugalula; ED: eastern domain; G: Gifurwe; GG: Gatumba-Gitarama; K: Kigali; Ka: Kabaya; Ku: Kuluti; L: Lutsiro; M: Mwaka; MN: Musha-Ntunga; N: Nyakabingo; R: Rutongo; Rw: Rwinkwavu; WD: western domain. After Fernandez-Alonso (2007) and Koegelenberg et al. (2015). Geochronological data from (1) Buchwaldt et al. (2008), (2) Tack et al. (2010), (3) Deblond et al. (2001), (4) Mäkitie et al. (2014), (5) Fernandez-Alonso et al. (2012), (6) Westerhof et al. (2014), (7) Maier et al. (2007).

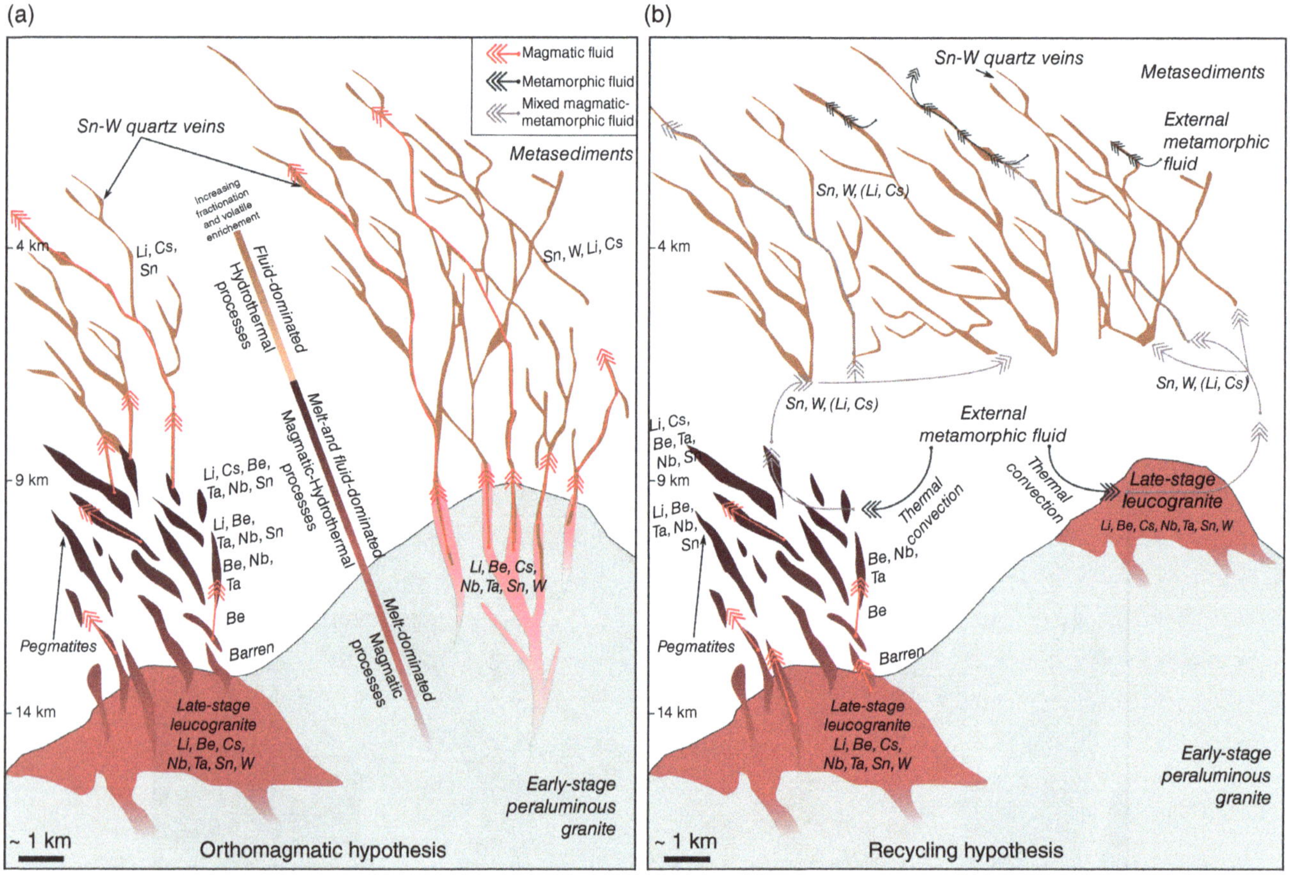

Figure 4.4 End-member metallogenic models for the development of granite-related Nb-Ta-Sn pegmatite-type and Sn-W quartz vein-type deposits. (a) The orthomagmatic hypothesis in which the granitic magma acted as the direct melt/fluid source of the ore metals. A continuous genetic relation can exist from a differentiated granite and pegmatite system towards the quartz veins (e.g., Varlamoff, 1972; Webster et al., 1997). Alternatively, the pegmatite and quartz vein systems could form disconnected but were both derived from a common granite source system (e.g., Audétat et al., 2000b; Chicharro et al., 2015; Hulsbosch et al., 2016; Moura et al., 2014). (b) The recycling hypothesis in which convection of external metamorphic fluids extracted ore metals from igneous and metasedimentary rocks within the circulation cell (e.g., Dewaele et al., 2010; Marignac & Cuney, 1999; Vindel et al., 1995).

associated with potential parental granites, or when they are, granite and pegmatites are slightly diachronous (Černý, 1991b; Müller et al., 2017; Shaw et al., 2016). In high-grade metamorphic areas, low-degree partial melting of quartzo-feldspathic or amphibolitic rocks can produce pegmatitic melts in the absence of felsic magmatism or generate pegmatites, which do not demonstrate a genetic relationships with spatially associated granitic plutons (Müller et al., 2015; Roda Robles et al., 1999; Shearer et al., 1992; Simmons & Webber, 2008). For example, in the Late-Proterozoic Sveconorwegian pegmatite province of southern Norway and southwest Sweden, more than 5000 rare-element pegmatites with a dominant niobium-yttrium-fluorine (NYF) signature are not related to a parental granite generation. Instead, they occur in areas of high-grade metamorphism and are the result of migmatization (Müller et al., 2017). In granitic source models, dominantly metasedimentary sequences are partially melted and the containing incompatible elements ultimately end up in a granitic melt which, in turn, can form a pegmatitic melt by extreme melt differentiation processes (e.g., > 99% of fractional crystallization; Černý, 1991c). In the anatectic source models, the same fluxing components and incompatible elements will be preferentially partitioned into a silicate melt derived from low-degree partial melting and that can directly form a pegmatite (Simmons & Webber, 2008; Turlin et al., 2017). Because the granites and pegmatites of the Kibara metallogenic province meet the criteria pointing toward a granitic source model (e.g., spatial relationships, age determinations, and chemical characteristics; see discussion in the following section 4.4), this chapter will not further discuss the specifics of anatectic source models for pegmatites in detail.

4.4. CASE STUDY: THE KIBARA METALLOGENIC PROVINCE

The Great Lakes region in Central Africa is marked by abundant Sn, W, Nb, and Ta deposits related to prominent granitic intrusions (Varlamoff, 1972). Geographically, this giant granite-related metallogenic province extends through the Katanga and Kivu-Maniema provinces of the Democratic Republic of Congo (DR Congo) in the southwest, over Rwanda and Burundi, up to southwest Uganda and northwest Tanzania in the northeast (Figs. 4.2 and 4.3). The Great Lakes region is known as the Kibara Metallogenic Province (Pohl et al., 2013) and comprises the Mesoproterozoic Karagwe-Ankole orogenic belt (KAB) in the northeast and the Kibara orogenic belt (KIB) in the southwest, both separated by the northwest-trending Paleoproterozoic Rusizian basement units of the Ubende belt (Figs. 4.2 and 4.3) (for a detailed discussion of the basement units, see Fernandez-Alonso et al., 2012; Tack et al., 2010).

Two contrasting tectono-magmatic models currently exist for the KAB and KIB region. The first set of models places the region in an intracratonic geodynamic setting. In this view, the KAB has been kinematically affected by different periods of extension and compression (Cahen et al., 1984; Klerkx et al., 1984; Klerkx et al., 1987) or alternatively by extensional detachments conditioned by transtensional strike-slip reactivations of shear zones in the Paleoproterozoic basement (Fernandez-Alonso & Theunissen, 1998). The most prominent tectono-magmatic event has set as a short-lived anorogenic event at 1375 Ma (Tack et al., 2010). The compressional deformation in the KAB is in this model set at ~1 Ga and interpreted as a far-field effect of the Irumide Orogen (Tack et al., 2010). The second set of models interprets the KAB region around 1375 Ma as an active margin, related to the Congo Craton–Tanzania/Bangweulu Block convergence, with localized extension in a back arc setting. The Kibara Sn (G4) granite magmatism at 1 Ga in the KAB suggests a renewed convergence and associated deformation, potentially related to the Rodinia amalgamation (Debruyne et al., 2015; Kampunzu et al., 1986; Koegelenberg & Kisters, 2014; Koegelenberg et al., 2015; Kokonyangi et al., 2004; Kokonyangi et al., 2007; Rumvegeri, 1991).

The primary rare metal mineralization of the Kibara Metallogenic Province is hosted in magmatic Nb-Ta-Sn pegmatites and magmatic-hydrothermal Sn or W quartz veins associated with the regionally defined "Kibara Sn granites" of an early Neoproterozoic age (Cahen et al., 1984; Fig. 4.2). Generally, Sn granites represent specialized granitic intrusions with high content of Sn (>15 ppm Sn; Barsukov, 1957), and spatially associated to Sn deposits (Lehmann, 1987). The mineralized systems consist of swarms of pegmatite dykes or quartz veins emplaced in the metasedimentary host rocks, or at the granitic cupola level, structurally situated above these Sn granites. Endogranitic deposits, such as breccia pipes, greisens, intragranitic veins, or stockscheider pegmatites, have not been described in the region. Estimates of historical mining concentrates are about 800,000 t cassiterite (SnO_2), 30,000 t Ta-Nb oxide ($(Nb,Ta)_2O_5$), 30,000 t wolframite ($(Fe,Mn)WO_4$), and 600 t gold. Pohl et al. (2013) report grades of 153 g/t Sn, 70 g/t Ta, and 82 g/t Nb for pegmatite deposits (i.e., Gatumba, Rwanda); 4.7 kg/t Sn for Sn quartz veins (Rutongo, Rwanda); and 1.5 to 2.3 g/t Au for Au quartz veins (Twangiza, DR Congo). Grades for W in quartz veins (Nyakabingo, Rwanda), are currently not published. At times, bismuth (Bi), molybdenite (MoS_2), beryl ($Be_3Al_2(SiO_3)_6$), and amblygonite ($(Li,Na)AlPO_4(F,OH)$) were by-products of mining (Pohl et al., 2013).

Individual pegmatite or quartz vein deposits of the Kibara Metallogenic Province have been intensely studied, especially at their type-localities in Rwanda; Gatumba-Gitarama, Rutongo, Nyakabingo, Musha, and Ntunga (Figs. 4.2, 4.3, and 4.5), which led to a vast amount of literature concerning the paragenesis, metallogenesis, and alteration of the ore bodies on a deposit scale (De Clercq et al., 2008; Dewaele et al., 2016a; Dewaele et al., 2010; Dewaele et al., 2016b; Dewaele et al., 2011; Günther & Ndutiye, 1990; Lehmann & Lavreau, 1987; Melcher et al., 2015; Pohl, 1994; Pohl & Günther, 1991; Pohl et al., 2013; Varlamoff, 1948, 1954, 1969, 1972). The evolution of the granite-pegmatite system and, especially, the pristine source of the mineralizing fluids and the metals for the peribatholitic quartz-vein deposits has been studied recently in depth (Dewaele et al., 2016b; Hulsbosch, 2016; Hulsbosch et al., 2016; Hulsbosch et al., 2014). Recent work suggests that the Kibara Sn granites, pegmatites, and quartz veins evolved spatiotemporally as one integrated metallogenic system operating at ca. 980 ± 20 Ma (Dewaele et al., 2016b; Hulsbosch et al., 2016; Hulsbosch et al., 2014; Lehmann et al., 2014; Pohl et al., 2013).

4.4.1. Kibara Sn (G4) Parental Granites

Major ilmenite-series, S-type granitic magmatism intruded the KAB (Günther et al., 1989; Lehmann & Lavreau, 1987; 1988) at 986 ± 10 Ma (SHRIMP U-Pb on zircon; Tack et al., 2010). Spatially, the rare-metal pegmatites and Sn-W vein systems are related to this granite event (Günther & Ndutiye, 1990; Lehmann & Lavreau, 1987; Pohl et al., 2013; Varlamoff, 1954). These Kibaran Sn granites (also called G4 granites in Rwanda) are leucocratic and show a typically equigranular and unfoliated texture (Cahen & Ledent, 1979). Their mineralogical composition consists of quartz, microcline, albite, and

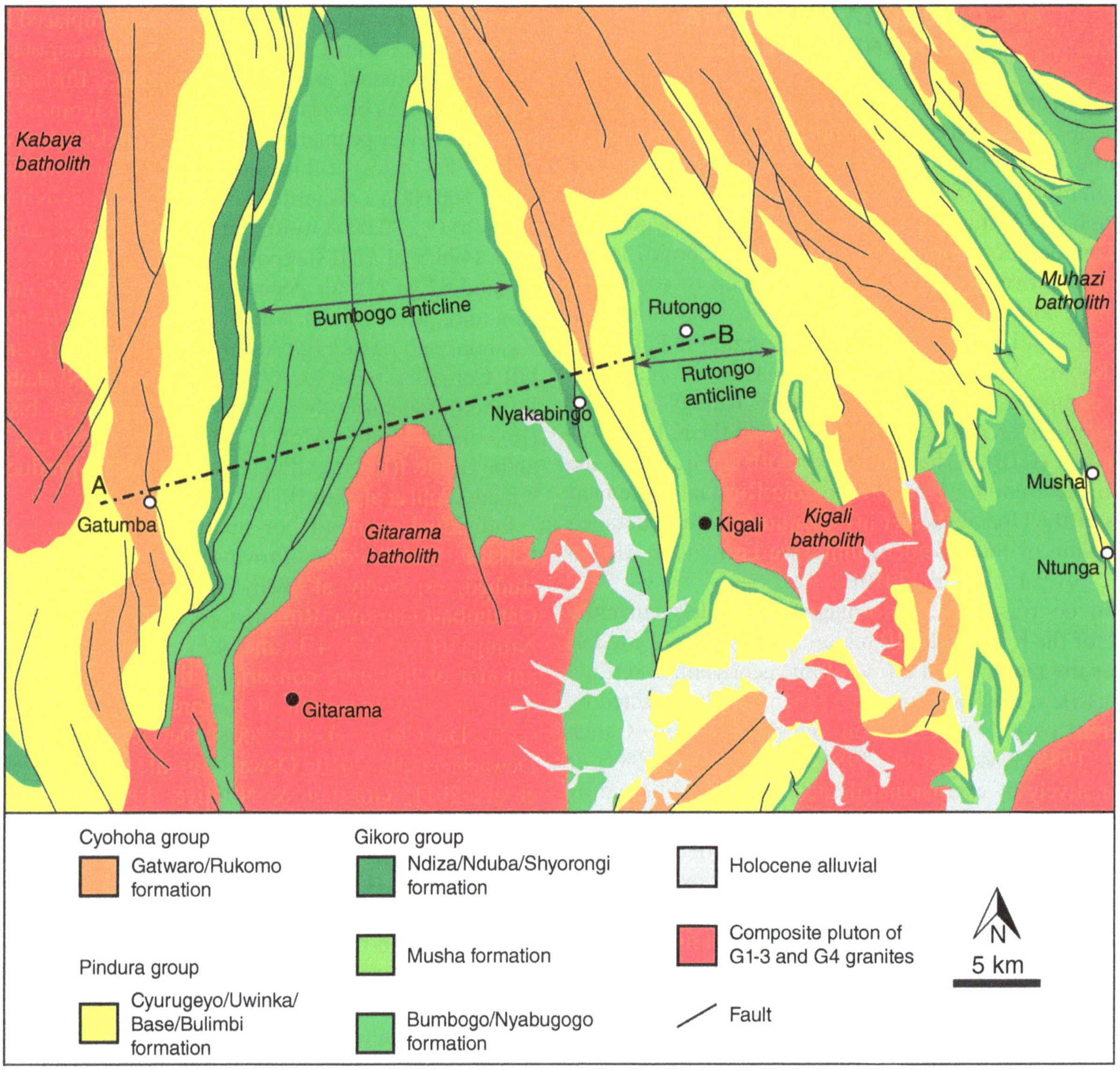

Figure 4.5 Geological map of central Rwanda with localization of the Gatumba-Gitarama, Nyakabingo, and Rutongo type-localities for Kibaran granite-related deposit. Profile A-B refers to Figure 4.13 (after Fernandez Alonso, 2007).

muscovite, with accessory apatite, garnet (almandine-spessartine), zircon, monazite, xenotime, and tourmaline (Cahen & Ledent, 1979). Many Kibara Sn granites exhibit aplitic and pegmatitic textures, and miaroles or veinlets of tourmaline and quartz or amethyst. Postsolidus hydrothermal alteration (albitization: see also Fig. 4.6d, muscovite/sericite formation, tourmalinization, and kaolinization) and poor exposure caused by deep lateritic weathering are characteristic (Günther et al., 1989; Pohl, 1994; Pohl et al., 2013).

Geochemically, the Kibara Sn granites can be classified as dominantly alkali-calcic to calc-alkalic, peraluminous leucogranites (Fig. 4.6a, b, c, and e). Their geochemical composition has been interpreted being a function of the variable degree of differentiation (Lehmann & Lavreau, 1988). The Kibara Sn granites are typically enriched in lithophile elements, such as Li, B, Rb, Cs, Nb, Ta, Th, U, Ga, Y, and W, and depleted in elements that are compatible, such as Sr, Ba, Ti, Co, Ni, Zr, and Eu (Pohl et al., 2013; see also Fig. 4.6f). A compilation of reported geochemical data of Kibara Sn granites in Rwanda (regionally called G4 granites), Burundi (regionally called Gr5 granites), and DR Congo (regionally called E-group granites) is given in Figure 4.6. They are characterized by high B ($n = 50$; 5–60 ppm; average 40 ppm) and relatively low F concentrations ($n = 50$: 315–2400 ppm; average

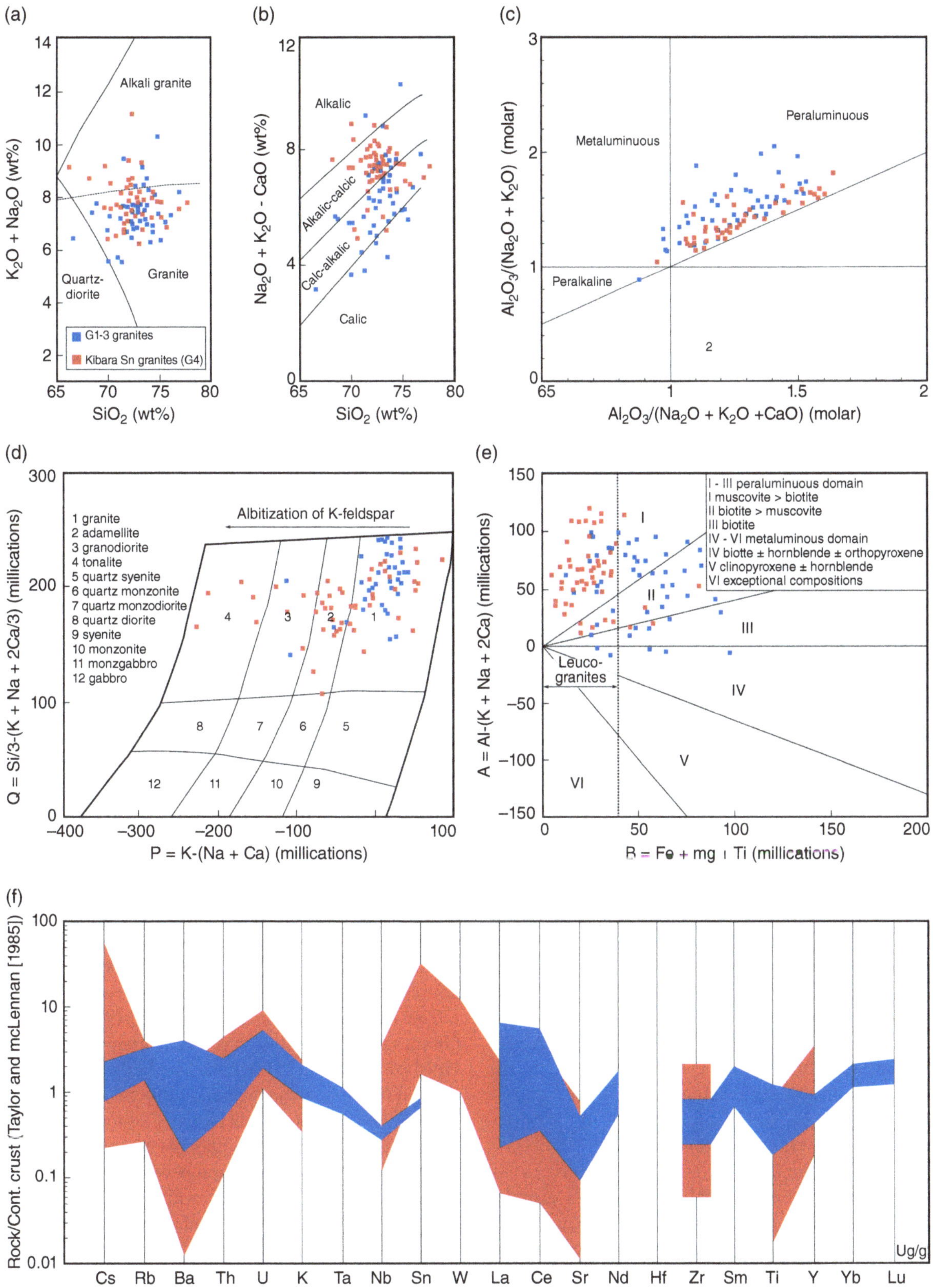

Figure 4.6 Compilation of reported geochemical data of G1-3 granites (blue; n = 51) and Kibara Sn granites (G4; red; n = 69) in Rwanda, Burundi, and DR Congo. (a) SiO_2 versus $Na_2O + K_2O$ variation diagram (after Wilson, 1989); (b) SiO_2 versus $Na_2O + K_2O - CaO$ variation diagram of Frost et al. (2001); (c) A/CNK versus A/NK variation diagram (after Shand, 1943), (d) and (e) P-Q and B-A variation diagrams of Debon and Le Fort (1983); and (f) Upper Continental Crust of Taylor and McLennan (1985) normalized multielement diagram. The colored fields correspond to reported ranges in elemental concentrations. Geochemical data compiled from Fernandez-Alonso (1981), Fernandez-Alonso et al. (1986), Lehmann and Lavreau (1987), Lehmann and Lavreau (1988), Günther (1990), and Hulsbosch et al. (2014).

1020 ppm) (De Clercq, 2012; Lehmann & Lavreau, 1988; Pohl et al., 2013) compared with average barren low-Ca granite abundances (B: 10 ppm; F: 850 ppm) (Turekian & Wedepohl, 1961). Sn concentrations in the Kibara Sn granites are between 4 and 80 ppm with an average of 19 ppm. The majority of the granite samples analyzed previously (24 out of 38 samples with reported Sn concentrations) are not enriched in Sn as typically observed for rare element granites (i.e., >15 ppm; De Clercq, 2012; Lehmann, 1990b). This has been ascribed to redissolution and redistribution of Sn during later hydrothermal/metasomatic events (Lehmann, 1990a). Reported W concentrations in the Kibara Sn granites are between 1.5 ppm and 11 ppm (Günther, 1990; Hulsbosch et al., 2014). The geochemical specialization of the Kibara Sn granites has been ascribed to prolonged magmatic differentiation (Lehmann, 1987) combined with hydrothermal fluid interaction (Lehmann & Lavreau, 1987) and fractional crystallization at depth (Lehmann & Lavreau, 1988) by crustal thickening and remelting of restites of older (1373 ± 6 Ma; SHRIMP U-Pb zircon; Tack et al., 2010) G1-3 peraluminous S-type granites (Günther et al., 1989; Pohl, 1994) and ~2 Ga Rusizian basement rocks (Tack et al., 2010). The widely distributed Kibara Sn granite magmatism occurs mainly as composite intrusions into the foliated batholiths of the G1-3 generation (Muchez et al., 2014; Pohl, 1994). The geochemistry of the G1-3 generation granites is variable, but within rather narrow limit, they are dominantly calc-alkalic to alkali-calcic, peraluminous, S-type, biotite-muscovite granites (Fig. 4.6a, b, c, d, and e; Fernandez-Alonso & Theunissen, 1998; Fernandez-Alonso et al., 1986). Their variable geochemical compositions have been interpreted to reflect the effects of partial melting of variably proportions of supracrustal metasedimentary rocks and crystalline basement rocks of presumably Paleoproterozoic age (i.e., Rusizian basement rocks), differentiation by fractional crystallization, and high degrees of contamination (Fernandez-Alonso et al., 1986; Tack et al., 2010). A geochemical comparison of the G1-3 granites with the Kibara Sn (G4) granites is given in Figure 4.6.

In certain regions of Rwanda, for example in the northeastern part of the Gitarama batholith (Fig. 4.5), G1-3 granitic and Kibara Sn (i.e., G4) granitic pluton types can be observed (Muchez et al., 2014). The G1-3 granites are characterized by a fine-grained quartz, plagioclase, K-feldspar, and biotite compositions that show a well-developed foliation. The Kibara tin granites outcrop as leucogranites with centimeter- to decimeter-size (pegmatitic) quartz, feldspar crystals. These leucogranites contain pods (up to meter-scale) of quartz- feldspar graphic intergrowths. The leucogranites and pods clearly cross-cut and thus postdate the foliated granite (Muchez et al., 2014). Reported Sn concentrations in the older G1-3 granites are ~2 ppm, that is, about the Clarke value of the Upper Continental Crust (Fig. 4.6f) (Lehmann & Lavreau, 1988).

The depth of emplacement of the Kibara Sn granites is constrained by metamorphic contact aureoles, which are characterized by andalusite, indicating a pressure of ≤4 kbar, and the presence of spodumene in the associated pegmatites, indicating a pressure of ≥2.5 kbar, which defines a depth range of 10 to 16 km under lithostatic regime (Lehmann et al., 2014). Pohl and Günther (1991) constrained the emplacement of the Kibara Sn granites at shallower depths of 6 to 12 km (1.6 to 3.2 kbar) based on fluid inclusion work. The depth of emplacement of the older G1-3 granites is estimated at a minimum of 16 km by metamorphic contact aureoles of quartz-kyanite-staurolite-garnet-chlorite-biotite-muscovite (≥4 kbar under lithostatic pressure; Lehmann et al., 2014). These pressure ranges correspond to the 2–5 kbar range of Fernandez-Alonso et al. (1986) derived from normative composition.

4.4.2. Nb-Ta-Sn Pegmatites (Gatumba-Gitarama)

The regional occurrence and mineralogy of the pegmatites of the KAB in Rwanda, Burundi, and the DR Congo have been studied in detail by N. Varlamoff (Varlamoff, 1954, 1960, 1961, 1963. 1969, 1972). These studies identified that the striking feature of these pegmatites is their well-developed regional zoning around the Kibara Sn granites. This regional zoning is defined as a succession of seven well-defined pegmatite types differing in their size and their mineral composition (Varlamoff, 1972). The regional pegmatite classification of Varlamoff (1972) is mainly based on minerals that can be easily identified in the field, such as quartz, micas, tourmalines, beryl, spodumene, amblygonite, apatite, cassiterite, columbite-group minerals, and pyrochlore, in combination with textures (e.g., graphic structures and crystal size).

The regional pegmatite zonation of the Gatumba-Gitarama type-locality has been mapped in detail by Hulsbosch et al. (2014) and Muchez et al. (2014) who showed that four major regional mineralogical zones could be identified in the field: (1) biotite, (2) biotite-muscovite (or two-mica), (3) muscovite, and (4) mineralized pegmatite zone. Zones 1 to 3 are common pegmatites belonging to the muscovite–rare-element class, while zone 4 is classified as rare-element class pegmatites (cf. classification of Černý & Ercit, 2005). The regional pegmatite zonation sequence can be observed starting from leucogranite feeder zones with Kibara Sn granite affinity evolving proximally in common pegmatites (zone 1 to 3) and culminating most distally in the mineralized pegmatites of zone 4. Regionally, the pegmatites of zones 1–3 occur both as intrabatholitic pegmatites and as exopegmatites in a 4.5 km wide aureole around the batholith at Gatumba-Gitarama. The mineralized pegmatites belonging to zone 4 at Gatumba-Gitarama occur only as exopegmatites in the most outward aureole of the pegmatite zonation sequence. (Hulsbosch et al., 2014). The regional zonation sequence is in fact a complex, superimposed distribution of individual pegmatite zonations starting from leucogranite extrusion sources. The magmatic crystallization age of the Gatumba-Gitarama pegmatites has been interpreted at ~975 Ma: 975±29 Ma (Rb-Sr on muscovite of zone 4; Cahen, 1964) and 975±8 Ma (U-Pb on columbite-tantalite of zone 4; Dewaele et al., 2011).

Biotite pegmatites (zone 1) are dominantly composed of microcline, plagioclase, biotite, and small amounts of quartz; and can developed abundant quartz-feldspar and quartz-tourmaline graphic textures. Biotite-muscovite pegmatites (zone 2) are also dominantly composed of quartz, microcline, plagioclase, biotite-muscovite intergrowths, in combination with quartz-tourmaline-(muscovite) graphic textures. Muscovite pegmatites (zone 3) show abundant muscovite, quartz, microcline, small and rare prisms of beryl, and a weak development of secondary albitization. Assemblages of quartz-tourmaline-(muscovite) graphic textures are often observed. Mineralized pegmatites (zone 4) are made up of pink K-feldspar (orthoclase), microcline, albite-oligoclase, quartz, Li-rich muscovite, spodumene, and elbaite. Accessory minerals include cassiterite, lepidolite, apatite, multicolored and white beryl, columbite-tantalite, rare phosphates, and other accessory phases (e.g., Daltry & Von Knorring, 1998). A famous example of a phosphate-rich pegmatite dyke in the Gatumba-Gitarama region is the Buranga pegmatite, which contains over one hundred mineral species, of which more than half are phosphates (Bertossa, 1965, 1968; Daltry & Von Knorring, 1998). The phosphates gatumbaite ($CaAl_2(PO_4)_2(OH)_2 \cdot H_2O$), burangaite ($NaFe^{2+}Al_5(PO_4)_4(OH)_6 \cdot 2H_2O$), and bertossaite ($(Li,Na)_2(Ca,Fe^{2+},Mn^{2+})Al_4(PO_4)_4(OH,F)_4$) have been first described in the Buranga dyke. At Buranga, the phosphates occur in an isolated zone next to the quartz core (Dewaele et al., 2011; Varlamoff, 1963). These mineralized pegmatites have been metasomatized (25–90 vol%), causing a large part of the primary mineralogy of the pegmatites to be altered to muscovite-quartz-(cassiterite) greisens (Varlamoff, 1963). Dewaele et al. (2008) and Dewaele et al. (2011) concluded, based on petrographic, fluid inclusion and stable oxygen and hydrogen isotope studies, that the Ta-Nb and cassiterite mineralization mainly formed during two distinct stages: the Ta-Nb and some of the cassiterite formed during an earlier stage of crystallization of the pegmatites prior to the albitization, while the major part of the cassiterite formed in quartz-muscovite-cassiterite greisen pockets during the later metasomatic-hydrothermal overprint. The greisen-hosted cassiterite is paragenetically associated with sericitization and muscovitization of the zone 4 pegmatites (Fig. 4.7). Nb and Ta mineralization in the pegmatites of the Kibara Metallogenic Province is regionally known as "coltan," but covers in fact a wide array of Ta-Nb oxide (TNO) mineralogical compositions. Most

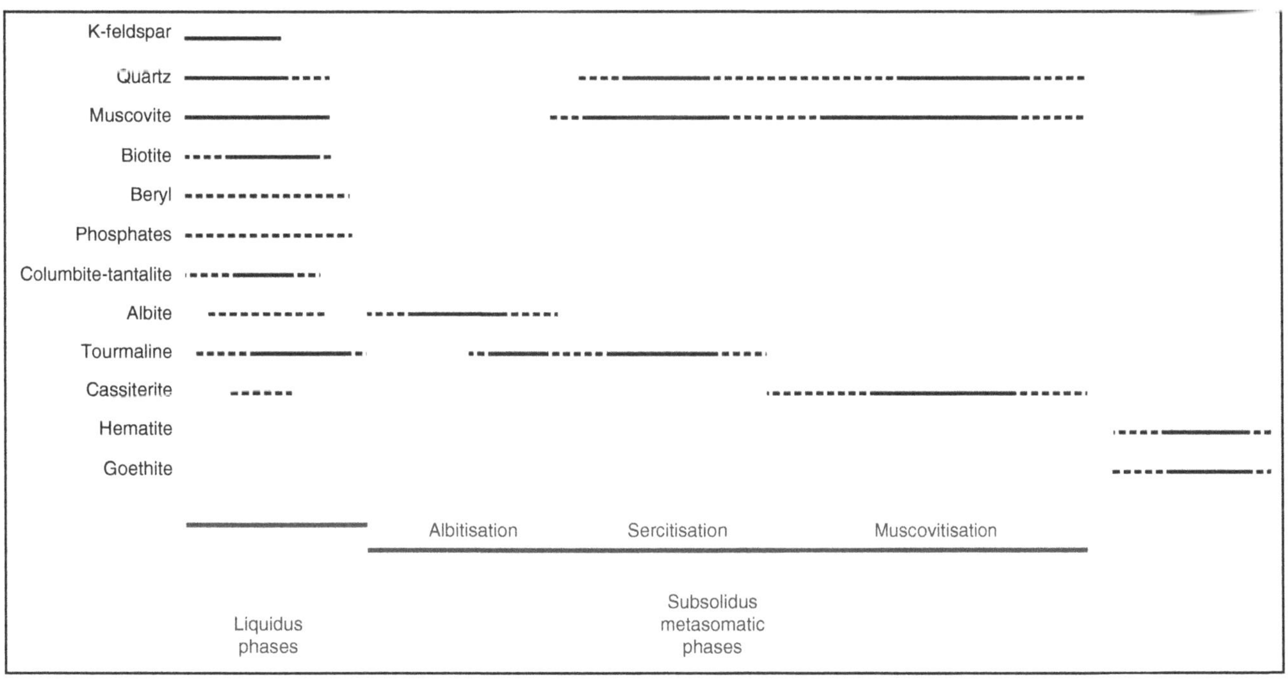

Figure 4.7 Paragenetic sequence of the pegmatite-hosted Nb-Ta-Sn mineralization at Gatumba-Gitarama (after Dewaele et al., 2011).

TNO mineralization in Rwanda (and KAB) comprises a solid solution of the columbite group minerals (CGM), that is, columbite-(Fe), columbite-(Mn), tantalite-(Fe), except for pure tantalite-(Mn) and the miscibility gap to tapiolite (Melcher et al., 2015). Apart from CGM (comprising on average 91% of the TNO), tapiolite (3% to 78%), wodginite-ixiolite (2% to 35%), pyrochlore supergroup minerals (mainly microlite; average 3% to 86%), and Ta-Nb rutile (1% to 40%) are reported by Melcher et al. (2015). Nb-Ta mineralization within the Gatumba-Gitarama pegmatites has been dated around 975 ± 8 Ma (LA-ICP-MS U-Pb on columbite-tantalite; Dewaele et al., 2011). On the basis of scanning electron microscope/mineral liberation analysis techniques, the non-TNO fraction of concentrates of the Gatumba-Gitarama mineralized pegmatites dominantly contains cassiterite and accessory ilmenite, zircon, monazite, xenotime, apatite, garnet, and wolframite, all in variable proportions (total fraction non-TNO: 5% up to 75%; Melcher et al., 2015).

The magmatic evolution of the four pegmatite zones at Gatumba-Gitarama from a leucogranitic Kibara Sn granite (G4) source has been successfully reconstructed by means of alkali metal and rare earth element (REE) geochemical modeling of rock-forming minerals (feldspars, micas, and tourmaline) sampled from dykes belonging to the four pegmatites zones (Hulsbosch et al., 2013, 2014). Fractionation modeling shows that the pegmatites adjacent to the parental Kibara Sn granite pluton are the least fractionated, and the distal pegmatites are the most fractionated. Biotite pegmatites (zone 1) form from 0% to 69% crystallization, two- mica pegmatites (zone 2) between 69% and 92% crystallization, and muscovite pegmatites (zone 3) from 92% crystallization onward. The extreme differentiation as observed in the rare-element pegmatites (zone 4) requires at least 98% fractionation of the initial Kibara Sn granite composition (Hulsbosch et al., 2014). Consequently, trace element modeling indicates that Rayleigh fractional crystallization governs the regional mineralogical and geochemical evolution from a granite source to muscovite–rare-element class and eventually rare-element class pegmatites. This mechanism shows that granitic pegmatites of the KAB are extremely fractionated melts, which are genetically and temporally connected to a main Kibara Sn granite body (Hulsbosch et al., 2014). During the end-stages of crystallization of individual dykes of zones 1–3 pegmatites of the Gatumba-Gitarama field, residual melt pools with a boro-aluminosilicate composition developed and caused the formation of mineralogical zonations within dyke, such as schorl-quartz-(muscovite) assemblage (Hulsbosch et al., 2017a).

Primary fluid inclusions in tourmaline of these less-differentiated pegmatites (zones 1–3) demonstrate the presence of low- to moderate-salinity, H_2O-NaCl-KCl-$MgCl_2$-complex salt (e.g., Rb-, Cs-salt) fluids, which started to exsolve at the Kibara Sn granite–pegmatite transition stage. LA-ICP-MS analysis of these primary fluid inclusions show significant W enrichment in these fluid phases (5–500 ppm) (Hulsbosch et al., 2016). The general absence of W mineralization in this magmatic suite, including the most differentiated columbite-tantalite mineralized pegmatites of the Gatumba-Gitarama area, emphasizes the efficiency of fluid saturation to extract crystal-melt incompatible W from the differentiating melt phase as tungstate species in aqueous solutions (Hulsbosch et al., 2016).

4.4.3. W Quartz Veins (Nyakabingo)

The Nyakabingo deposit (Figs. 4.3 and 4.5) has been studied as a representative locality for W quartz veins in the KAB (De Clercq, 2012; De Clercq et al., 2008; Dewaele et al., 2016b; Goldmann et al., 2013; Pohl & Günther, 1991; Pohl et al., 2013). The quartz vein system is typically located in the cores and flanks of second- to fourth-order complex anticlinal folds of the regional Bumbogo anticline (Fig. 4.5). Recent dating of selvage muscovite from the Nyakabingo W quartz veins provided an age of 992.4 ± 1.5 Ma, which has been interpreted as the formation age of the veins (Ar-Ar on muscovite; Dewaele et al., 2016b). The veins are hosted in sandstones and black, organic-rich metapelitic rocks of the Nduba-Shyorongi and Bumbogo formation belonging to the Gikoro Group (Fig. 4.8; Dewaele et al., 2016b). Two types of W-mineralized quartz veins are exposed in the field: bedding-parallel veins with a thickness between 0.1 and 2 m (exceptionally up to 9 m thick) and bedding-crosscutting quartz veins that are at a high angle to the bedding, and which have a thickness of a few centimeters up to several meters (Dewaele et al., 2016b). Metapelitic host rock around the veins has been altered to muscovite, tourmaline, and minor biotite early during vein opening and fluid-rock interactions. Centimeter large muscovite sheets precipitated in this contact zone between the veins and the surrounding metapelitic host rock during early vein formation (Fig. 4.9). During the main vein stage, the first quartz phase precipitated together with euhedral arsenopyrite, pyrite, scheelite, massive ferberite, and molybdenite (De Clercq, 2012; Goldmann et al., 2013). Often, the massive ferberite contains aligned scheelite inclusions, indicating ferberite after scheelite (Fig. 4.9). Subsequently, the scheelite crystals recrystallized and ferberite pseudomorphs (i.e., reinite variant) formed. The reinite pseudomorphs have been fractured and filled by an assemblage of quartz, muscovite, and xenotime (De Clercq, 2012; Goldmann et al., 2013). Afterward, the quartz veins have been crosscut by a sulfide phase

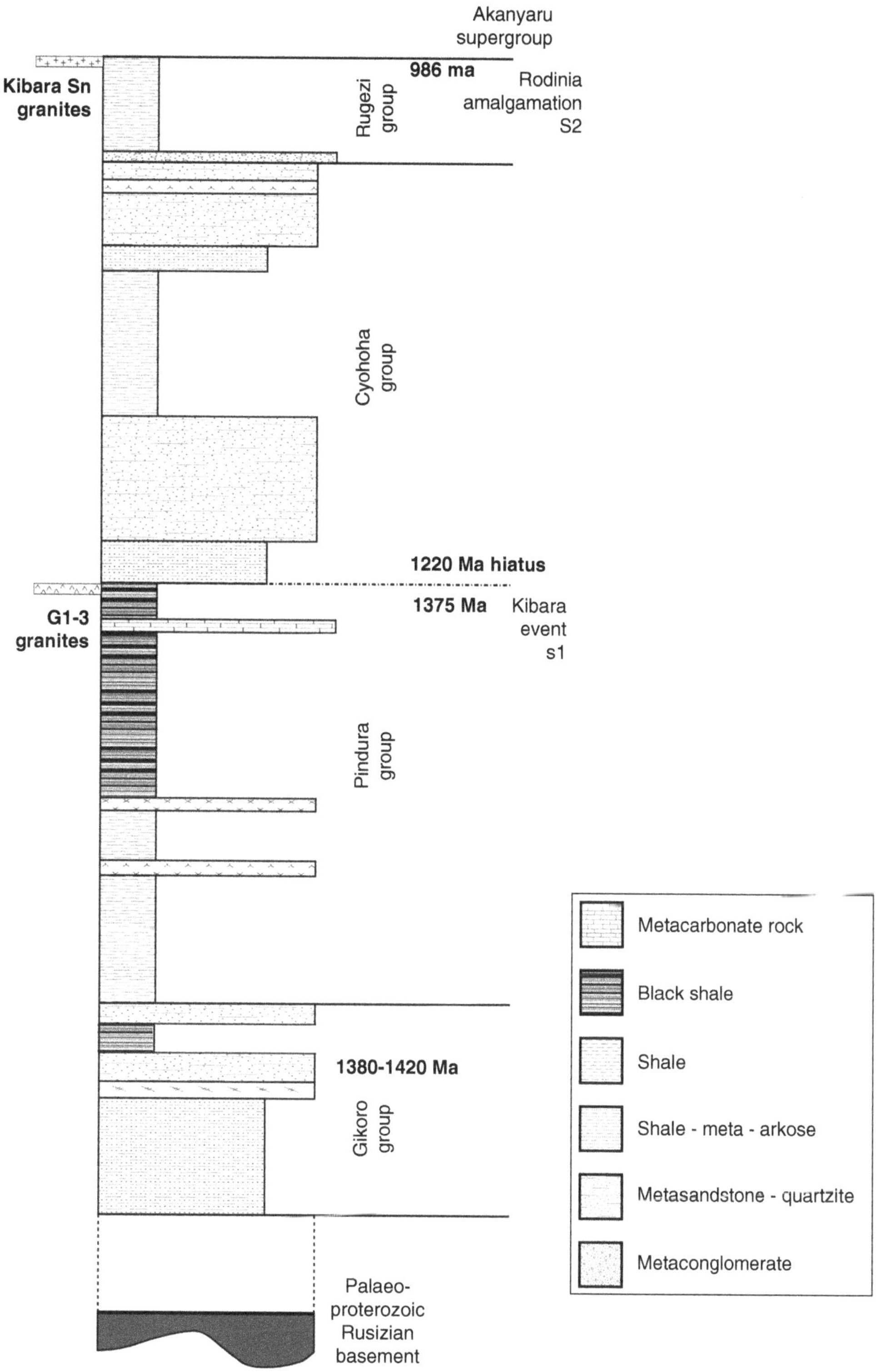

Figure 4.8 Synthetic litho- and chronostratigraphic reconstruction of sedimentation and magmatism in the western domain of the Karagwe-Ankole Belt (after Fernandez-Alonso et al., 2012).

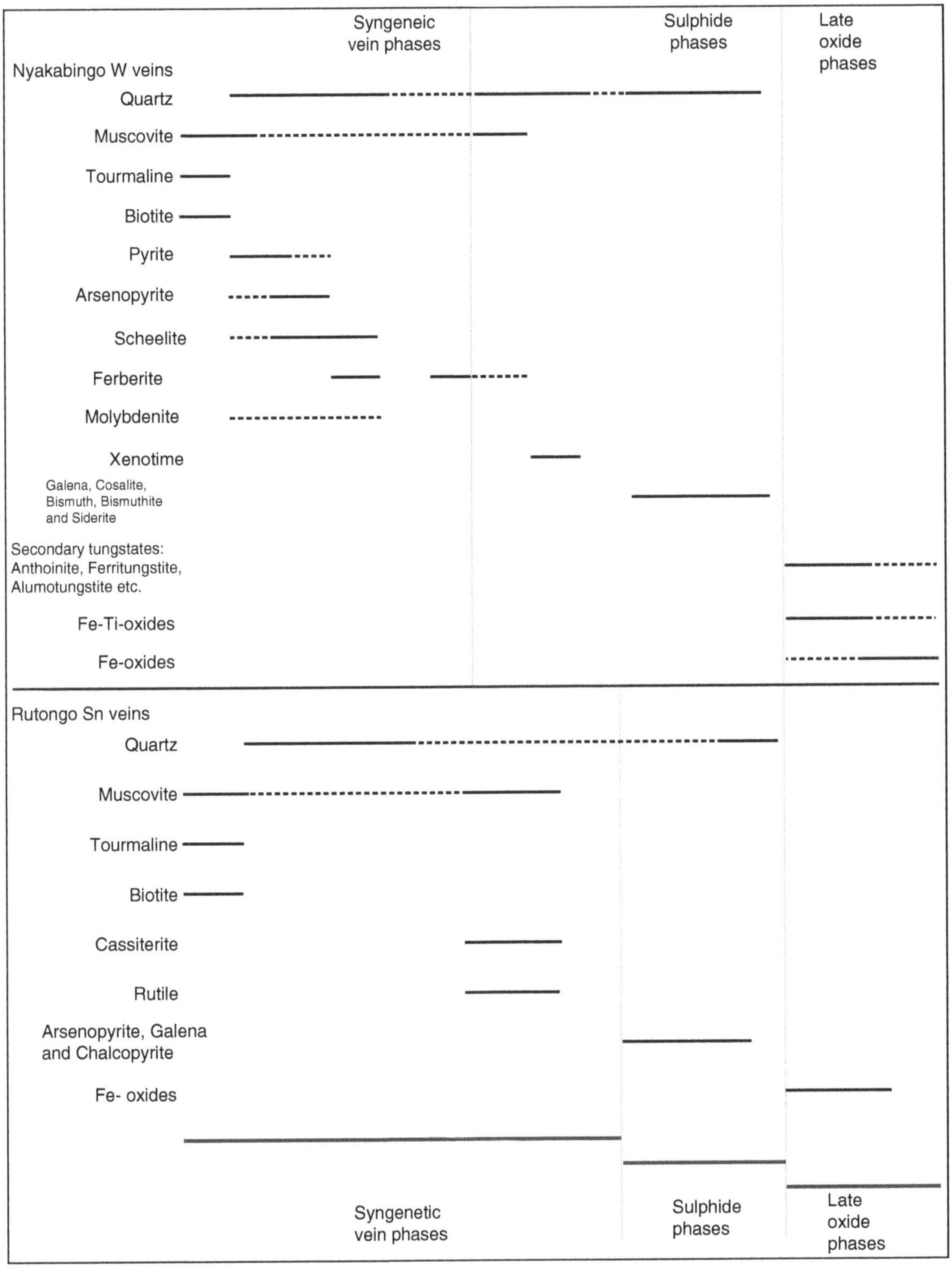

Figure 4.9 Simplified paragenetic sequences of the vein-hosted W and Sn mineralization at Nyakabingo and Rutongo, respectively (after De Clercq, 2012).

consisting of galena, cosalite, bismuth and bismuthite, and siderite and a late oxide phase (De Clercq, 2012).

The W ore minerals are interpreted to have precipitated from an aqueous-gaseous H_2O-CO_2-CH_4-N_2-NaCl-(KCl) fluid, with a low to moderate salinity (0.6–13.8 eq. wt% NaCl) and minimal trapping temperatures between 247 and 344°C based on fluid inclusion studies (Dewaele et al., 2016b). Based on LA-ICP-MS analyses of individual fluid inclusions by De Clercq (2012), the composition of the mineralizing fluid is characterized as dominantly Na rich, with lesser amounts of Mg (220–3800 ppm), K (850–1700 ppm), Ca (930–4710 ppm), Mn (230–1030 ppm), and Fe (160– 1080 ppm). Trace amounts of Sr (10–180 ppm), Cs (20–50 ppm), Ba (300–1300 ppm), W (50 ppm), and Pb (10–160 ppm) were also measured (De Clercq, 2012). Intense fluid–host rock interactions controlled the scheelite-ferberite precipitation during vein formation and caused the associated selvage contact alteration by muscovitization and tourmalinization reactions (see also Lecumberri-Sanchez et al., 2017). This alteration overprinted the fluid composition, causing a change in the gaseous composition with enrichment in N_2 and CH_4 likely derived from the host rock (6–44 mol% N_2 and 5–10 mol% CH_4; Dewaele et al., 2016a). Moreover, the quartz stable isotope record, with $\delta^{18}O_{quartz}$ values of +14.4‰ to +15.6‰ V-SMOW and δD_{quartz} values of −33‰ to −64 ‰ V-SMOW, would favor a dominant "metamorphic" signature of the mineralizing fluid, which is, however, likely caused by intense fluid-metapelite interactions (Fig. 4.12 in Dewaele et al., 2016b; Hulsbosch et al., 2016). This is supported by the typical association of the tungsten vein-type deposits with organic-rich black shales (Dewaele et al., 2016b). The latter are typically altered to a tourmaline-muscovite-(biotite) selvage at the contact with the quartz vein. These black shales acted also as a source of iron and calcium for the formation of tungstate minerals (Dewaele et al., 2016b; Lecumberri-Sanchez et al., 2017). For example, De Clercq (2012) and Günther and Ndutiye (1990) reported abundant cubic cavities (2–3 mm) of boxwork pyrite in the black shales and sandstones of Nyakabingo indicating Fe leaching. Sulfide dissolution is also observed on the vein selvages at contact with the enclosing metapelite. Additionally, during fluid-metapelite interaction, the mineralizing fluid could have been mixed with diluted "metamorphic" fluids, which could have caused a decrease in chloride concentration of the mineralizing fluid and induced a substantial decrease in the W solubility (Hulsbosch et al., 2016). The ^{40}Ar–^{39}Ar age of vein muscovite from Nyakabingo (992 ± 2 Ma; Dewaele et al., 2016b) overlaps with the 986 ± 10 Ma intrusion age of the Kibara Sn granites (SHRIMP U-Pb on zircon; Tack et al., 2010). At the deposit scale, there are no field observations indicating a direct spatial link between the W-bearing veins and the Kibara Sn granites. However, based on magnetometric data, a hidden granitic body is inferred to be present at an estimated depth of 0.5–2 km (Günther & Ndutiye, 1990).

4.4.4. Sn Quartz Veins (Rutongo)

One of the largest Sn districts of the KAB, reasonably covered by more recent scientific studies (Dewaele et al., 2007; Dewaele et al., 2010; Günther, 1990; Pohl, 1994; Pohl & Günther, 1991; Pohl et al., 2013), is the Rutongo mining district in central Rwanda (Figs. 4.3 and 4.5). The Rutongo Sn veins are also located in secondary anticlinal structures of the regional Rutongo anticline (Fig. 4.5). In contrast to the W deposits, the Sn deposits are typically hosted in a number of massive sandstone and/or quartzite units that are separated by units of alternating metapelitic rocks and sandstones (De Clercq, 2012; Pohl et al., 2013). The Sn deposits are mainly hosted in metasediments, which belong to the Gikoro Group (Fig. 4.8; Baudet et al., 1988). The Sn-mineralized quartz veins occur in swarms of several hundreds of subparallel crosscutting veins that are oriented perpendicular to the fold axes. The individual quartz veins have a thickness that varies between a few cm to several meters (more than 5 m), with an average between 0.5 and 1 m. They can be several hundred meters long (Dewaele et al., 2010).

The emplacement age of the Rutongo veins is estimated at 965 ± 29 Ma by muscovite Rb-Sr age dating (muscovite associated with cassiterite; Monteye-Poulaert et al., 1962). Recent dating of cassiterite from the Rutongo Sn quartz veins provided ages of 957 ± 21 Ma (U-Pb; Zhang et al., 2016), that is, within the same time interval uncertainty as the Rb-Sr age of Monteye-Poulaert et al. (1962). Structural analysis indicated a relatively late formation of the Sn-mineralized veins in the general evolution of the Rutongo anticline (De Clercq, 2012). More specifically, in the southeastern part of the Rutongo anticline, granitic rocks of the Kigali granite can be found. This muscovite-bearing leucogranite is interpreted to be a Kibara Sn granite (Bertossa et al., 1964). In the contact zone between the Kigali granite intrusion and the Rutongo host rocks, it is reported that the intrusion postdates microfolds in the host rocks (Lhoest, 1957a, b). The Kigali granite is as such interpreted to have been emplaced during a late phase or postfolding of the rocks (Lhoest, 1957a; b; Peeters, 1956; Pohl, 1974). Consequently, the Rutongo Sn veins are interpreted to be formed syn- to post-Kigali granite emplacement.

The quartz veins are associated with intense wall-rock alteration, comprising silicification, tourmalinization, muscovitization, and precipitation of biotite (Fig. 4.9; De Clercq, 2012; Dewaele et al., 2010). These reactions are typical for the Sn vein-type deposits in the central part of Rwanda (Dewaele et al., 2010). Cassiterite formed in a late stage of the evolution of the mineralized veins and is intimately associated with muscovite crystals in fractures in the quartz veins. Afterward, the quartz veins were crosscut by a sulfide phase, consisting of arsenopyrite, galena, and chalcopyrite, and late Fe oxides (De Clercq, 2012). Cassiterite is interpreted to have precipitated from an aqueous-gaseous fluid with a low to moderate salinity (6.2–15.4 eq. wt% NaCl) and a minimal trapping temperature between 225°C and 349°C, based on fluid inclusion studies (De Clercq, 2012). The gas phase is composed of CO_2 (50–78 mol%), N_2 (11–40 mol%), and smaller amounts of CH_4 (10–15 mol%). Na is the dominant cation in solution, with lesser amounts of K (1444–3377 ppm), Li (149–284 ppm), and Cs (446 ppm) (De Clercq, 2012). The $\delta^{18}O_{quartz}$ values of the Sn-mineralized quartz veins are +13.6 and +14.1‰ V-SMOW. The δD_{quartz} show a large spread between -47 and -67‰ V-SMOW. The oxygen and deuterium isotopic values of the fluids are interpreted to reflect metamorphic, possibly magmatic signatures (De Clercq, 2012; Fig. 4.5 in Dewaele et al., 2010). Both oxygen and deuterium isotopic values of cassiterite are relatively homogeneous (7.5 and 7.9‰ V-SMOW δD_{quartz}, and −90 and −105‰ $\delta^{18}O_{quartz}$) and indicate that the Sn-mineralizing fluid interacted with metamorphic rocks (De Clercq, 2012). As for the W vein-type deposits, intense water-rock interactions have overprinted the pristine source signature.

4.5. NB-TA-SN-W FRACTIONATION MECHANISMS

The co-occurrence of Nb-Ta-Sn pegmatite-type and Sn-W quartz vein-type mineralization at multiple locations in the KAB region of the Kibara Metallogenic Province provides a first-class and, moreover, relatively rare opportunity to investigate the geochemical evolution of a global granite-related ore system and to evaluate the element distribution between pegmatites and quartz veins during the magmatic-hydrothermal transition. The next sections provides a discussion and interpretation of the metallogenesis of the Nb-Ta-Sn-W granite-related ore deposits and its application on the Rwandese, and to some extent the Kibara Metallogenic Province, granite-related ore deposits. Main fractionation processes responsible for element enrichment will be discussed. As a conclusion, a general granite-related metallogenic model will be presented for the B-rich, F-poor KAB metallogenic system.

4.5.1. Processes Related to Disseminated Magmatic Mineralization

4.5.1.1. Generalities

Disseminated rare-metal, Nb-Ta-Sn-W-Li-Be-(Zr-U-REE), mineralization is restricted to shallow albite-rich, peraluminous granites (Černý et al., 2005). The crustal emplacement is generally from subvolcanic to miarolitic (<5 km) and mineralization occurs often in isolated bodies, or rhyolitic to microgranitic dykes (Cuney et al., 1992; Raimbault et al., 1995). Albite-rich granites represent most-fractionated end members of layered sequences in upwardly differentiated plutons (Candela, 1992, 1997).

Various classifications of rare-metal granites have been proposed based on their geochemistry. However, only the peraluminous, high phosphorus rare-metal granite type is associated with Nb-Ta-Sn-W mineralization (Černý et al., 2005; Linnen & Cuney, 2005). A well-studied example is the Sn-Ta-Nb-Rb-Cs-Li-rich, late-Variscan Beauvoir granite in the French Massif Central (Cuney et al., 1992; Raimbault et al., 1995), which has been interpreted to be intruded at a pressure range of 1.0 to 1.2 kbar, which corresponds to emplacement depths of <4.5 km. (Aissa et al., 1987). In general, these low-seated peraluminous rare-metal granites are spatially associated with large muscovite-biotite leucogranite plutons at depth that formed during postorogenic relaxation in continental collisions (Bouchot et al., 2005). Lenharo et al. (2003) constrained the PT evolution of several granite types associated with disseminated rare-metal mineralization in Pitinga, Amazonas, Brazil. Parental, barren quartz monzonites have been interpreted to intrude at ~930°C and 5 kbar while the last and mineralized albite granite formed below 650°C and 1 kbar. This suggests that (1) initial crystallization of the early intrusive phases occurred at around 15 km depth under lithostatic regime and (2) the final emplacement of the more differentiated rare-metal granites took place at a depth of 0.5–1 km (Lenharo et al., 2003).

In general, these peraluminous granites are mineralized in the form of microlite, CGM, or Ta-rich cassiterite (Linnen et al., 2014). They are characterized by low concentrations of REE (close to chondrititic values) and contain Li-rich muscovite, amblygonite, or lepidolite (Cuney et al., 1992; Kovalenko & Yarmolyuk, 1995). P, Rb, and Cs levels are generally high (Černý et al., 2005). Most metallogenic studies identified that disseminated mineralization in rare-metal granites is the product of primary magmatic crystal-melt fractionation processes (Candela, 1997; Černý et al., 2005; Kovalenko & Yarmolyuk, 1995; Linnen, 1998; Webster et al., 2004) due to the (very) low crystal/melt distribution coefficients of Nb (0.05–0.09), Ta (0.02–0.05), W (0.04–0.4), and Sn (~0.3) (Antipin et al., 1981; Bartels et al., 2010; Candela, 1992; Kovalenko et al., 1988; Lehmann, 1990b; Linnen, 1998; Linnen & Cuney, 2005; Linnen et al., 2014; Raimbault et al., 1995). However, extensive hydrothermal reworking of the magmatic mineralization is documented by the formation of Li-rich, Sn-W greisens (Webster et al., 2004). Based on bulk Ta enrichment trends from increasingly more fractionated granite batches, that is,

≤ 1 ppm in less-fractionated up to >100 ppm in the most fractionated units, Černý et al. (2005) proposed a degree of fractionation ≥ 0.99 of the parental melt. The associated enrichment of the crystal-incompatible fluxes, such as F, Li, B, and P, during fractionation causes modification of the physicochemical properties of the melt (lowering of the viscosity and solidus temperature), which facilitates crystal-melt separation and favors protracted crystal-melt fractionation, promoting extreme rare-metal enrichment (Bartels et al., 2013; Černý et al., 2005; Dingwell, 1988; Linnen, 1998; Pichavant, 1981). In addition, Raimbault et al. (1995) proposed that melt-melt and melt-aqueous fluid exsolution processes can occur as additional enrichment mechanisms, superimposed on the general crystal-melt fractionation trends. These processes can potentially lead to a immiscible hydrosaline liquid that could transport elements such as Li, Rb, Cs, Sn, W, Nb, and Ta upward through the magma toward the roof of the granite bodies (Badanina et al., 2004; Harlaux et al., 2017b; Reyf et al., 2000; Thomas et al., 2000; Veksler et al., 2002). However, the process of liquid-liquid immiscibility in the formation of orthomagmatic deposits, such as the composition and dynamics of liquid-liquid immiscibility or the metal distribution and solubility in immiscible phases, is not well understood (cf. Černý et al., 2005; Kamenetsky & Kamenetsky, 2011; Linnen et al., 2014).

4.5.1.2. Consequences for the KAB

The B and Cl-rich, F-poor leucocratic and calc-alkaline, peraluminous Kibara Sn granites are spatiotemporally related to the rare-metal pegmatites and Sn-W vein mineralization systems in the KAB (operated in a ca. 980 ± 20 Ma range; Lehmann & Lavreau, 1988; Pohl, 1994; Pohl et al., 2013; Tack et al., 2010). These major ilmenite-series granites are, however, reported as barren (Günther et al., 1989; Lehmann & Lavreau, 1987, 1988). The depth of emplacement is interpreted at 10–16 km (Lehmann et al., 2014) (see section 4.4.1). The depth of emplacement of the Kibara Sn granites appears in contrast with the shallow rare-metal granites, which are generally emplaced at ~4 km (e.g., Aissa et al., 1987; Lenharo et al., 2003).

The Kibara Sn granites are mainly derived by anatexis of supracrustal metasedimentary rocks and crystalline basement rocks (Fernandez-Alonso et al., 1986; Pohl, 1994; Tack et al., 2010) and such granitic melts are, in general, hydrated (water-saturated) and can contain water in the range of 7–8 wt% H_2O (Burnham, 1997). Typically, these peraluminous S-type magmas, like the Kibara Sn system, can maximally rise upward in the crust until they intersect the water-saturated granite solidus (cf. Holtz et al., 2001). By this time, typically at 15–18 km (4–5 kbar under lithostatic pressure), they would have become completely solid and could not intrude upward any farther (Robb, 2013; Strong, 1988). H_2O-fluid saturation in these magmas at these crustal levels will eventually occur by second boiling and eventually yield to the formation of pegmatites and associated hydrothermal systems like quartz veins (e.g., Strong, 1988). In addition, F-enriched granitic magmas have the potential to rise upward to higher crustal levels in comparison to F-poor systems due to the positive effective of F in reducing the viscosity and the minimum liquidus temperature of granitic melts (Manning, 1981; Pichavant & Manning, 1984). F also significantly delays fluid/vapor saturation of granitic melts (Holtz et al., 1993; Manning, 1981; Strong, 1988). The absence of F in the Kibara Sn granite could also explain its more deep-seated nature compared with granites with disseminated rare-metal mineralization. Consequently, the deeper-seated nature of the Kibara Sn granites, at crustal depths of ~10–16 km, is probably caused by the high initial water content, and by the exsolution of a magmatic aqueous fluid phase by second boiling during crystallization until solidus conditions (cf. Whitney, 1988), that is, at the granite-pegmatite transition stage in case of the Gatumba-Gitarama pegmatites (Hulsbosch et al., 2016).

Rare elements Nb, Ta, Sn, and W are generally incompatible in peraluminous melt systems (Linnen et al., 2014). The lack of magmatic disseminated mineralization in Kibara Sn granites can be consequently interpreted to result from "an unsuccessful alignment and combination" (cf. Richards, 2013) of magmatic to hydrothermal fractionation processes. A competing interplay can be expected in a differentiating magma system with an exsolved fluid phase between (1) the enrichment of an incompatible ore element in the melt phase up to ore mineral saturation and precipitation, versus (2) the fractionation of the incompatible but soluble ore element and partition into the exsolved fluid phase. At each infinitesimal step of melt differentiation, there is as a result a competition between the further enrichment of the incompatible element in the melt versus the depletion of the element in the melt due to the combination of its soluble behavior and the dominance of the fluid phase in the system (i.e., mass fraction of fluid, which is related to the water content of the melt). For example, solubility data of Sn in silicic melts strongly suggest that crystallization of magmatic cassiterite from a melt containing Sn at the ppm level is unlikely (Lehmann, 1990c). This is particularly unlikely for reduced, peraluminous granites, such as the Kibara Sn granites, even with their Sn content at the 100 ppm level (Štemprok, 1990). These estimated values are exceedingly lower than the Sn saturation value for peraluminous melts, which ranges between ca. 0.4 and 0.8 wt% Sn (Linnen et al., 1996). However, cassiterite saturation may occur in extremely fractionated pegmatite phases (Štemprok, 1990). Experimental W solubility studies by

Che et al. (2013) show that the crystallization of magmatic wolframite is possible in flux-rich subaluminous to peraluminous melts, such as Kibara Sn granites, at approximately 500°C. The lack of magmatic wolframite in the Kibara Sn granites and associated pegmatites has been correlated to the exsolution of a mobile fluid phase and the subsequent fractionation of soluble W toward this fluid phase (Hulsbosch et al., 2016). Fractionation of W toward an early aqueous fluid phase hampers further enrichment of incompatible W in the melt phase because the W content of the melt is depleted by fluid scavenging of W. If the felsic magma system is not fluid saturated, incompatible W concentrates in the melt and is forced to saturate at the appropriate W-mineral solubility products (e.g., Che et al., 2013). Based on average W concentrations in magmatic muscovite from the Gatumba-Gitarama pegmatites, which range from ~30 ppm in the pegmatites of zone 2 to ~7 ppm in the pegmatites of zone 3 and 4 (Hulsbosch et al., 2013, 2014), and the W muscovite-melt distribution coefficient (i.e., 1.6; Kovalenko et al., 1988), calculated W melt concentration decreases from 18 ppm to 4 ppm in the pegmatites of zone 2 to zone 4. These W melt concentrations are well below experimentally derived solubility products for wolframite in peraluminous melts, for example, ~4600 ppm W for ferberite and ~1100 ppm for hubnerite at 800°C (calculated from data of Che et al., 2013).The solubility behavior of Nb and Ta in peraluminous melts is probably related to the formation of bonds with Al; Ta is preferentially bonded to nonbridging oxygens with Al as first neighbors (Van Lichtervelde et al., 2010). In sub- and peraluminous melts tantalite solubility is higher than that of columbite resulting in a systematic decrease of Nb/Ta during the crystallization of these melts (Linnen et al., 2014; Van Lichtervelde et al., 2010). The effect of fluxing compounds on Nb-Ta-solubilities is somewhat controversial (Linnen et al., 2014): Li increases the solubility of columbite and tantalite (Linnen, 1998) while F does not increase its solubility (Fiege et al., 2011). Notwithstanding, flux-rich granitic melts can dissolve weight percent levels of Nb and Ta (Bartels et al., 2010). As with the rare elements W and Sn, the solubility of columbite and tantalite is strongly temperature dependent, especially in the case of peraluminous melt compositions (Linnen & Keppler, 1997; Linnen & Cuney, 2005; Van Lichtervelde et al., 2010). Experimental Ta solubilities systematically decrease from ~25000 ppm at 1050°C to ~8000 ppm at 800°C (at an ASI of 1.0; Linnen & Keppler, 1997). Taking the above-mentioned fractionation experiments into account, Nb-Ta and Sn magmatic disseminated mineralization can be expected only at magmatic end-stages (Linnen & Cuney, 2005) characterized by very high degrees of crystal-melt fractionation (~F ≥ 0.99), such as observed in the mineralized pegmatites of the Gatumba-Gitarama area. In contrast, the (early) magmatic disseminated W mineralization can be more easily expected in flux-rich, volatile undersaturated subaluminous to peraluminous melts, which are characterized by significant degrees of crystal-melt saturation in the absence of a mobile aqueous fluid phase (Hulsbosch et al., 2016).

4.5.2. Processes Related to Pegmatitic Rare-Element Mineralization

4.5.2.1. Generalities

Generally, the granite intrusions generating rare-element class pegmatites range in morphology from small stocks and sheets to medium-sized plutons and diapiric cupolas up to batholithic masses (Černý, 1991a,b). Intrusion depths for the granite and pegmatite aureoles have been estimated at 5 to 2 kbar (London, 2008). Homogenous biotite to two-mica granites have been identified to be spatially associated with low- to medium-fractionated pegmatites (Beus, 1966; Černý et al., 2005). However, granites parental to highly fractionated pegmatites are leucocratic and heterogeneous two-mica granites with local pods of graphic-textured pegmatitic leucogranites (Beus, 1966). Older studies have unsuccessfully attributed fractionation processes such as the Soret effect (i.e., thermogravitational diffusion: heavy elements migrate toward the colder end and lighter ones to the hotter end of the melt gradient), differential breakdown of divers complexes, aqueous fluid separation, and late aqueous fluid overprint for the extensive fractionation of Nb and Ta (and other HFSE) in pegmatitic melts (cf. review in Beus, 1966). Recent experiments, however, confirmed that crystal-melt fractionation and differential solubility of Nb and Ta are mainly responsible for rare-metal fractionation in pegmatitic melts, in combination with depolymerization of the fluxed melt (review in Linnen et al., 2014). The extreme solubility of rare metals particularly in conjugate/immiscible peralkaline melt fractions is also demonstrated for Ta by melt inclusions studies (Thomas et al., 2011a; Thomas et al., 2011b). The systematics of rare-metal solubility and melt fractionation in the case of peraluminous lithium-cesium-tantalum (LCT family) pegmatite melts is in concept similar to the behavior in granitic systems, in particular in high P peraluminous granites (section 4.5.1).

4.5.2.2. Consequences for the KAB

The regional pegmatite zonation in Gatumba-Gitarama is mineralogically characterized by an evolution from muscovite–rare-element class (zone 1 to 3) to rare-element class pegmatites (zone 4), which is generated by magmatic crystal-melt differentiation from parental leucocratic Kibara Sn granite (Hulsbosch et al., 2014). Moreover, the four pegmatite zones intruded metasedimentary units of relatively low-pressure, greenschist metamorphic facies (cf. Gérards, 1965), which is above the regional depth for an anatectic origin (i.e., upper amphibolite facies, cf.

Brown, 1994). The culmination of Li-Be-Cs-Ta-Nb mineralization in pegmatites of zone 4 (i.e., end-member units of fractional crystallization of Kibara Sn source granites) is in line with the most recent experimental work, which showed that crystal-melt fractionation and differential solubility of Nb and Ta are the main differentiation processes acting in crystallizing pegmatitic melts (Linnen et al., 2014). Extreme crystal-melt fractionation of peraluminous granitic melt system ($F > 98\%$; Hulsbosch et al., 2014) caused (1) the large-scale, regional pegmatite melt evolution and (2) the magmatic enrichment of incompatible Nb, Ta, Sn, and W at the Gatumba-Gitarama pegmatite field. Phase separation by aqueous fluid immiscibility is observed in the mineral and inclusion record (e.g., Hulsbosch et al., 2016; Sirbescu & Nabelek, 2003; Thomas & Davidson, 2013; Thomas et al., 2012) and can cause fluid partitioning of fluid soluble elements, like W, which can form tungstate species in aqueous solutions (Wood & Samson, 2000). Fluid immiscibility processes become increasingly more important at high degrees of fractional crystallization of B-rich, F-poor hydrous granite-pegmatite melts (Hulsbosch et al., 2017a).

4.5.3. Processes Related to Hydrothermal/ Metasomatic Mineralization

4.5.3.1. Generalities

Hydrothermal vein and greisen systems are, in contrast to disseminated magmatic rare-metal mineralization in granites and pegmatites, dominantly driven by fluid-melt equilibria, fluid mixing, and aqueous geochemistry (Černý et al., 2005; Heinrich, 1990; Wood, 2005; Wood & Samson, 2000). The Kibara Sn magmatic system is distinctly F-poor, B- and Cl-rich (Pohl et al., 2013), consequently fluid-melt partitioning will be mainly discussed for B- and Cl-dominated fluid–peraluminous melt systems. Moreover, F partitions generally in favor of the melt (Carroll & Webster, 1994) with reported fluid/melt distribution coefficients of < 0.5 for peraluminous granite-pegmatite systems (London et al., 1988). In addition, the solubility of water increases with increasing B content (Pichavant, 1981), an effect also observed with increasing Al content in peraluminous melts (Dingwell et al., 1984).

Fluid distribution of Nb and Ta is very low with experimental fluid-melt partition coefficients ranging from ~0 to 0.1, and from ~0.001 to 0.08, respectively (Borodulin et al., 2009; Chevychelov et al., 2004; Linnen & Cuney, 2005). However, it has been demonstrated that Nb and Ta can be soluble as hydroxyl-fluorine species in concentrated fluoride solutions ($(HF) > 10^{-3}$ mol/kg) under reducing conditions (Timofeev et al., 2015; Zaraisky et al., 2010). Chloride and carbonate hydrothermal solutions are ineffective in transporting Ta and Nb and are not capable of transferring these metals in the amounts sufficient for the formation of economic viable ore deposits (Zaraisky et al., 2010). The lack of Nb or Ta enrichment in geochemical aureoles above most rare-element granite and pegmatite cupolas (e.g., the Beauvoir granite or KAB system), except for some Li-F type granites and pegmatites (Zaraisky et al., 2010), supports an extremely low-fluid affinity of these elements in Cl-dominated magmatic fluids (cf. Cuney et al., 1992; Pohl, 1994; Varlamoff, 1972).

Experimental determination of Sn fluid-partitioning is problematic due to the reaction of cassiterite with precious metals making up the experiment capsule, especially at reduced conditions (Linnen et al., 1995). At oxygen fugacities near the NNO buffer and with a fluid chlorinity less than 1 $mol \cdot l^{-1}$, Sn partitions experimentally in favor of the melt (Taylor, 1988 as cited in Linnen & Cuney, 2005). Keppler and Wyllie (1991) report lower fluid-melt partition coefficients ranging from 0.005 to 0.08 for fluid HCl concentrations of 0 to 2 $mol \cdot l^{-1}$. Experimental data of Hu et al. (2008) and Hu et al. (2016) constrained the influence of melt composition on the fluid partitioning of Sn. They showed that the fluid-melt partition coefficient increases with increasing aluminum saturation index, $K_{fm,Sn} = 0.1886 \cdot ASI - 0.1256$, giving absolute coefficients for peraluminous melts alike to the Kibara Sn granites between 0.06 and 0.18 (ASI = 1.0–1.6; Fig. 4.6c). Given that Sn is readily complexed by Cl in aqueous solutions (Wilson & Eugster, 1990), it is expected that Sn will partition more preferentially in favor of the fluid at very high Cl concentrations (Linnen & Cuney, 2005). However, experiments of Wilson and Eugster (1990) and Taylor and Wall (1993) showed that significant amounts of Sn can partition into the chloride-bearing acidic magmatic fluid phase (100s–10,000s ppm Sn; Taylor & Wall, 1993). Based on Sn partitioning between coexisting fluid and silicate melt inclusions, Zajacz et al. (2008) found no systematic variation of the fluid-melt partition coefficients of Sn with the chlorinity of the fluid. In the low chlorinity range (1–3 $mol \cdot kg^{-1}$), partition coefficients vary between 1.5 and 6.7. A contrasting behavior is observed in high chlorinity fluids with reported partition coefficients ranging from 0.26 and 2.11, up to ~42 (Zajacz et al., 2008).

W fluid distribution differs from Sn in that W forms oxyacid complexes like HWO_4^-, and as such, complexes with alkalis, such as $NaHWO_4$ (Wood & Samson, 2000). As a consequence, the observed positive correlation between the W fluid-melt partition coefficient and the NaCl content of the fluid may be caused by Na ion pairing rather than Cl complexes (cf. Linnen & Cuney, 2005). There is, as such, some variation in experimental fluid-melt partition coefficients. For halogen-free experiments, fluid-melt partition coefficients range from ~0.1

(Manning & Henderson, 1984) to ~4 (Schäfer et al., 1999) and ~5 (Keppler & Wyllie, 1991). For low-salinity (NaCl) fluids, fluid-melt partition coefficients of ~0.3 to 1.0 have been derived (Chevychelov & Chevychelova, 1997, as stated in Linnen & Cuney, 2005). For HCl-rich fluids, coefficients of ~0.4 to 1.0 have been reported (Keppler & Wyllie, 1991). Natural fluid-melt partition coefficients of W show a positive correlation with the chlorinity of the fluid phase between 1 and 3 $mol \cdot kg^{-1}$ (Zajacz et al., 2008). For high-chlorinity natural fluids, the scatter of the natural data is high (Zajacz et al., 2008). Variations in W fluid-melt partition coefficients with chlorinity have been interpreted on the basis of different mass transfer reactions of W from the silicate melt to the exsolved aqueous fluid (Keppler & Wyllie, 1991; Manning & Henderson, 1984; Zajacz et al., 2008):

$$WO_3(melt) + 2NaCl(aq) \leftrightarrow WO_2Cl_2(aq) + Na_2O(melt) \quad \text{(Reaction 1)}$$

$$WO_3(melt) + H_2O(aq) \leftrightarrow H_2WO_4(aq) \quad \text{(Reaction 2)}$$

$$WO_3(melt) + 2NaCl(aq) + H_2O(aq) \leftrightarrow Na_2WO_4(aq) + 2HCl(aq) \quad \text{(Reaction 3)}$$

Fractionation according to reaction 1 would result in a positive correlation between the chlorinity of the fluid phase and W fluid-melt partition coefficient, while reaction 2 would probably result in a negative correlation due to decreasing water activity with increasing salinity. Also fractionation according to reaction 3 would result in a positive correlation between the salinity of the fluid and the fluid-melt partition coefficient (Zajacz et al., 2008). Gibert et al. (1992) demonstrated that the presence of volatiles (N_2, CH_4, CO_2) in W mineralizing fluids have an effect on W speciation and scheelite solubility. Their modeling showed that the addition of N_2 (≤10 mol %) to the mineralizing fluids results in an increase in $H_2WO_4^0$ and HWO_4^- concentrations relative to WO_4^- and in a decrease in scheelite solubility. This mechanism favors scheelite precipitation and illustrates the commonly observed association of W deposits with black shales, which generates mixed volatile fluids and acts as sources for Fe (e.g., ferberite) or and Ca (e.g., scheelite) during host-rock interaction and W mineralization (Chicharro et al., 2016; Dewaele et al., 2016b; Hulsbosch et al., 2016; Lecumberri-Sanchez et al., 2017; Polya et al., 2000).

4.5.3.2 Consequences for the KAB

Magmatic Nb-Ta and Sn mineralization in the KAB is only observed at the magmatic end-stages of melt differentiation (F ≥ 0.98; i.e., the pegmatites of zone 4). Based on primary fluid inclusions analyses in tourmaline, Hulsbosch et al. (2016) observed that magmatic-hydrothermal fluids, originating from the differentiating Kibara Sn granite-pegmatite system, contain both W (~5–500 ppm) and Sn (~10–400 ppm). Depending on the host-rock lithology, these fluids can exclusively precipitate W or Sn in hydrothermal quartz veins. W mineralization exclusively occurs in the KAB in the peribatholitic quartz vein hosted in black shales (e.g., Nyakabingo) while hydrothermal Sn-mineralization is observed in quartz–muscovite veins hosted in muscovite–feldspar-rich quartzites (e.g., Rutongo). Dewaele et al. (2016b) and Hulsbosch et al. (2016) demonstrate that in case of W mineralization in the KAB, the black shale host rocks are typically altered to a tourmaline-muscovite-(biotite) selvage at the contact with the quartz vein. During host-rock interaction, the black shales acted as a source of iron and calcium for the formation of tungstate minerals (Dewaele et al., 2016b). For example, De Clercq (2012) reported abundant leaching of pyrite in the black shales and sandstones of Nyakabingo. Additionally, during fluid-black shale interaction, the mineralizing fluid could have been mixed with diluted "metamorphic" fluids causing a decrease in chloride concentration of the mineralizing fluid and inducing a substantial decrease in the W solubility (Hulsbosch et al., 2016). Selective cassiterite precipitation from a W- and Sn-containing hydrothermal fluid is favored by oxidation (with H_2 consumption), acid neutralization, or decreasing salinity (Heinrich, 1990). H_2 is consumed by vapor separation and in reactions of the mineralizing fluids with aluminosilicate host rocks. Acidity is consumed during interaction of the mineralizing fluid with host rocks rich in feldspars, which results in the formation of muscovite and quartz assemblages. Cassiterite and cogenetic muscovite occur mainly at the contact zone between the quartz vein and the quartzite host rock due to the selective precipitation mechanism of Sn by acid-consuming induced by feldspar consumption. As such, in the KAB, a magmatic-hydrothermal fluid containing both W and Sn, which originated from the differentiating Kibara Sn granite–pegmatite system, can exclusively precipitate W or Sn depending on the host-rock lithology and associated host-rock reactions (Hulsbosch et al., 2016).

In the Kibara Sn granite-pegmatite system, only Sn can precipitate from these fluids due to the oxidation of Sn^{2+} to Sn^{4+} by hydrolysis during greisenization reactions in the zone 4 pegmatites (see also reactions 6 and 7). This general behavior is in line with the experimentally derived fluid-melt partition coefficients stated above: extremely low fluid affinity for Nb and Ta (i.e., retained generally in the melt), insoluble behavior for Sn (i.e., dominantly retained in the melt), and moderately insoluble to soluble

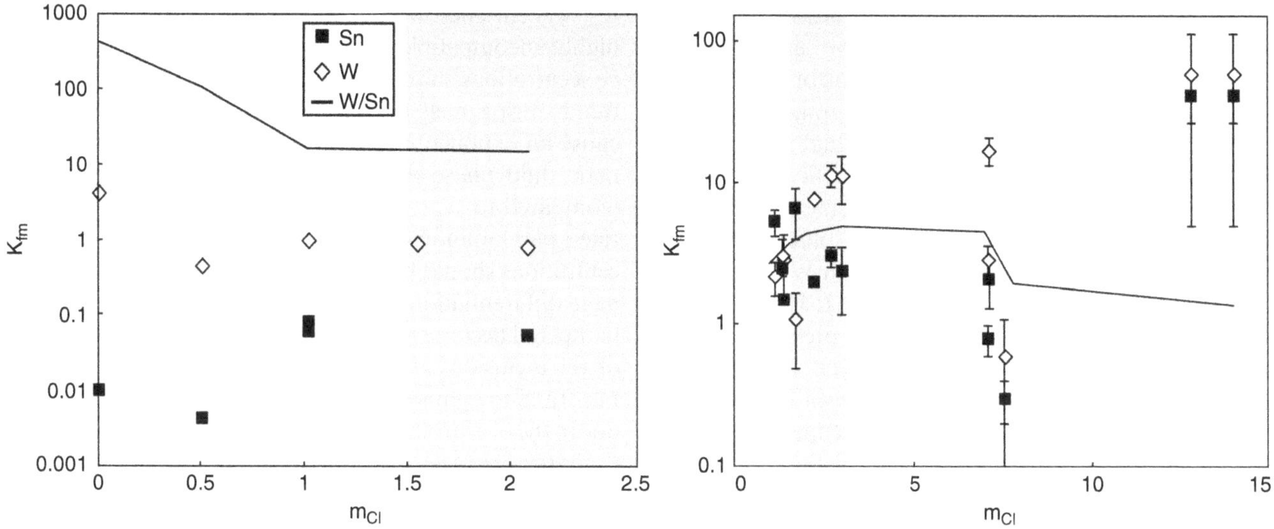

Figure 4.10 Comparison of reported fluid-melt partition coefficient (K_{fm}) for both Sn and W as a function of the molinity of Cl in the fluid phase (moles of Cl/kg of solution). (a) Experimental partitioning in haplogranitic-H_2O-HCl system at 2 kbar, 750°C and NNO buffer conditions (adapted and recalculated from Keppler & Wyllie, 1991). (b) Natural partitioning based on LA-ICP-MS analysis of coexisting fluid and silicate melt inclusions in various granitic systems (peralkaline to peraluminous in composition) at oxygen fugacities ranging from NNO-1.7 to NNO+4.5 (adapted from Zajacz et al., 2008). Grey shaded areas correspond to calculated molinity of Cl in the fluid phase of type I and type II fluid inclusions in the pegmatites of Gatumba-Gitarama (Hulsbosch et al., 2016). Black lines correspond to calculated ratios of the W over Sn fluid-melt partition coefficients.

behavior for W (i.e., partitioning into the aqueous fluid). Notwithstanding the high melt affinity for Sn, magmatic cassiterite is observed only as an accessory mineral phase in the pegmatites of zone 4. The high solubility of Sn^{2+} in these reduced peraluminous melts (cf. Johnston, 1965; Lehmann, 1990b) is surmounted only by extreme degrees of melt fractionation ($F > 0.98$; Hulsbosch et al., 2014). In these highly fractionated pegmatites of zone 4, oversaturation relatively to SnO_2 can cause direct cassiterite precipitation from the melt when high solubility limits have been exceeded (1000–2000 ppm Sn at the NNO buffer, and at 1.5 kbar and 750°C; Ryabchikov et al., 1978). Bhalla et al. (2005) determined for an H_2O-satured peraluminous melt (~1 wt% F and at the NNO buffer) the following relationship: $C_{SnO_2} = 3.17 - \frac{3.51 \cdot 10^3}{T}$ for the temperature-dependent (T, Kelvin) cassiterite saturation (C_{SnO_2}, wt%). For a peraluminous melt, cassiterite solubility at 750°C amounts ~5500 ppm SnO_2 (Bhalla et al., 2005). Based on Sn concentrations in magmatic muscovite from zone 4 pegmatites, which range from ~600 to ~1000 ppm (Hulsbosch et al., 2013; 2014), and Sn muscovite-melt distribution coefficients (1.1 at 720°C and 4.9 at 580°C; Kovalenko et al., 1988), calculated Sn concentration in the highly fractionated pegmatitic melt ($F > 0.98$) are between 125 and 204 ppm (at 580°C) and between 545 and 909 ppm (at 720°C). Sn melt contents for the Kibara Sn granites have been estimated between 4–80 ppm (concentrations can be affected by hydrothermal redistribution; Lehmann & Lavreau, 1987), and Sn melt contents in the biotite to muscovite pegmatites of zone 1 to 3 have been calculated to vary between 7 and 62 ppm (at 580°C) and between 30 and 270 ppm (at 720°C), which are all significantly lower than the cassiterite saturation limit of Ryabchikov et al. (1978) and Bhalla et al. (2005).

In order to evaluate consistently the hydrothermal-metasomatic mineralization potential for both Sn and W, data from studies that determined fluid-melt partition coefficient for Sn and W will be applied. However, studies reporting coefficients for both Sn and W are scarce, with the exception of the experimental study of Keppler and Wyllie (1991) and the natural study of Zajacz et al. (2008). Although, absolute values for fluid-melt partition coefficients can vary over more than one order in magnitude between these studies, overall general conclusions for hydrothermal element fractionation can still be drawn (Fig. 4.10).

Based on the literature review above and the experimental and natural fluid partitioning data of Sn and W in Figure 4.10, the metallogenic processes for hydrothermal Sn and W quartz vein-hosted deposits of an orthomagmatic origin for the KAB can be proposed. (1) A first stage of enrichment of Sn and W in the Kibara Sn granitic up to the initial pegmatitic melt stages by incompatible crystal-melt fractionation. (2) The highly differentiated peraluminous B-rich, F-poor Kibara Sn granite systems will reach water saturation at relative deeper crustal levels in comparison with F-dominated and/or metaluminous melts (see section 4.5.1.). First indications of fluid saturation are observed just prior to or at the granite-pegmatite transition stage. The relative late-stage fluid saturation of the Kibara Sn system is particularly beneficial for hydrothermal W enrichment because it enables sufficient accumulation of the otherwise highly soluble W in the melt phase. (3) Based on the overall high W to Sn fluid-melt partition coefficient ratio ($k_{fm,W/Sn} > 1$; Fig. 4.10), the coexisting aqueous fluid phase will preferentially be enriched in W compared to Sn. However, absolute fluid concentrations depend also on the concentration in the Kibara Sn melt systems, which have been estimated between 4 and 80 ppm for Sn (bulk rock data; Lehmann & Lavreau, 1987) and 7.5 ppm for W (Hulsbosch et al., 2016). (4) In the case of W, the high fluid affinity ($K_{fm} = 1$–5) (i.e., natural systems; Hulsbosch et al., 2016; Zajacz et al., 2008) in combination with the moderately incompatible melt affinity (K_{cm} ~0.4) (Hulsbosch et al., 2016) causes preferential enrichment in the coexisting fluid phase during melt differentiation. However, based on W concentrations in primary pegmatitic fluid inclusions and melt-fluid-crystal fractionation calculations (Hulsbosch et al., 2016), W fluid concentrations do not demonstrate a typical exponential enrichment trend during melt differentiation. This is a direct consequence of the depletion of W in the melt phase during differentiation by the dominant partitioning of W into the fluid phase ($K_{fm} > 1$) (Hulsbosch et al., 2016). As a result, the mineralization potential of W from hydrothermal fluids can be expected already high at the initial stages of magmatic fluid saturation. Protracted melt fractionation in equilibrium with an exsolved fluid phase will only further deplete W from the melt. The lithological control of black shales (e.g., Fe-source; Dewaele et al., 2016b; Hulsbosch et al., 2016) on W precipitation is, moreover, a critical factor for hydrothermal W mineralization in quartz veins (reactions 4 and 5).

$$WO_4^{2-} + MCl_n^{2-n} \leftrightarrow MWO_4 + nCl^- \quad \text{(Reaction 4)}$$

$$HWO_4^{-} + MCl_n^{2-n} \leftrightarrow MWO_4 + H^+ + nCl^- \quad \text{(Reaction 5)}$$

with $M = (Fe,Mn)^{2+}$ derived from black shales
$n = 0$ to 2.

(5) In the case of Sn, it preferentially partitions into the melt phase ($k_{fm,W/Sn} > 1$; Fig. 4.10) and gets more enriched toward the latest magmatic units of the zone 4 pegmatites where melt oversaturation relatively to SnO_2 may cause cassiterite precipitation. In combination with incompatible crystal-melt fractionation, Sn also (but not preferential; $K_{fm} < 1$) enriches in the coexisting fluid phase. Due to its highly incompatible behavior(cf. Taylor, 1979), Sn melt concentrations should describe an exponential enrichment trend during melt differentiation, which in turn, should cause an exponential fluid enrichment trend in the cogenetic fluid phase due to the relatively low fluid affinity (compared to W; e.g., Fig. 4.10). Ultimately, in H_2O-saturated peraluminous melt systems, very high Sn fluid concentrations should be expected at high ($F > 0.98$) degrees of melt differentiation. However, verification is, up to now, hampered because no LA-ICP-MS compositional analyses of the inclusions in minerals from the highly fractionated, Nb–Ta–Sn pegmatites of zone 4 could not be performed due to the overprinting of the inclusion record by secondary fluids (cf. Dewaele et al., 2008). As a result, the mineralization potential of Sn from hydrothermal fluids can be expected to increase with melt differentiation in equilibrium with an exsolved fluid. Hydrothermal Sn mineralization is consequently expected to culminate at the highest degrees of melt differentiation. Moreover, due to the capability of both pegmatites and sandstones/quartzites to act as lithologically controlled precipitation horizons by feldspar hydrolysis reactions (reactions 6 and 7; Dewaele et al., 2016b; Hulsbosch et al., 2016) and due to the higher Sn/W ratios in Kibara Sn granites (e.g., ~16 based on data of Lehmann & Lavreau, 1987), Sn mineralization is more widespread in the KAB granite-related ore system.

$$Sn(II)Cl_x^{2-x} + 2H_2O \leftrightarrow Sn(IV)O_2 + 2H^+ + xCl^- + H_2 \quad \text{(Reaction 6)}$$

$$3KAlSi_3O_8 + 2H^+ + 2Cl^- \leftrightarrow KAl_3Si_3O_{10}(OH)_2 + 6SiO_2 + 2K^+ + 2Cl^- \quad \text{(Reaction 7)}$$

4.6. W-SN ORE POTENTIAL OF GRANITE-RELATED SYSTEMS

In section 4.5.3., it is concluded that W and Sn have an opposing hydrothermal mineralization potential in water-saturated, peraluminous granite-related ore systems. The potential of W to form hydrothermal deposits can be expected already optimally high at the initial stages of magmatic fluid saturation (i.e., a melt at relatively lower degrees of differentiation). Protracted melt fractionation in equilibrium with an exsolved fluid phase will only further deplete W in the melt and in turn in the coexisting fluid phase due the partitioning of W into the fluid phase. The mineralization potential of Sn from hydrothermal fluids increases with melt differentiation in equilibrium with an exsolved fluid. Hydrothermal Sn mineralization is consequently expected to culminate at the highest degrees of melt differentiation. In order to conceptualize this opposing fractionation behavior, two theoretical parameters are proposed to quantify the potential from magmatic and hydrothermal Sn and W ores, that is, the Magmatic Ore Potential (MOP) and Hydrothermal Ore Potential (HOP), shown in equations (4.1)–(4.4):

$$MOPW = \frac{C_{m,W}}{C_{m,\mathrm{Sn}}} = C^0_{m,W/Sn} \frac{\left(\frac{f_m}{f_m^0}\right)^{K_{cm,W}\left(1-\frac{K_{fm,W}}{K_{cm,W}}\right)\Phi_{H_2O}+K_{cm,W}-1}}{\left(\frac{f_m}{f_m^0}\right)^{K_{cm,Sn}\left(1-\frac{K_{fm,Sn}}{K_{cm,Sn}}\right)\Phi_{H_2O}+K_{cm,Sn}-1}} \tag{4.1}$$

$$MOP\,Sn = \frac{C_{m,\mathrm{Sn}}}{C_{m,W}} = C^0_{m,Sn/W} \frac{\left(\frac{f_m}{f_m^0}\right)^{K_{cm,Sn}\left(1-\frac{K_{fm,Sn}}{K_{cm,Sn}}\right)\Phi_{H_2O}+K_{cm,Sn}-1}}{\left(\frac{f_m}{f_m^0}\right)^{K_{cm,W}\left(1-\frac{K_{fm,W}}{K_{cm,W}}\right)\Phi_{H_2O}+K_{cm,W}-1}} \tag{4.2}$$

$$HOPW = \frac{C_{f,W}}{C_{f,\mathrm{Sn}}} = C^0_{m,W/Sn} \frac{K_{fm,W}\left(\frac{f_m}{f_m^0}\right)^{K_{cm,W}\left(1-\frac{K_{fm,W}}{K_{cm,W}}\right)\Phi_{H_2O}+K_{cm,W}-1}}{K_{fm,Sn}\left(\frac{f_m}{f_m^0}\right)^{K_{cm,Sn}\left(1-\frac{K_{fm,Sn}}{K_{cm,Sn}}\right)\Phi_{H_2O}+K_{cm,Sn}-1}} \tag{4.3}$$

$$HOP\,Sn = \frac{C_{f,\mathrm{Sn}}}{C_{f,W}} = C^0_{m,Sn/W} \frac{K_{fm,Sn}\left(\frac{f_m}{f_m^0}\right)^{K_{cm,Sn}\left(1-\frac{K_{fm,Sn}}{K_{cm,Sn}}\right)\Phi_{H_2O}+K_{cm,Sn}-1}}{K_{fm,W}\left(\frac{f_m}{f_m^0}\right)^{K_{cm,W}\left(1-\frac{K_{fm,W}}{K_{cm,W}}\right)\Phi_{H_2O}+K_{cm,W}-1}} \tag{4.4}$$

With C_m = trace element concentration in the melt phase, C_f = trace element concentration in the fluid phase, C_m^0 = initial trace element concentration in the parental melt, K_{fm} = bulk fluid-melt partition coefficient, K_{cm} = bulk crystal-melt partition coefficient, f_m = mass fraction of melt, $f_m^0 = 1$ = melt is water-saturated at the liquidus, Φ_{H_2O} = the negative of the solubility of water in the melt (see details in Spera et al., 2007).

In Equations (4.1) to (4.4), the partitioning of the ore elements among coexisting crystals, melt, and supercritical fluid melt (i.e., C_m and C_f) during crystallization are calculated on the basis of equations of Spera et al. (2007) (see also Fig. 4.11). The MOP and HOP parameters express the normalized (i.e., $C^0_{m,W} = C^0_{m,\mathrm{Sn}} = 1$ at $f_m = 1$) potential of a melt or hydrothermal fluid to be relatively enriched in W or Sn in function of melt differentiation (i.e., the mass fraction of melt f_m). The evolution of the adimensional MOP or HOP parameters in function of mass fraction of melt provides a semiquantitative evaluation of the ore elements Sn and W and their potential to form magmatic or hydrothermal granite-related deposits with protracting fractionation of the melt. In the case of magmatic deposits (cf. section 4.5.1, i.e., MOP), enrichment of an incompatible element in the melt and subsequent saturation-precipitation of the respective ore mineral are the main mineralization processes while for hydrothermal deposits fluid-mineral fractionation, solubility in aqueous phase, and precipitation processes dominate (cf. section 4.5.3, i.e., MOP).

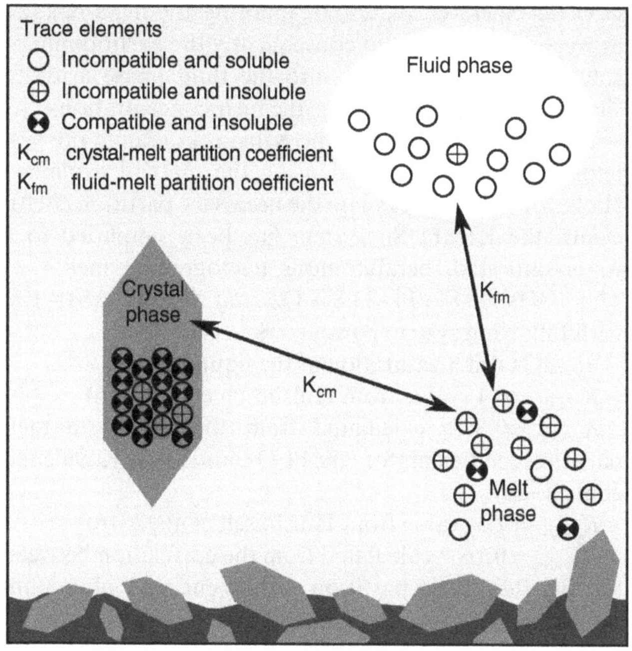

Figure 4.11 Conceptual sketch demonstrating the partitioning of trace elements among coexisting crystals, melt, and fluid, and illustrating the chemical behavior of compatible versus incompatible, and soluble versus insoluble elements as quantified by their crystal-melt and fluid-melt partition coefficients. The distribution of elements among these three phases forms the backbone of the Magmatic Ore Potential (MOP) and Hydrothermal Ore Potential (HOP) calculation.

By applying the Spera et al. (2007) formulation to calculate the melt (C_m) and fluid (C_f) concentrations, the essential role of the fluid phase is considered by implementing a mass-balance expression of water (ϕ_{H_2O}) to the traditional fractional crystallization expressions. This is the only valid procedure for multiphase crystal-fluid-melt systems, especially if water solubilities exceed ~5 wt%

H_2O (cf. Spera et al., 2007). Ignoring the mass-balance expression of water and consequently the partitioning of soluble elements, like W, into the fluid phase leads to significant miscalculation of the melt concentrations.

The MOP and HOP of the Kibara Sn granite ore systems have been evaluated using the method explained above. In order to calculate the necessary partition coefficients, the Kibara Sn system has been simplified to a water-saturated, peraluminous haplogranitic melt (i.e., 47.5% Kfs, 23.8% Pl, 23.8% Qz, and 5% Ms; ASI = 1.5) with following system parameters:

$f_m^0 = 1$ = water saturation at the liquidus

$K_{cm,W} = 0.4$ (value from Hulsbosch et al., 2016)

$K_{cm,\mathrm{Sn}} = 0.3$ = calculated from the Sn crystal-melt partition coefficients for Kfs, Pl, Qz, and Ms of Kovalenko et al. (1988)

$K_{fm,W} = 2.5$ (value from Hulsbosch et al., 2016)

$K_{fm,\mathrm{Sn}} = 0.16$ = calculated from the correlation between the Sn fluid-melt partition coefficient and aluminum saturation index of Hu et al. (2008)

The MOP and HOP parameters of W and Sn for the peraluminous and water-saturated Kibara Sn granite are given in Figure 4.12. Based on the MOP and HOP parameters of W (Fig. 4.12 a and b), a distinct preference of W to fractionation toward the immiscible fluid phase is observed at the expense of the melt with an approximate factor of 10 (HOP >> MOP). With protracted melt fractionation, both the MOP and HOP decrease due to high solubility of W. This behavior can especially be noticed by evaluating the MOP and HOP parameters calculated for different water solubilities (Fig. 4.12 a, b: 5, 10, and 15 wt% H_2O) showing a general decrease with increasing water solubilities. The potential to form magmatic or hydrothermal deposits decrease over several orders of magnitude with melt differentiation from $f_m = 1$ to $f_m = 0.01$. The MOP and HOP parameters quantify and visualize the proposed fractionation behavior of W in sections 4.5.1 and 4.5.3: the potential of W to form magmatic disseminated and hydrothermal deposits decreases exponentially with protracted melt differentiation and is

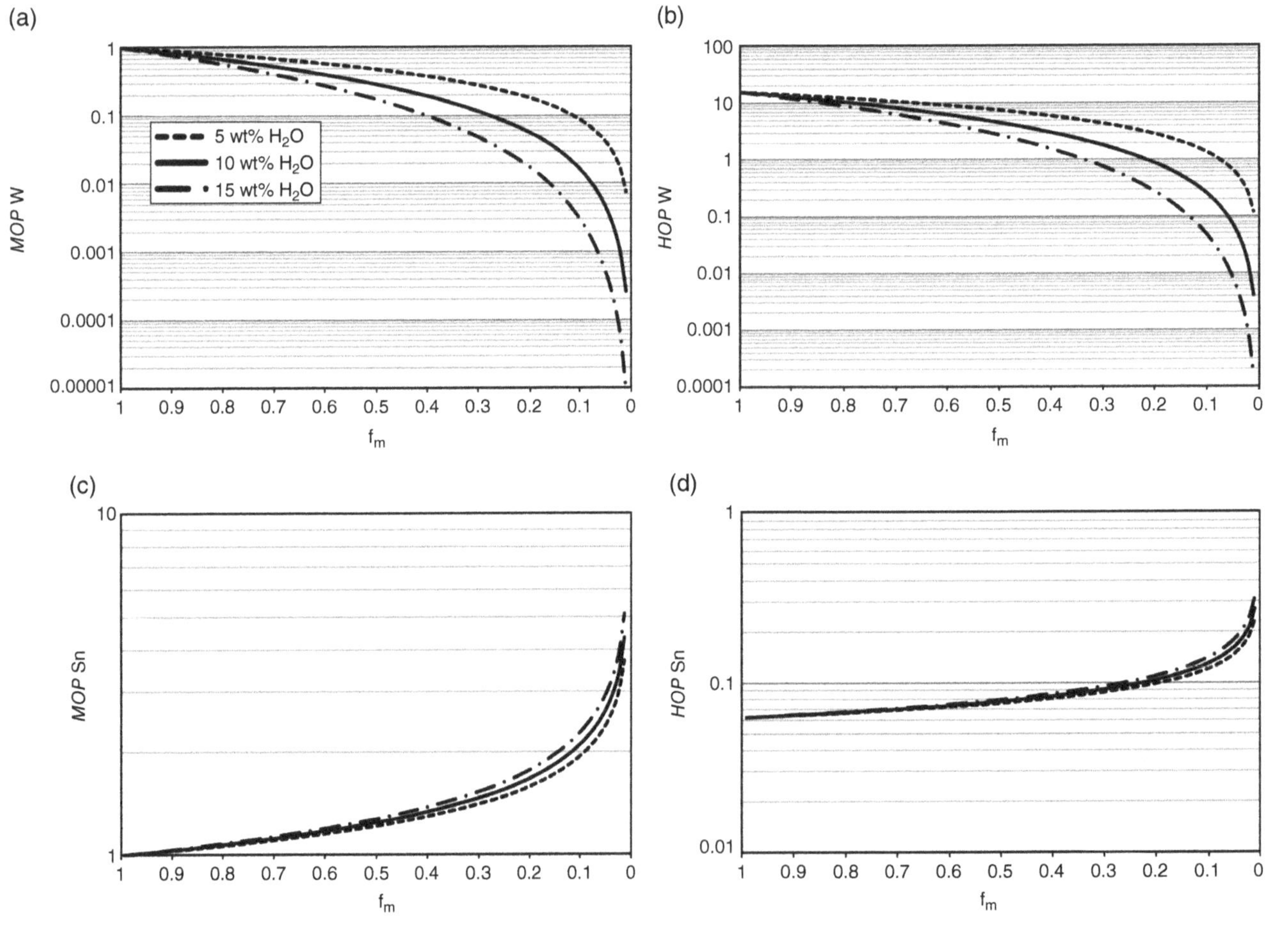

Figure 4.12 The relative Magmatic Ore Potential (MOP) and Hydrothermal Ore Potential (HOP) parameters for W (a) and (b), and Sn (c) and (d) versus the mass fraction of melt (f_m), respectively, and for different water solubilities (5, 10, 15 wt.%). Calculations stopped at $f_m = 0.01$. For calculation and interpretation see text.

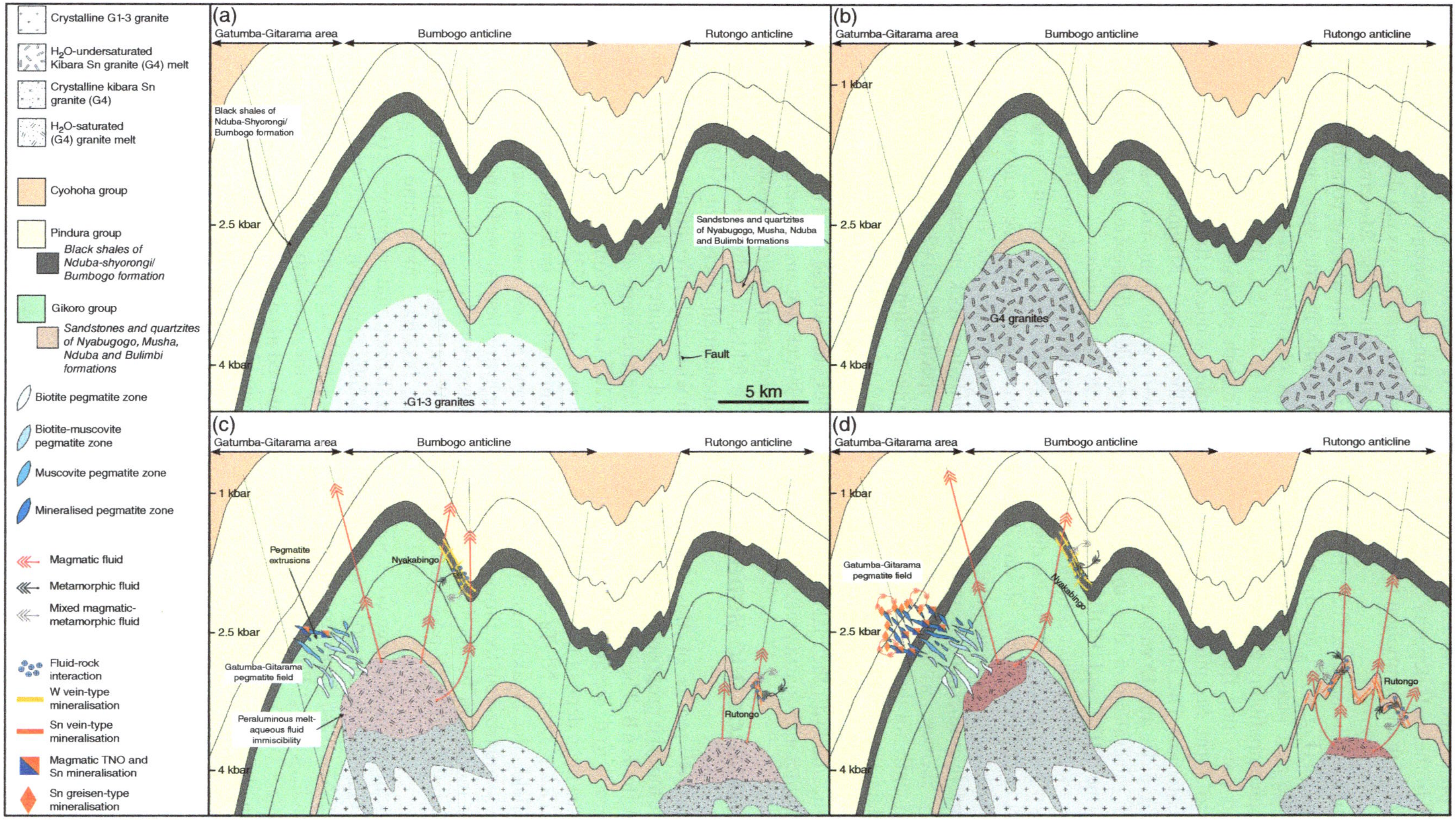

Figure 4.13 Metallogenic model of the granite-related Nb-Ta-Sn-W deposits in the KAB. The location of cross section A-B is shown in Figure 4.5. (a) Prior to G4 intrusions (~1375 Ma-1000 Ma), (b) G4 intrusion (1000 Ma-986 Ma), (c) W vein-type and magmatic Nb-Ta-Sn pegmatite-type mineralization (~992 Ma-975 Ma), and (d) Sn vein-type and greisen-hosted pegmatite-type mineralization (<975 Ma-966 Ma). See text for details.

negatively affected by protracted W fluid fractionation. The potential for hydrothermal deposits is high at relatively low degrees of melt differentiation (f_m between 1 and 0.25).

A contrasting evolution can be deduced from the MOP and HOP parameters of Sn (Fig. 4.12 c and d). A general preference of Sn for the melt phase is observed (MOP >>HOP). Due to the incompatibility and insolubility of Sn, both the melt and coexisting fluid phase show an increasing potential to form Sn mineralization with protracted melt differentiation. The amount of exsolved fluids (Fig. 4.12 c and d: 5, 10, and 15 wt% H_2O) is a less crucial factor for Sn mineralization due to the low Sn solubility. Again, the MOP and HOP parameters quantitatively describe the fractionation behavior of Sn as proposed in sections 4.5.1 and 4.5.3. Magmatic disseminated Sn mineralization can occur only at very high degrees of melt differentiation (i.e., oversaturation) and the potential to form hydrothermal or metasomatic deposits exponentially increase with protracted melt differentiation. Consequently, for the KAB, the MOP and HOP parameters explain the formation of only hydrothermal mineralization in quartz veins for W, whereas Sn can form both magmatic and hydrothermal/metasomatic mineralization in pegmatites and quartz veins.

4.7. KIBARA METALLOGENIC MODEL

Based on the presented results of this study, an integrated metallogenic model is proposed for the granite-related ore deposits in the KAB. The model is based on four evolutionary stages (a to d, see also Figs. 4.13 a–d) and focused on the three main deposit types of the KAB as they occur in Central Rwanda: the Nb-Ta-Sn Gatumba-Gitarama pegmatite field, the Nyakabingo W quartz vein, and the Rutongo Sn quartz-vein localities (Figs. 4.5 and 4.8). This generic model can be applied to all ore deposit types associated with B-rich, F-poor granite systems (Fig. 4.13).

1. Intrusion and crystallization of barren G1-3 granitic magmas at ca. 1.375 Ga (Tack et al., 2010) in a basin sequence composed of Akanyaru Supergroup metasediments (see also Fig. 4.8; see also Fig. 4.7; Fernandez-Alonso et al., 2012).

2. Regional intrusion of the late-orogenic Kibara Sn or so-called G4 granite generation at ~986 Ma (Tack et al., 2010) as composite intrusions within the G1–3 generation in a folded and foliated basin sequence. Initial differentiation due to Rayleigh fractionation causes incompatible elements Sn, W, Nb, and Ta to progressively concentrate in the melt (Hulsbosch et al., 2014).

Subsequent second boiling of this relatively deep-seated system (≤4 kbar; Lehmann et al., 2013) initiated saturation of a magmatic medium saline fluid (Hulsbosch et al., 2016). W partitions strongly into the fluid phase as (alkali-) tungstate complexes. Especially the lack of F in this specific granite system delays the exsolution of the W-bearing fluid phase. Moreover, Sn is retained preferentially in the melt at these conditions and will selectively partition only into the fluid phase at elevated Cl concentrations. Notwithstanding that this magmatic fluid will contain W and Sn, the fluid's ore potential for W will decline with protracting melt differentiation and fluid exsolution, while the fluid's ore potential for Sn will exponentially increase. Magmatic fluid expulsion and fluid flow will occur along pre-existing planes of structural weakness like bedding, cleavage, and fractures (Hulsbosch et al., 2017b; Muchez et al., 2014). Once these initial W-(Sn) magmatic fluids are channeled out of the differentiating Kibara Sn granite system, the fluid source signature is modified due to mixing with external meteoric or "metamorphic" fluid reservoirs and overprinting by fluid–host rock interactions (e.g., De Clercq, 2012). More specifically, black shales controlled the scheelite-ferberite precipitation in hydrothermal quartz veins (992.4 ± 1.5 Ma at Nyakabingo; ^{40}Ar-^{39}Ar on muscovite) by a combination of decreasing temperature, addition of volatiles, an increase in pH through wall-rock reaction, and wall-rock element leaching. Hydrothermal cassiterite-containing quartz veins will occur in quartzite-sandstone units by feldspar hydrolysis reaction. However, based on the exponentially increasing MOP and HOP parameters for Sn with melt differentiation, a culmination of hydrothermal Sn mineralization in quartz veins is expected near the end-stage of the magmatic Kibara Sn system (965 ± 29 Ma at Rutongo; Rb-Sr on muscovite; Monteye-Poulaert et al., 1962). Both the lithological control of different host-rock units and contrasting fluid evolution of W and Sn will cause a decoupling of these elements during hydrothermal ore formation.

Contemporaneous and cogenetically with the formation of a W-enriched fluid phase, the residual melt-fluid system becomes more and more differentiated by Rayleigh fractionation processes (Hulsbosch et al., 2016). Eventually, expulsion of the highly evolved pegmatite-forming melts is triggered. The magmatic crystallization age of the Gatumba-Gitarama pegmatites has been determined at ca. 975 Ma (i.e., zone 4): 975 ± 29 Ma (Rb-Sr on muscovite; Cahen, 1964) and 975 ± 8 Ma (U-Pb on columbite-tantalite; Dewaele et al., 2011). The residual pegmatitic melt is extremely enriched in incompatible melt structure modifiers (e.g., Nb, Ta, Sn B, P, alkali metals, carbonates, sulfates, borates, phosphates, and H_2O), causing the melt viscosity to be several orders of magnitude lower than the crystal mush matrix (Thomas & Davidson, 2013). These chemical and physical conditions between a low-viscous, residual melt and its parental melt are favorable conditions to extract the pegmatitic liquid from the granitic system. The pegmatite-forming melt system will increasingly become more differentiated by Rayleigh fractionation processes with the formation of biotite, biotite-muscovite, muscovite, and mineralized pegmatite zones. Eventually Ta-Nb oxide and cassiterite saturation initiates at the most differentiated and late magmatic end-stages, that is, the mineralized pegmatite zone, and a magmatic Nb-Ta and Sn mineralization forms in the residual pegmatite system (ages between 975 ± 8 Ma and 966 ± 9 M; U-Pb on CGM) (Dewaele et al., 2011; Melcher et al., 2015).

3. Late-stage reduced acidic and saline magmatic fluids in both the granitic and pegmatitic system (Thomas & Davidson, 2013; Veksler, 2005) transported Sn as Sn^{2+}-chloride complexes (Černý et al., 2005). Intra- and peribatholitic Sn deposits form by the interaction of this Sn-bearing fluid with host (pegmatite) and wall rock (sandstone-quartzite), triggering acid-consuming reactions (cf. muscovitization), and eventually precipitation of the major phase of cassiterite.

4.8. CONCLUSION

The results of this study strongly suggest an orthomagmatic origin for granite-related Nb-Ta-Sn-W mineralization, especially at their type-localities in Central Rwanda. It classifies the KAB as a granite-related ore system genetically related to Kibara Sn granites (i.e., G4 granites). Rayleigh-type fractional crystallization acted as the main mechanism by which pegmatitic magmas differentiate from a parental leucogranitic melt and by which incompatible elements Nb, Ta, Sn, and W initially are enriched. However, early aqueous fluid immiscibility has been identified to occur during differentiation of this B-rich, F-poor melt system, which greatly affects further W enrichment in the melt by preferential partitioning of W into the mobile fluid phase. Element-specific melt-fluid-crystal fractionation together with element-specific and lithological-controlled precipitation conditions are the key enrichment processes and are all responsible for the decoupling of Nb-Ta, Sn, and W and their subsequent precipitation in pegmatite, hydrothermal quartz vein, and greisen deposits.

ACKNOWLEDGMENTS

Sincere thanks to Philippe Muchez and Stijn Dewaele for all their advice, guidance, and insights. Research of the author is funded by a postdoctoral fellowship of the Research Foundations–Flanders (FWO). Helpful comments to the manuscript made by two AGU reviewers are greatly appreciated. Last, thanks to editors Sophie Decrée and Laurence Robb for the opportunity to write this chapter and for their editorial handling.

REFERENCES

Aissa, M., Weisbrod, A., & Marignac, C. (1987). Caractéristiques chimiqies et thermodynamiques des circulations hydrothermales du site d'Echassières. *Géologie de la France*, *2–3*, 335–350.

Antipin, V. S., Kovalenko, V. I., Kuznetsova, A. I., & P. L. A. (1981). Tin and tungsten behavior in ore-bearing acidic magmatic rocks according to analyses of the partition coefficients. *Geokhimiya*, *2*, 163–177.

Audétat, A., Günther, D., & Heinrich, C. A. (2000a). Magmatic-hydrothermal evolution in a fractionating granite: A microchemical study of the Sn-W-F-mineralized Mole Granite (Australia). *Geochimica et Cosmochimica Acta*, *64*(19), 3373–3393.

Audétat, A., Gunther, D., & Heinrich, C. A. (2000b). Causes for large-scale metal zonation around mineralized plutons: Fluid inclusion LA-ICP-MS evidence from the Mole Granite, Australia. *Econ. Geol. Bull. Soc. Econ. Geol.*, *95*(8), 1563–1581.

Badanina, E. V., Veksler, I. V., Thomas, R., Syritso, L. F., & Trumbull, R. B. (2004). Magmatic evolution of Li-F, rare-metal granites: a case study of melt inclusions in the Khangilay complex, Eastern Transbaikalia (Russia). *Chemical Geology*, *210*(1–4), 113–133.

Baker, T., Pollard, P. J., Mustard, R., Markl, G., & Graham, J. L. (2005). A comparison of granite-related tin, tungsten, and gold-bismuth deposits: Implications for exploration. *SEG Newsletter*, *61*, 5–17.

Barsukov, V. L. (1957). The geochemistry of tin. *Geochemistry*, *1*, 41–52.

Bartels, A., Holtz, F., & Linnen, R. L. (2010). Solubility of manganotantalite and manganocolumbite in pegmatitic melts. *American Mineralogist*, *95*(4), 537–544.

Bartels, A., Behrens, H., Holtz, F., Schmidt, B. C., Fechtelkord, M., Knipping, J., Crede, L., et al. (2013), The effect of fluorine, boron and phosphorus on the viscosity of pegmatite forming melts. *Chemical Geology*, *346*(0), 184–198.

Baudet, D., Hanon, M., Lemonne, E., Theunissen, K., Buyagu, S., Dehandschutter, J., Ngizimana, W., et al. (1988). Lithostratigraphie du domaine sédimentaire de la chaîne Kibarienne au Rwanda. *Annales de la Société Géologique de Belgique*, *112*, 225–246.

Bertossa, A. (1965). La pegmatite de Buranga. *Bulletin de la Société belge de Rwanda*, *2*, 1–5.

Bertossa, A. (1968). Inventaire des minéraux du Rwanda. *Bulletin du Service Géologique de Rwanda*, *4*, 25–46.

Bertossa, A., Gérards, J., & Petricec, V. (1964). Géologie de la région de Kigali. *Bulletin du Service Géologique de Rwanda*, *1*, 3–12.

Beus, A. A. (1966). *Geochemistry of Beryllium And Genetic Types of Beryllium Deposits*. W. H. Freeman, San Fransisco.

Beus, A. A. (1986). *Geology of Tungsten*. UNESCO Publishing.

Bhalla, P., Holtz, F., Linnen, R. L., & Behrens, H. (2005). Solubility of cassiterite in evolved granitic melts: effect of T, fO2, and additional volatiles. *Lithos*, *80*(1–4), 387–400.

Borodulin, G. P., Chevychelov, V. Y., & Zaraysky, G. P. (2009). Experimental study of partitioning of tantalum, niobium, manganese, and fluorine between aqueous fluoride fluid and granitic and alkaline melts. *Doklady Earth Sciences*, *427*(1), 868–873.

Bouchot, V., Ledru, P., Lerouge, C., Lescuyer, J. L., & Milesi, J. P. (2005). Late Variscan mineralizing systems related to orogenic processes: The French Massif Central. *Ore Geology Reviews, 27*(1–4), 169–197.

Breiter, K., Forster, H. J., & Seltmann, R. (1999). Variscan silicic magmatism and related tin-tungsten mineralization in the Erzgebirge-Slavkovsky les metallogenic province. *Mineralium Deposita, 34*(5–6), 505–521.

Breiter, K., Škoda, R., & Uher, P. (2007). Nb-Ta-Ti-W-Sn-oxide minerals as indicators of a peraluminous P- and F-rich granitic system evolution: Podlesí, Czech Republic. *Mineralogy and Petrology, 91*(3–4), 225–248.

Brown, M. (1994). The generation, segregation, ascent and emplacement of granite magma: the migmatite-to-crustally-derived granite connection in thickened orogens. *Earth-Science Reviews, 36*(1–2), 83–130.

Buchwaldt, R., Toulkeridis, T., Todt, W., & Ucakuwun, E. K. (2008). Crustal age domains in the Kibaran belt of SW-Uganda: Combined zircon geochronology and Sm-Nd isotopic investigation. *Journal of African Earth Sciences, 51*(1), 4–20.

Burnham, C. W. (1997). *Magmas and hydrothermal fluids*, 3rd ed. John Wiley & Sons.

Cahen, L. (1964). État de la géochronologie du Rwanda. *Bulletin du Service Géologique de Rwanda, 1*, 35–38.

Cahen, L., & Ledent, D. (1979). Précisions sur l'age, la pétrogénèse et la position stratigraphique des granites à étain de l'est de l'Afrique Centrale. *Bulletin de la Sociétébelge de Géologie, 88*, 33–49.

Cahen, L., Snelling, N. J., Delhal, J., Vail, J. R., Bonhomme, M., & Ledent, D. (1984). *The Geochronology and Evolution of Africa*. Oxford University Press.

Candela, P. A. (1992). Controls on ore metal ratios in granite related ore systems: an experimental and computational approach. *Transactions of the Royal Society of Edinburgh: Earth Sciences, 83*, 317–326.

Candela, P. A. (1997). A Review of Shallow, Ore-related granites: textures, volatiles, and ore metals. *Journal of Petrology, 38*(12), 1619–1633.

Carroll, M. R., & Webster, J. D. (1994). Solubilities of sulfur, noble gases, nitrogen, chlorine, and fluorine in magmas. *Volatiles in Magmas, 30*, 231–279.

Černý, P. (1991a). Rare-element granitic pegmatites. Part I. Anatomy and internal evolution of pegmatite deposits. *Geoscience Canada, 18*, 49–67.

Černý, P. (1991b). Fertile granites of Precambrian rare-clement pegmatite fields: Is geochemistry controlled by tectonic setting or source lithologies? *Precambrian Research, 51*(1–4), 429–468.

Černý, P. (1991c). Rare-element granitic pegmatites. Part II. Regional to global environments and petrogenesis. *Geoscience Canada, 18*, 68–81.

Černý, P., & Ercit, T. S. (2005). The classification of granitic pegmatites, *The Canadian Mineralogist, 43*(6), 2005–2026.

Černý, P., Blevin, P. L., Cuney, M., & London, D. (2005). Granite-Related Ore Deposits. *Economic Geology 100th Anniversary volume*, 337–370.

Che, X. D., Linnen, R. L., Wang, R. C., Aseri, A., & Thibault, Y. (2013). Tungsten solubility in evolved granitic melts: An evaluation of magmatic wolframite. *Geochimica et Cosmochimica Acta, 106*, 84–98.

Chevychelov, V., Zaraisky, C., Borisovsky, S., & Borkov, D. (2004). Partitioning of Ta and Nb between magmatic melt and aqueous (K,Na,H)F-containing fluid. Effects of temperature and chemical composition of the melt. *Lithos, 73*(1–2), S17–S17.

Chicharro, E., Boiron M.-C., López-García, J. Á., D. N. Barfod, & Villaseca, C. (2016). Origin, ore forming fluid evolution and timing of the Logrosán Sn–(W) ore deposits (Central Iberian Zone, Spain). *Ore Geology Reviews, 72, Part 1*, 896–913.

Chicharro, E., Martín-Crespo, T., Gómez-Ortiz, D., López-García, J. Á., Oyarzun, R., & Villaseca, C. (2015). Geology and gravity modeling of the Logrosán Sn–(W) ore deposits (Central Iberian Zone, Spain). *Ore Geology Reviews, 65, Part 1*(0), 294–307.

Cuney, M., Marignac, C., & Weisbrod, A. (1992). The Beauvoir topaz-lepidolite albite granite (Massif Central, France); the disseminated magmatic Sn-Li-Ta-Nb-Be mineralization. *Economic Geology, 87*(7), 1766–1794.

Daltry, V. D. C., & Von Knorring, O. (1998). Type-mineralogy of Rwanda with particular reference to the Buranga pegmatite. *Geologica Belgica, 1*, 9–15.

De Clercq, F. (2012). *Metallogenesis of Sn and W vein-type deposits in the Karagwe-Ankole belt (Rwanda)*, KU Leuven, Department of Earth and Environmental Sciences.

De Clercq, F., Muchez, P., Dewaele, S., & Boyce, A. (2008). The tungsten mineralisation at Nyakabingo and Gifurwe (Rwanda): Preliminary results. *Geologica Belgica, 11*(3–4), 251–258.

Deblond, A., Punzalan, L. E., Boven, A., & Tack, L. (2001). The Malagarazi supergroup of southeast Burundi and its correlative Bukoba supergroup of northwest Tanzania: neo- and mesoproterozoic chronostratigraphic constraints from Ar-Ar ages on mafic intrusive rocks. *Journal of African Earth Sciences, 32*(3), 435–449.

Debon, F., & Le Fort, P. (1983). Chemical-mineralogical classification of common plutonic rocks and associations. *Transactions of the Royal Society of Edinburgh: Earth Sciences, 73*, 135–149.

Debruyne, D., Hulsbosch, N., Van Wilderode, J., Balcaen, L., Vanhaecke, F., & Muchez, P. (2015). Regional geodynamic context for the Mesoproterozoic Kibara Belt (KIB) and the Karagwe-Ankole Belt: Evidence from geochemistry and isotopes in the KIB. *Precambrian Research, 264*, 82–97.

Dewaele, S., Muchez, P., Burgess, R., & Boyce, A. (2015). Geological setting and timing of the cassiterite vein type mineralization of the Kalima area (Maniema, Democratic Republic of Congo). *Journal of African Earth Sciences, 112*, 199–212.

Dewaele, S., Tack, L., Fernandez, M., Boyce, A., & Muchez, P. (2007). *Cassiterite mineralization in vein-type deposits of the Kibara orogen (Central Africa): Nyamiumba (Rutongo area, Rwanda)*, 1007–1010.

Dewaele, S., Hulsbosch, N., Cryns, Y., Boyce, A., Burgess, R., & Muchez, P. (2016a). Geological setting and timing of the world-class Sn, Nb-Ta and Li mineralization of Manono–Kitotolo (Katanga, Democratic Republic of Congo). *Ore Geology Reviews, 72, Part 1*, 373–390.

Dewaele, S., De Clercq, F., Muchez, P., Schneider, J., Burgess, R., Boyce, A., & Fernandez Alonso, M. (2010). Geology of the cassiterite mineralisation in the Rutongo area, Rwanda (Central Africa): current state of knowledge. *Geologica Belgica*, *13*(1–2), 91–111.

Dewaele, S., De Clercq, F., Hulsbosch, N., Piessens, K., Boyce, A., Burgess, R., & Muchez, P. (2016b). Genesis of the vein-type tungsten mineralization at Nyakabingo (Rwanda) in the Karagwe–Ankole belt, Central Africa. *Mineralium Deposita*, *51*(2), 283–307.

Dewaele, S., Tack, L., Fernandez Alonso, M., Boyce, A., Muchez, P., Schneider, J., Cooper, G., & Wheeler, K. (2008). Geology and mineralisation of the Gatumba area, Rwanda: present state of knowledge. *Etudes Rwandaises*, *16*, 6–24.

Dewaele, S., Henjes-Kunst, F., Melcher, F., Sitnikova, M., Burgess, R., Gerdes, A., Fernandez, M. A., Clercq, F. D., Muchez, P., & Lehmann, B. (2011). Late Neoproterozoic overprinting of the cassiterite and columbite-tantalite bearing pegmatites of the Gatumba area, Rwanda (Central Africa). *Journal of African Earth Sciences*, *61*(1), 10–26.

Dingwell, D. B. (1988). The structure and properties of fluorine-rich magmas: a review of experimental studies, in R. P. Taylor & D. F. Strong, *Recent advances in the geology of granite-related mineral deposits* (pp. 1–12). Canadian Institute of Mining and Metallurgy.

Dingwell, D. B., Harris, D. M., & Scarfe, C. M. (1984). The solubility of H2O in melts in the system: SiO2- Al2O3-Na2O-K2O at 1 to 2 Kbars. *Journal of Geology*, *92*(4), 387–395.

EC (2017). *Study on the review of the list of critical raw materials: Critical raw materials factsheets*, European Commision, Directorate-General for Internal Market, Industry, Entrepreneurship and SMEs.

Fernandez-Alonso, M. (1981). Données préliminaires sur la géochimie de granites protérozoiques du Rwanda et du Burundi. *Rapport Annuel de la Musée royal de l'Afrique Centrale*, 195–207.

Fernandez-Alonso, M., & Theunissen, K. (1998). Airborne geophysics and geochemistry provide new insights in the intracontinental evolution of the Mesoproterozoic Kibaran belt (Central Africa). *Geological Magazine*, *135*(2), 203–216.

Fernandez-Alonso, M., Lavreau, J., & Klerkx, J. (1986). Geochemistry and geochronology of the Kibaran granites in Burundi, Central Africa: Implications for the Kibaran orogeny. *Chemical Geology*, *57*(1–2), 217–234.

Fernandez-Alonso, M., Cutten, H., De Waele,B., Tack, L., Tahon, A., Baudet, D., & Barritt, S. D. (2012). The Mesoproterozoic Karagwe-Ankole Belt (formerly the NE Kibara Belt): The result of prolonged extensional intracratonic basin development punctuated by two short-lived far-field compressional events. *Precambrian Research*, *216–219*, 63–86.

Fernandez-Alonso, M. (2007). *Geological Map of the Mesoproterozoic Northeastern Kibara Belt, Royal Museum for Central Africa*, Tervuren, Belgium.

Fiege, A., Kirchner, C., Holtz, F., Linnen, R. L., & Dziony, W. (2011). Influence of fluorine on the solubility of manganotantalite (MnTa2O6) and manganocolumbite (MnNb2O6) in granitic melts: An experimental study. *Lithos*, *122*(3–4), 165–174.

Förster, H. J., Romer, R. L., Gottesmann, B., Tischendorf, G., & Rhede, D. (2009). Are the granites of the Aue-Schwarzenberg Zone (Erzgebirge, Germany) a major source for metalliferous ore deposits? A geochemical, Sr-Nd-Pb isotopic, and geochronological study. *Neues Jahrbuch für Mineralogie, Abhandlungen*, *186*(2), 163–184.

Frost, B. R., Barnes, C. G., Collins, W. J., Arculus, R. J., Ellis, D. J., & Frost, C. D. (2001). A geochemical classification for granitic rocks. *Journal of Petrology*, *42*(11), 2033–2048.

Gast, P. W. (1972). The chemical composition and structure of the moon. *The Moon*, *5*(1), 121–148.

Gérards, J. (1965). Géologie de la région de Gatumba. *Bulletin du Service Géologique Rwandaise*, *2*, 31–42.

Gibert, F., Moine, B., Schott, J., & Dandurand, J.-L. (1992). Modeling of the transport and deposition of tungsten in the scheelite-bearing calc-silicate gneisses of the Montagne Noire, France. *Contributions to Mineralogy and Petrology*, *112*(2), 371–384.

Goldmann, S., Melcher, F., Gäbler, H.-E., Dewaele, S., Clercq, F., & Muchez, P. (2013). Mineralogy and Trace Element Chemistry of Ferberite/Reinite from Tungsten Deposits in Central Rwanda. *Minerals*, *3*(2), 121–144.

Günther, M. A. (1990). Flüssigkeitseinschlüsse und geologisches Umfeld zentralafrikanischer Sn-, W- und Au-Lagerstätten (Rwanda and Burundi), technischen Universität Carolo-Wilhelmina zu Braunschweig, Braunschweig.

Günther, M. A., & Ndutiye, W. (1990). The tungsten deposit of Nyakabingo, Rwanda. *International Geological Correlation Programme, Project n° 255, TU Braunschweig-MRAC Tervuren*, *3*, 31–36.

Günther, M. A., Dulski, P., Lavreau, J., Lehmann, B., Möller, P., & Pohl, W. (1989). The Kibaran tin granites: Hydrothermal alteration versus plate tectonic setting. *International Geological Correlation Programme, Project n° 255, TU Braunschweig-MRAC Tervuren*, *2*, 21–27.

Harlaux, M., Romer, R. L., Mercadier, J., Morlot, C., Marignac, C., & Cuney, M. (2017a). 40 Ma of hydrothermal W mineralization during the Variscan orogenic evolution of the French Massif Central revealed by U-Pb dating of wolframite. *Mineralium Deposita*.

Harlaux, M., Mercadier, J., Bonzi, W., Marc-Emile, D., Kremer, V., Marignac, C., & Cuney, M. (2017b). Geochemical Signature of Magmatic-Hydrothermal Fluids Exsolved from the Beauvoir Rare-Metal Granite (Massif Central, France): Insights from LA-ICPMS Analysis of Primary Fluid Inclusions. *Geofluids*, *2017*, 25.

Heinrich, C. A. (1990). The chemistry of hydrothermal tin(-tungsten) ore deposition. *Economic Geology*, *85*(3), 457–481.

Holtz, F., Dingwell, D. B., & Behrens, H. (1993). Effects of F, B2O3 and P2O5 on the solubility of water in haplogranite melts compared to natural silicate melts. *Contributions to Mineralogy and Petrology*, *113*(4), 492–501.

Holtz, F., Johannes, W., Tamic, N., & Behrens, H. (2001). Maximum and minimum water contents of granitic melts generated in the crust: a reevaluation and implications. *Lithos*, *56*(1), 1–14.

Hu, X., Bi, X., Hu, R., Shang, L., & Wenling, F. (2008). Experimental study on tin partition between granitic silicate melt and coexisting aqueous fluid. *Geochemical Journal*, *42*(2), 141–150.

Hu, X., Bi, X., Hu, R., Cai, G., & Chen, Y. (2016). Tin partition behavior and implications for the Furong tin ore formation associated with peralkaline intrusive granite in Hunan Province, China. *Acta Geochimica*, *35*(2), 138–147.

Hulsbosch, N. (2016). Chemical fractionation in granite-related ore systems: evidence from Nb-Ta-Sn pegmatite-type and Sn-W quartz vein-type mineralisation in the Karagwe-Ankole Belt (Rwanda). University of Leuven, KU Leuven, Leuven.

Hulsbosch, N., Boiron, M.-C., Dewaele, S., & Muchez, P. (2016). Fluid fractionation of tungsten during granite-pegmatite differentiation and the metal source of peribatholitic W quartz veins: Evidence from the Karagwe-Ankole Belt (Rwanda). *Geochimica et Cosmochimica Acta*, *175*, 299–318.

Hulsbosch, N., Hertogen, J., Dewaele, S., André, L., & Muchez, P. (2013). Petrographic and mineralogical characterisation of fractionated pegmatites culminating in the Nb-Ta-Sn pegmatites of the Gatumba area (western Rwanda). *Geologica Belgica*, *16*(1–2), 105–117.

Hulsbosch, N., Hertogen, J., Dewaele, S., André, L., & Muchez, P. (2014). Alkali metal and rare earth element evolution of rock-forming minerals from the Gatumba area pegmatites (Rwanda): Quantitative assessment of crystal-melt fractionation in the regional zonation of pegmatite groups. *Geochimica et Cosmochimica Acta*, *132*, 349–374.

Hulsbosch, N., Thomas, R., Boiron, M.-C., Dewaele, S., & Muchez, P. (2017a). Direct Observation of Boro-Aluminosilicate Melt Compositions: Insights From Raman Spectroscopy of Melt Inclusions In Pegmatitic Tourmaline of the Gatumba-Gitarama Area (Rwanda). *The Canadian Mineralogist*, *55*(3), 377–397.

Hulsbosch, N., Van Daele, J., Reinders, N., Dewaele, S., Jacques, D., & Muchez, P. (2017b). Structural control on the emplacement of contemporaneous Sn-Ta-Nb mineralized LCT pegmatites and Sn bearing quartz veins: Insights from the Musha and Ntunga deposits of the Karagwe-Ankole Belt, Rwanda. *Journal of African Earth Sciences*, *134*, 24–32.

Johnston, W. D. (1965). Oxidation-reduction equilibria in molten Na2O-SiO2 glass. *Journal of the American Ceramic Society*, *48*, 184–190.

Kamenetsky, V. S., & Kamenetsky, M. B. (2011). Magmatic fluids immiscible with silicate melts: examples from inclusions in phenocrysts and glasses, and implications for magma evolution and metal transport, in B. Yardley, C. Manning, & G. Garven, *Frontiers in Geofluids* (pp. 293–311). Wiley-Blackwell.

Kampunzu, A. B., Rumvegeri, B. T., Kapenda, D., Lubala, R. T., & Caron, J. P. H. (1986). Les Kibarides d'Afrique centrale et orientale: Une chaîne de collision. *UNESCO. Geology for Development Newsletter*, *5*, 125–137.

Keppler, H., & Wyllie, P. (1991). Partitioning of Cu, Sn, Mo, W, U, and Th between melt and aqueous fluid in the systems haplogranite-H2O − HCl and haplogranite-H2O − HF. *Contributions to Mineralogy and Petrology*, *109*(2), 139–150.

Klerkx, J., Liégeois, J. P., Lavreau, J., & Theunissen, K. (1984). Granitoides kibariens précoces et tectonique tangentielle au Burundi: magmatisme bimodal lié a une distention crustale, in J. Klerkx & J. Michot, *African Geology, a Volume in Honour of L. Cahen* (pp. 29–46). Royal Museum for Central Africa, Tervuren.

Klerkx, J., Liegeois, J. P., Lavreau, J., & Claessens, W. (1987). Crustal evolution of the northern Kibaran Belt, eastern and central Africa, in A. Kröner, *Proterozoic Lithospheric Evolution* (pp. 217–233). American Geophysical Union and the Geological Society of America.

Koegelenberg, C., & Kisters, A. F. M. (2014). Tectonic wedging, back-thrusting and basin development in the frontal parts of the Mesoproterozoic Karagwe-Ankole belt in NW Tanzania. *Journal of African Earth Sciences*, *97*(0), 87–98.

Koegelenberg, C., Kisters, A. F. M., Kramers, J. D., & Frei, D. (2015). U-Pb detrital zircon and 39Ar-40Ar muscovite ages from the eastern parts of the Karagwe-Ankole Belt: Tracking Paleoproterozoic basin formation and Mesoproterozoic crustal amalgamation along the western margin of the Tanzania Craton. *Precambrian Research*, *269*, 147–161.

Kokonyangi, J., Armstrong, R., Kampunzu, A. B., Yoshida, M., & Okudaira, T. (2004). U-Pb zircon geochronology and petrology of granitoids from Mitwaba (Katanga, Congo): Implications for the evolution of the Mesoproterozoic Kibaran belt. *Precambrian Research*, *132*(1–2), 79–106.

Kokonyangi, J. W., Kampunzu, A. B., Armstrong, R., Arima, M., Yoshida, M., & Okudaira, T. (2007). U-PbSHRIMP dating of detrital zircons from the Nzilo Group (Kibaran belt): Implications for the source of sediments and mesoproterozoic evolution of central Africa. *Journal of Geology*, *115*(1), 99–113.

Kovalenko, V. I., & Yarmolyuk, V. V. (1995). Endogenous rare metal ore formations and rare metal metallogeny of Mongolia. *Econ. Geol. Bull. Soc. Econ. Geol.*, *90*(3), 520–529.

Kovalenko, V. I., Ryabchikov, I. D., & Antipin, V. S. (1988). Temperature dependence of the distribution coefficients for Sn, W, Pb, and Zn in magmatic systems. *Geochemistry International*, *25*(1), 1–10.

Lecumberri-Sanchez, P., Vieira, R., Heinrich, C. A., Pinto, F., & Walle, M. (2017). Fluid-rock interaction is decisive for the formation of tungsten deposits. *Geology*, *45*(7), 579–582.

Lehmann, B. (1987). Tin granites, geochemical heritage, magmatic differentiation. *Geologische Rundschau*, *76*(1), 177–185.

Lehmann, B. (1990a). Regional element distribution patterns and the problem of pregranitic tin enrichments, in *Metallogeny of Tin*. Springer, Berlin-Heidelberg-New York.

Lehmann, B. (1990b). Magmatic enrichment of tin, in *Metallogeny of Tin*, 45–85. Springer, Berlin-Heidelberg-New York.

Lehmann, B. (1990c). Model of tin ore formation, in *Metallogeny of Tin*. Springer, Berlin-Heidelberg-New York.

Lehmann, B., & Lavreau, J. (1987). Tin granites of the northern Kibaran belt, Central Africa (Kivu/Zaire, Rwanda, Burundi), in G. Matheis & H. Schandelmeier, *Current Research in African Earth Sciences* (pp. 33–36). Balkema, Rotterdam.

Lehmann, B., & Lavreau, J. (1988). Geochemistry of the tin granites from Kivu (Zaire), Rwanda and Burundi. *255 Newsletter/Bulletin*, *1*, 43–46.

Lehmann, B., Halder, S., Ruzindana Munana, J., de la Paix Ngizimana, J., & Biryabarema, M. (2013). The geochemical signature of rare-metal pegmatites in Central Africa: Magmatic rocks in the Gatumba tin-tantalum mining district, Rwanda. *Journal of Geochemical Exploration*.

Lehmann, B., Halder, S., Ruzindana Munana, J., de la Paix Ngizimana, J., & Biryabarema, M. (2014). The geochemical signature of rare-metal pegmatites in Central Africa: Magmatic rocks in the Gatumba tin-tantalum mining district, Rwanda. *Journal of Geochemical Exploration*, *144, Part C*, 528–538.

Lenharo, S. L. R., Pollard, P. J., & Born, H. (2003). Petrology and textural evolution of granites associated with tin and rare-metals mineralization at the Pitinga mine, Amazonas, Brazil. *Lithos*, *66*(1–2), 37–61.

Lhoest, A. (1957a). Les différents types de filons de la concession Somuki à Rutongo. *Annales de la Société Géologique de Belgique*, *80*, 503–530.

Lhoest, A. (1957b). Note préliminaire sur la géologie de la région Kigali-Rutongo dans le Ruanda. *Bulletin de la Société belge de Géologie*, *66*, 190–198.

Linnen, R. L. (1998). The solubility of Nb-Ta-Zr-Hf-W in granitic melts with Li and Li + F: Constraints for mineralization in rare metal granites and pegmatites. *Econ. Geol. Bull. Soc. Econ. Geol.*, *93*(7), 1013–1025.

Linnen, R. L., & Williams-Jones, A. E. (1995). Genesis of a Magmatic Metamorphic Hydrothermal System: The Sn-W Polymetallic Deposits at Pilok, Thailand. *Economic Geology*, *90*, 1148–1166.

Linnen, R. L., & Keppler, H. (1997). Columbite solubility in granitic melts: consequences for the enrichment and fractionation of Nb and Ta in the Earth's crust. *Contributions to Mineralogy and Petrology*, *128*(2–3), 213–227.

Linnen, R. L., & Cuney, M. (2005). Granite-Related Rare-Element Deposits and Experimental Constraints on Ta-Nb-W-Sn-Zr-Hf, in R. L. Linnen & I. M. Samson, *Rare-Earth Geochemistry and Mineral Deposits* (pp. 45–68). Geological Association of Canada.

Linnen, R. L., Pichavant, M., & Holtz, F. (1996). The combined effects of f(O2) and melt composition on SnO2 solubility and tin diffusivity in haplogranitic melts. *Geochimica Et Cosmochimica Acta*, *60*(24), 4965–4976.

Linnen, R. L., Pichavant, M., Holtz, F., & Burgess, S. (1995). The effect of *f*O2 on the solubility, diffusion, and speciation of tin in haplogranitic melt at 850°C and 2 kbar. *Geochimica et Cosmochimica Acta*, *59*(8), 1579–1588.

Linnen, R. L., Samson, I. M., Williams-Jones, A. E., & Chakhmouradian, A. R. (2014). Geochemistry of the Rare-Earth Element, Nb, Ta, Hf, and Zr Deposits, 543–568.

London, D. (2008). *Pegmatites*. Mineralogical Association of Canada.

London, D. (2014). A petrologic assessment of internal zonation in granitic pegmatites. *Lithos*, *184–187*, 74–104.

London, D., Hervig, R., & Morgan, G. V. I. (1988). Melt-vapor solubilities and elemental partitioning in peraluminous granite-pegmatite systems: experimental results with Macusani glass at 200 MPa. *Contributions to Mineralogy and Petrology*, *99*(3), 360–373.

Maier, W. D., Peltonen, P., & Livesey, T. (2007). The ages of the Kabanga North and Kapalagulu intrusions, Western Tanzania: A Reconnaissance Study. *Economic Geology*, *102*(1), 147–154.

Mäkitie, H., Data, G., Isabirye, E., Mänttäri, I., Huhma, H., Klausen, M. B., Pakkanen, L., & Virransalo, P. (2014). Petrology, geochronology and emplacement model of the giant 1.37Ga arcuate Lake Victoria Dyke Swarm on the margin of a large igneous province in eastern Africa. *Journal of African Earth Sciences*, *97*, 273–297.

Manning, D. A. C. (1981). The effect of fluorine on liquidus phase relationships in the system Qz-Ab-Or with excess water at 1 kb. *Contributions to Mineralogy and Petrology*, *76*(2), 206–215.

Manning, D. C., & Henderson, P. (1984). The behaviour of tungsten in granitic melt-vapour systems. *Contributions to Mineralogy and Petrology*, *86*(3), 286–293.

Marignac, C., & Cuney, M. (1999). Ore deposits of the French Massif Central: insight into the metallogenesis of the Variscan collision belt. *Mineralium Deposita*, *34*(5–6), 472–504.

Marignac, C., & Cathelineau, M. (2010). *The Nature of Ore-forming Fluids in Peri-Batholitic Sn-W Deposits and a Classification*. James Cook University, Townsville, 245–247.

Melcher, F., Graupner, T., Gäbler, H.-E., Sitnikova, M., Henjes-Kunst, F., Oberthür, T., Gerdes, A., & Dewaele, S. (2015). Tantalum–(niobium-tin) mineralization in African pegmatites and rare metal granites: Constraints from Ta-Nb oxide mineralogy, geochemistry and U-Pb geochronology. *Ore Geology Reviews*, *64*(0), 667–719.

Melcher, F., Graupner, T., Gäbler, H.-E., Sitnikova, M., Oberthür, T., Gerdes, A., Badanina, E., & Chudy, T. (2017). Mineralogical and chemical evolution of tantalum-(niobium-tin) mineralisation in pegmatites and granites. Part 2: Worldwide examples (excluding Africa) and an overview of global metallogenetic patterns. *Ore Geology Reviews*, *89*, 946–987.

Monteye-Poulaert, G., Delwiche, R., & Cahen, L. (1962). Âges de minéralisations pegmatitiques et filoniennes du Rwanda et du Burundi. *Annales de la Société Géologique de Belgique*, *71*, 210–221.

Moura, A., Dória, A., Neiva, A. M. R., Gomes, C. L., & Creaser, R. A. (2014). Metallogenesis at the Carris W-Mo-Sn deposit (Gerês, Portugal): Constraints from fluid inclusions, mineral geochemistry, Re-Os and He-Ar isotopes. *Ore Geology Reviews*, *56*, 73–93.

Muchez, P., Hulsbosch, N., & Dewaele, S. (2014). Geological mapping and implications for Nb-Ta, Sn and W prospection in Rwanda. *Mededelingen Zittingen Koninklijke Academie Overzeese Wetenschappen*, *60*(3–4), 515–530.

Mulja, T., Williams-Jones, A. E., Wood, S. A., & Boily, M. (1995a). The rare-element-enriched monzogranite-pegmatite-quartz vein systems in the Preissac-Lacorne Batholith, Quebec. II. Geochemistry and Petrogenesis. *Can. Mineral.*, *33*, 817–833.

Mulja, T., Williams-Jones, A. E., Wood, S. A., & Boily, M. (1995b). The rare-element-enriched monzogranite-pegmatite-quartz vein systems in the Preissac-Lacorne Batholith, Quebec. I. Geology and Mineralogy. *Can. Mineral.*, *33*, 793–815.

Müller, A., Romer, R. L., & Pedersen, R.-B. (2017). The Sveconorwegian Pegmatite Province–Thousands of pegmatites without parental granites. *The Canadian Mineralogist*, *55*(2), 283–315.

Müller, A., Ihlen, P. M., Snook, B., Larsen, R. B., Flem, B., Bingen, B., & Williamson, B. (2015). The Chemistry of Quartz in Granitic Pegmatites of Southern Norway: Petrogenetic and Economic Implications. *Economic Geology*, *110*(7), 1737–1757.

Peeters, L. (1956). Contribution à la géologie des terrains anciens du Ruanda-Urundi et du Kivu. *Annales du Musée Royale du Congo belge, Tervuren, série in 8°. Sciences Géologiques*, *16*, 1–197.

Pettke, T., Audétat, A., Schaltegger, U., & Heinrich, C. A. (2005). Magmatic-to-hydrothermal crystallization in the

W-Sn mineralized Mole Granite (NSW, Australia). *Chemical Geology*, *220*(3–4), 191–213.

Pichavant, M. (1981). An Experimental Study of the Effect of Boron on a Water Saturated Haplogranite at 1 Kbar Vapour Pressure: Geological Applications. *Contributions to Mineralogy and Petrology*, *76*(4), 430–439.

Pichavant, M., & Manning, D. (1984). Petrogenesis of tourmaline granites and topaz granites; the contribution of experimental data. *Phys. Earth Planet. Inter.*, *35*(1–3), 31–50.

Pohl, W. (1974). *Rapport préliminaire: Mission géologique Rutongo*.

Pohl, W. (1994). Metallogeny of the northeastern Kibara belt, Central Africa: Recent perspectives. *Ore Geology Reviews*, *9*(2), 105–130.

Pohl, W., & Günther, M. A. (1991). The origin of Kibaran (late Mid-Proterozoic) tin, tungsten and gold quartz vein deposits in Central Africa: a fluid inclusions study. *Mineralium Deposita*, *26*(1), 51–59.

Pohl, W. L., Biryabarema, M., & Lehmann, B. (2013). Early Neoproterozoic rare metal (Sn, Ta, W) and gold metallogeny of the Central Africa Region: a review. *Applied Earth Science*, *122*(2), 66–82.

Polya, D. A., Foxford, K. A., Stuart, F., Boyce, A., & Fallick, A. E. (2000). Evolution and paragenetic context of low δD hydrothermal fluids from the Panasqueira W-Sn deposit, Portugal: new evidence from microthermometric, stable isotope, noble gas and halogen analyses of primary fluid inclusions. *Geochimica et Cosmochimica Acta*, *64*(19), 3357–3371.

Raimbault, L., Cuney, M., Azencott, C., Duthou, J. L., & Joron, J. L. (1995). Geochemical Evidence for a Multistage Magmatic Genesis of Ta-Sn-Li Mineralization in the Granite at Beauvoir, French Massif Central. *Econ. Geol. Bull. Soc. Econ. Geol.*, *90*(3), 548–576.

Reyf, F. G. (1997). Direct evolution of W-rich brines from crystallizing melt within the Mariktikan granite pluton, west Transbaikalia. *Mineralium Deposita*, *32*(5), 475–490.

Reyf, F. G., Seltmann, R., & Zaraisky, G. P. (2000). The role of magmatic processes in the formation of banded Li,F-enriched granites from the Orlovka tantalum deposit, Transbaikalia, Russia: Microthermometric evidence. *Can. Mineral.*, *38*, 915–936.

Richards, J. P. (2013). Giant ore deposits formed by optimal alignments and combinations of geological processes. *Nature Geosci*, *6*(11), 911–916.

Robb, L. (2013). *Introduction to Ore-Forming Processes*, Wiley.

Roda Robles, E., Pesquera Perez, A., Velasco Roldan, F., & Fontan, F. (1999). The granitic pegmatites of the Fregeneda area (Salamanca, Spain); characteristics and petrogenesis. *Mineralogical Magazine*, *63*(4), 535–558.

Romer, R., & Kroner, U. (2015). Sediment and weathering control on the distribution of Paleozoic magmatic tin-tungsten mineralization. *Mineralium Deposita*, *50*(3), 327–338.

Romer, R. L., & Kroner, U. (2016). Phanerozoic tin and tungsten mineralization—Tectonic controls on the distribution of enriched protoliths and heat sources for crustal melting. *Gondwana Research*, *31*, 60–95.

Rumvegeri, B. T. (1991). Tectonic significance of Kibaran structures in Central and eastern Africa. *Journal of African Earth Sciences*, *13*(2), 267–276.

Ryabchikov, I. D., Durasova, N. A., Barsukov, V. L., & Efimov, A. S. (1978). Redox potentials as a factor of metalliferous ability of acid magmas. *Geokhimiya*(*6*), 832–834.

Schäfer, B., Frischknecht, R., Gunther, D., & Dingwell, D. B. (1999). Determination of trace-element partitioning between fluid and melt using LA-ICP-MS analysis of synthetic fluid inclusions in glass. *European Journal of Mineralogy*, *11*(3), 415–426.

Shand, S. J. (1943). *The Eruptive Rocks*. John Wiley, New York.

Shaw, R. A., Goodenough, K. M., Roberts, N. M. W., Horstwood, M. S. A., Chenery, S. R., & Gunn, A. G. (2016). Petrogenesis of rare-metal pegmatites in high-grade metamorphic terranes: A case study from the Lewisian Gneiss Complex of north-west Scotland. *Precambrian Research*, *281*, 338–362.

Shearer, C. K., Papike, J. J., & Jolliff, B. L. (1992). Petrogenetic links among granites and pegmatites in the Harney Peak rare-element granite-pegmatite system, Black Hills, South Dakota. *The Canadian Mineralogist*, *30*(3), 785–809.

Simmons, W. B. S., & Webber, K. L. (2008). Pegmatite genesis: state of the art. *European Journal of Mineralogy*, *20*(4), 421–438.

Sirbescu, M.-L. C., & Nabelek, P. I. (2003). Crystallization conditions and evolution of magmatic fluids in the Harney Peak Granite and associated pegmatites, Black Hills, South Dakota: Evidence from fluid inclusions. *Geochimica et Cosmochimica Acta*, *67*(13), 2443–2465.

Spera, F. J., Bohrson, W. A., Till, C. B., & Ghiorso, M. S. (2007). Partitioning of trace elements among coexisting crystals, melt, and supercritical fluid during isobaric crystallization and melting. *American Mineralogist*, *92*(11–12), 1881–1898.

Štemprok, M. (1987). Greisenization (A Review). *Geologische Rundschau*, *76*(1), 169–175.

Štemprok, M. (1990). Solubility of tin, tungsten and molybdenum oxides in felsic magmas. *Mineralium Deposita*, *25*(3), 205–212.

Strong, D. F. (1988). A review and model for granite-related mineral deposits, in R. P. Taylor & D. F. Strong, *Recent Advances in the Geology of Granite-related Mineral Deposits* (pp. 424–445). Canadian Institute of Mining Metallurgy.

Tack, L., Wingate, M. T. D., De Waele, B., Meert, J., Belousova, E., Griffin, B., Tahon, A., & Fernandez-Alonso, M. (2010). The 1375 Ma "Kibaran event" in Central Africa: Prominent emplacement of bimodal magmatism under extensional regime. *Precambrian Research*, *180*(1–2), 63–84.

Taylor, J. R., & Wall, V. J. (1993). Cassiterite solubility, tin speciation, and transport in a magmatic aqueous phase, *Economic Geology*, *88*(2), 437–460.

Taylor, R. G. (1979). *Geology of tin deposits*. Elsevier, Amsterdam, Oxford and New York.

Taylor, S. R., & McLennan, S. M. (1985). *The Continental Crust: Its Composition and Evolution*. Blackwell Scientific Publication, Carlton.

Thomas, R., & Davidson, P. (2013). The missing link between granites and granitic pegmatites. *Journal of Geosciences*, 183–200.

Thomas, R., Webster, J. D., & Heinrich, W. (2000). Melt inclusions in pegmatite quartz: complete miscibility between silicate melts and hydrous fluids at low pressure. *Contributions to Mineralogy and Petrology*, *139*(4), 394–401.

Thomas, R., Davidson, P., & Beurlen, H. (2011a). Tantalite-(Mn) from the Borborema Pegmatite Province, northeastern

Brazil: Conditions of formation and melt- and fluid-inclusion constraints on experimental studies. *Mineralium Deposita*, *46*(7), 749–759.

Thomas, R., Davidson, P., & Schmidt, C. (2011b). Extreme alkali bicarbonate- and carbonate-rich fluid inclusions in granite pegmatite from the Precambrian Rønne granite, Bornholm Island, Denmark. *Contributions to Mineralogy and Petrology*, *161*(2), 315–329.

Thomas, R., Davidson, P., & Beurlen, H. (2012). The competing models for the origin and internal evolution of granitic pegmatites in the light of melt and fluid inclusion research. *Mineralogy and Petrology*, *106*(1–2), 55–73.

Timofeev, A., Migdisov, A. A., & Williams-Jones, A. E. (2015). An experimental study of the solubility and speciation of niobium in fluoride-bearing aqueous solutions at elevated temperature. *Geochimica et Cosmochimica Acta*, *158*(0), 103–111.

Turekian, K., & Wedepohl, K. H. (1961). Distribution of the Elements in Some Major Units of the Earth's Crust. *GSA Bulletin*, *72*(2), 175–192.

Turlin, F., André-Mayer, A.-S., Moukhsil, A., Vanderhaeghe, O., Gervais, F., Solgadi, F., Groulier, P.-A., & Poujol, M. (2017). Unusual LREE-rich, peraluminous, monazite- or allanite-bearing pegmatitic granite in the central Grenville Province, Québec. *Ore Geology Reviews*, *89*, 627–667.

Van Lichtervelde, M., Holtz, F., & Hanchar, J. M. (2010). Solubility of manganotantalite, zircon and hafnon in highly fluxed peralkaline to peraluminous pegmatitic melts. *Contributions to Mineralogy and Petrology*, *160*(1), 17–32.

Varlamoff, N. (1948). Gisements de cassitérite de la région de Kalima (Maniema, Congo belge). *Annales de la Société Géologique de Belgique*, *71*, 194 – 237.

Varlamoff, N. (1954). Répartition des types de pegmatites autour de la partie nord-ouest du grand massif granitique de Nyanza. *Annales de la Société Géologique de Belgique*, *78*, 1–21.

Varlamoff, N. (1960). Habitus des cristaux de cassitérite des pegmatites stannifères de Kirengo (région de Katumba, Ruanda). *Annales de la Société Géologique de Belgique*, *84*, 169–177.

Varlamoff, N. (1961). Pegmatites à amblygonite et à spodumène et pegmatites fortement albitisées à spodumène et à cassitérite de la région de Katumba (Ruanda). *Annales de la Société Géologique de Belgique*, *84*, 257–278.

Varlamoff, N. (1963). Les phénomènes de greisenification, d'albitisation et de lépidolitisation et leurs relations spatiales avec les granites et les pegmatites granitiques d'Afrique. *Annales de la Société Géologique de Belgique*, *86*, 285–322.

Varlamoff, N. (1969). Transitions entre les filons de quartz et les pegmatites stannifères de la région de Musha-Ntunga (Ruanda). *Annales de la Société Géologique de Belgique*, *92*, 193–213.

Varlamoff, N. (1972). Central and West African Rare-Metal Granitic Pegmatites,. Related Aplites, Quartz Veins and Mineral Deposits. *Mineralium Deposita*, *7*(2), 202–216.

Veksler, I. A. (2005). Element Enrichment and Fractionation by Magmatic Aqueous Fluids: Experimental Constraints on Melt-Fluid Immiscibility and Element Partitioning in R. L. Linnen & I. M. Samson, *Rare-Earth Geochemistry and Mineral Deposits* (pp. 69–85). Geological Association of Canada.

Veksler, I. V., Thomas, R., & Schmidt, C. (2002). Experimental evidence of three coexisting immiscible fluids in synthetic granitic pegmatite. *American Mineralogist*, *87*(5–6), 775–779.

Vindel, E., Lopez, J. A., Boiron, M. C., Cathelineau, M., & Prieto, A. C. (1995). P-V-T-X -f02 evolution from wolframite to sulphide depositional stages in intragranitic W-veins. An example from the Spanish Central System. *European Journal of Mineralogy*, *7*(3), 675–688.

Webster, J. (2004). The exsolution of magmatic hydrosaline chloride liquids. *Chemical Geology*, *210*(1–4), 33–48.

Webster, J., Thomas, R., Förster, H.-J., Seltmann, R., & Tappen, C. (2004). Geochemical evolution of halogen-enriched granite magmas and mineralizing fluids of the Zinnwald tin-tungsten mining district, Erzgebirge, Germany. *Mineralium Deposita*, *39*(4), 452–472.

Webster, J. D., Thomas, R., Rhede, D., Forster, H. J., & Seltmann, R. (1997). Melt inclusions in quartz from an evolved peraluminous pegmatite: Geochemical evidence for strong tin enrichment in fluorine-rich and phosphorus-rich residual liquids. *Geochimica Et Cosmochimica Acta*, *61*(13), 2589–2604.

Westerhof, A. B., Härmä, P., Isabirye, E., Katto, E., Koistinen, T., Kuosmanen, E., Lehto, T., et al. (2014). Geology and geodynamic development of uganda with explanation of the 1: 1,000,000 scale geological map. *Geological Survey of Finland Special Paper*, *55*.

Whitney, J. A. (1988). The origin of granite: The role and source of water in the evolution of granitic magmas. *GSA Bulletin*, *100*(12), 1886–1897.

Wilkinson, J. J. (1990). The role of metamorphic fluids in the development of the Cornubian orefield: fluid inclusion evidence from south Cornwall. *Mineralogical Magazine*, *54*, 219–230.

Wilson, G. A., & Eugster, H. P. (1990). Cassiterite solubility and tin speciation in supercritical chloride solutions, in R. J. Spencer & I. M. Chou, *Geochemical Society Special Publication* (pp. 179–195).

Wilson, M. (1989). *Igneous Petrogenesis*. Springer, Dordrecht.

Wood, S. A. (2005). The Aqueous Geochemistry of Zirconium, Hafnium, Niobium and Tantalum, in *Rare-Element Geochemistry and Mineral Deposits*, edited by R. L. Linnen & I. M. Samson, pp. 217–268, Geological Association of Cananda.

Wood, S. A., & Samson, I. M. (2000). The Hydrothermal Geochemistry of Tungsten in Granitoid Environments: I. Relative Solubilities of Ferberite and Scheelite as a Function of T, P, pH, and mNaCl. *Economic Geology*, *95*(1), 143–182.

Zajacz, Z., Halter, W. E., Pettke, T., & Guillong, M. (2008). Determination of fluid/melt partition coefficients by LA-ICPMS analysis of co-existing fluid and silicate melt inclusions: Controls on element partitioning. *Geochimica et Cosmochimica Acta*, *72*(8), 2169–2197.

Zaraisky, G. P., Korzhinskaya, V., & Kotova, N. (2010). Experimental studies of Ta2O5 and columbite-tantalite solubility in fluoride solutions from 300 to 550°C and 50 to 100 MPa. *Mineralogy and Petrology*, *99*(3–4), 287–300.

Zhang, R.-Q., Sun, W.-D., Lehmann, B., Seltmann, R., & Li, C. Y. (2016). Multiple tin mineralization events in Africa: Constraints by in-situ LA-ICPMS cassiterite U-Pb ages in *P. N. 2798. 35th International Geological Congress*, Cape Town, South Africa.

5

The Southern Breccia Metasomatic Uranium System of the Great Bear Magmatic Zone, Canada: Iron Oxide-Copper-Gold (IOCG) and Albitite-Hosted Uranium Linkages

E.G. Potter[1], J.-F. Montreuil[2], L. Corriveau[3], and W.J. Davis[1]

ABSTRACT

The Southern Breccia metasomatic uranium (U) showings are located 1 km south of the NICO deposit, an iron oxide-copper-gold (IOCG) deposit, in the Great Bear magmatic zone of Canada. The timing of both occurrences is tightly constrained to 1873–1868 Ma, linking formation of the albitite-hosted U to development of IOCG mineralization. During this period, regional iron oxide and alkali-calcic metasomatism formed: Na (albite), high-temperature Ca-Fe (amphibole + magnetite), high-temperature Ca-K-Fe (amphibole + magnetite + biotite ± K-feldspar), high-temperature K-Fe (K-feldspar/biotite + magnetite), K (K-feldspar ± biotite), and low-temperature K-Fe-Mg (K-feldspar + hematite + chlorite) assemblages. Primary uraninite and brannerite occur within high-temperature K-Fe alteration composed of magnetite + K-feldspar ± biotite-cemented breccias developed in earlier albitite. The chemistry of primary uraninite supports precipitation from high-temperature, magmatic-derived fluids, as previously proposed for the NICO deposit. Field relationships, petrography, whole-rock geochemistry, and geochronology indicate that alteration of the Southern Breccia corridor host rocks was coeval with early alteration at NICO, whereas U mineralization postdated Au-Co-Bi at NICO. The linkage of the Southern Breccia to the regional iron-oxide and alkali-calcic alteration system that generated the NICO deposit presents a new driver for formation of albitite-hosted U deposits and highlights an exploration target in IOCG districts globally.

5.1. INTRODUCTION

Renowned for its legacy vein-type U and Ag deposits, the Great Bear magmatic zone (GBMZ; Fig. 5.1) of the Northwest Territories is currently one of the most prospective terranes for iron oxide-copper-gold (IOCG) deposits in Canada, with a high potential for undiscovered IOCG deposits and affiliated mineral deposits (Corriveau, 2007; Corriveau et al., 2010a, 2016). IOCG and affiliated deposits such as iron oxide-apatite and albitite-hosted U, form within regional metasomatic systems that have the largest hydrothermal footprints of known ore systems, commonly altering country rocks for several hundred square kilometers (Williams et al., 2005; Mumin et al., 2010; Porter, 2010). In the GBMZ, within such large alteration systems, regional zones of early high to low temperature Na alteration form along fault zones or above subvolcanic intrusions and create zones of porous rocks dominated by secondary albite with minor quartz, zircon, titanite, rutile, and iron-oxides (Corriveau et al., 2010a, b; Montreuil et al., 2015; De Toni, 2016). The porosity can be transient, annihilated through mineral growth

[1] *Geological Survey of Canada, Natural Resources Canada, Ottawa, ON, Canada*

[2] *Formerly Institut National de la Recherche Scientifique, Québec, QC, Canada; Red Pine Exploration Inc., Toronto, ON, Canada*

[3] *Geological Survey of Canada, Natural Resources Canada, Québec, QC, Canada*

Ore Deposits: Origin, Exploration, and Exploitation, Geophysical Monograph 242, First Edition.
Edited by Sophie Decrée and Laurence Robb.

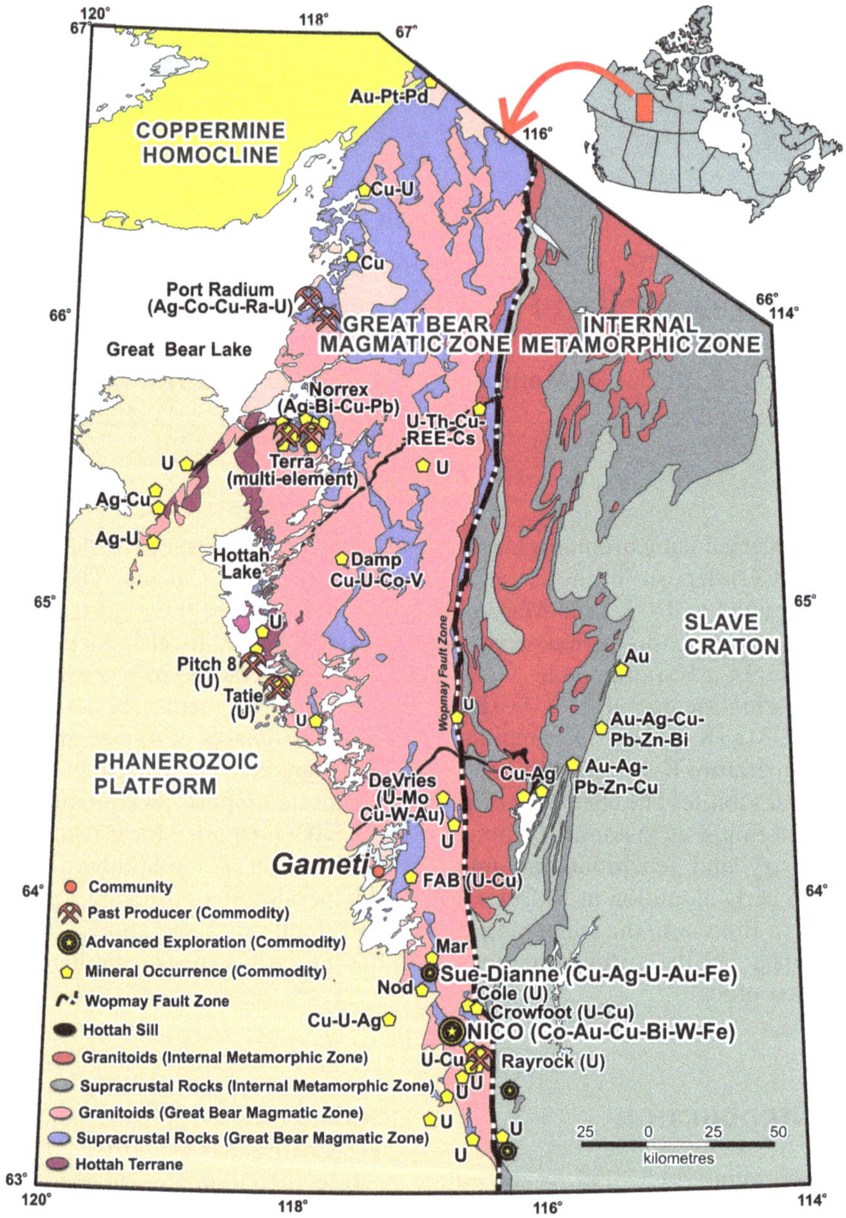

Figure 5.1 Location of the Great Bear magmatic zone (GBMZ) and historic mineral occurrences, in particular the NICO deposit situated in the southern GBMZ. Geology after Hoffman and Hall (1993) and mineral occurrences from the NORMIN database (www.nwtgeoscience.ca/normin).

within pores, grain coarsening, and recrystallization, and even propagate as waves as fluid pressures change and fluid-rock (interface-coupled dissolution-reprecipitation reactions) progress (Putnis, 2015; Omlin et al., 2017). Albitites above subvolcanic intrusions are also prone to extensive recrystallization (Corriveau et al., 2010b). Along fault zones, deformation-induced albitite breccia zones significantly increase the permeability of the structures for fluid circulation (Poulet, 2012; Montreuil et al., 2012, 2015).

Prior to discovery of the Southern Breccia corridor in 2010 (Corriveau et al., 2011), the typically "barren" albitized zones associated with IOCG deposits were only hypothesized to have potential for albitite-hosted or Na-metasomatic U (Williams et al., 2005; Wilde, 2013). This study presents new petrography and mineral chemistry data and uses descriptions of the Southern Breccia alteration assemblages, geochemical signatures of the alteration, and evolution of the regional metasomatic system (Montreuil et al., 2015), coupled with recent geochronology results (Gandhi et al., 2001; Montreuil et al., 2016a) to further support a common genetic linkage between the Southern Breccia and NICO deposit already proposed by field relationships, geophysics, and geochemistry

(Enkin et al., 2012; Montreuil et al., 2015, 2016b; Corriveau et al., 2016; Hayward et al., 2016). This linkage presents a new driver for the formation of albitite-hosted U deposits and potential exploration targets in IOCG districts worldwide. In turn it expands the mineral potential of settings with known albitite corridors, albitite-hosted U, and albitite-hosted Au-Co-U for IOCG deposits.

5.2. IRON OXIDE-ALKALI-CALCIC ALTERATION (IOAA) SYSTEMS

As proposed by Porter (2010) and revisited by Corriveau et al. (2016), the regional iron oxide and alkali-calcic alteration (IOAA) hosting IOCG deposits can encompass not only IOCG deposits (magnetite, magnetite-hematite, and hematite groups), their alteration, and breccia zones, but also the wide spectrum of affiliated deposits that form within such systems such as iron oxide-apatite (IOA) deposits (Williams, 2010a, b), iron oxide-uranium (IOU; Hitzman & Valenta, 2005; Skirrow et al., 2011), some skarns (Gandhi, 2003; Williams, 2010a; Corriveau et al., 2016), as well as alkaline intrusion-hosted IOCG deposits (Corriveau, 2007; Groves et al., 2010; Porter, 2010), and epithermal type veins (Mumin et al., 2010; Kreiner & Barton, 2011).

5.2.1. Iron Oxide-Copper-Gold (IOCG) Deposits

While still contentious in the literature, the definition of IOCG deposits used herein is polymetallic hydrothermal mineral occurrences that contain elevated Cu concentrations, with or without Au among abundant hydrothermal low-Ti Fe-oxide (magnetite, hematite) gangue minerals or associated alteration (cf. Williams, 2010a). Ore zones have moderate contents of sulphides consisting of low-sulphur base-metal sulphides and arsenides, such as chalcopyrite, bornite, chalcocite, pyrrhotite, and arsenopyrite, and native elements such as gold and bismuth (Williams et al., 2005; Corriveau, 2007; Corriveau & Mumin, 2010; Williams, 2010a). IOCG deposits range widely in age, but many prospective terranes are Proterozoic orogens formed at the margins of Archean cratons (Corriveau, 2007; Groves et al., 2010; Skirrow, 2010).

As proposed by Corriveau et al. (2010b, 2016), all IOCG deposits occur within broad-scale, chemically and mineralogically diverse haloes of intensely altered rocks and breccia zones with a systematic evolution of regional to deposit-scale alteration facies (Fig. 5.2). These facies have diagnostic paragenetic sets: Na (± skarns); high-temperature (HT) Ca-Fe (± Na); HT K-Fe (± K-skarns,

Depth	Alternation facies	Mineralization
Near surface or distal ≤ 350°C	6. LT silicification (epithermal K-Al alteration)	Epithermal style
250°C | 350°C	5. LT K-Fe and LT Ca-Fe (K-feldspar/sericite-hematite-carbonate-chlorite-epidote-sulphides)	Hem-group IOCG ±U metasomatic U
350°C | 450°C	4. 'Felsite' or skarns (K-feldspar or clinopyroxene-garnet-sulphides)	Mag-to hem-group IOCG, K-skarns, metasomatic U
	3. HT K-Fe (K-feldspar/biotite-magnetite-sulphides)	Mag-group IOCG (Cu, Au, Co, Bi, ...)
450°C | 800°C	2. Ca-Fe (±Na) (amphibole-magnetite ± apatite)	(IOA) ± REE, Fe-skarns
300 – 600°C	1. Na (albite - scapolite)	Ground preparation
4 – 10 km		

Figure 5.2 IOAA ore system model, modified from Corriveau et al. (2010b, 2016). The alteration facies are listed from the roots of the system to surface following their prograde reaction path but telescoping of alteration facies due to tectonic activity or reenergizing of thermal system can lead to overprinting (e.g., albitite-hosted U deposits in LT K-Fe alteration). HT = high temperature, LT = low temperature.

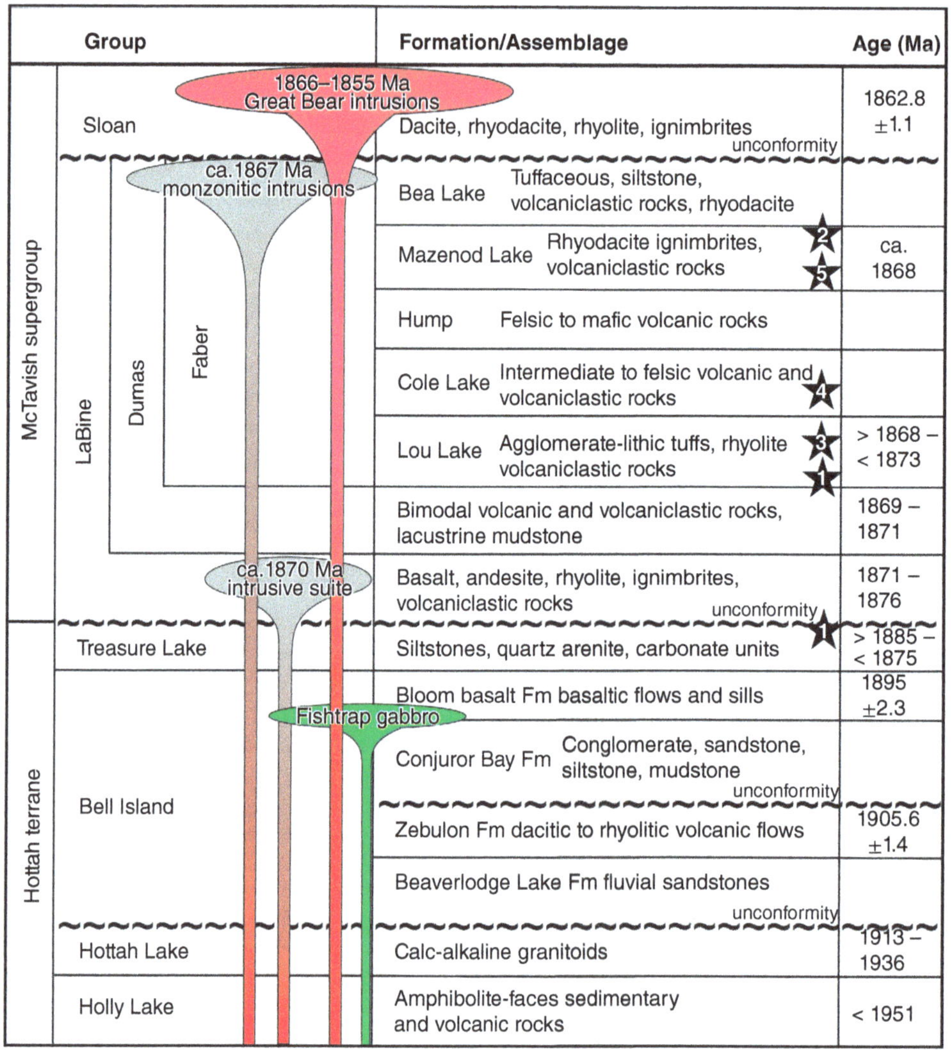

Figure 5.3 Generalized GBMZ stratigraphy and age constraints. "<" denotes ages defined by detrital zircon populations (adapted from Ootes et al., 2015; with geochronology from Bowing, 1984, Reichenbach, 1991, Gandhi et al., 2001, Gandhi & van Breemen, 2005, Bennett & Rivers, 2006, Ootes et al., 2015, and Montreuil et al., 2016a). Hosts to IOCG deposits shown by stars: 1. NICO and Southern Breccia (at Treasure Lake Group, Lou Lake Formation unconformity); 2. Sue Dianne (Bea Lake–Mazenod Lake unconformity); 3. Nod, Brooke, and Mar; 4. Cole; and 5. Fab. Note that the age of the host is not representative of the age of the deposits.

K-felsites); low-temperature (LT) K-Fe; LT Ca-Fe-Mg; and epithermal-style veins (Fig. 5.3; Corriveau et al., 2010b, 2016). These alteration facies produce several lithological, mineralogical, chemical, and geophysical characteristics that can be useful in regional exploration (Belperio et al., 2007; Benavides et al., 2008; Corriveau et al., 2010a, b, 2016, 2017; McMartin et al., 2011; Montreuil et al., 2013; Hayward et al., 2013, 2016; Enkin et al., 2016; Normandeau et al., 2018). Collectively, the alteration facies, their replacement zones, vein networks, and breccias, and the mineralization form the ore systems. When the ore systems are subjected to orogenic events, renewed magmatism, or incursion of basin brines, the metals can be remobilized into vein systems like the five-element veins of the northern GBMZ (Kissin, 1992; Gandhi et al., 2018). In most cases, the metal associations in veins and the host alteration facies remain similar (Corriveau et al., 2016). However, mobility of hexavalent U can be significant leading to decoupling of U from the original alteration assemblages.

The Great Bear magmatic zone (GBMZ) hosts two main belts of IOCG and affiliated deposits/showings that parallel the eastern and western magmatic arc margins if late northeast-trending fault movements are removed and the belt is reconstructed to its original configuration (Fig. 5.1; Hayward & Corriveau, 2013).

Two of the occurrences are the only IOCG deposits known in Canada with estimated reserves or resources: the magnetite-group NICO deposit and the Sue-Dianne magnetite- to hematite-group IOCG deposit. Current National Instrument 43-101 compliant proven and probable reserves at NICO are 33,077 Kt at 1.03 g/t Au, 0.11% Co, 0.14% Bi, and 0.04% Cu (Burgess et al., 2014). The Cu-Ag-(Au) Sue-Dianne deposit characterized by K-feldspar + magnetite breccias that evolve to K-feldspar magnetite-hematite and K-feldspar hematite breccias toward the surface (Mumin et al., 2010), contains National Instrument 43-101 compliant indicated resources of 8.4 Mt grading at 0.80% Cu, 0.07 g/t Au, and 5.2 g/t Ag (Hennessey & Puritch, 2008). Elevated U concentrations are also common among the mineralized GBMZ occurrences (e.g., Sue Dianne, Fab, Cole, Damp, DeVries; Gandhi, 1988, 1989; Gandhi & Prasad, 1993, 1995; Camier, 2002; Corriveau et al., 2010b, 2015, 2016, 2017; Mumin et al., 2010; Ootes et al., 2010, 2013; Potter et al., 2013a, b; Montreuil et al., 2016b). Whole-rock geochemical data and coordinates of the GBMZ samples, mineral occurrences, host rocks, and alteration discussed in this study are presented in Corriveau et al. (2015).

5.2.2. Albitite-Hosted Uranium Deposits

"Na-metasomatic" or albitite-hosted U deposits occur in regional-scale structural corridors characterized by multiple stages of Na (dominantly albite but also scapolite, riebeckite, and aegirine), Ca + Mg + Fe (amphibole, pyroxene, chlorite, carbonate, hematite, magnetite), and K-metasomatism (K-feldspar, biotite) formed under high-temperature, syndeformational settings (Benavides et al., 2008; Cuney & Kyser, 2008; Polito et al., 2009; Skirrow et al., 2009; Cuney et al., 2012). The Na-metasomatic zones are typically narrow (e.g., < 1 m to 700 m wide) but can extend up to several thousand meters along strike. The dominant U minerals are uraninite, brannerite, davidite, and coffinite, and are frequently accompanied by secondary hexavalent U mineral species (Cuney & Kyser, 2008). Key examples include Valhalla in Australia (Polito et al., 2009), Lagoa Real in Brazil (Porto Da Silveira, 1991; Lobato et al., 2015), Aricheng South in Guyana (Alexandre, 2010), the Michelin deposit in Labrador (Gandhi, 1978; Corriveau et al., 2010a; Sparkes, 2017), the Turamdih and other U-Cu-(Fe) deposits in the Singhbhum Shear Zone, India (Pal & Chaudhuri, 2016), and deposits in the central Ukraine shield (Cuney et al., 2012). Some of these albitite-hosted deposits occur within districts with IOCG-U, IOCG, and Cu mineralization (Hitzman & Valenta, 2005; Williams, 2005; Wilde, 2013) and exhibit similar alteration assemblages and ore mineralogy.

5.3. GEOLOGICAL SETTINGS

5.3.1. The Great Bear Magmatic Zone

The GBMZ forms a curvi-linear magmatic belt at the northwest margin of the exposed Canadian Shield. It is interpreted as an Andean-style continental magmatic arc developed at the western margin of the Wopmay orogen (Fig. 5.1), an orogen that formed in response to accretion of the composite Hottah terrane on the western margin of the Slave craton (Hildebrand et al., 2010a; Ootes et al., 2015, and references therein). The Hottah is a 1.93 to 1.89 Ga volcanic arc that likely developed south of the Slave craton, but translated northward along the proto-Wopmay fault to its present location on the western margin of the Slave craton in response to ca. 1.91–1.88 Ga rifting (Davis et al., 2015; Ootes et al., 2015). Following translation, renewed subduction at ca. 1880 Ma drove onset of the ca. 1875–1855 Ma dominantly high-K calc-alkaline volcanic and plutonic rocks of the GBMZ (Bowring, 1984; Gandhi et al., 2001; Hildebrand et al., 2010b; Davis et al., 2015; Ootes et al., 2015; Montreuil et al., 2016a). The GBMZ is exposed for 450 km north-south, although geophysical surveys have indicated an additional 500 km may lie beneath Paleozoic rocks to the south (Coles et al., 1976; Hildebrand et al., 1987; Cook et al., 1999) and 300 km beneath Proterozoic sedimentary rocks to the northwest (Hildebrand et al., 2010b).

The 1875–1861 Ma, predominantly intermediate to felsic suites of subvolcanic, volcanic, and volcaniclastic rocks of the GBMZ (shoshonitic to high-K calc-alkaline series) have been ascribed to the McTavish Supergroup (Fig. 5.3; Hildebrand et al., 2010b; Azar, 2007; Montreuil et al., 2016a). The Supergroup has been subdivided into the Faber, LaBine, Dumas, and Sloan groups that range in composition from basalt through andesite, dacite to rhyolite (Fig. 5.3). The Faber Group is only exposed in the southern part of the belt, where it unconformably overlies a platform-type sedimentary sequence termed the Treasure Lake Group. Detrital zircons in a meta-quartzite of the Treasure Lake Group constrain its maximum age to 1886 ± 8 Ma (Gandhi & van Breemen, 2005), while a crosscutting 1878 ± 4 Ma syndeformation granitic dyke defines the minimum age (Bennett & Rivers, 2006). The Treasure Lake Group was metamorphosed at greenschist to amphibolite facies (grade increases toward the Wopmay Fault Zone), deformed, and tilted before extrusion of the Faber Group volcanic sequence (Gandhi & van Breemen, 2005; Montreuil et al., 2016a). The Dumas Group occurs only adjacent to the Wopmay fault (Fig. 5.1) but recent geochronology indicates that the ca. 1870 Ma volcanogenic strata of the Faber and Dumas groups may be equivalent (Ootes et al., 2015). The Treasure Lake and Faber groups are the principal hosts of the IOCG deposits and prospects of the southern GBMZ (e.g., Sue Dianne, NICO, Brooke, Mar, Nod, Fab; Fig. 5.3). The McTavish arc volcanic suite was intruded by A-type granitic plutons from 1866–1855 Ma (Gandhi, 1988; Gandhi et al., 2001; Bennet & Rivers, 2006; Ootes et al., 2015; Montreuil et al., 2016a).

5.3.2. The NICO Magnetite-Group IOCG Deposit

The NICO deposit formed within the greenschist facies metasedimentary Treasure Lake Group at and below a regional angular unconformity with the overlying volcanic rocks of the basal and central part of the Faber Group (Fig. 5.4; Gandhi & Lentz, 1990; Goad et al., 2010a, b). The Faber Group volcanic rocks consist of basal massive to flow-banded rhyolitic-rhyodacitic flows overlain by a succession of volcaniclastic units dominated by "ash flow tuff with minor porphyritic tuff and lapilli tuff" that are also pervasively metasomatized (K-feldspar dominant; Gandhi & Lentz, 1990; Gandhi, 1994; Montreuil et al., 2016b).

Current drilling has defined an ore zone ("Bowl zone") that extends from the near surface to a depth of 300 m and strikes parallel to the 120°-trending and moderate to steeply north-dipping, thick-bedded subarkosic wackes and interlayered laminar and thin-bedded siltstones (Goad et al., 2000a, b; Hennessey et al., 2007). All the metasedimentary units are pervasively and intensely replaced by stratabound alteration zones dominated by amphibole and magnetite that evolve into localized discordant replacement-type alteration, veining, and brecciation, some of which being K-feldspar and magnetite dominant (Sidor, 2000; Acosta-Góngora et al., 2015a, b; Montreuil et al., 2016b). Collectively, the mineralization and alteration zones, veins networks, and breccias constitute the northern exposed part of the Lou IOAA system.

Along the shear zones spatially associated with the NICO deposit, brittle-ductile deformation is common and most shears are hydrothermally altered by magnetite and magnetite-biotite-amphibole assemblages, indicating a tectonically active environment during IOAA metasomatism. The extent of the highly magnetic alteration facies has been modeled by Craven et al. (2013), Hayward (2013) and Hayward et al. (2016) to > 6 km with a core of higher susceptibility extending from near surface to a depth of ~1500 m along and slightly discordant with respect to original bedding now dipping to the northeast. A low resistivity anomaly extends from a depth of ~250 m (interpreted as currently defined ore zone) to a depth of 1500 m while exploration drilling reached only about 300 m. Another highly resistive region occurs to the west of the NICO deposit at a depth of ~300 m (Craven et al., 2013). This anomaly is truncated by a southwest-dipping discontinuity interpreted as a thrust sheet that juxtaposed the Southern Breccia against the NICO footprint (Hayward et al., 2016). A low-resistivity region, dipping southwest below the NICO deposit, extends to upper mantle depths and represents a potential crustal-scale fault zone (Craven et al., 2013).

The dominant IOAA assemblages at NICO are early skarn (clinopyroxene of hedenbergite + augite varieties), and mild albitization followed by many cycles of high-temperature Ca-Fe and K-Fe (including transitional Ca-Fe-K), manifested in the development of amphibole (actinolite, ferrohornblende), amphibole + magnetite, amphibole + biotite, magnetite, amphibole + magnetite + biotite ± K-feldspar, and K-feldspar (Goad et al., 2000a, b; Sidor, 2000; Corriveau et al., 2011, 2016; Acosta-Góngora et al., 2014, 2015a, b). Carbonate, albite, tourmaline, and earthy hematite alteration are minor in extent within the ore zone and biotite-hematite is locally present. Intense Na (albite) alteration is locally present in the footwall of the deposit within amphibole-bearing stratabound-altered quartz arenite units and increases in intensity in the siltstone units below the deposit (Montreuil et al., 2016b).

Previous studies have documented multiple cycles of stratabound to discordant replacement-type alteration, veining, and brecciation that involved repeated influx of fluids under physicochemical conditions that could produce high-temperature magnetite-bearing Ca-Fe and K-Fe alteration types as well as Au-Co-Bi-Cu mineralization (Fig. 5.5; Goad et al., 2000a, b; Corriveau et al., 2011, 2016; Acosta-Góngora et al., 2014, 2015a, b; Montreuil et al., 2016b). The main mineralization event is characterized by the formation of cobaltite, Co-rich loellingite, and Co-rich arsenopyrite with accessory bismuthinite at relatively high temperatures in a gangue composed of ferro-actinolite, ferro-hornblende, and biotite (>400 °C, Acosta-Góngora et al., 2015a; > 470 °C, and possibly up to 600 °C, Sidor, 2000). During Au-Co-Bi sulpharsenide mineralization, bedding-parallel alteration, veins, and tectonic foliation indicate synmineralization brittle-ductile deformation (Montreuil et al., 2016b).

Subsequent to Au-Co-Bi mineralization, chalcopyrite precipitated in a K-Fe gangue assemblage composed of either magnetite or pyrrhotite with Fe-rich biotite and K-feldspar in the NICO deposit and overlying Summit Peak showing (Fig. 5.4), as (Sidor, 2000; De Toni, 2016; Montreuil et al., 2016b). This mineralization cuts the Au-Co-Bi sulpharsenide mineralization at NICO and principally occurs as veins and zones of brittle brecciation.

5.3.3. The Southern Breccia Corridor

Located less than 1 km southeast of the NICO IOCG deposit, the Southern Breccia corridor was previously mapped as hornfelsed siltstones (Goad et al., 2000a, b;

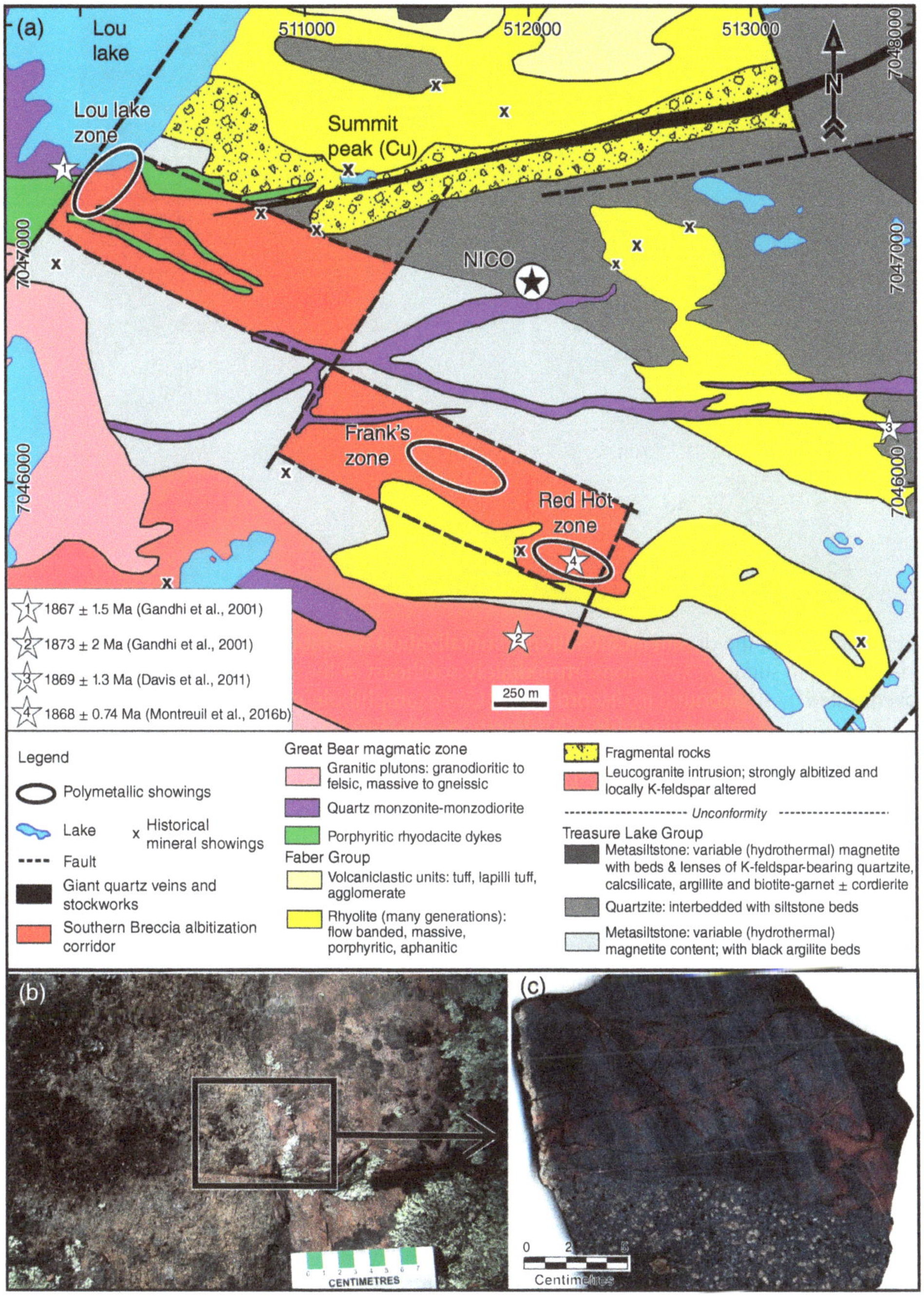

Figure 5.4 (a) Geology and geochronological constraints of the Lou IOAA system, including location of the NICO deposit and Southern Breccia mineral showings. The Southern Breccia showings are within a 3 km long, southeast-trending albite alteration corridor (modified from Montreuil et al., 2015). (b, c) Unaltered 1868 ± 0.74 Ma porphyritic quartz monzonite dyke that cuts alteration and mineralization at the Red Hot zone. Outcrop station 11PUA-002. Photographs courtesy of Geological Survey of Canada.

Mumin et al., 2010). It is currently defined for 3 km along strike based on the series of white albitite breccias, distinctive bright red alteration zones, and U-rich, polymetallic (U ± Th-Cu-Mo-Bi-Au-REE) showings (Fig. 5.4). The 500 m wide albitite corridor is bound by local-scale faults and is hosted within metasiltstones of the Treasure Lake Group. It lies at the transition from high magnetic and gravity anomalies to a northwest-trending regional gravity low along a southwest-dipping fault zone models to at least 1.5 km depth (Hayward et al., 2016). The albitized corridor strikes roughly 120°, parallel to the tectonic foliations in the shear zones observed in the altered

Figure 5.5 Photographs of the dominant alteration-mineralization assemblages in Treasure Lake metasiltstones at the NICO deposit, from surface expressions immediately southeast of the deposit. (a) Stratabound Na (albite; Ab) alteration overprinted by stratabound to discordant HT Ca-Fe (amphibole-magnetite; Am, Mgt) alteration; all overprinted by K-Fe (K-feldspar, magnetite stable; Kfs) alteration with K_2O reaching 6 wt. % based on gamma-ray spectrometer measurements on outcrop. (b) Intense stratabound to discordant HT Ca-Fe (amphibole-magnetite) alteration, overprinted by later K-Fe (biotite, K-feldspar, magnetite; Bt, Kfs, Mgt). (c) Stratabound Na (albite) and HT Ca-Fe (amphibole-magnetite) alteration cut by HT Ca-Fe (Am) and subsequently HT K-Fe (Kfs-Mgt) veins with haloes of stratabound alteration. All photos from station 11PUA-028, courtesy of the Geological Survey of Canada.

Treasure Lake metasedimentary rocks of the NICO footprint, as well as bedding at the regional scale. In contrast, at outcrop scale, veining, replacement-type alteration, and brecciation follow and cut the metasiltstone bedding. The center of the Southern Breccia corridor is also dextrally offset by 045°-trending faults visible as well defined geophysical (airborne and ground magnetic; Hayward et al., 2013, 2016) and topographic lineaments (e.g., LANDSAT images and air photos).

An albitized granitic intrusion truncates the metasedimentary rocks to the south and provides a maximum age of 1873 ± 2 Ma for both NICO and the Southern Breccia (Gandhi et al., 2001). An unaltered, porphyritic quartz monzonite dyke that cuts all mineralization and alteration in the Southern Breccia corridor (Fig. 5.4b,c) was dated at 1868 ± 0.74 Ma, confining the minimum age of the Southern Breccia development (Montreuil et al., 2016a). Texturally similar and geochemically indistinguishable porphyry dykes also cut the ore at NICO. Finally, a monzodioritic porphyritic dyke that is part of a regionally extensive, composite dyke swarm and which cuts the NICO alteration zones, was dated at 1869 ± 1.3 Ma (Davis et al., 2011, p.97). Considering age uncertainties, the NICO deposit and Southern Breccia corridor formed within a 8 m.y. window from 1875–1867 Ma. Acosta-Góngora et al. (2015a) also presented Re-Os molydbenite ages from both occurrences that lie within this time frame: 1865 ± 9 Ma for the NICO deposit and 1877 ± 8 Ma for the Southern Breccia. A quartz monzonite porphyry pluton located just immediately west of the Southern Breccia, yielded an age of 1867 ± 1.5 Ma (Gandhi et al., 2001). This pluton is geochemically distinct from the porphyry dykes and further highlights the abundant intrusive activity during and subsequent to formation of the ore deposit and corridor.

5.4. METHODS

Thick (~200 μm) polished thin sections were prepared from sample slabs and the resulting sections were examined using a Zeiss EVO 50 series Scanning Electron Microscope (SEM) equipped with a Backscattered Electron Detector at the Geological Survey of Canada (GSC) in Ottawa. The Oxford energy dispersive spectrometry (EDS) system includes the X-MAX 150 Silicon Drift Detector, the INCA Energy 450 software, and the latest AZtec microanalysis software. The SEM was operating at 20 kV with a beam current of 400 pA to 1 nA. Once suitable uraninite grains were located using a petrographic microscope and SEM, major elements (i.e., wt.% in oxides of Mg, Al, Si, P, Ca, Ti, V, Mn, Fe, Y, Zr, La, Ce, Pr, Nd, Sm, Pb, Th, U; Table 5.1) were

Table 5.1 Chemistry of Uraninite Determined by Electron Microprobe (wt.%) and LA-ICP-MS (ppm) from the Southern Breccia Corridor.

	SBx -Primary (n=164)				SBx - secondary (n=5)			
wt.%	Average	1 σ	Min	Max	Average	1 σ	Min	Max
MgO	0.01	0.03	0.00	0.24	0.12	0.10	0.00	0.27
Al_2O_3	0.02	0.04	0.00	0.29	0.11	0.11	0.00	0.28
SiO_2	0.06	0.16	0.00	1.19	1.02	0.34	0.50	1.39
P_2O_5	0.02	0.22	0.00	2.80	0.02	0.04	0.00	0.09
CaO	0.29	0.20	0.00	0.78	5.62	2.93	0.38	7.02
TiO_2	0.03	0.05	0.00	0.25	0.21	0.15	0.01	0.36
V_2O_3	0.01	0.02	0.00	0.19	0.01	0.01	0.00	0.03
MnO	0.03	0.05	0.00	0.22	0.26	0.21	0.00	0.48
FeO	0.14	0.21	0.00	1.49	1.26	0.30	0.86	1.61
Y_2O_3	2.21	1.08	0.00	4.72	0.22	0.17	0.06	0.49
ZrO_2	0.31	0.23	0.00	1.03	0.16	0.22	0.00	0.50
La_2O_3	0.16	0.13	0.00	0.59	0.38	0.08	0.25	0.44
Ce_2O_3	1.17	0.42	0.00	2.19	1.11	0.42	0.43	1.51
Pr_2O_3	0.15	0.14	0.00	0.64	0.00	0.01	0.00	0.02
Nd_2O_3	1.01	0.32	0.08	1.78	0.51	0.30	0.00	0.77
Sm_2O_3	0.32	0.14	0.00	0.69	0.09	0.08	0.00	0.17
PbO	19.15	1.16	14.92	22.50	3.05	0.54	2.43	3.66
ThO_2	4.30	2.76	0.25	13.03	0.02	0.04	0.00	0.10
UO_2	64.63	2.42	60.10	70.96	78.21	1.49	76.84	80.70
ppm	(n=42)				(n=8)			
La	1258	390	740	2629	4121	218	3836	4451
Ce	10159	1659	6587	12303	9215	1880	7700	12418
Pr	1914	302	1290	2436	1088	201	905	1425
Nd	11253	1620	7968	13212	5431	592	4761	6321
Sm	4012	452	2822	4536	1459	182	1259	1859
Eu	62	20	41	126	347	48	290	434
Gd	4073	346	2900	4549	1212	137	1080	1509
Tb	780	71	531	902	135	26	104	183
Dy	6010	526	4133	6779	692	150	498	963
Ho	1283	115	963	1483	110	26	76	152
Er	4392	427	2798	4939	282	65	192	386
Tm	592	73	382	718	35	8	23	47
Yb	3949	501	2681	4762	237	54	161	314
Lu	346	51	207	431	30	7	20	40

analyzed with a Cameca SX50 electron microprobe at the GSC and a JEOL 8230 electron microprobe at the University of Ottawa. Operating conditions were 20 kV accelerating voltage and 10 nA current, with 20 s on peak and 10 s off-peak counting times. A mixture of natural and synthetic pure metal, simple oxides, and simple compounds were used as standards.

Trace element concentrations (i.e., ppm of Mg, Al, Si, Ca, Sc, Ti, V, Mn, Fe, Co, Sr, Y, Zr, Nb, Cs, Ba, REE, Hf, Ta, Pb, Bi) were analyzed in-situ on polished thin sections at the Geological Survey of Canada using a Photon-Machines Analyte.193 excimer laser ablation system (LA-ICP-MS) ($\lambda = 193$ nm) with Helex ablation cell and an Agilent 7700x quadrupole ICP-MS. The laser was operated at 10 Hz, at 50% of 5 mJ for polished sections 200 μm thick whereas standard polished thin sections (~30 μm thick) were analyzed at a lower frequency of 5 Hz to reduce the ablation rate. Helium gas was used to transport ablated material from the ablation cell and was mixed with Ar plasma (flow rate of 1.05 mL/min) before entering the ICP-MS. Data acquisition cycles were 100 s in length, including 40 s of background signal monitoring prior to ablation. Standards were analyzed using a 52 μm spot size (NIST 610), whereas samples were ablated with a 26 μm diameter spot size. Element concentrations were determined using the GLITTER program (version 4.4.3, Macquarie Research Ltd.), with data acquisition adjusted to exclude data from inclusions and/or ablation of surrounding minerals. In addition to standards, runs were bracketed by analyses of GSE-1G plus BCR2 and normalized to uranium contents determined by EMPA, following methods outlined in Jackson (2008).

5.5. RESULTS

5.5.1. Field Relationships and Petrography

The U mineralized zones of the Southern Breccia corridor were differentially uplifted and tilted (Enkin et al., 2012) and record several cycles of stratabound to discordant replacement alteration, veining, and brecciation. Detailed descriptions of the alteration assemblages, including the geochemical signatures of each alteration type, are presented in Montreuil et al. (2015) and are summarized as follows. The first generation of alteration consists of white to cream-colored, stratabound and selective Na alteration that albitized Treasure Lake Group metasedimentary rocks to form white albitites when intense (Fig. 5.6a). This barren Na-alteration was overprinted by

Figure 5.6 Photographs of progressive hydrothermal alteration in the Southern Breccia corridor: (a) early, regional, stratabound, creamy albite (Ab) alteration of the metasedimentary rocks; (b) rose-colored, stratatransgressive albite that transitions into (c) and (d) precursor texture-destructive, whole-rock albite (Ab2) replacements with bright red K-Fe veinlets (arrow) and breccias (stations 11PUA-018 and 11PUA-017); (e) polished sample illustrating rose-colored albite cut by pyrite and arsenopyrite-bearing amphibole + magnetite veinlets (arrows) and variegated K-feldspar overprint along northern margin of the corridor (sample 11PUA-511G2); (f) uraninite-bearing (weathered to bright yellow secondary U phases on surface), K-feldspar + magnetite veins and breccias developed within intensely albitized metasedimentary rocks from the Red Hot zone (station 11PUA-002); and (g) polished, bright red albitite breccia from Frank's zone and autoradiograph of the same sample (h) illustrating U-bearing K-Fe (magnetite) veinlets, breccia, and disseminations overprinted by, and evolving into, LT K-Fe (hematite; hem) veinlets (arrowed) (sample 11PUA-018C2). Photographs courtesy of Geological Survey of Canada.

a second generation of stratabound to discordant Na alteration that completely transformed the albitized metasiltstone and albitite into rose-colored albitite (Fig. 5.6b, c). Next, along the northern margin of the corridor, a weak HT Ca-Fe overprint generated amphibole and magnetite veins that cut the early Na alteration generations (Fig. 5.6e). The HT Ca-Fe alteration increases in intensity to the north toward the NICO deposit but does not extend more than a few meters into the main albitite corridor west of the geophysically modeled, unexposed fault zone that defines the northern boundary of the Southern Breccia corridor. Finally, two generations of K-Fe alteration imparted the diagnostic bright red color to the showings of the Southern Breccia corridor (Fig. 5.6d, f, g). The U and polymetallic mineralization took place during the latest stage of K-Fe (magnetite) veining and brecciation that transition into a hematite-bearing K-Fe assemblage (Fig. 5.6g). Later U remobilization resulted in the formation of earthy hematite + chlorite + K-feldspar veins trending parallel to the Lou Lake fault (~045°) near the southeast shoreline of Lou Lake (Fig. 5.4).

The main U showings in the corridor are referred to as the Lou Lake, Frank's, and Red Hot zones (Fig. 5.4). In the Red Hot and Frank's zones, U is systematically associated with the presence of K-feldspar in crosscutting K-Fe veins and all the showings occur within the bright red alteration fronts. The Red Hot zone showings are characterized by U and Th enrichment with low base metal (sulphide) contents. Uraninite is the dominant U mineral and is in sulphide-poor magnetite + K-feldspar + biotite veins and breccias with accessory to minor chlorite, ilmenite, rutile, apatite, thorite, galena, titanite, and zircon (Fig. 5.7).

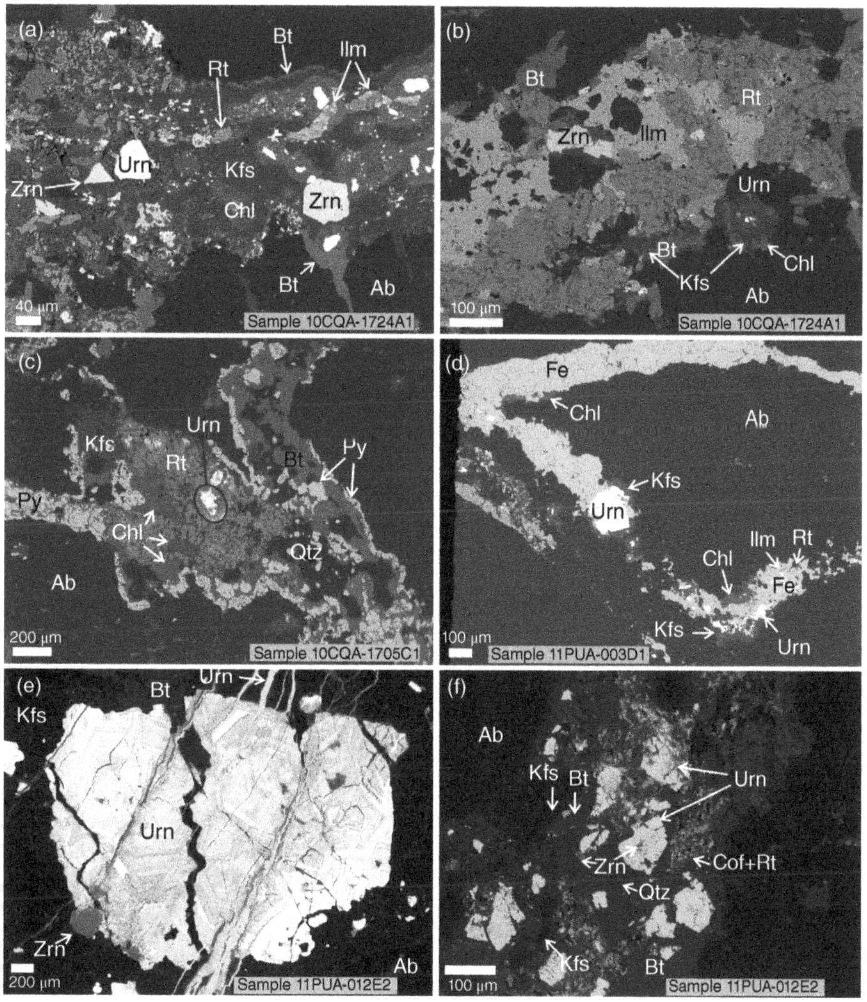

Figure 5.7 Representative backscatter electron (BSE) images of the ore and alteration mineral assemblages from the Frank's and Red Hot zones of Southern Breccia albite-altered (Ab) corridor. (a–e) Uraninite and (f) coffinite are hosted within K-feldspar (Kfs), biotite (Bt), and magnetite (Fe) veins with accessory pyrite (Py), apatite (Ap), thorite (Th), chlorite (Chl), ilmenite (Ilm), rutile (Rt), zircon (Zrn), bornite (Brn), and trace quartz (Qtz). Sample and station numbers indicated by file name. Images courtesy of Geological Survey of Canada.

Trace minerals present include: monazite, allanite, arsenopyrite, barite, wolframite, xenotime, ferrocolombite, and titanian ferrocolumbite. Less abundant U minerals include brannerite and coffinite, which occur in trace to minor concentrations as primary subhedral crystals and as alteration products of uraninite, respectively. Mimicking the evolution in Fe-oxides from magnetite to hematite, zoning of trace modal Cu minerals in the halo of the U showings also records a progressive evolution from reducing to oxidizing conditions, with bornite cores rimmed by chalcocite, covellite, and chalcopyrite (Montreuil et al., 2015).

The Frank's zone showings are characterized by U and Th enrichment accompanied by high base metal and REE concentrations. In this zone, uraninite, brannerite, and coffinite occur in sulphide-rich magnetite + K-feldspar/biotite veins and breccias with the same accessory and trace minerals as the Red Hot zone. However, Frank's zone is characterized by a greater abundance of the sulphide minerals pyrite, chalcopyrite, bornite, and molybdenite (Fig. 5.7).

The Lou Lake zone located on the northwestern edge of the corridor is distinct from both the Frank's and Red Hot zones in terms of metal endowment and alteration. Within this zone, pitchblende, coffinite, becquerelite, and fine-grained uranium-potassium-silicates (possibly compreignacite, boltwoodite, or weeksite) are hosted in late, 045°-trending earthy hematite + chlorite + K-feldspar veins that cut the albitized host rocks (Fig. 5.8). The same fault orientation, without any veining or alteration, dextrally offsets the Southern Breccia corridor at its

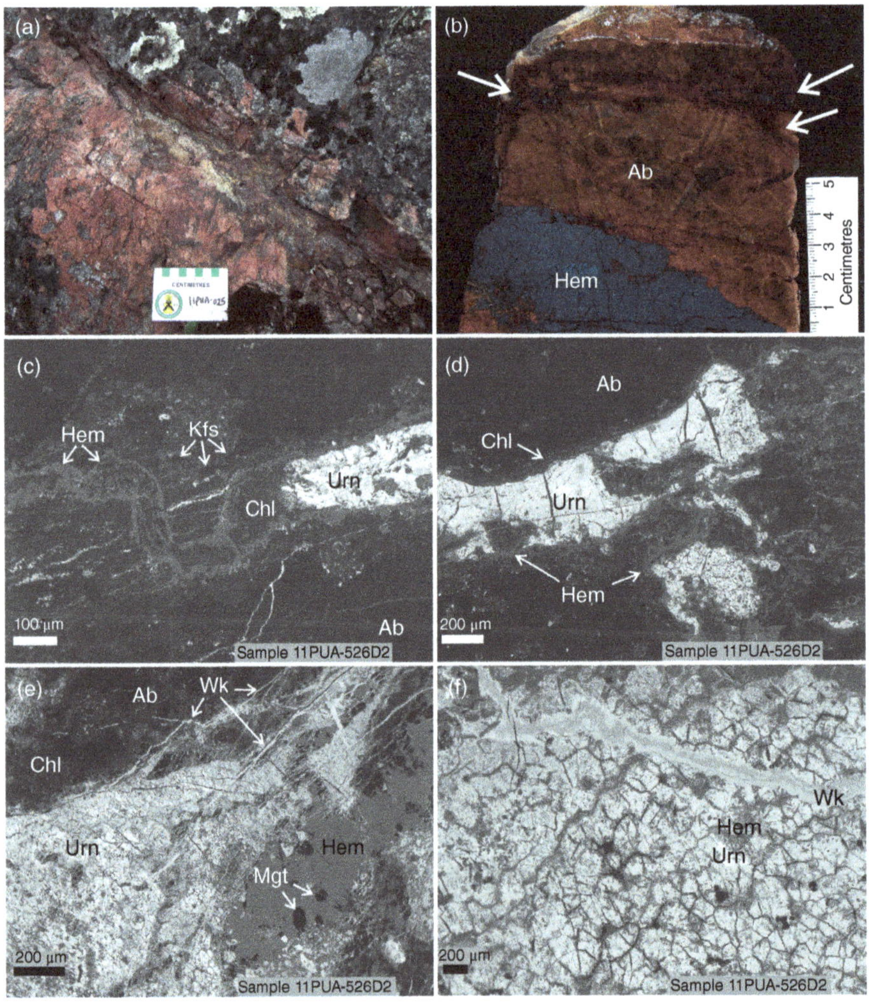

Figure 5.8 (a, b) Representative photographs of U-bearing, K-feldspar + hematite + chlorite veinlets (arrows) cutting albitized metasedimentary rocks from station 11PUA-025. A barren (no U, Th or REE) specular hematite vein is also visible in (b) formed during LT K-Fe alteration. (c–f) BSE images of U-bearing and alteration phases from the Lou Lake zone. Uraninite (Urn) and its alteration products (possibly weeksite; Wk) occur within K-feldspar (Kfs) + hematite (Hem) + chlorite (Chl) veinlets cutting the abilitized (Ab) host rocks. Images courtesy of Geological Survey of Canada.

currently exposed midpoint by ~250 m (Fig. 5.4). This zone hosts elevated Au (Gandhi & Lentz, 1990) and Bi, La, and Ce (2, 1490, 93, and 340 ppm, respectively; Corriveau et al., 2015).

5.5.2. Chemistry of Uraninite from the Southern Breccia Corridor

Analyses of primary, unaltered uraninite from the Frank's and Red Hot zones (Fig. 5.9a) yield high concentrations of rare earth elements (REE) with relatively flat chondrite-normalized REE patterns dominated by strong negative Eu anomalies. Analyses of uraninite from the Lou Lake zone yield distinct chemistries with lower REE, Th, and Y contents and fractionation of the chondrite-normalized REE pattern (Fig. 5.9b). The uraninite analyses from the Lou Lake zone also contain lower concentrations of Pb (e.g., 3.05 versus 19.15 wt. % PbO in the Frank's/Red Hot zones; Table 5.1, Fig. 5.10a), reflecting a younger crystallization age of the U-bearing minerals. This was empirically confirmed through calculation of chemical U-Th-Pb age dates on the uraninite grains, using the method

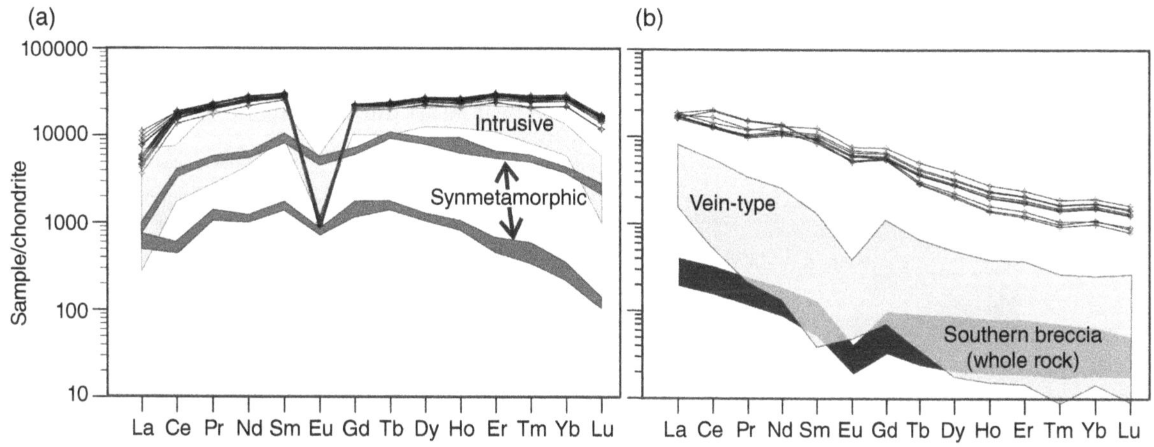

Figure 5.9 Chondrite-normalized plots of REE concentrations in uraninite from (a) Frank's and Red Hot zones and (b) the Lou Lake zone, with reference to published ranges from Mercadier et al. (2011; intrusive, synmetamorphic, and vein-type) and this study. Southern Breccia whole rock data from Corriveau et al. (2015). Chondrite normalization values from McDonough and Sun (1995).

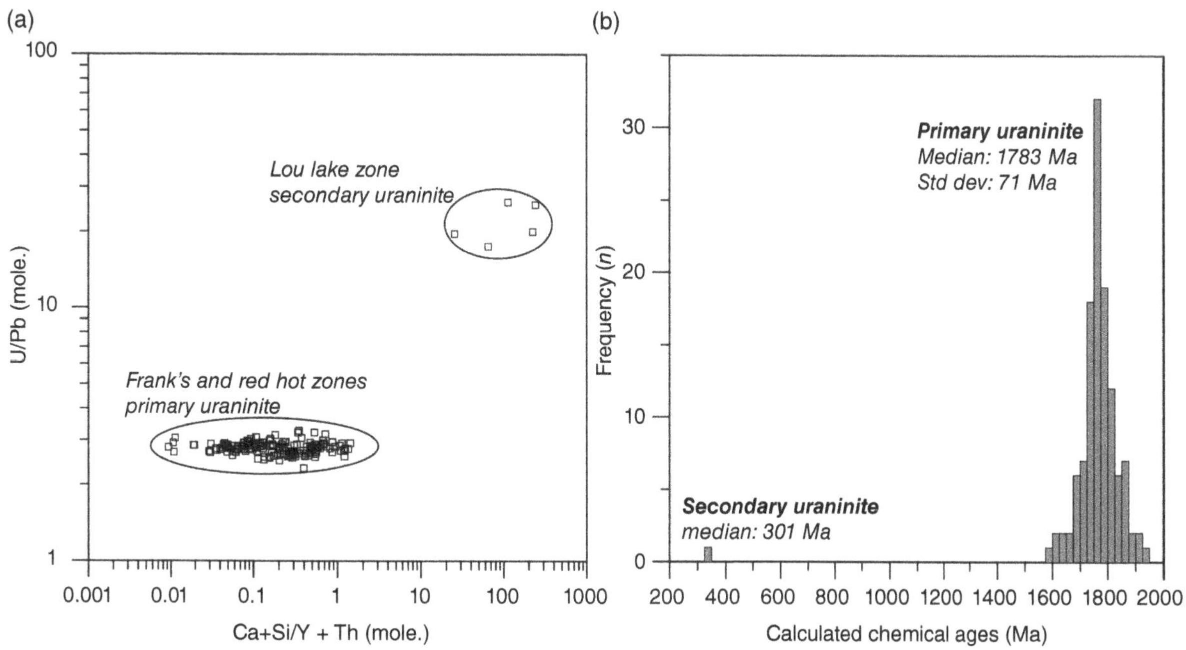

Figure 5.10 (a) Discrimination of uraninite populations based on Ca, Si, Y, Th, U, and Pb contents. (b) Summary of chemical U-Th-Pb ages calculated using the Bowles (2015) method.

of Bowles (2015), with a distinct separation of the uraninite populations: 1592–2093 Ma and 230–344 Ma (Fig. 5.10b). While these calculated chemical ages do not fall within the tightly constrained time frame defined by precise U-Pb geochronology of the host rocks, they do support the presence of two generations of uranium mineralization: a primary event associated with K-Fe alteration and a secondary remobilization along later structures.

5.6. DISCUSSION

5.6.1. Magmatism and the Great Bear Magmatic Zone IOAA Systems

The NICO and Southern Breccia corridor formed in an active tectono-magmatic-hydrothermal setting. This activity is reflected in the presence of numerous generations of porphyry and aplite dykes that were emplaced along trend of, or that cut, the extensive breccia zones in the region (Figs. 5.4, 5.11). Early dykes are also faulted and magnetite-bearing assemblages infill fault zones. The early dykes observed in the Southern Breccia corridor have synmagmatic folds and elongate boudins with tapered ends, are emplaced along the breccia fabric, were syntectonically emplaced, and are themselves invaded by subsequent K-Fe alteration (Fig. 5.11a–d; Montreuil et al., 2016a). Similarly, some early porphyry dykes display attributes of emplacement coeval with magnetite + K-feldspar alteration within the NICO deposit: all features critical for interaction between magmatic-hydrothermal and externally sourced hydrothermal fluids (basinal, meteoric) and heat influx during formation of IOCG deposits.

Figure 5.11 Field photographs of (a, b) porphyry dykes emplaced synductile and postductile deformation and IOAA alteration near the NICO deposit (station 11PUA-008); (c, d) Ductile deformed Na and Ca- Fe± K altered Treasure Lake metasedimentary rocks associated with the NICO deposit (station 11PUA-028); and (e, f) brittle breccia that formed during K-Fe alteration and overprinted albitite from the Southern Breccia corridor (stations 11PUA-018 and 11PUA-019). Photographs courtesy Geological Survey of Canada.

U-Pb zircon geochronology of host rocks and crosscutting dykes from IOCG and related mineralization throughout the GBMZ constrain the main IOAA events in the GBMZ to a narrow window between 1875 and >1866 Ma (Davis et al., 2011, p. 97; Montreuil et al., 2016a, c). This confirms that IOAA development and IOCG plus related mineralization are intimately associated with high-level intrusive activity that precedes, is coeval, and follows alteration plus mineralization. This period quickly follows initiation of the Great Bear magmatic arc after accretion of the Hottah basement to Slave craton ca. 1880 Ma (Bowring & Grotzinger, 1992; Davis et al., 2015; Ootes et al., 2015). During this period, host caldera filled complexes and stratovolcanoes were intruded by coeval and cogenetic, 1.87 Ga laccoliths and porphyries of the McTavish Supergroup and were subsequently deroofed during caldera collapse associated with intrusion of resurgent plutons (Hildebrand, 1986; Gandhi et al., 2001; Hildebrand et al., 2010). As such, the GBMZ rocks record the presence of a high-temperature regime with extreme temperature and chemical gradients that were reenergized by heat and fluids exsolved from intrusions, which ultimately facilitated metal mobility during formation of the IOAA systems.

The linkage to magmatism and mineralization is supported by the trace element concentrations of primary uraninite, with high REE contents and negative Eu anomalies reflecting precipitation from high-temperature magmatic-hydrothermal fluids that were involved in formation of the Na, HT Ca-Fe, and HT K-Fe alteration assemblages developed within the Lou IOAA system (Fig. 5.9a). As shown in Figure 5.3, Na, transitional Na Ca-Fe, and HT Ca-Fe alteration assemblages are inferred to have precipitated from 300 °C to 800 °C in IOCG districts (De Jong & Williams, 1995; Mark & Foster, 2000; Pollard, 2001; Williams et al., 2005; Oliver et al., 2006; Niiranen et al., 2007), whereas high-temperature K-Fe assemblages are reported to have formed at temperatures ranging from 350 °C to 450 °C (Beane, 1974; Bastyrakov & Skirrow, 2007; Baker et al., 2008; Skirrow, 2010; Williams, 2010b). The resulting REE signatures in uraninite are similar in shape and concentration to those of magmatic-derived U systems, a comparison also supported by moderate-to-high Th contents (up to 13.02 wt. % ThO_2; Table 5.1) in the uraninite, typical of magmatic or intrusive systems (Cuney, 2010; Mercadier et al., 2011; Frimmel et al., 2014). Secondary U-minerals from the Lou Lake zone yielded LREE-enriched patterns similar to vein-type systems where the REE patterns mirror the host rocks, such as the whole-rock data of the Southern Breccia mineralized samples (Fig. 5.9b). This interpretation is supported by lower Th and Y concentrations in uraninite. The distinct separation of the chemical U-Th-Pb uraninite age populations (Fig. 5.10b) support the presence of two generations of uranium mineralization, a primary event associated with K-Fe alteration and a secondary low temperature remobilization along later structures.

As reported by Acosta-Góngora et al. (2015a, 2018), sulphur isotope data for the GBMZ IOCG deposits and $\delta^{18}O$ values of the fluids precipitating magnetite and Co-rich arsenopyrite at NICO are also consistent with magmatic-dominated fluid origin, whether from a magmatic-hydrothermal fluid or dissolution/redeposition in a closed system. However, the authors note that $\delta^{65}Cu$ and the $\delta^{34}S$ data may record minor crustal S and Cu, reflecting crustal remobilization during the build up of the IOAA systems (i.e., metal leaching during Na alteration).

5.6.2. Linkages Between NICO and the Southern Breccia

Field relationships, whole-rock geochemistry, and geochronology all indicate that the formation of the Southern Breccia corridor was coeval with the NICO deposit (Fig. 5.12; Gandhi et al., 2001; Gandhi & van Breemen, 2005; Davis et al., 2011, p. 97; Acosta-Góngora et al., 2015a; Montreuil et al., 2015, 2016a, b; Corriveau et al., 2016; Hayward et al., 2016). In fact, the discovery of the Southern Breccia was facilitated by applying the IOAA ore system model (Fig. 5.3; Corriveau et al., 2010b, 2011, 2016) to the NICO deposit, which highlighted the absence of regional Na and low temperature K-Fe alteration in previous studies (Goad et al., 2000a, b; Sidor, 2000). The presence of regional Na alteration is considered essential in the buildup of IOAA systems and the genesis of IOCG deposits (Corriveau et al., 2016), as Na alteration zones record massive metal leaching, mass losses, and influx of voluminous, highly saline fluids at regional scales (Yadav et al., 2000; Oliver et al., 2004; Engvik et al., 2008; Kaur et al., 2012; Montreuil et al., 2012, 2013, 2015). Geochemical signatures, isotopic studies, and mass balance calculations of alteration zones associated with evolution of IOAA systems globally also support the cogenetic relationship among alteration facies at regional scales: metals that are leached during early, regional Na alteration are reprecipitated during Ca-Fe and K-Fe alteration as the fluid chemistry and temperature evolves (Mark & Foster, 2000; Oliver et al., 2004; Clark et al., 2005; Yang, 2008; Corriveau et al., 2010a, b, 2016; Skirrow, 2010; Mumin et al., 2010; Montreuil et al., 2015, 2016a, b, c; Acosta-Góngora et al., 2018). As such albitite corridors serve as metal sources and ground preparation for subsequent mineralization.

At NICO, the ores are enriched in metals that are generally not abundant in the Southern Breccia corridor

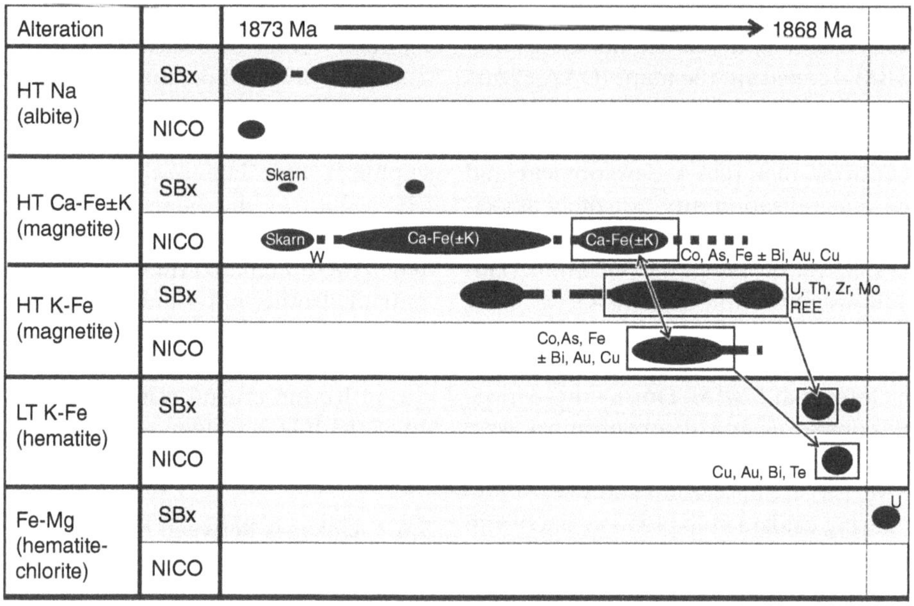

Figure 5.12 Simplified sequence of alteration and ore assemblages from the NICO deposit and Southern Breccia corridor in context of the IOAA ore system model (Fig. 5.3).

(i.e., Ni, Co, Au, Te, etc.; Fig. 5.12). Conversely, metals that are enriched in the Southern Breccia corridor are not abundant in the NICO ores (i.e., U, Th, Zr, Mo, REE, Y, etc.; Fig. 5.12). Metals present in both components of the system (mainly Cu), are generally minor phases that precipitated in HT K-Fe (NICO; Southern Breccia) alteration assemblages. Although the overall ore chemistry differs, the IOCG mineralization at NICO consists of chalcopyrite-magnetite veins while mineralization in the Southern Breccia occurs in magnetite-uraninite-chalcopyrite-pyrite veins, both of which formed during brittle deformation, crosscutting the brittle-ductile shear zones, and are cut by ca. 1868 Ma unaltered and undeformed porphyritic dyke (Montreuil et al., 2013, 2015). Metals present in hematite + chlorite + K-feldspar veining in the Lou Lake zone (i.e., U, Bi, Co, Au, and LREE) may reflect remobilization from both NICO and Southern Breccia rocks during late faulting or channeling of ore-forming fluids through the Lou Lake fault during late stages in the evolution of the regional IOAA system. This is supported by trace element patterns in uraninite from the Lou Lake zone that mirror those of whole-rock data from the Southern Breccia zone (Fig. 5.9). When viewed in the context of the IOAA sequence (Fig. 5.12), both systems have complementary alteration assemblages: Na, and K-Fe dominant in the Southern Breccia, and HT Ca-Fe plus K-Fe alteration dominant at NICO. At surface, the only HT Ca-Fe alteration recognized in the Southern Breccia corridor occurs along the northern deformed margin nearest to the NICO deposit (Fig. 5.6e; Corriveau et al., 2016).

In order to account for the lack of intense HT Ca-Fe alteration in the Southern Breccia corridor and absence of the regional carbonate sequence in the Treasure Lake Group, Montreuil et al. (2015, 2016a, b) and Hayward et al. (2016) proposed that the Southern Breccia corridor was initially located at greater depths, closer to the magmatic heat and fluid(s) sources responsible for Na alteration. Following early regional Na alteration, ductile-brittle deformation and Au-Co-Bi mineralization in the NICO deposit, progressive, and differential exhumation of the Southern Breccia corridor is interpreted to have led to widespread brittle fracturing and brecciation associated with IOCG mineralization in the NICO deposit and magnetite-uraninite veins that formed albitite-hosted U mineralization in the Southern Breccia. This relationship is supported by:

- the fault bounded nature of the albitite corridor (Fig. 5.4);
- the intensity and extent of Na alteration in the Southern Breccia corridor, which is unparalleled in exposures of altered rocks in the region;
- the IOCG imprint in the magnetite-uraninite veins of the Southern Breccia increases to the north with increasing proximity to the center of IOCG mineralization spatially centered around the NICO deposit in the Lou IOAA system;
- the transition from brittle-ductile shearing to dominantly brittle fracturing and brecciation during IOCG mineralization in NICO and uraninite-magnetite mineralization in the Southern Breccia (Figs. 5.5, 5.6, 5.11);
- post-IOCG in NICO and post-magnetite-uraninite in the Southern Breccia evolution of magnetite to hematite at the scale of the Lou IOAA system (Fig. 5.6); and

• the overall decreasing abundance of Lou assemblage volcanic rocks to the southeast (e.g., rhyolite) and exhumation of the apical part of the leucogranite.

In the Southern Breccia, the preferential formation of K-Fe alteration and U mineralization in the albitites was likely enhanced by the rheological attributes of the albitites. Compared with the surrounding anisotropic siltstones composed of alternating biotite-rich and weakly Na-altered beds, the albitites were likely mechanically weakened by the porosity formation during albitization (cf. Putnis & Austrheim, 2010) and are relatively isotropic. These rheological attributes of the albitites are interpreted to have preferentially partitioned brittle fracturing and brecciation to form albitite-hosted U mineralization where albitization and U mineralization occurred as two separate events.

5.7. CONCLUSIONS, EXPLORATION MODEL, AND TARGETS

The main phase of U exploration in the GBMZ was completed in an era (pre-1960) that focused on polymetallic vein systems, exemplified by the Port Radium, Terra, and Rayrock mines (Fig. 5.1; Kidd & Haycock, 1935; Campbell, 1955; Ruzicka & Thorpe, 1996). This era came to an abrupt end due to changing U market conditions and subsequent discovery of higher grade quartz-pebble conglomerate deposits in Elliot Lake, Ontario, and then unconformity-related deposits in the Athabasca Basin, Saskatchewan (Beck, 1969; Hoeve & Sibbald, 1978; Roscoe, 1996). While the GBMZ and related vein systems did contain localized zones of high-grade U, the limited tonnage and commonly erratic ore grade detracted from their economic feasibility.

The results of this study highlight the potential of a new U target in the GBMZ and global IOCG districts, related to development of an IOAA system that hosts an IOCG deposit(s). Critical in this updated model is the systematic precipitation of U and other metals driven by dominantly magmatic fluids during evolution of the IOAA systems, relationships proposed by Corriveau et al. (2010a, b, 2016) and further supported by field work of Mumin et al. (2010), Potter et al. (2013b), Ootes et al. (2017), and analytical results of Montreuil et al. (2013, 2016a, b), Acosta-Góngora et al. (2018), and this study. Field mapping and literature reviews document at least thirty IOAA systems in the GBMZ, some of which contain elevated U in albitized host rocks (e.g., Cole Lake and Damp showings, Fig. 5.1; Acosta-Góngora et al., 2015a; Corriveau et al., 2015). The Cole Lake U showing located 16 km to the northeast of NICO contains uraninite in magnetite, K-feldspar, amphibole, and apatite veins within an intensely Na-altered, brecciated volcanic host. These veins also contain trace chlorite, zircon, monazite, chalcopyrite, titanite, and rutile. The Damp Cu-Co-U-V showing consists of V-bearing, K-Fe-cemented hydrothermal breccias rich in U within an extensively albitized volcanic precursor (Na_2O at 9.4 wt. %; Gandhi & Prasad, 1995; Acosta-Góngora et al., 2011, 2015a).

Recognition of the commonly cryptic Na and K alteration may open a new era of U exploration in the GBMZ focused on the numerous IOAA systems, and the presence of remobilized U veins can be used as a vector for IOCG deposits within regional IOAA systems worldwide. This study also supports exploration interests for IOCG and affiliated deposits in districts that host albitite corridors such as those of Scandinavia (Kuusamo district, Bamble sector; Eilu, 2012; Slack, 2013) and India (Pal et al., 2009).

ACKNOWLEDGMENTS

This is a contribution to the Targeted Geoscience Initiative (TGI) program of Natural Resources Canada (NRCan), building on knowledge gained during the 2008–2013 Geomapping for Energy and Minerals (GEM) program. The GEM project was undertaken in collaboration with the Northwest Territories Geoscience Office, the Community Government of Gamètì, and the local Tlicho and Sahtu governments. Field research was permitted under the Aurora Research Institute Scientific Research licenses 14548, 14649, and 14548; land use permits W2009C0001 and W2010J0004 granted by the Wek'èezhìi Land and Water Board (WLWB) in accordance with the Mackenzie Valley Resource Management Act; and NWT archaeological sites database agreement # DR2009-335 and # DR2010-390. The authors would like to acknowledge the significant logistical support provided by Polar Continental Shelf program and Fortune Minerals Limited to study the entire southern Great Bear magmatic zone. Technical support from Pat Hunt (SEM), Katherine Venance (EMPA), Zhaoping Yang (LA-ICP-MS), and Simon Jackson (LA-ICP-MS) at the Geological Survey of Canada made this study possible. This contribution benefitted from a constructive review by Charlie Jefferson. Finally, the authors would like to acknowledge the Natural Sciences and Engineering Research Council and the "Fond québécois de la recherche sur la nature et les technologies" that supported the PhD studies of JFM. ESS Contribution number 20130267.

REFERENCES

Acosta-Góngora, P., Gleeson, S. A., Ootes, L., Jackson, V. A., Lee, M., & Samson, I. (2011). Preliminary observations on the IOCG mineralogy at the Damp, Fab, and Nori showings and Terra-Norex mines, Great Bear magmatic zone. NWT Geoscience Office, NWT Open File Report 2011-01.

Acosta-Góngora, P., Gleeson, S. A., Samson, I., Corriveau, L., Ootes, L., Taylor, B. E., Creaser, R. A., & Muehlenbachs, K.

(2015a). Genesis of the Paleoproterozoic NICO Iron-Oxide-Cobalt-Gold-Bismuth deposit, Northwest Territories, Canada: Evidence from isotope geochemistry and fluid inclusions. *Precambrian Research*, *268*, 168–193.

Acosta-Góngora, G. P., Gleeson, S. A., Samson, I., Ootes, L., & Corriveau, L. (2014). Trace element geochemistry of magnetite and its relationship to Cu-Bi-Co- Au-Ag-U-W mineralisation in the Great Bear magmatic zone, NWT, Canada. *Economic Geology*, *109*, 1901–1928.

Acosta-Góngora, G. P., Gleeson, S. A., Samson, I., Ootes, L., & Corriveau, L. (2015b). Gold refining by bismuth melts in the iron oxide-dominated NICO Au-Co-Bi (± Cu ± W) deposit, NWT, Canada. *Economic Geology*, *110*, 291–314.

Acosta-Góngora, P., Gleeson, S. A., Samson, I. M., Corriveau, L., Ootes, L., Jackson, S. E., Taylor, B. E., & Girard, I. B. (2018). Origin of sulfur and crustal recycling of copper in polymetallic (Cu-Au-Co-Bi-U ± Ag) iron-oxide-dominated systems in the Great Bear Magmatic Zone, NWT, Canada. *Mineralium Deposita*, *53*, 353–376.

Alexandre, P. (2010). Mineralogy and geochemistry of the sodium metasomatism-related uranium occurrence of Aricheng South, Guyana. *Mineralium Deposita*, *45*, 351–367

Azar, B. (2007). The lithogeochemistry of volcanic and subvolcanic rocks of the Fab Lake area, Great Bear magmatic zone, Northwest Territories, Canada. Bachelor thesis, University of Toronto.

Baker, T., Mustard, R., Fu, B., Williams, P., Dong, G., Fisher, L., Mark, G., & Ryan, C. (2008). Mixed messages in iron oxide-copper-gold systems of the Cloncurry district, Australia: Insights from PIXE analysis of halogens and copper in fluid inclusions. *Mineralium Deposita*, *43*, 599–608.

Bastrakov, E. N., & Skirrow, R. G. (2007). Fluid evolution and origins of iron oxide-Cu-Au prospects in the Olympic Dam district, Gawler Craton, South Australia. *Economic Geology*, *102*, 1415–1440.

Beane, R. E. (1974). Biotite stability in the porphyry copper environment. *Economic Geology*, *69*, 241–256.

Beck, L. S. (1969). Uranium deposits of the Athabasca region, Saskatchewan. Saskatchewan Department of Mineral Resources, Report 126.

Belperio, A., Flint, R., & Freeman, H. (2007). Prominent Hill: A Hematite-dominated, iron oxide copper-gold system. *Economic Geology*, *102*, 1499–1510.

Benavides, J., Kyser, T. K., Clark, A. H., Stanley, C., & Oates, C. (2008). Exploration guidelines for copper-rich iron oxide-copper-gold deposits in the Mantoverde area, northern Chile: The integration of host-rock molar element ratios and oxygen isotope compositions. *Geochemistry: Exploration, Environment, Analysis*, *8*, 343–367.

Bennett, V., & Rivers, T. (2006). U-Pb ages of zircon primary crystallization and inheritance for magmatic rocks of the southern Wopmay orogen, Northwest Territories. Northwest Territories Geoscience Office, Open Report 2006-006.

Bowles, J. F. W. (2015). Age dating from electron microprobe analyses of U, Th, and Pb: Geological advantages and analytical difficulties. *Microscopy and Microanalysis*, *21*, 1114–1122.

Bowring, S. A. (1984). Uranium-lead zircon geochronology of Early Proterozoic Wopmay Orogen, Northwest Territories, Canada: An example of rapid crustal evolution. Doctoral thesis, University of Kansas.

Bowring, S. A., & Grotzinger, J. P. (1992). Implications of new chronostratigraphy for tectonic evolution of Wopmay orogen, northwest Canadian Shield. *American Journal of Sciences*, *292*, 1–20.

Burgess, H., Gowans, R. M., Hennessey, B. T., Lattanzi, C. R., Puritch, E. (2014). Technical report on the feasibility study for the NICO gold-cobalt-bismuth-copper deposit, Northwest Territories, Canada. Fortune Minerals Ltd., NI 43-101 Technical Report No. 1335. Available at www.sedar.com.

Camier, W. J. (2002). The Sue-Dianne Fe-oxide Cu-Ag-Au breccia complex, southern Great Bear magmatic zone, Northwest Territories, Canada. Master thesis, University of Western Ontario.

Campbell, D. D. (1955). Geology of the pitchblende deposits of Port Radium, Great Bear Lake, Northwest Territories. Doctoral thesis, California Institute of Technology at Berkeley.

Clark, C., Schmidt Mumm, A., & Faure, K. (2005). Timing and nature of fluid flow and alteration during Mesoproterozoic shear zone formation, Olary Domain, South Australia. *Journal of Metamorphic Geology*, *23*, 147–164.

Coles, R. L., Haines, G. V., & Hannaford, W. (1976). Large scale magnetic anomalies over western Canada and the Arctic: A discussion. *Canadian Journal of Earth Sciences*, *13*, 790–802.

Cook, F. A., van der Velden, A. J., & Hall, K. W. (1999). Frozen subduction in Canada's Northwest Territories: Lithoprobe deep lithospheric reflection profiling of the western Canadian Shield. *Tectonics*, *18*, 1–24.

Corriveau, L. (2007). Iron oxide copper-gold deposits: A Canadian perspective. In W. D. Goodfellow (Ed.), *Mineral Deposits of Canada: A synthesis of major deposit-types, district metallogeny, the evolution of geological provinces and exploration methods* (pp. 307–328). Geological Association of Canada, Mineral Deposits Division, Special Publication 5.

Corriveau, L., & Mumin, A. H. (2010). Exploring for iron oxide copper-gold deposits: The need for case studies, classifications and exploration vectors. In L. Corriveau & A. H. Mumin (Eds.), *Exploring for Iron Oxide Copper-Gold Deposits: Canada and global analogues* (pp. 1–12). Geological Association of Canada, Short Course Notes 20.

Corriveau, L., Lauzière, K., Montreuil, J.-F., Potter, E.G., Prémont, S., & Hanes, R. (2015). Dataset of new lithogeochemical analysis in the Great Bear magmatic zone, Northwest Territories, Canada. Geological Survey of Canada, Open File 7643.

Corriveau, L., Montreuil, J.-F., & Potter, E. G. (2016). Alteration facies linkages among IOCG, IOA and affiliated deposits in the Great Bear magmatic zone, Canada. In J. Slack, L. Corriveau, & M. Hitzman (Eds.), *Proterozoic Iron Oxide-Apatite (± REE) and Iron Oxide-Copper-Gold and Affiliated Deposits of Southeast Missouri, USA, and the Great Bear Magmatic Zone, Northwest Territories, Canada* (pp. 2045–2072). *Economic Geology, 111*.

Corriveau, L., Mumin, A. H., & Montreuil, J.-F. (2011). The Great Bear magmatic zone: The IOCG spectrum and related deposit types. *Proceedings of the 11th SGA Biennial Meeting*, Antofagasta, Chile (26–29 September 2011), 524–526.

Corriveau, L., Mumin, A. H., & Setterfield, T. (2010a). IOCG environments in Canada: Characteristics, geological vectors to ore and challenges. In T. M. Porter (Ed.), *Hydrothermal iron oxide copper-gold and related deposits. A global perspective, volume 4* (pp. 311–344). Adelaide: Porter Geoscience Consultancy Publishing.

Corriveau, L., Potter, E. G., Acosta-Góngora, P., Blein, O., Montreuil, J.-F., De Toni, A. F., Day, W., Slack et al. (2017). Petrological mapping and chemical discrimination of alteration facies as vectors to IOA, IOCG, and affiliated deposits within Laurentia and beyond. *Proceedings of the 14th SGA Biennial Meeting*, Québec City (20–23 August 2017), 851–855.

Corriveau, L., Williams, P. J., & Mumin, A. H. (2010b). Alteration vectors to IOCG mineralisation: From uncharted terranes to deposits. In L. Corriveau & A. H. Mumin (Eds.), *Exploring for Iron Oxide Copper-Gold Deposits: Canada and Global Analogues* (pp. 89–110). Geological Association of Canada, Short Course Notes 20.

Craven, J. A., Roberts, B. J., Hayward, N., Stefanescu, M., & Corriveau, L. (2013). A magnetotelluric survey and preliminary geophysical inversion and visualization of the NICO IOCG deposit, Northwest Territories. Geological Survey of Canada, Open File 7465.

Cuney, M. (2010). Evolution of uranium fractionation processes through time: Driving the secular variation of uranium deposits. *Economic Geology*, *105*, 553–569.

Cuney, M., & Kyser, T. K. (2008). Deposits related to Na-metasomatism and high-grade metamorphism. In M. Cuney & T. K. Kyser (Eds.), *Recent and Not-So-Recent Developments in Uranium Deposits and Implications for Exploration* (pp. 97–116). Mineralogical Association of Canada, Short Course Series 39.

Cuney, M., Emtz, A., Mercadier, J., Mykchaylov, V., Shunko, V., & Yuslenko, A. (2012). Uranium deposits associated with Na-metasomatism from central Ukraine: A review of some of the major deposits and genetic constraints. *Ore Geology Reviews*, *44*, 82–106.

Davis, W. D., Corriveau, L., van Breemen, O., Bleeker, W., Montreuil, J.-F., Potter, E. G., & Pelleter, E. (2011). Timing of IOCG mineralising and alteration events within the Great Bear magmatic zone. *Proceedings of 39th Annual Yellowknife Geoscience Forum*. NWT Geological Survey, Yellowknife, Canada (15–18 November 2011), 97.

Davis, W. J., Ootes, L., Newton, L., Jackson, V., & Stern, R. A. (2015). Characterization of the Paleoproterozoic Hottah terrane, Wopmay Orogen using multi-isotopic (U-Pb, Hf and O) detrital zircon analyses: An evaluation of linkages to northwest Laurentian Paleoproterozoic domains. *Precambrian Research*, *269*, 296–310.

De Jong, G., & Williams, P. J. (1995). Giant metasomatic system formed during exhumation of mid-crustal Proterozoic rocks in the vicinity of the Cloncurry Fault, northwest Queensland. *Australian Journal of Earth Sciences*, *42*, 281–290.

De Toni, A. F. (2016). Les paragénèses à magnétite des altérations associées aux systèmes à oxydes de fer et altérations en éléments alcalins, zone magmatique du Grand lac de l'Ours. Master thesis, Institut National de la Recherche Scientifique.

Eilu, P. (Ed.) (2012). Mineral deposits and metallogeny of Fennoscandia. Geological Survey of Finland, Special Paper 53.

Engvik, A., Putnis, A., Fitz Gerald, J. D., & Austrheim, H. (2008). Albitization of granitic rocks: The mechanism of replacement of oligoclase by albite. *The Canadian Mineralogist*, *46*, 1401–1415.

Enkin, R., Corriveau, L., & Hayward, N. (2016). Metasomatic alteration control of petrophysical properties in the Great Bear magmatic zone (Northwest Territories, Canada). *Economic Geology*, *111*, 2073–2085.

Enkin, R., Montreuil, J. F., & Corriveau, L. (2012). Differential exhumation and concurrent fluid flow at the NICO Au-Co-Bi-Cu deposit and Southern Breccia U-Th-REE-Mo anomaly, Great Bear magmatic zone, NWT - A paleomagnetic and structural record. *Program with Abstracts, Geological Association of Canada, Mineralogical Association Canada*, St. John's, Canada (27–29 May 2012).

Frimmel, H. E., Schedel, S., & Bratz, H. (2014). Uraninite chemistry as forensic tool for provenance analysis. *Applied Geochemistry*, *48*, 104–121.

Gandhi, S. S. (1978). Geological setting and genetic aspects of uranium occurrences in the Kaipokok Bay–Big River area, Labrador. *Economic Geology*, *73*, 1492–1522.

Gandhi, S. S. (1988). Volcano-plutonic setting of U-Cu bearing magnetite veins of FAB claims, southern Great Bear magmatic zone, Northwest Territories. Geological Survey of Canada Paper 88-1C, 177–187.

Gandhi, S. S. (1989). Rhyodacite ignimbrites and breccias of the Sue-Dianne and Mar Cu-Fe-U deposits, southern Great Bear Magmatic Zone, Northwest Territories. Geological Survey of Canada Paper 89-1C, 263–273.

Gandhi, S. S. (1994). Geological setting and genetic aspects of mineral occurrences in the southern Great Bear Magmatic Zone, Northwest Territories. *Geological Survey of Canada Bulletin*, *475*, 63–96.

Gandhi, S. S. (2003). An overview of the Fe oxide-Cu-Au deposits and related deposit types. Canadian Institute of Mining and Metallurgy Conference, Montreal Canada.

Gandhi, S. S., & Lentz, D. R. (1990). Bi-Co-Cu-Au-As and U occurrences in the Snare Group metasediments and felsic volcanics of the southern Great Bear Magmatic Zone, Lou Lake, Northwest Territories. Geological Survey of Canada Paper 90-1C, 239–253.

Gandhi, S. S., & Prasad, N. (1993). Regional metallogenic significance of Cu, Mo, and U occurrences at De Vries lake, southern Great Bear magmatic zone, Northwest Territories. Geological Survey of Canada Paper 93-1C, 29–39.

Gandhi, S. S., & Prasad, N. (1995). Geological setting of Bode copper and Damp polymetallic prospects, central Great Bear magmatic zone, Northwest Territories. Geological Survey of Canada Paper 1995-C, 201–212.

Gandhi, S. S., & van Breemen, O. (2005). SHRIMP U-Pb geochronology of detrital zircons from the Treasure Lake Group; new evidence for Paleoproterozoic collisional tectonics in the southern Hottah Terrane, northwestern Canadian Shield. *Canadian Journal of Earth Sciences*, *42*, 833–845.

Gandhi, S. S., Mortensen, J. K., Prasad, N., & van Breemen, O. (2001). Magmatic evolution of the southern Great Bear continental arc, northwestern Canadian Shield, Geochronological constraints. *Canadian Journal of Earth Sciences*, *38*, 767–785.

Gandhi, S. S., Potter, E. G., & Fayek, M. (2018). New constraints on genesis of the polymetallic veins at Port Radium, Great Bear Lake, Northwest Canadian Shield. *Ore Geology Reviews*, *96*, 28–47

Goad, R. E., Mumin, A. H., Duke, N. A., Neale, K. L., & Mulligan, D. L. (2000a). Geology of the Proterozoic iron oxide-hosted, NICO cobalt-gold-bismuth, and Sue Dianne copper-silver deposits, southern Great Bear magmatic zone, Northwest Territories, Canada. In T. M. Porter (Ed.), *Hydrothermal iron oxide copper-gold and related deposits. A global perspective, volume 1* (pp. 249–267). Adelaide: Porter Geoscience Consultancy Publishing.

Goad, R. E., Mumin, A. H., Duke, N. A., Neale, K. L., Mulligan, D. L., & Camier, W. J. (2000b). The NICO and Sue-Dianne Proterozoic, iron oxide-hosted, polymetallic deposits, Northwest Territories. Application of the Olympic Dam model in exploration. *Exploration Mining Geology*, *9*, 123–140.

Groves, D. I., Bierlein, F. P., Meinert, L. D., & Hitzman, M. W. (2010). Iron oxide copper-gold (IOCG) deposits through Earth history. Implications for origin, lithospheric setting, and distinction from other epigenetic iron oxide deposits. *Economic Geology*, *105*, 641–654.

Hayward, N. (2013). 3D Magnetic Inversion of Mineral Prospects in the Great Bear Magmatic Zone, Northwest Territories. Geological Survey of Canada Open File 7421.

Hayward, N., & Corriveau, L. (2014). Fault reconstructions using aeromagnetic data in the Great Bear magmatic zone, Northwest Territories, Canada. *Canadian Journal of Earth Sciences*, *51*, 1–16.

Hayward, N., Corriveau, L., Craven, J., & Enkin, R. (2016). Geophysical signature of alteration and mineralisation envelope at the Au-Co-Bi-Cu NICO deposit, NT, Canada. In J. Slack, L. Corriveau, & M. Hitzman (Eds.), *Proterozoic Iron Oxide-Apatite (± REE) and Iron Oxide-Copper-Gold and Affiliated Deposits of Southeast Missouri, USA, and the Great Bear Magmatic Zone, Northwest Territories, Canada* (pp. 2087–2110). *Economic Geology*, *111*.

Hayward, N., Enkin, R. J., Corriveau, L., Montreuil, J-F., & Kerswill, J. (2013). The application of rapid potential field methods for the targeting of IOCG mineralisation based on physical property data, Great Bear Magmatic Zone, Canada. *Journal of Applied Geophysics*, *94*, 42–58.

Hennessey, B. T., & Puritch, E. (2008). A technical report on a mineral resource estimate for the Sue-Dianne deposit, Mazenod Lake area, Northwest Territories, Canada. Fortune Minerals Limited, National Instrument 43-101 Technical Report.

Hennessey, B. T., Puritch, E., Ward, I. R., Konigsmann, K. V., Hayden, A. S., Bocking, K. A., & Rougier, M. (2007). Technical report on the bankable feasibility study for the NICO cobalt-gold-bismuth deposit, Mazenod Lake area, Northwest Territories, Canada. Micon International Limited National Instrument 43-101 Technical Report.

Hildebrand, R. S. (1986). Kiruna-type deposits: Their origin and relationship to intermediate subvolcanic plutons in the Great Bear magmatic zone, Northwestern Canada. *Economic Geology*, *81*, 640–659.

Hildebrand, R. S., Hoffman, P. F., & Bowring, S. A. (1987). Tectono-magmatic evolution of the 1.9 Ga Great Bear magmatic zone, Wopmay Orogen, Northwestern Canada. *Journal of Volcanology and Geothermal Research*, *32*, 99–118.

Hildebrand, R. S., Hoffman, P. F., & Bowring, S. A. (2010a). The Calderian orogeny in Wopmay orogen (1.9 Ga), northwest Canadian Shield. *Geological Society of America Bulletin*, *122*, 794–814.

Hildebrand, R. S., Hoffman, P. F., Housh, T., & Bowring, S. A. (2010b). The nature of volcano-plutonic relations and shapes of epizonal plutons of continental arcs as revealed in the Great Bear magmatic zone, northwestern Canada. *Geosphere*, *6*, 812–839.

Hitzman, M. W., & Valenta, R. K. (2005). Uranium in iron oxide-copper-gold (IOCG) systems. *Economic Geology*, *100*, 1657–1661.

Hoeve, J., & Sibbald, T. I. I. (1978). On the genesis of Rabbit Lake and other unconformity-type uranium deposits in northern Saskatchewan. *Economic Geology*, *73*, 1450–1473.

Hoffman, P. F., & Hall, L. (1993). Geology, Slave craton and environs, district of Mackenzie, Northwest Territories. Geological Survey of Canada, Open File 2559.

Jackson, S. E. (2008). Calibration strategies for elemental analysis by LA-ICP-MS. In P. Sylvester (Ed.), *Laser Ablation-ICP-Mass Spectrometry in the Earth Sciences: Current Practices and Outstanding Issues* (pp. 169–188). Mineralogical Association of Canada, Short Course Series 40.

Kaur, P., Chaudhri, N., Hofmann, A. W., Raczek, I., Okrusch, M., Skora, S., & Baumgartner, L. P. (2012). Two-stage, extreme albitization of A-type granites from Rajasthan, NW India. *Journal of Petrology*, *53*, 919–948.

Kidd, D. F., & Haycock, M. H. (1935). Mineralography of the ores of Great Bear Lake. *Geological Society of America Bulletin*, *46*, 879–960.

Kissin, S. A. (1992). Five-element (Ni-Co-As-Ag-Bi) veins. *Geoscience Canada*, *19*, 113−124.

Kreiner, D., & Barton, M. D. (2011). High-level alteration in iron-oxide (-Cu-Au) (IOCG) vein systems, examples near Copiapó, Chile. Society for Geology Applied to Mineral Deposits, 11th, Antofagasta, Chile, Extended Abstracts, 497–499.

Lobato, L. M., Pimentel, M. M., Cruz, S. C. P., Machado, N., Noce, C. M., & Alkmim, F. F. (2015). U-Pb geochronology of the Lagoa Real uranium district, Brazil: Implications for the age of the uranium mineralization. *Journal of South American Earth Sciences*, *58*, 129–1006.

Mark, G., & Foster, D. R. W. (2000). Magmatic-hydrothermal albite-actinolite-apatite-rich rocks from the Cloncurry district, NW Queensland, Australia. *Lithos*, *51*, 223–245.

McDonough, W. F., & Sun, S-S. (1995). The composition of the Earth. *Chemical Geology*, *120*, 223–253.

McMartin, I., Corriveau, L., & Beaudoin, G. (2011). An orientation study of the heavy mineral signature of the NICO Co-Au-Bi deposit, Great Bear magmatic zone, Northwest Territories, Canada. *Geochemistry: Exploration, Environment, Analysis*, *11*, 293–307.

Mercadier, J., Cuney, M., Lach, P., Boiron, M-C., Bonhoure, J., Richard, A., Leisen, M., & Kiser, P. (2011). Origin of uranium deposits revealed by their rare element signature. *Terra Nova, 23*, 264–269.

Montreuil, J. -F., Corriveau, L., & Davis, W. (2016a). Tectonomagmatic evolution of the southern Great Bear magmatic zone (Northwest Territories, Canada), Implications on the genesis of iron oxide alkali-altered hydrothermal systems. In J. Slack, L. Corriveau & M. Hitzman (Eds.), *Proterozoic Iron Oxide-Apatite (± REE) and Iron Oxide-Copper-Gold and Affiliated Deposits of Southeast Missouri, USA, and the Great Bear Magmatic Zone, Northwest Territories, Canada* (pp. 2111–2138). *Economic Geology, 111*.

Montreuil, J. -F., Corriveau, L., & Grunsky, E. C. (2013). Compositional data analysis of IOCG systems, Great Bear magmatic zone, Canada: To each alteration types its own geochemical signature. *Geochemistry: Exploration, Environment, Analysis, 13*, 229–247.

Montreuil, J. -F., Corriveau, L., & Long, B. (2012). Porosity in albitites and the development of albitite-hosted U deposits: Insights from x-ray computed tomography. CT Scan workshop: developments in non-medical environments, Québec, Canada.

Montreuil, J. -F., Corriveau, L., & Potter, E. G. (2015). Formation of albitite-hosted uranium within IOCG systems: The Southern Breccia, Great Bear magmatic zone, Northwest Territories, Canada. *Mineralium Deposita, 50*, 293–325.

Montreuil, J.-F., Corriveau, L., Potter, E. G., & De Toni, A. F. (2016b). On the relation between alteration facies and metal endowment of iron oxide-alkali-altered systems, southern Great Bear magmatic zone (Canada). In J. Slack, L. Corriveau, & M. Hitzman (Eds.), *Proterozoic Iron Oxide-Apatite (± REE) and Iron Oxide-Copper-Gold and Affiliated Deposits of Southeast Missouri, USA, and the Great Bear Magmatic Zone, Northwest Territories, Canada* (pp. 2139–2168). *Economic Geology, 111*.

Montreuil, J.-F., Potter, E. G., Corriveau, L., & Davis, W. J. (2016c). Element mobility patterns in magnetite-group IOCG systems: The Fab IOCG system, Northwest Territories, Canada. *Ore Geology Reviews, 72*, 562–584.

Mumin, A. H., Somarin, A. K., Jones, B., Corriveau, L., Ootes, L., & Camier, J. (2010). The IOCG-porphyry-epithermal continuum of deposits types in the Great Bear magmatic zone, Northwest Territories, Canada. In L. Corriveau & A. H. Mumin (Eds.), *Exploring for Iron Oxide Copper-Gold Deposits: Canada and Global Analogues* (pp. 59–78). Geological Association of Canada, Short Course Notes 20.

Niiranen, T., Poutiainen, M., & Mänttäri, I. (2007). Geology, geochemistry, fluid inclusion characteristics, and U-Pb age studies on iron-oxide-Cu-Au deposits in the Kolari region, northern Finland. *Ore Geology Reviews, 30*, 75–105.

Normandeau, P. X., Harlov, D. E., Corriveau, L., Paquette, J., & McMartin, I. (2018). Characterization of fluorapatite within iron oxide alkali-calcic alteration systems of the Great Bear magmatic zone: A potential metasomatic process record. *The Canadian Mineralogist*, in press.

Oliver, N. H. S., Mark, G., Pollard, P. J., Rubenach, M. J., Bastrakov, E., Williams, P. J., Marshall, L. C., et al. (2004). The role of sodic alteration in the genesis of iron oxide-copper-gold deposits: Geochemistry and geochemical modelling of fluid-rock interaction in the Cloncurry district, Australia. *Economic Geology, 99*, 1145–1176.

Oliver, N. H. S., Rubenach, M. J., Baker, B. F., Blenkinsop, T. G., Cleverley, J. S., Marshall, L. J., & Ridd, P. J. (2006). Granite-related overpressure and volatile release in the mid crust: Fluidized breccias from the Cloncurry district, Australia. *Geofluids, 6*, 346–358.

Omlin, S., Malvoisin, B., & Podladchikov, Y. Y. (2017). Pore fluid extraction by reactive solitary waves in 3-D. *Geophysical Research Letters, 44*, 9267–9275.

Ootes, L., Davis, W. J., Jackson, V. A., & van Breemen, O. (2015). Chronostratigraphy of the Hottah terrane and Great Bear magmatic zone of Wopmay Orogen, Canada, and exploration of a terrane translation model. *Canadian Journal of Earth Sciences, 52*, 1062–1092.

Ootes, L., Goff, S., Jackson, V. A., Gleeson, S. A., Creaser, R. A., Samson, I. M. S., Evensen, N., Corriveau, L., & Mumin, A. H. (2010). Timing and thermo-chemical constraints on multi-element mineralization at the Nori/RA Cu-Mo-U prospect, Great Bear magmatic zone, Northwest Territories, Canada. *Mineralium Deposita, 45*, 549–566.

Ootes, L., Harris, J., Jackson, V. A., Azar, B., & Corriveau, L. (2013). Uranium-enriched bedrock in the central Wopmay orogen: Implications for uranium mineralization. In E. G. Potter, D. Quirt, & C. W. Jefferson (Eds.), *Uranium in Canada: Geological environments and exploration developments* (pp. 85–103). *Exploration and Mining Geology, 21*.

Ootes, L., Snyder, D. B., Davis, W., Acosta-Gongora, P., Corriveau, L., Mumin, A. H., Gleeson, S. A., Samson, I. M., Montreuil, J.-F., Potter, E. G., & Jackson, V. A. (2017). A Paleoproterozoic Andean-type iron oxide copper-gold environment, the Great Bear magmatic zone, Northwest Canada. *Ore Geology Reviews, 81*, 123–139.

Pal, D. C., & Chaudhuri, T. (2016). Radiation damage-controlled localization of alteration haloes in albite: Implications for alteration types and patterns vis-à-vis mineralization and element mobilization. *Mineralogy and Petrology, 110*, 823–843.

Pal, D. C., Barton, M. D., & Sarangi, A. K. (2009). Deciphering multistage history affecting U-Cu (-Fe) mineralization in the Singhbhum Shear Zone, eastern India using pyrite textures and compositions in the Turamdih U-Cu (-Fe) deposit. *Mineralium Deposita, 44*, 61–80.

Polito, P. A., Kyser, T. K., & Stanley, C. (2009). The Proterozoic, albitite-hosted, Valhalla uranium deposit, Queensland, Australia: A description of the alteration assemblage associated with uranium mineralization in diamond drill hole V39. *Mineralium Deposita, 44*, 11–40.

Pollard, P. J. (2001). Sodic (-calcic) alteration in Fe oxide-Cu-Au districts: An origin via unmixing of magmatic H_2O-CO_2–NaCl ± $CaCl_2$-KCl fluids. *Mineralium Deposita, 36*, 93–100.

Porter, T. M. (2010). Current understanding of iron oxide associated-alkali altered mineralised systems. Part 1: An overview. In T. M. Porter (Ed.), *Hydrothermal iron oxide copper-gold and related deposits: A global perspective, volume 3* (pp. 5–32). Adelaide: Porter Geoscience Consultancy Publishing.

Porto da Silveira, C. L., Schorscher, H. D., & Miekeley, N. (1991). The geochemistry of albitization and related uranium mineralization, Espinharas, Paraiba (PB), Brazil. *Journal Geochemistry Exploration*, *40*, 329–347.

Potter, E. G., Corriveau, L., & Montreuil, J. -F. (2013b). Iron oxide copper-gold ± uranium in the Great Bear magmatic zone: Nature of uranium in IOCG systems. Geological Survey of Canada, Open File 7254.

Potter, E. G., Montreuil, J. -F., Corriveau, L., & De Toni, A. (2013a). Geology and hydrothermal alteration of the Fab Lake region, Northwest Territories. Geological Survey of Canada, Open File 7339.

Poulet, T., Karrech, A., Regenauer-Lieb, K., Fisher, L., & Schaubs, P. (2012). Thermal-hydraulic-mechanical-chemical coupling with damage mechanics using ESCRIPTRT and ABAQUS. *Tectonophysics*, *526–529*, 124–132.

Putnis, A. (2015). Transient porosity resulting from fluid-mineral interaction and its consequences. *Reviews in Mineralogy and Geochemistry 80*, 1–23.

Putnis, A., & Austrheim, H. (2010). Fluid-induced processes: Metasomatism and metamorphism. *Geofluids*, *10*, 254–269.

Reichenbach, I. G. (1991). The Bell Island bay Group, remnant of an early Proterozoic ensialic marginal basin in Wopmay orogeny, District of Mackenzie. Geological Survey of Canada Paper 88–28.

Roscoe, S. M. (1996). Paleoplacer uranium, gold. In O. R. Eckstrand, W. D. Sinclair, & R. I. Thorpe (Eds.), *Geology of Canadian mineral deposit types* (pp. 10–23). Geological Survey of Canada, Geology of Canada no. 8.

Ruzicka, V., & Thorpe, R. I. (1996). Arsenide vein silver, uranium. In O. R. Eckstrand, W. D. Sinclair, & R. I. Thorpe (Eds.), *Geology of Canadian mineral deposit types* (pp. 296–306). Geological Survey of Canada, Geology of Canada no. 8.

Sidor, M. (2000). The origin of black rock alteration overprinting iron-rich sediments and its genetic relationship to disseminated polymetallic sulphide ores, Lou Lake, Northwest Territories, Canada. Master Thesis, University of Western Ontario.

Skirrow, R. (2010). "Hematite-group" IOCG ± U ore systems. Tectonic settings, hydrothermal characteristics, and Cu-Au and U mineralizing processes. In L. Corriveau & A. H. Mumin (Eds.), *Exploring for iron oxide copper-gold deposits: Canada and global analogues* (pp. 39–58). Geological Association of Canada, Short Course Notes 20.

Skirrow, R. G., Creaser, R., & Hore, S. B. (2011). Mt Gee-Armchair U-REE deposits, South Australia. In R. G. Skirrow (Ed.), Uranium mineralisation events in Australia: Geochronology of the Nolans Bore, Oasis, Kintyre, Mt Gee-Armchair, and Maureen uranium deposits (pp. 36 58). Geoscience Australia Record 2011/12.

Skirrow, R. G., Jaireth, S., Huston, D. L., Bastrakov, E. N., Schofield, A., van der Wielen, S. E., & Barnicoat, A. C. (2009). Uranium mineral systems: Processes, exploration criteria and a new deposit framework. Geoscience Australia Record 2009/20.

Slack, J. (2013). Descriptive and geoenvironmental model for cobalt-copper-gold deposits in metasedimentary rocks. U.S. Geological Survey Scientific Investigations Report 2010-5070-G.

Sparkes, G. W. (2017). Uranium mineralization within the Central Mineral Belt of Labrador: A summary of the diverse styles, settings and timing of mineralization: Government of Newfoundland and Labrador, Department of Natural Resources, Geological Survey, St. John's, Open File LAB/1684.

Wilde, A. (2013). Towards a model for albitite-type uranium. *Minerals*, *3*, 36–48.

Williams, P. J. (2010a). Classifying IOCG deposits. In L. Corriveau & A. H. Mumin (Eds.), *Exploring for iron oxide copper-gold deposits: Canada and global analogues* (pp. 13–22). Geological Association of Canada, Short Course Notes 20.

Williams, P. J. (2010b). "Magnetite-group" IOCGs with special reference to Cloncurry (NW Queensland) and Northern Sweden. Settings, alteration, deposit characteristics, fluid sources, and their relationship to apatite-rich iron ores. In L. Corriveau & A. H. Mumin (Eds.), *Exploring for iron oxide copper-gold deposits: Canada and global analogues* (pp. 23–38). Geological Association of Canada, Short Course Notes 20.

Williams, P. J., Barton, M. D., Johnson, D. A., Fontboté, L., de Haller, A., Mark, G., Oliver, N. H. S., et al. (2005). Iron oxide copper-gold deposits; geology, space-time distribution, and possible modes of origin. *Economic Geology*, *100*, 371–406.

Yadav, O. P., Jain, R. B., Singh, R., Singh, G., Sharma, D. K., & Fahmi, S. (2000). Geology and geochemistry of uraniferous albitites of the middle Proterozoic Delhi Supergroup, Rajasthan, India. In K. C. Udaipur Gyani, & P. Katariya (Eds.), *Tectonomagmatism, geochemistry and metamorphism of Precambrian terrains* (pp. 303–320). Mohanlal Sukhadia University.

Yang, C. L. (2008). Progressive albitisation in the "Migmatite Creek" region, Weekeroo Inlier, Curnamona. Master thesis, University of Adelaide.

Section II
New Methods for Mineral Exploration

6

Cathodoluminescence Applied to Ore Geology and Exploration

Jean-Marc Baele[1], Sophie Decrée[2], and Brian Rusk[3]

ABSTRACT

The cathodoluminescence (CL) of minerals reflects with great sensitivity the physiochemical conditions in mineralized systems and their evolution through time. CL textures, colors, and spectra provide Earth scientists with unique signatures that can decipher multistage mineralization, fingerprint ore deposit type, relate ore deposition to specific minerals or mineralizing episodes, thus enhancing our understanding of ore deposits and improving exploration strategies. CL also enables fast identification of luminescent minerals and represents an excellent, if not required, complementary technique to in-situ geochemistry, geothermometry, and geochronology. Reports of the application of CL to ore deposits from magmatic, hydrothermal, and sedimentary affinities are briefly presented, including alkaline and carbonatite complexes, iron-oxide copper gold (IOCG) and Kiruna-type deposits, porphyry–Cu-Mo-W, and epithermal, Granite-related (Sn-W and U), orogenic and Carlin-type gold deposits, volcanic massive sulfides (VMS), sedimentary-exhalative (Sedex), Mississippi-Valley type (MVT), low-T fluorite deposits, and phosphates in sedimentary environments. In these ore deposits, quartz and apatite appear as extremely versatile minerals, with high potential for CL research, because they can form in a wide range of geological environments, during virtually all evolutionary stages in a mineralized system. Other minerals, such as fluorite, calcite/dolomite, zircon, feldspars, scheelite, cassiterite, and sphalerite, are also of great interest but in more specific contexts or for more specific applications.

6.1. INTRODUCTION

Cathodoluminescence (CL) has a wide range of applications in the field of ore geology. This technique is particularly useful for revealing the internal crystal texture, growth history of minerals, and other features that are not or not easily visible using other analytical methods (e.g., Götze, 2012). In addition, it provides essential information to reconstruct the processes and formation environment of various minerals in a wide range of geological contexts. CL allows a rapid visualization and identification of luminescent minerals, even when they are present as accessory phases. This provides Earth scientists with a powerful method for fingerprinting and provenance analysis (e.g., Brokus et al., 2015; Götze et al., 1999; Scholonek & Augustsson, 2016; Baele et al., 2016).

Together with CL imaging, CL spectroscopy (spectral measurement of the CL emission) yields information about the chemistry of minerals, which typically present various CL emission lines (and resulting CL colors) depending on their trace element content (e.g., Marshall, 1988; Barbarand & Pagel, 2001; Götze, 2012; Coulson, 2014). Hence, CL can bridge the gap between petrography and geochemistry. This is not always straightforward, however, since CL activation is a complex phenomenon, which also involves crystal defects. This topic is not completely understood in various aspects, even for very common minerals such as quartz.

[1] *Department of Geology and Applied Geology, University of Mons, Mons, Belgium*

[2] *Geological Survey of Belgium, Royal Belgian Institute of Natural Sciences, Brussels, Belgium*

[3] *Department of Geology, Western Washington University, Bellingham, Washington, USA*

Ore Deposits: Origin, Exploration, and Exploitation, Geophysical Monograph 242, First Edition.
Edited by Sophie Decrée and Laurence Robb.

In a number of geological studies, CL images nicely document heterogeneities in minerals, and provide critical textural context for assisting and interpreting in-situ geochemical, geothermometrical, and geochronological microanalyses, especially with the LA-ICP-MS (Laser ablation-Inductively coupled Plasma-Mass Spectrometry) and the EPMA (Electron-Probe Micro-Analysis).

After some generalities about cathodoluminescence, this paper discusses a few key topics: fingerprinting and provenance analysis, alteration and geochemical guides, and radiation damage in quartz. Then a series of case studies where CL was applied to ore deposits are briefly presented according to a widely accepted typology of mineralized systems, from magmatic to hydrothermal to sedimentary affinities. Although we have chosen to focus on metallic ores and associated nonmetallic substances such as fluorite and phosphates, CL may also have applications to other geomaterial resources, such as industrial minerals (quartz and silica in general, andalusite, talc, etc.) and building materials (aggregates, lime industry, dimensional stones).

6.2. CATHODOLUMINESCENCE (CL)

6.2.1. Principle of CL

Cathodoluminescence is the emission of light, which occurs in insulating and semiconducting solids when they are bombarded (excited) with high-energy electrons (Marshall, 1988; Yacobi & Holt, 1990). Commonly reported excitation conditions of electron beams used in CL work are 10 to 20 kV acceleration voltage and 5 to 15 μA/mm^2 current density. The electrons of the material are excited by incident electrons and their subsequent deexcitation (or relaxation) is accompanied with the emission of photons (radiative deexcitation). Excitation/deexcitation processes can involve the valence and conduction energy bands, which are separated by an energy band gap, as well as any other energy level due to the presence of defects in the crystal (Fig. 6.1). Direct deexcitation from the conduction to the valence band results in a CL emission mainly in the UV range because the amount of energy released is high. Deexcitation from/to intermediate energy levels (called traps) caused by crystal defects or chemical impurities involves the transfer of less energy than direct deexcitation and mostly produces visible CL. With multiple traps, which is the general situation, deexcitation cascades where the smallest steps are nonradiative, that is, they do not produce photons but phonons (vibrational energy transferred to the host lattice, which physically expresses as heat). Intrinsic CL refers to direct band-to-band or defect-induced deexcitation. The CL that is produced by the additional energy levels of substitutional chemical impurities (influenced or not by the host lattice, cf. crystal field effect below) is called extrinsic CL and such impurities act as luminescence centers. Chemical impurities in either substitution or insertion can also cause lattice distortion or enhance certain defects that favor CL emission but without acting, themselves, as luminescence center. One interest of CL is thus to make it possible to detect and characterize the defects in minerals, including chemical impurities, and interpret

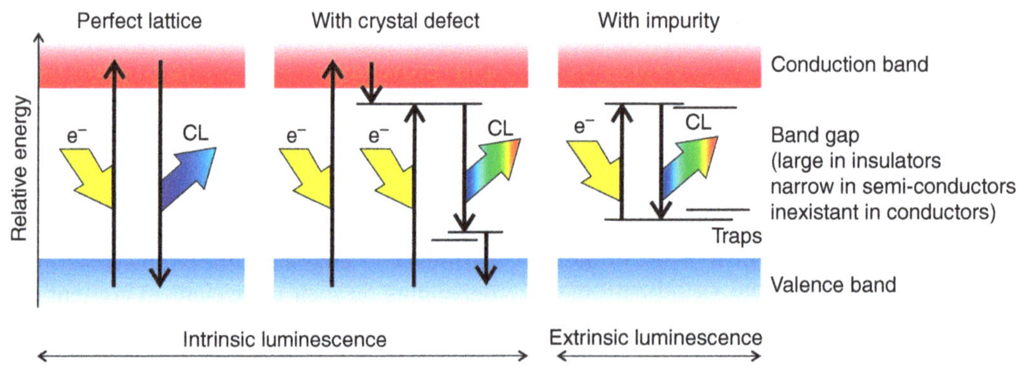

Figure 6.1 Simplified sketch of excitation/deexcitation processes responsible for CL emission in insulator and semiconductor materials. When bombarded by energetic electrons (e-), electrons from the material may be promoted to a higher energy (excited) state. Ground states are in the valence band and excited states in the conduction band. The direct, band-to-band deexcitation is accompanied by the emission of photons with relatively large energy, that is, in the UV-violet range. This is the only luminescence mechanism in perfectly pure and defect-free materials. Crystal defects introduce additional energy levels (electron traps), which lower the deexcitation energy and results in CL emission in the visible or near infrared. Smaller deexcitation steps (bare arrows) produce non-CL vibrational energy (phonons). Chemical impurities substituting for major elements in the crystal lattice introduce more energy levels and new ground states according to their own electronic structure (which may be slightly modified by crystal field effect). Only one or two possible paths are shown here for the sake of clarity.

them in terms of formation environment or use it as a fingerprint of a specific environment or physiochemical condition of formation.

6.2.2. CL Activation

In minerals, lattice defects and chemical impurities always occur in variable concentration and some are CL-active at trace level (starting at the low-ppm range; Marshall, 1988). A variety of point defects associated with CL emission have been identified in quartz and are still under investigation (e.g., Steven-Kalceff et al., 2000; Götze, 2010; Götze et al., 2015). Chemical impurities substituting for major elements in a host mineral have different CL behaviors depending on their electronic structure and concentration. CL activators act as luminescence centers owing to their favorable electronic structure. Typical activators are Mn^{2+}, Fe^{3+}, Ti^{4+}, Cr^{3+}, $REE^{3+/2+}$, and so on (Marshall, 1988). Some CL-activators, such as REE^{3+} and Cr^{3+}, have approximately the same emission wavelengths whatever their host. On the contrary, other activators such as Mn^{2+} luminesce with different wavelengths depending on the host mineral because their electronic orbitals and the associated energy levels involved in the CL process are influenced or deformed by the crystal environment (crystal field effect). As a result, Mn^{2+} activates yellow/green CL in aragonite, orange in calcite, and red in magnesite and most dolomite. However, yellow/green Mn-activated CL also occurs in other minerals such as apatite and plagioclase.

Spectroscopy is needed to reliably identify a specific activator, especially if the resulting global CL color is dominated by other activators. For example, the brown CL, which is typical of plagioclase, is due to both Mn^{2+} activation in the green and Fe^{3+} in the red region of the spectrum (around 570 and 710 nm, respectively, depending on the composition; see for example Kayama et al., 2009). In general, spectroscopic CL is most beneficial to the determination of CL-activators and evaluating their impact on the resulting CL color, which can be misleading because a similar CL color can be produced by different emission spectra while, in contrast, different CL colors can result from slight changes in a spectrum.

There is no simple relationship between the concentration of an activator and its CL intensity, which prevents the use of CL for chemometry, except in very particular cases such as the quantification of Mn in carbonate, which is possible with detection limit as low as 0.1–0.3 ppm (Habermann et al., 2000). In addition, the electron beam is not a passive probe, as it may also induce heating, ionic diffusion, decomposition (amorphization), and the creation of new color centers (e.g., Townsend & Rowlands, 2000; Stevens-Kalceff et al., 2000).

Temporal changes in the intensity of some CL emissions are observed in some samples. A typical example is quartz, where its common blue CL shifts to red with bombardment time (Marshall, 1988, and references therein). This is due to a decrease of the 450 nm (blue) peak emission and an increase of the 650 nm (red) peak upon irradiation (Zinkernagel, 1978), which is also observed in alpha-irradiated quartz (e.g., Rusk et al., 2006; Götze, 2010). While this process may take more than 10 s to be noticeable under classical excitation conditions (i.e., 15 kV acceleration voltage and 10 $\mu A/mm^2$ beam current density), some other emissions may extinguish within only a few seconds upon electron bombardment. A buildup of the heat generated by the electron beam also contributes to a gradual decrease in intensity of many CL emissions (Marshall, 1988).

When they are present in the same crystal structure, not all the CL activators will show up in emission spectra because they have different luminescence efficiencies and they may interact with each other. The interaction involves energy transfer mechanisms, which can weaken (quenchers, see below) or enhance (sensitizers) the CL emission of those activators that transfer/receive energy to/from other ions or lattice defects. REE^{3+} are a striking example in this respect. Each REE has a characteristic CL emission spectrum as evidenced by systematic analysis of doped synthetic minerals (Blanc et al., 2000; Mitchell, 2014). However, in REE-bearing natural minerals, only a few REE are ubiquitously observed in CL spectra, namely Ce^{3+}, Eu^{2+}, Dy^{3+}, Sm^{3+}, Nd^{3+}, and, more rarely, Tb^{3+}, Er^{3+}, and Eu^{3+} (the CL of Ce^{3+} and Nd^{3+} may not be observable depending on the instrument used as they luminesce in the UV and NIR regions, respectively). Thus, not all the REE that are present in a mineral can be detected by CL but useful information can still be retrieved such as an evaluation of the relative light to heavy REE enrichment using the relative intensity of the emission lines of Sm^{3+} (or Nd^{3+}) and Dy^{3+} in apatite (Roeder et al., 1987) and in fluorite (Baele et al., 2012).

6.2.3. CL Quenching

In too high concentration, some activators may have adverse effects on CL emission, which decreases in intensity when a certain concentration threshold is exceeded. This process is known as self or concentration quenching. For Mn^{2+} in calcite, for instance, concentration quenching starts at the 0.1–1% level (Marshall, 1988). An example of concentration quenching with Nd^{3+} in monazite is given in section 6.4.4. The reason behind self-quenching is that when present in sufficiently high concentration, activator ions are in such close proximity in the host crystal that they can absorb CL photons emitted by their neighbors (Marshall, 1988). This mutual

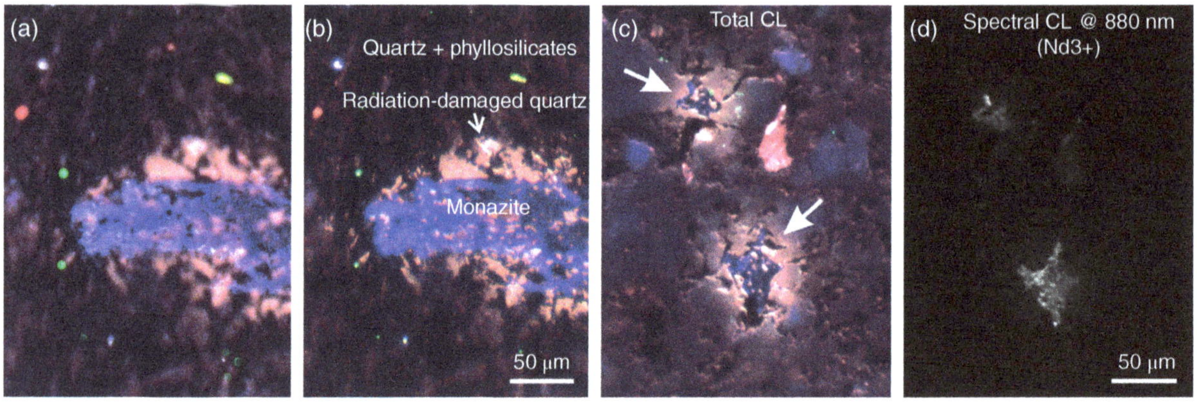

Figure 6.2 Cathodoluminescence imaging of low-grade metamorphic monazite ("grey monazite") from Lower Paleozoic shales of Ardenne (a) and (b) and Brabant (c) and (d), Belgium. (a) Monazite, which luminesces blue, is in sharp focus while the background is out of focus. (b) Same as (a) but with the microscope set to focus on the background, which comprises quartz (brown CL) and phyllosilicates (dark). The observed difference in focus is likely caused by a different electron penetration depth and/or activation energy. Note the bright luminescence of the quartz surrounding monazite due to alpha-particle damages (radiation-induced CL halo). (c) Small monazite aggregates with dark-blue CL are readily detected by their radiation-induced halo in quartz. (d) The spectral imaging of the Nd^{3+} emission at ca. 880 nm highlights the monazite grains shown in (c). Note that the tail of the emission from the irradiated quartz and the red-luminescent albite grain is captured as well.

interaction of the activators increases the internal energy cycling and results in a reduction of the CL intensity.

Substitutional foreign ions such as Fe^{2+}, Co^{2+}, and Ni^{2+} also decrease the CL intensity at concentration levels similar to activators and are called quenchers or inhibitors (Marshall, 1988). The quenching of Mn^{2+} activation by Fe^{2+} is ubiquitous in carbonates but the interplay with self-quenching and other factors including sensitizing (coactivation) by other ions such as Pb^{2+} or Ce^{3+} is complicated and not well understood (Machel, 1985; Machel, 2000). Quenching is an important limiting factor to the applicability of CL because minerals containing Fe^{2+} as a major constituent are not luminescent.

6.2.4. Sample Preparation

A variety of sample types can be studied under CL although uncovered, polished thin-sections and polished slabs are the most common. CL images are sharper and crisper than conventional optical images in transmission mode because incident electrons penetrate only a few μm into the material depending on beam acceleration voltage and the mean atomic number of the target material. However, this results in an enhancement of surface imperfections, especially if the electron beam is at an angle with the sample surface, which creates a shadowing effect. In addition, surface irregularities scatter light and this degrades the quality of CL images. Therefore, the quality of polishing is a critical issue in CL work. Also, since the depth of field of many microscope objectives is very narrow, blurring in CL images often results from the lack of planarity and/or horizontality of the observed sample surface. Note that even under good surface and geometrical conditions of the sample, some minerals can be blurred (out of focus) while the other are in sharp focus. This effect, which was notably observed with monazite (Fig. 6.2) and albite, is possibly due to the influence of the mean atomic number of the target material on electron penetration depth and the excitation energy that is required for the different activators.

6.2.5. Instrumentation and Methods

For detailed information on CL instrumentation, the reader is directed to Marshall (1988), Boggs and Krinsley (2000), and Götze and Kempe (2008). CL instruments fall into two basic categories depending on the excitation source, namely cold and hot cathode electron guns. With cold cathode electron guns, the electron beam is an electrical discharge sustained in moderate vacuum conditions (low pressure gas such as air, helium, or argon) that are achieved with a single pumping stage. Hot cathode electron guns work under higher vacuum by thermal emission from a tungsten filament, LaB6 single crystal cathode or newer field emission guns, which all require two or three pumping stages. Cold cathode guns are simpler and cheaper than hot cathode and they are mounted on an optical microscope (OM-CL), which can alternatively be equipped with a hot cathode gun (Neuser et al., 1995). Hot cathode guns produce higher excitation conditions (up to 30 kV or more) and a more stable beam. However, they are more expensive and require the application of a conductive coating on the specimen. The guns in scanning electron microscopes (SEM) and electron microprobes (EPMA) are widely used for hot cathode CL work (SEM-CL), which takes advantage of the analytical capability and flexibility of these instruments (BSE, EDS, WDS, EBSD, etc.). High magnification and access to the UV emission are other advantages of SEM-CL over OM-CL, although in many applications, having a large field of view is desirable (cf. the frequent use of stitching of SEM-CL micrographs).

While this review is largely based on literature, most data in the figures were acquired in our lab at University of Mons, Belgium, with a cold cathode CITL CL system (Cambridge Image Technology model Mk5, UK) operated at 15 kV and 500 μA acceleration voltage and beam current, respectively. With these settings, the unfocused beam has a current density of 8 $\mu A/mm^2$. Helium is used instead of air in order to improve beam stability and to help reduce specimen heating. CL images were captured with a Peltier-cooled digital color camera (Lumenera model Infinity 4, Canada), set from 0.5 s to a few seconds exposure time depending on CL intensity. Multiple frame averaging is routinely used to reduce noise unless short-lived CL is expected. Color calibration of the camera (white balance) is regularly performed using the blue-filtered, tungsten-halogen light source of the microscope, which may result in CL colors that are slightly different from other equipment (especially around the yellow band, which is narrow) but ensures more or less standardized observation conditions. Panchromatic CL images of hydrothermal quartz were acquired at Western Washington University, using a Vega TS 5136 SEM equipped with a Deben Centaurus panchromatic CL detector. Acceleration voltage was 15kV, and beam current typically varies between 1 and 10 nA, depending on CL intensity.

Inserting optical bandpass filters with appropriate central wavelengths in the lightpath provides a simple method for CL spectral imaging of specific activators (Fig. 6.2). The camera is then set to monochrome mode by acquiring in 2 x 2 binning mode, which also increases sensitivity but at the expense of resolution. Spectra should be examined before doing spectral imaging as this method does not discriminate the CL of interest from other potentially overlapping emissions. This method is thus different from "true" hyperspectral imaging, where a complete emission spectrum is acquired from every point in a raster image (which is only possible with scanning electron microscopes and EPMA), and where peak fitting can be used to correct for overlapping peaks (e.g., Vasyukova et al., 2013).

CL spectra were recorded using a CITL COS 8200 system equipped with a Avantes fiber spectrometer with 350–1100 nm and 4 nm spectral range and resolution, respectively. The detector is a Peltier-cooled 1024 pixels CCD array. The spectrometer is wavelength-calibrated using a Hg-Ar lamp but not corrected for relative intensity (system response).

6.3. CATHODOLUMINESCENCE APPLIED TO ORE EXPLORATION

6.3.1. Fingerprinting and Provenance Analysis

Cathodoluminescence allows for the observation of a variety of primary and secondary textures in crystals such as growth and sector zoning, alteration, fractures and other strain textures, and radiation-damage halos that are either not visible with other techniques or would necessitate more analytical efforts and costs (Figs. 6.3 and 6.4). These characteristics reflect the specific genetic environment and constitute a unique record of the complex processes that lead to the formation of the deposit. Therefore, they can be used for fingerprinting a specific deposit or deposit type (Fig. 6.5). Minerals derived from the primary mineralization in exploration materials such as soil, stream, and shore sediments can then be traced with CL. In this context, heavy/resistate minerals survive erosion and weathering, keeping their typical primary CL signature. For example, zircon luminescence can be used for diamond exploration (Belousova et al., 1998, 2002), and apatite and scheelite for other deposits (e.g., Kempe & Götze, 2002; Poulin et al., 2016).

Pegmatitic quartz shows a blue-green emission at 500 nm and a homogeneous texture, constituting a unique signature for this type of quartz (Götze et al., 2005). Hydrothermal quartz derived from deposits as various as epithermal, porphyry-type and orogenic Au can be distinguished based on their CL textures (Fig. 6.5) (Rusk, 2012). Since quartz is ubiquitous and quite resistant against weathering, the CL textures preserved in quartz are reliable fingerprints of the type of ore deposits from which it originates, with clear implications for exploration. Similarly, scheelite is another luminescing mineral than can be used to trace the derivation of sediments from mineralized rocks in heavy mineral concentrates. Poulin et al. (2016) showed that the CL zonation patterns in scheelite are specific in mineralized environments as diverse as orogenic Au, skarn, VMS, and greisen deposits. In porphyry copper deposits, the CL of apatite reveals characteristics that allow one to distinguish apatite from altered and unaltered rocks (Bouzari et al., 2016). The CL color of resistate apatite in soils thus yields insight into the nature of basement rocks and their potential as a source for mineralization. CL is also a useful technique to evaluate the potential in REE of placers and heavy mineral concentrates, by distinguishing weakly luminescent monazite crystals from other minerals

Figure 6.3 Examples of CL textures in quartz associated with mineralization. (a) Crossed polarized light micrograph of massive quartz from Saint-Yriex gold mine, France. (b) Same as (a) viewed under cathodoluminescence, showing a complex pattern of healed microfractures that are not visible under polarizing microscopy but may be faintly underlined by inclusion trails. (c) and (d) Quartz with green, blue, and grey CL exhibiting growth and sector zoning. (c) As-Ag sulfide vein, Sainte-Marie aux Mines, France. (d) Uraniferous quartz vein, Limousin, France. The small speckles in (d) are due to poor quality of the polishing.

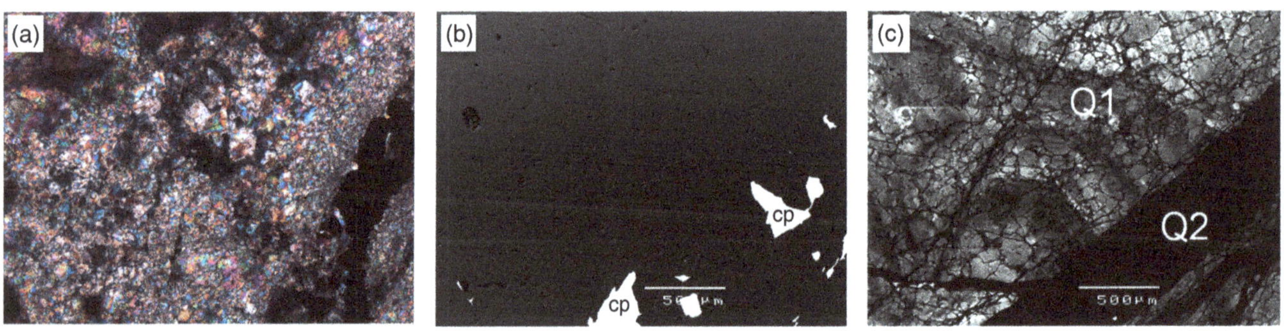

Figure 6.4 (a) Transmitted light, (b) Back scattered electron, and (c) SEM-CL image of a quartz-sulfide vein from the Chuquicamata porphyry copper deposit, Chile. The SEM-CL image reveals textures not visible by the other techniques. These textures are critical to gathering and evaluating data to understand ore genesis and improve exploration models.

(as zircon), which exhibit otherwise very similar optical characteristics (Richter et al., 2006; Cobert et al., 2015).

6.3.2. Alteration Processes and Geochemical Guides

Alteration processes can dramatically affect the CL characteristics of minerals. Since the identification and mapping of alteration related to mineralization are key aspects for exploration, CL can be used to highlight alteration zones in various mineralized environments and demarcate prospective areas. This was well demonstrated in deposits related to alkaline magmatism *sensu lato*. In this environment, for instance, the observation of red-CL (Fe^{3+}-activated) in K-feldspar indicates the alkaline affinity of the context and allows delineating altered zones such as fenites (e.g., Mariano & King, 1975; Hagni, 1984;

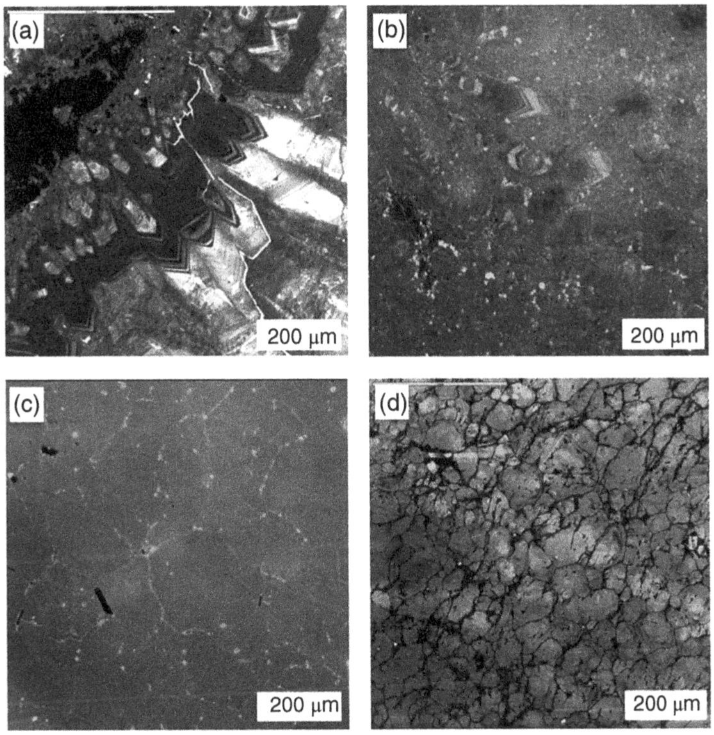

Figure 6.5 SEM-CL images can be used to fingerprint ore deposit types, based on the textures observed in vein quartz. (a) epithermal quartz is characterized by euhedral quartz and is dominated by primary growth textures, typically with little overprinting (Butte, MT, USA). (b) Quartz from the West Shasta, CA, massive sulfide deposit shows double-terminated euhedral quartz crystals, suggesting the quartz grew while suspended in a soft substrate. (c) Quartz in orogenic Au veins typically displays homogenous or faint, mottled CL textures (Sao Bento orogenic Au deposit, Brazil). (d) Porphyry deposits display a wide range in CL textures. Early quartz is typically characterized by mosaics of CL-bright sugary grains. Quartz crystals may range from sharply growth zoned, to wavy concentric zonation to mottled or homogenous internal CL textures, depending on the history of the sample (Los Pelambres, Chile).

Marshall, 1988; Decrée et al., 2013; Corriveau et al., 2016; Westhues et al., 2016).

If a majority of sulfide minerals are nonluminescent because they are conductive materials or contain Fe^{2+}, their oxidation within the supergene alteration zone can generate brightly luminescing secondary minerals such as smithsonite, willemite, hemimorphite, and hydrozincite in zinc deposits (Hagni, 1984). This is of particular interest as surface exploration often deals with the upper, weathered region of mineralized systems and CL provides an efficient technique to detect zinc mineralization and to decipher multistage oxidation in nonsulfide Zn deposits (Coppola et al., 2008).

The acquisition of CL spectra coupled to CL imaging may reveal trace-element enrichment in mineralized systems, highlighting the (sub)economic potential of elements that could be retrieved as by-products of the main mineralization. For instance, in the alkaline syenites at Thor Lake (Canada), zoned albite shows a rim with a distinct CL that is correlated with a strong Ga enrichment (up to 4000 ppm), which reflects Ga activity at the time of albite crystallization (De St. Jorre & Smith, 1988). In this deposit, it could be worthy to exploit Ga from albite (beside other rare-metal deposits in the area), if new extractive technologies are developed.

In the phosphate deposit associated with the Matongo carbonatite (Burundi), bright blue luminescent apatite indicates REE enrichment of this mineral (Decrée et al., 2016). Similarly, the CL characteristics of fluorite can be correlated to its REE content, as shown in fluorite from carbonatites of Ambadongar, India (Singh & Tiwari, 2010) and Motzfeldt alkaline complex, Greenland (Schönenberger et al., 2008), and in hydrothermal fluorite from Kerio Valley, Kenya (Ogola et al., 1994). It is thus expected that the occurrence of such REE-rich minerals are indicative of higher economical potential.

Hydrothermal influences can also be detected in anhydrite associated with mineralization using its CL signature. In the IOCG deposit of Nefza, Tunisia (Ben Abdallah et al., 2013), the anhydrite associated with

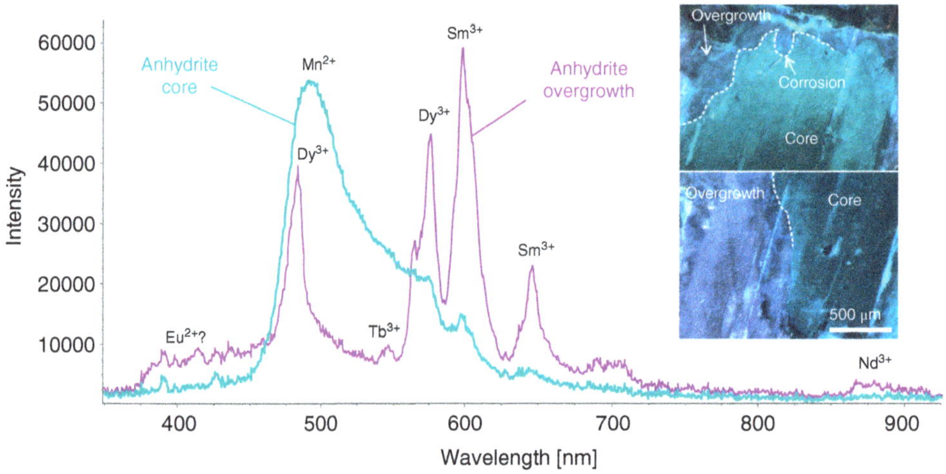

Figure 6.6 Cathodoluminescence imaging and spectroscopy of hydrothermally-influenced anhydrite associated with Miocene IOCG mineralization (Nefza, Tunisia). The greenish-blue core is a relic of country anhydrite showing corrosion and predominant Mn^{2+}-activation. The pale violet anhydrite overgrowth is REE^{3+}-activated (activator assignment after Marshall, 1988).

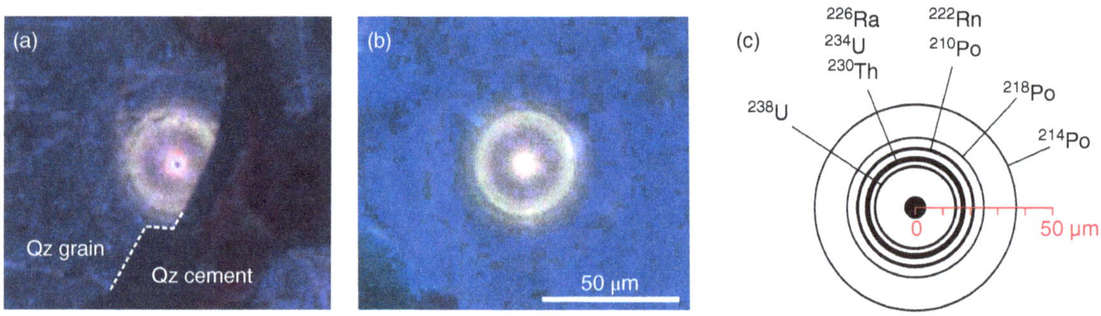

Figure 6.7 Cathodoluminescence induced by alpha-particle radiation damage in quartz. (a) & (b) Luminescent rings induced by a micron-size radioactive inclusion (visible in (a)) in quartz grains from Thanetian sandstone, south Belgium. Note in (a) the absence of radiation-induced luminescence in the dark-luminescent quartz cement. (c) Sketch based on the micrograph in (b) showing the different alpha-emitting radionuclides in the ^{238}U decay chain that are responsible for the luminescent rings (radionuclide assignment after Dill & Weber, 2010).

mineralization shows a core with blue-green, mostly Mn-activated CL (presumably a relic of country evaporite formation) and an overgrowth with violet, REE-activated CL (Fig. 6.6). The overgrowth deposited after an episode of corrosion. A similar observation was made for the Luohe-Xiaobaozhuang Iron-Oxide-Apatite deposit (Zeng, pers. com.), which is similar to the Taocun IOA deposit in the Middle-Lower Yangtze River Valley Metallogenic Belt, China (Zeng et al., 2016).

6.3.3. Radiation-Damage in Quartz

Alpha particles damage the crystal lattice of quartz, which activates a red- to yellow-luminescence (Hu et al., 2008, and references therein). This phenomenon is clearly visible when quartz contains micron-sized uraninite inclusions or other minute radioactive minerals (Fig. 6.7). In this case, a set of concentric luminescent rings is created by the alpha particles emitted by the different radionuclides in the ^{238}U decay series. Dill and Weber (2010) have observed similar patterns of radiation-induced damage in the so-called fetid fluorite from Nabburg-Wölsendorf, SW Germany. In fluorite, radiation-induced damage is visible in optical microscopy but in quartz, they require cathodoluminescence to be observed. Each luminescent ring corresponds to an alpha particle of specific energy, which determines the travel range (stopping power) of the particle. Most CL-active crystal defects are generated at the end of the course of the alpha particles and produce the observed ring pattern.

When the radioactive inclusions are large or the radioactive source is a gas or a liquid flowing through the porosity of the rock, no distinctly separated luminescent rings are observed because alpha particles are no longer radiating from a single point but from more diffuse locations. Instead, luminescent halos of 20–25 μm thick are formed in quartz (Fig. 6.2). The CL induced by radiation damages can be used as an exploration guide for uranium deposits as quartz is a widespread mineral in many different types of rocks (Hu et al., 2008). These features are best developed in quartz from unconformity boundaries, faults, and fracture, as illustrated for the Athabaska basin (Botis et al., 2005, 2006). Similarly, radiation damage in minerals can delineate the zones or discriminate the evolutionary stages that were influenced by radioactive mineralizing fluids (Dill & Weber, 2010). Radiation-induced luminescent halos in quartz are also useful for screening radioactive minerals such as monazite and xenotime in shales, which are potential source rocks for placers (Baele, 2014; Cobert et al., 2015). Lanthanide-bearing minerals, which are often actinide-bearing as well, show no or weak luminescence due to their high trace-element concentration, but the radiation-induced CL in adjacent quartz is readily observed (Fig. 6.2; see also section 6.4.4).

6.4. CATHODOLUMINESCENCE APPLIED TO THE STUDY OF ORE DEPOSITS

6.4.1. Deposits Related to Alkaline Complexes and with Alkaline Affinity

The use of cathodoluminescence in exploration and analysis of carbonatites has been widely documented. Distinct CL colors are indeed observed in minerals from this type of environment (e.g., Mariano & Ring, 1975; Mariano, 1979; Marshall, 1988). Some of these characteristics can be extended to other deposits related to magmas with an alkaline affinity, such as the Kiruna-type apatite-iron ore and iron-oxide copper gold deposits (IOCG).

6.4.1.1. Alkaline and Carbonatite Complexes

Among the typical minerals that are observed in alkaline and carbonatite complexes, calcite and dolomite usually luminesce with varying intensity in yellow-orange and red, respectively (e.g., Decrée et al., 2015; Maki et al., 2016; Tremblay et al., 2017). As in sedimentary carbonates, the relative concentrations of Mn^{2+} and Fe^{2+} control the intensity of CL in carbonatite carbonates, with Mn acting as an activator and Fe as a quencher (Hayward & Jones, 1991). Despite the abundance of REE in carbonatitic environments and their uptake by calcite when it crystallizes, the CL of calcite is almost invariably activated by Mn^{2+} (Marshall, 1988). REE-activated calcite has been observed in very rare cases, such as in weathered carbonatites, where oxidized Mn cannot substitute for Ca in the calcite lattice but precipitates in separate oxide minerals.

Feldspars are other common minerals in carbonatites but in this environment they show typical CL colors, especially a dull to bright red luminescence due to Fe^{3+}-activation (Marshall, 1988). In the Matongo Carbonatite, Burundi, both albite and K-feldspar have a distinct red CL (Decrée et al., 2015). This is commonly observed in feldspars occurring in fenitized rocks as well (e.g., Hagni & Shivdasan, 2001). The CL of a few minerals in carbonatite is illustrated in Fig. 6.8.

Apatite is almost ubiquitously luminescent as it is a good host for CL activators such as Mn^{2+} and $REE^{3+/2+}$, but not for Fe^{2+}, which is the most common CL quencher. In addition, apatite may crystallize at any stage in mineralized systems, from orthomagmatic to late hydrothermal and even during subsequent supergene mineralization. It is therefore a mineral of particular interest for investigating mineralization processes and timing in many ore deposits including carbonatites and phosphorites. Fluorapatite in carbonatites is predominantly blue under CL (e.g., Mariano & Ring, 1975; Marshall, 1988; Kempe & Götze, 2002; Waychunas, 2002; Decrée et al., 2016). This CL color has been assigned to (mainly light-)REE activation. From a more genetic perspective, apatite can record the evolution from silicate to carbonatitic melts as shown by Wang et al. (2014), who used CL to visualize the internal structures of apatite in the Kaiserstuhl Miocene carbonatite complex in Germany.

A great diversity of apatite CL colors is encountered in the phosphate deposit associated with the Matongo carbonatite, Burundi (Decrée et al., 2016). The phosphate-rich breccias, which form the main deposit, exhibit angular to subrounded clasts. A number of these clasts are made of primary (magmatic) CL-zoned fluorapatite. The core of apatite crystals exhibits blue-green CL and the outer rim a blue-violet CL, with a stronger activation by Nd^{3+} as revealed by spectral imaging in the near infrared (Fig. 6.9). This zonation reflects the change from Mn/REE to predominant REE activation with time. The breccia matrix comprises two different types of carbonate-fluorapatite formed in supergene conditions. The first type is a fine-grained, dark-green to brown luminescent groundmass of tabular crystals with a CL intensity that is lower than primary apatite. The second type is nonluminescent, which indicates the absence of activator elements ($REE^{3+/2+}$ and/or Mn^{2+}). Such nonluminescent fluorapatite has been reported in surficial environments associated with other carbonatites (De Toledo et al., 2004).

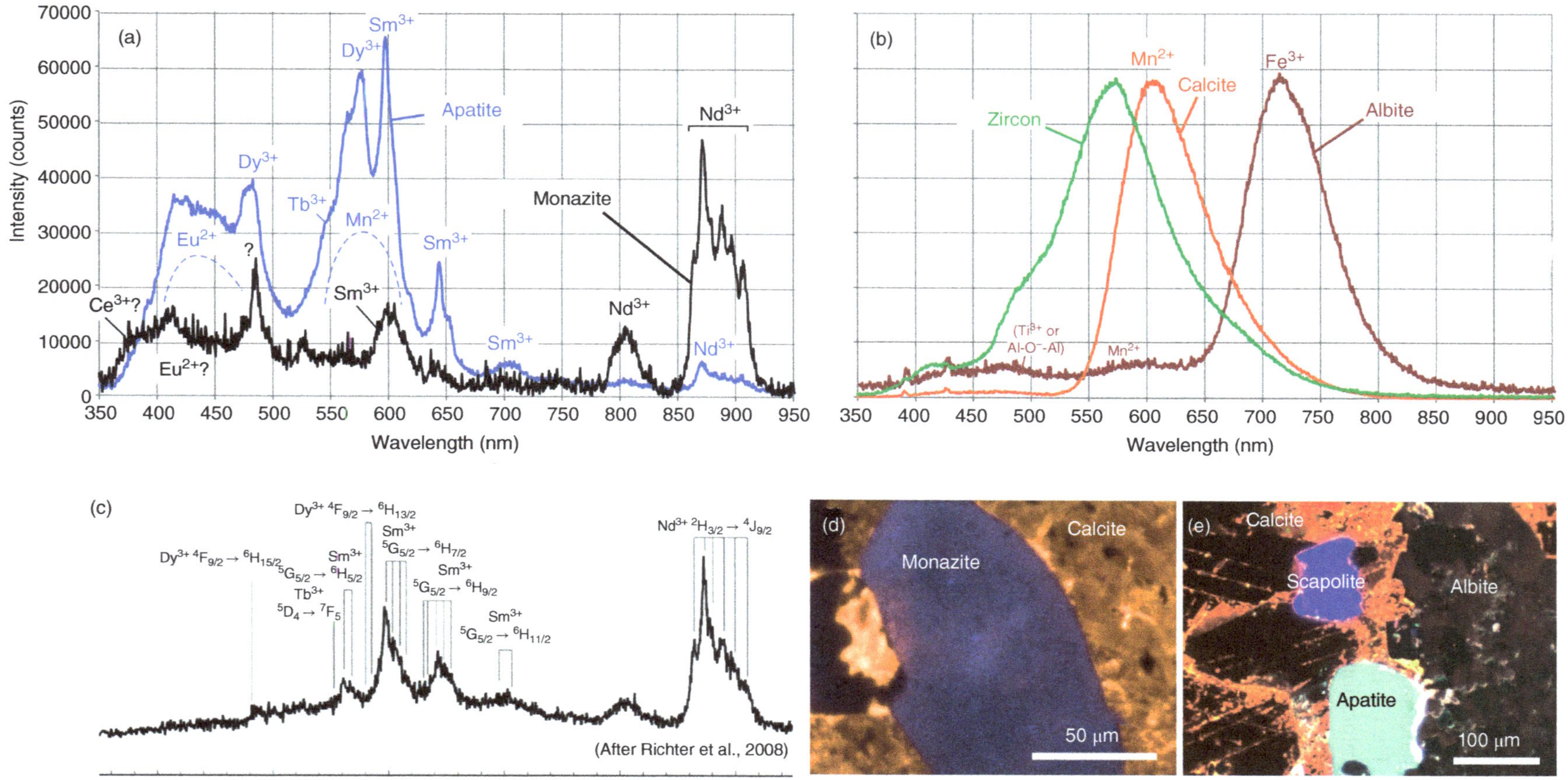

Figure 6.8 Cathodoluminescence characteristics of some carbonatite minerals in the Matongo carbonatite, Burundi. (a) CL spectra from violet-blue luminescent monazite shown in (d) and light blue luminescent apatite with noticeable Mn^{2+} and $REE^{3+/2+}$ activation, shown in (e). Note the strong Nd^{3+} emission of monazite in the near-infrared. The peak at ca. 480 nm is not assigned to Dy^{3+} as this activator always produces an additional peak around 580 nm, which is lacking here (Pr^{3+} could be a candidate as observed in synthetic doped apatite; Mitchell, 2014). The shoulder at 380 nm could indicate a strong Ce^{3+} activation in the UV as sensitivity of the optical system drops dramatically below this point. (b) CL spectra of calcite, albite and zircon showing activation by Mn^{2+}, Fe^{3+} and radiation-induced defects, respectively (zircon luminescence is yellow-green). (c) CL spectra of detrital monazite (after Richter et al., 2008) shown for comparison with (a) (same wavelength scale). This type of monazite luminesces dark-green and is likely sourced from granitic-gneissic rocks. (d) and (e) CL imaging of some minerals from the Matongo carbonatite: monazite (blue-violet CL), calcite (dark to yellow-red CL), apatite (light blue CL), scapolite (deep blue CL) and albite (dull brownish-red CL).

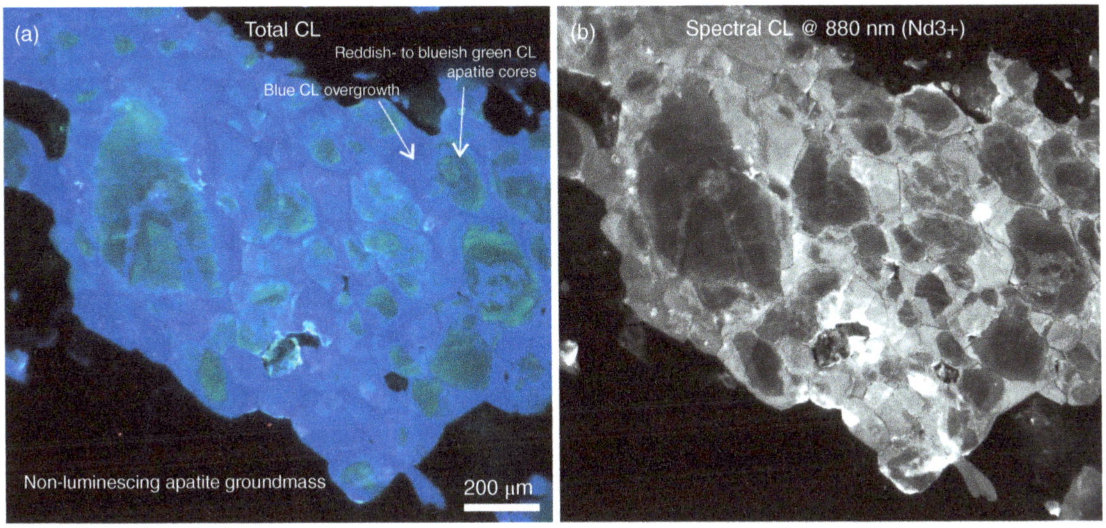

Figure 6.9 Cathodoluminescence imaging of apatite in supergene phosphate ore developed at the expense of the Matongo carbonatite, Burundi. (a) Total (color) CL micrograph showing primary (magmatic) apatite with a reddish- to blueish-green CL due to both Mn^{2+} and $REE^{3+/2+}$ activation. The blue CL in apatite overgrowths indicates a stronger REE activation. (b) Spectral CL of the Nd^{3+} region (880 nm) showing an increase in Nd-activation in apatite overgrowth.

A last generation of bright green luminescent fibrous fluorapatite, which is Mn^{2+}-activated still formed in supergene environment, fills the remaining voids (Decrée et al., 2016).

Campbell and Henderson (1997) have documented the diversity of CL textures and colors of apatite in REE ore from Bayan Obo (China). Changes in REE geochemistry can be recorded with great sensitivity in apatite, which makes this mineral an ideal candidate to investigate mineralization episodes in REE deposits. CL has enabled distinguishing four fluorapatite types at Bayan Obo: yellow-green discrete grains (due to Mn^{2+} activation), anhedral pink-grey REE-rich apatite (with Eu^{3+} or Sm^{3+} as possible activators), yellow-cream REE-Sr-bearing apatite in veins (Dy^{3+} or Mn^{2+} as activators), and cream-grey (due to a variety of activators) pure apatite associated with fluorite in veins. Strontianite can be used for a similar purpose as its CL is predominantly REE-activated in REE-rich environments such as carbonatites (Marshall, 1988).

Fluorite cathodoluminescence has been well investigated in ore deposits related to carbonatites and alkaline complexes. Fluorite commonly exhibits a bright bluish-violet luminescence (Hagni & Shivdasan, 2001; Graupner et al., 2015), which is mainly related to activation by Eu^{2+} (Marshall, 1988; Blanc et al., 2000). Green-luminescent fluorite has also been described in carbonatite rocks of Ambadongar, India (Singh & Tirawi, 2010), and in the alkaline complex of Ivigtut, Greenland (Schönenberger et al., 2008). Green-luminescent fluorite shows strong Dy^{3+}-activation (Baele et al., 2012; Götze, 2012). The intensity of fluorite CL is seemingly directly related to its trace-element content (Schönenberger et al., 2008; Baele et al., 2012). The CL of fluorite has also been used indirectly to investigate the Nb mineralization associated with the Saint-Honoré alkaline complex (Canada), where typical blue fluorite (and yellow calcite) are present as inclusions in Nb-bearing minerals (Tremblay et al., 2017). Because fluorite, apatite, and calcite are luminescent minerals, their identification and relationships are readily obtained with CL. Hagni (2012) used CL imaging to analyze the carbonatite-hosted fluorite deposit at Okorusu, Namibia, in relation to beneficiation problems. The author was able to discriminate the different types of fluorite ores according to their content in phosphorus (which is a contaminant) and their host formation as fluorite replaced predominantly the carbonatite (P-rich ore) and less the adjacent marbles and fenites (P-poor ores).

The CL of zircon is a helpful indicator of carbonatitic processes (Tichomirowa et al., 2006, 2013). The "yellow" broad emission band between 400 and 700 nm in zircon, which originates from radiation-induced defects (Nasdala et al., 2010), decreases in intensity where there is a very high concentration of defects. This, together with the CL textures of zircon, allowed Tichomirowa et al. (2013) to reconstruct multistage carbonatitic processes, which influenced the primary geochemical and isotope signatures. In general, the CL of zircon is blue to bluish-white or yellow-green, due to REE, especially Dy^{3+} activation, and the above- mentioned broadband yellow-green emission peaking at ca. 560 nm, respectively (e.g., Marshall, 1988; Nasdala et al., 2003, and references therein). Zircon commonly shows a well-developed concentric zoning but also various CL textures that are routinely used for conducting geochronological investigation and interpreting the resulting data (e.g., Kempe et al., 2000).

6.4.1.2. Iron-Oxide Copper Gold (IOCG) and Kiruna-type Deposits

IOCG deposits and Kiruna-type apatite-iron oxide ores are affiliated with (variably) alkali granites, porphyries, and even carbonatites (Groves & Vielreicher, 2001). They are often associated with pervasive large-scale alkali alteration (Williams et al., 2005). The alteration footprints in IOCG allow defining prospective areas with high mineralization potential (Corriveau et al., 2007, 2010).

Erickson (2011) used CL to characterize the alkali alteration in the vicinity of IOCG deposits in Elk Mountain (Colorado, USA). This study focused on feldspar, emphasizing that red-CL albite has partly replaced the core of igneous plagioclase in the most altered zones. In the IOCG-type Oued Belif breccia, Nefza district, Tunisia (Decrée et al., 2013), K-metasomatism is associated with mineralization, as evidenced by the replacement of blue-luminescent K-feldspar by red-luminescent K-feldspar, and the growth of euhedral K-feldspar with alternating blue and red luminescence. This suggests an increase in alkalinity of metasomatic fluids, which stabilizes Fe^{3+} species and allows their incorporation in feldspar and CL activation in the red region of the spectrum (Marshall, 1988).

Comparable observations have been made in Kiruna-type magnetite-apatite deposits of the Lyon Mountain Granite, USA (Valley et al., 2011). In rocks affected by sodic alteration, albite mostly luminesces red. In the Kiruna district, the metasomatically altered feldspar exhibits a dull to red luminescence, which contrasts with the blue luminescence of unaltered K-feldspar (Westhues et al., 2016).

Apatite also helps in deciphering alteration/metasomatism processes affecting IOCG and Kiruna-type deposits. In the Kiruna district (Sweden), apatite from the altered zones shows a green to grey REE-activated luminescence (Westhues et al., 2016). In the Kiruna-type magnetite-apatite deposits of the Lyon Mountain Granite, USA, apatite from albite granite and microcline granite, which has experienced sodic and potassic alteration, respectively, luminesces purple to orange due to REE activation (Valley et al., 2011). This is rather uncommon for primary magmatic apatite in felsic rocks, which typically shows a yellow-green CL activated by Mn^{2+} (Marshall, 1988).

In the Sossego IOCG deposit, Brazil, apatite associated with the copper-gold mineralization exhibits altered zones delineated by a yellow luminescence while nonaltered apatite luminesces green (Monteiro et al., 2008; Moreto et al., 2015). Apatite from the IOCG mineralized system of the Great Bear magmatic zone, Canada, shows a variety of CL colors and textures (Lypaczewski et al., 2013). In mineralized samples, the CL is characterized by a distinct blue to yellow-green luminescence, which is related to REE and Mn^{2+} substituting for Ca^{2+}, respectively (Marshall, 1988; Kempe & Götze, 2002). In addition, complex zonation patterns occur in zones close to the hydrothermal source (Rae et al., 1996).

Quartz is not abundant in IOCG deposits. When it is present, its relationship to mineralization is not always clear. Although few studies of CL of quartz from IOCG deposits exist, CL images from Ernest Henry IOCG in the Cloncurry District of Queensland, Australia, show that a variety of quartz CL textures are present, supporting a complex and multigenerational origin of this variety of ore deposit (Fig. 6.10). CL textures vary from homogenous to highly fractured to internally zoned.

6.4.2. Magmatic-Hydrothermal Ore Deposits

6.4.2.1. Cu-Mo-W Porphyry and Epithermal Deposits

Quartz is a critical mineral used to reconstruct complex vein generations and stages of mineralizing systems (and subsequently to discuss the conditions and evolution of ore-forming fluids) in porphyry to epithermal systems (e.g., Landtwing & Pettke, 2005; Rusk et al., 2006; Pudack et al., 2009; Rusk, 2012; Vasyukova et al., 2013; Stefanova et al., 2014; Frelinger et al., 2015; Wang et al., 2017). Quartz textures observed under CL offer a wealth of information on growth history of minerals, vein generations, and formation conditions. Magmatic to hydrothermal quartz may indeed exhibit a wide variety of textures, which result from primary and secondary processes like growth and sector zoning, dissolution, overgrowth, and recrystallization (Rusk et al., 2006; Müller et al., 2010; Rusk, 2012; Frelinger et al., 2015) (Fig. 6.11).

Quartz CL reflects the physiochemical conditions of the formation environment and its evolution through time. In porphyry copper hydrothermal systems, a decrease in quartz luminescence intensity, together with a decrease in titanium concentration is widely observed

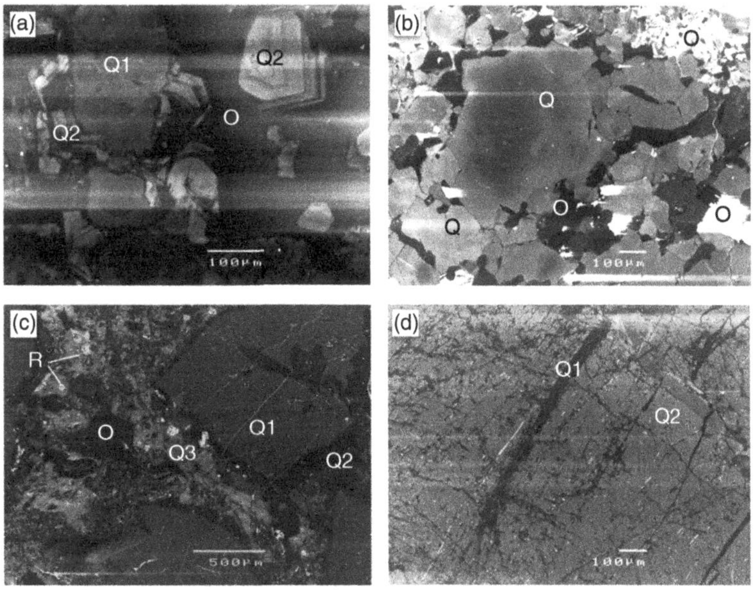

Figure 6.10 SEM-CL images of quartz from the Ernest Henry IOCG deposit, Cloncurry District, Queensland, Australia. Horizontal CL-bright streaks in all images are caused by bits of apatite and calcite intergrown with the quartz. The common occurrence of these bright-luminescing minerals makes imaging such quartz difficult. (a) Early mottled-CL quartz (Q1) overgrown by later zoned quartz (Q2), intergrown with non-luminescing sulfides and magnetite labeled "O" for "other mineral". (b) CL-grey homogenous to mottled quartz (Q) intergrown with a variety of non-luminescing and brightly luminescing minerals (O). (c) Brecciated CL-grey homogenous quartz fragments (Q1) overgrown by CL-dark overgrowths (Q2), and then infilled by CL-bright complex and zoned quartz-rich matrix (Q3). Small bits and pieces of other minerals (O) are also present in the matrix. Several radiation halos, only visible by CL caused by the presence of small grains of monazite are labeled R. (d) Coarse and intensely fractured CL-grey quartz. Entire field of view is quartz.

from the earliest generation of quartz to the latest owing to a decrease in temperature of the crystallizing fluids with time (Landtwing & Pettke, 2005; Rusk et al., 2006; Müller et al., 2010; Rusk et al., 2011; Rusk, 2012). Commonly, the latest, low-temperature generation of quartz in these deposits, which forms under dominantly epithermal conditions, contains elevated concentrations of Al (up to 4000 ppm), Li (up to 300 ppm), and sometimes other trace elements (Rusk et al., 2008; Maydagan et al., 2015; Mao et al., 2017). There is a correlation in porphyry quartz between increased CL intensity (predominantly in the blue wavelengths) and elevated Ti concentration. The incorporation of Ti in quartz is temperature and pressure dependent and it can be used as a thermobarometer (e.g., Wark & Watson, 2006; Thomas et al., 2010; Huang & Audetat, 2012), especially when combined with CL and fluid inclusion microanalysis. Quantitative mapping of Ti in quartz is possible using CL and calibration via EPMA or LA-ICP-MS (Rusk et al., 2008; Donovan et al., 2011; Leeman et al., 2012). Because of the relation between Ti concentration in quartz and pressure and temperature, such maps document the changing conditions of the hydrothermal system through time. CL textures also provide information about the environment of quartz formation. Dissolution textures that are frequently observed in porphyry copper systems likely result from the cooling of magmatic/hydrothermal fluids through the zone of retrograde quartz solubility, in a temperature range of 460°C–550°C at pressures below 800 bars (e.g., Rusk & Reed, 2002; Landtwing & Pettke, 2005; Müller et al., 2010). Peculiar CL textures such as healed fractures and spiders (splatter and cobweb textures of Rusk & Reed, 2002) could be regarded as good indicators of high-temperature formation environments as (1) they are observed in quartz from plutonic and some high-grade metamorphic environments as well as in some porphyry systems; and (2) they could originate from retrograde dissolution (cf. above), which, in addition, may have occurred in microfractures left by beta- to alpha-quartz inversion upon cooling (Boggs & Krinsley, 2006, and references therein). In contrast, sector zoning, which produces simple wedge shapes to very complex textures superimposed on growth zoning is typical in quartz in epithermal deposits and other low-temperature environments (e.g., Rusk, 2012). In porphyry systems, quartz frequently displays well-developed concentric zoning under

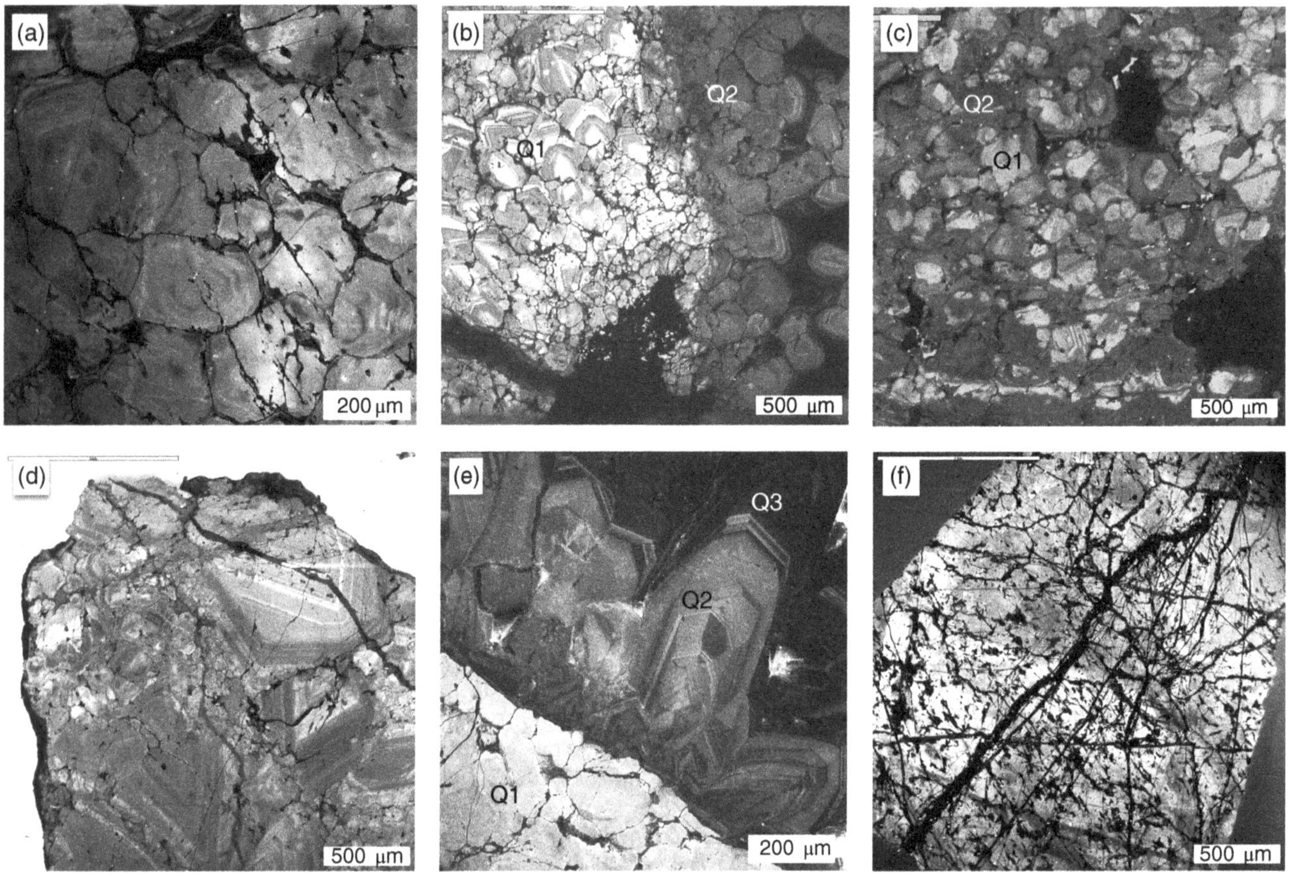

Figure 6.11 Textures typical of porphyry-Cu (Mo-Au) deposits. (a) Wavy concentric zonations in CL-bright quartz mosaics (Butte, MT). (b) Early zoned CL-bright quartz cut by later zoned CL-darker quartz intergrown with chalcopyrite (Grasberg, Indonesia). (c) Early CL-bright quartz that has been resorbed and subsequently overgrown by a later generation of CL-darker quartz related to sulfides (El Abra, Chile). (d) Euhedral quartz crystals with no strong orientation intergrown with mosaics of fine grained quartz (Los Pelambres, Chile). (e) Early CL-bright quartz cut and infilled by later strongly euhedral quartz projecting into the vein center (Butte, MT). (f) CL-bright quartz that is highly fractured and displays strong splatter and cobweb texture (El Teniente, Chile).

CL but not sector zoning. To date, no explanation has been given for this observation. A plausible hypothesis could be that only low-temperature (alpha) quartz is able to widely develop sector zoning because of its lower degree of symmetry compared with high-temperature (beta) quartz. This means that alpha-quartz has several symmetrically nonequivalent crystals faces, which, owing to their different surface structure, will tend to partition trace elements and defects during crystal growth and this may result in sector zoning. High-temperature (beta) quartz, with its higher, sixfold symmetry, would produce either no sector zoning as the six pyramid faces all have the same surface properties, or a simple zoning if prism faces form as well. Well-developed sector zoning in quartz observed under CL could then be interpreted as an evidence for direct precipitation of alpha-quartz, which is stable at < 573°C under atmospheric pressure.

In many hydrothermal ore deposits, CL of quartz can be used to fingerprint different mineralization stages. In the porphyry-Cu-Mo deposit in Butte, Montana, USA, quartz from quartz-molybdenite veins has a high luminescence intensity, while the CL of quartz associated with pyrite and chalcopyrite in later veins is weaker (Rusk et al., 2006). Similar characteristics are shared by other porphyry deposits, such as the Dabaoshan Mo deposit in China (Mao et al., 2017). In that case, four generations of quartz veins (V1 to V4) were described, with a decreasing luminescence intensity from the earliest (V1) to the latest (V4). Quartz associated with base metals (V4) has a weaker CL than quartz in veins hosting the Mo mineralization (V2).

Similarly, the following sequence is observed in the Altar Cu-(Au-Mo) porphyry deposit, Argentina (Maydagán et al., 2015): (1) quartz from early quartz-chalcopyrite-pyrite and quartz-molybdenite veins exhibits bright CL and homogeneous texture, (2) quartz from quartz-molybdenite veins shows growth zones with moderate oscillatory CL intensity, and (3) quartz from later quartz-pyrite veins has lower CL intensity. The dark CL of the latter type of quartz, rich in sulfides and sulfosalts, is typical of epithermal deposits.

In Oyu Tolgoi and Zesen Uul, southern Mongolia, two porphyry systems of different age and located 100 km apart show a similar evolutionary sequence of quartz CL (Müller et al., 2010). Each of the four recognized quartz stages have a distinct CL and trace-element characteristics but together they record a trend of decreasing temperature. The earliest stage comprises barren quartz showing the highest CL intensity, Ti content, and temperature formation (>600°C). The main mineralization is associated with the second quartz stage, which exhibits a dull CL and precipitated under acidic fluid conditions. Subsequent quartz stages, barren or sulfide-rich, are also weakly luminescent to nonluminescent. The quartz adjacent to sulfides commonly shows elevated content in Fe, which could explain its dark CL (Penniston-Dorland, 2001; Müller et al., 2010). The association of weakly luminescent quartz with sulfide mineralization was also described in the following Cu and/or Mo porphyry-type deposits: Tongcun, South China (Ni et al., 2017), Grasberg Igneous Complex, Irian Jaya, Indonesia (Penniston-Dorland, 2001), and Red Hills, Texas, USA (Frelinger et al., 2015).

The cathodoluminescence of quartz has also been used, among other techniques, to support the magmatic nature of the gold mineralization at the Bilihe deposit, China (Yang et al., 2015), and to detail the paragenetic sequence of the Asachinskoe epithermal gold deposit, Kamchatka, Russia (Takahashi et al., 2008).

Apatite and scheelite are other minerals that can be used for studying porphyry-type deposits under CL. Similar to apatite in a majority of felsic rocks, apatite has a predominant yellow-green, Mn^{2+}-activated CL in porphyry-type deposits (Marshall, 1988). Bouzari et al. (2016) showed that it is possible to distinguish apatite from altered and unaltered rocks in several porphyry copper deposits in British Columbia based on their luminescence. Apatite luminesces green in K silicate-altered rocks, whereas it shows a characteristic yellow/brown-luminescence in unaltered rocks. This could be applied in exploration, by investigating heavy minerals in regoliths and stream sediments. Scheelite, which exhibits a blue CL with common oscillatory zoning in porphyry systems, could be used for similar applications (Poulin et al., 2016; Wang et al., 2017). The CL in scheelite is due to activation by both WO^{4-} groups, which produce a broadband blue emission, and REE, with a series of narrow emission lines (Marshall, 1988; Gaft et al., 1999; Blanc et al., 2000).

A study of the CL of anhydrite from various geological environments shows that REE activation in this mineral is typical of hydrothermal systems, including porphyry-copper deposits (Marshall, 1988).

6.4.2.2. Granite-related (Sn-W and U) Deposits

In the Piaotang granite-related W-Sn deposit, China, zoned cassiterite with a dark CL core and a bright CL rim reveals different formation temperature during the hydrothermal evolution of the system (Zhang et al., 2017). The CL color of cassiterite ranges from yellow-green to blue with Ti and W as activators, respectively, and Nb, Ta, and Fe as quenchers (Fig. 6.12; Marshall, 1988; Farmer et al., 1991). In the South Crofty granite-related Sn mineralization, Cornwall, UK, successive generations of cassiterite are distinguished by their distinct CL colors and growth textures, including sector zoning. These CL characteristics support the hypothesis that two contrasting hydraulic regimes succeeded during Sn mineralization (Farmer et al., 1991).

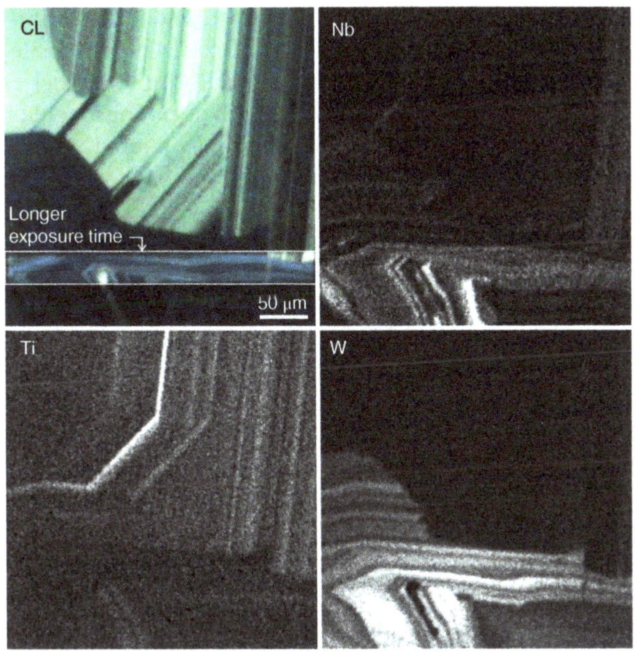

Figure 6.12 Cathodoluminescence and WDS maps in a single crystal of cassiterite from Rwanda. The cassiterite crystal exhibits both growth (fine bands) and sector zoning. The CL is green in the two upper sectors while it is dark in the lower sector. The CL in the dark sector is actually blue, as revealed by imaging with a longer exposure time (inset), which was not used for the entire photograph because it would have overexposed the green CL. A comparison between CL and WDS maps shows that green and blue CL in cassiterite is activated by Ti and W, respectively. However, Nb acts as a quencher as demonstrated by the lower CL intensity where this element is in higher concentration.

Quartz CL has been used to address the issue of (late) magmatic processes and the transition from magmatic-dominated to hydrothermal-dominated processes leading to metal concentration in ore-bearing granite. Breiter et al. (2017) showed that alteration and purification of primary magmatic quartz during late magmatic processes in the Cinovec/Zinnwald Sn-W-Li deposit, Czech Republic, led to a decrease of CL intensity due to a decrease in trace-element concentration including Ti. In the Land's End pluton, Cornwall, UK, the CL colors and textures of quartz, coupled with in-situ geochemical analysis, document the transition from magmatic to hydrothermal environment with a decrease in CL intensity and an increase in Al and Fe concentration (Müller et al., 2006). A decrease in Ti concentration was also observed from blue-CL magmatic quartz to red-brown hydrothermal quartz.

Apatite from granite, greisen, and pegmatite typically exhibits a strong Mn^{2+}-activated yellow-green luminescence (Marshall, 1988; Kempe & Götze, 2002; Waychunas, 2002; Müller et al., 2006). The luminescence of apatite has been used as a tool to trace fluids in the U mineralized system associated with the emplacement of peraluminous leucogranites in the Pontivy-Rostrenen composite intrusion, France (Ballouard et al., 2017). In these rocks, primary (magmatic) apatite is unzoned or exhibits regular zonation under CL while hydrothermally altered apatite has irregular CL textures and zoning. As^{5+} incorporation (which quenches the CL) in red-pink CL apatite overgrowths in episyenite points to oxidizing hydrothermal conditions. Interestingly, Eu^{3+} activation is observed in the associated altered apatite, which, in contrast to the view of Roeder et al. (1987), could support the interpretation of Eu^{3+} activation as indicating oxidizing conditions for the mineralizing fluids.

Scheelite from granite/greisen-related deposits luminesces from bright to dark blue, as shown by scheelite from the greisen deposit of Zinnwald-Cinovec, Germany-Czech Republic. In proximal intrusion systems such as porphyries, skarns, and greisens, scheelite is strongly zoned under CL while it is homogenous in orogenic Au deposits, which may have implications for exploration (Poulin et al., 2016).

The cathodoluminescence of fluorite has been used to provide an essential petrographic framework for in-situ geochemistry and data analysis of the Mount Pleasant Sn-W-Mo deposits, New Brunswick, Canada (Assadzadeh et al., 2017). From this study, it appears that fluorite associated with molybdenite shows complex CL textures. By contrast, fluorite related to the base-metal sulfide mineralization is quite homogenous or displays only simple zoning patterns under CL, which reveals that sulfides formed in a less complex fluid environment compared to Mo mineralization.

Zircon coeval with wolframite in the Nanling Range tungsten mineralization, Southeast China, exhibits weak luminescence and oscillatory zoning, which (among other characteristics) points to a hydrothermal origin (Wang et al., 2016).

6.4.3. Hydrothermal Deposits

6.4.3.1. Orogenic Gold Deposits

Cathodoluminescence of quartz is helpful to recognize the different generations of veins and associated mineralization in orogenic gold deposits. Quartz CL was investigated in order to determine the paragenetic position of microshears associated with a third phase of vein-hosted gold mineralization at Curraghinalt, Northern Ireland (Rice et al., 2016).

In Sheba and Fairview gold deposits, South Africa, the CL of quartz documents both the structural and fluid contexts leading to Au mineralization (Agangi et al., 2014). Mineralized quartz veins show euhedral growth zones under CL, suggesting precipitation in open space generated by late brittle deformation. Its bright luminescence can be correlated to Al content, which is typically observed in low- to moderate-temperature hydrothermal systems (Götze et al., 2004). Since pH may control the solubility of Al in fluids (Rusk et al., 2008), the authors related the CL of quartz with pH changes and gold/sulfide deposition.

The study of the Charmitan gold(-tungsten) mineralization, Uzbekistan, provides additional information about quartz and scheelite luminescence in the context of orogenic Au deposits (Graupner et al., 2010). Four generations of quartz and two generations of scheelite have been defined based on their luminescence. The blue or green-yellow CL colors of quartz from the gold-bearing quartz veins support a hydrothermal origin, the bluish color being mostly preserved in the core of the quartz grains. When quartz is associated with sulfides, its luminescence is weak. Two generations of scheelite were determined. The first is characterized by a weak luminescence and shows narrow oscillatory growth zoning. This scheelite is partly altered into a second scheelite generation, which luminesces brightly with a rather homogeneous texture, and is associated with economic gold mineralization. The CL color of scheelite is mainly due to intrinsic, broad-band emission of the WO_4^{2-} group in the blue range and subordinate REE^{3+} activation.

As a whole, scheelite from orogenic Au deposits luminesces blue and commonly shows no oscillatory zoning under CL, which would correspond to primary precipitation features (e.g., Poulin et al., 2016; Grzela, 2017). Such characteristics could help to detect the presence of orogenic gold deposits by examining scheelite in residual deposits and stream sediments (Poulin et al., 2016).

Similarly, Rusk (2012) showed that orogenic Au quartz is commonly homogenous under CL, likely due to annealing of the primary CL textures during metamorphism. Orogenic gold quartz is thus texturally distinct from the quartz in porphyry copper and epithermal deposits, which has potential applications for exploration.

Zircon CL textures yield insight into the genesis of the Dongping orogenic Au deposit, Dongping deposit, North China (Bao et al., 2014). Hydrothermal zircons disseminated in altered syenite and occurring in auriferous quartz veins show vague zoning or no internal texture, whereas zircon from unaltered syenite shows oscillatory zoning that indicates a magmatic origin.

6.4.3.2. Carlin-Type Gold Deposits

Most cathodoluminescence studies of Carlin-type Au deposits focus on alteration associated with mineralization processes. Emsbo et al. (2003) studied the hydrothermal alteration and textures of carbonate rocks in the Meikle deposit, Carlin trend, Nevada, USA, which directly relate to formation of high-grade ore. In that system, the underlying limestone is pervasively dolomitized prior to mineralization, with the formation of "zebra" texture characterized by red-brown to orange-yellow CL colors. The luminescence of dolomite varies according to its Mn and Fe content. Quartz is a minor constituent during ore-stage mineralization. It is mostly nonluminescent, but exhibits an increasing luminescence from early to late mineralization stages, which could reflect the increasing contribution from a second mineralizing fluid. Late ore-stage drusy quartz exhibits prominent concentric growth zoning grading from bright blue-green to yellow-grey luminescence. Finally, late ore-stage calcite luminesces bright red-orange.

Vaughan et al. (2016) proposed a model for zoned alteration of carbonates in the Banshee Carlin-type Au deposit, Carlin trend, taking into account the spatial distribution of low-temperature alterations, with a particular attention on calcite veining linked to mineralization. At Banshee, carbonate cement from distal samples is non- to dull-luminescent, whereas it becomes brighter in areas close to alteration zones. Similarly, calcite from early veins, which are abundant in altered rocks and less common in the vicinity of the mineralization, is dull-luminescent, while late calcite veins within altered areas are zoned and brightly luminescent. Hydrothermal calcite veins at the Banshee deposit thus provide an indicator for the proximity of ore-bearing zones. Cathodoluminescence, together with other techniques, therefore constitutes an effective exploration tool for Carlin-type Au deposits (Vaughan et al., 2016).

A CL study was carried out in the Osiris, Isis, and Isis East Carlin-type Au showings, Yukon, USA (Beaton, 2015). In these systems, the micritic matrix of the host limestone underwent dolomitization, with the formation of a dull CL dolomite during a pre-ore stage. The base metal sulfide mineralization is cemented by a dull brown luminescent dolomite with local red/orange zoning. Finally, the deposition of a distinctive bright red CL dolomite is associated with the main ore Au-bearing stage. As expected, the brightness of dolomite CL is associated with high Mn/Fe.

Lubben et al. (2012) examined the CL of quartz in the Betze-Post deposit, Carlin trend, and distinguished four main generations of quartz. Au-bearing pyrite is spatially associated with ore-stage jasperoid and late-ore drusy quartz, both being nonluminescent and with high Al contents. This contrasts with the two-stage post-ore quartz overgrowth, which is characterized by zoning, bright luminescence, and low trace-element contents. Identical trends in CL texture and trace elements were recognized in quartz infill at the Jerritt Canyon Carlin deposits (Rusk et al., 2008). This indicates that CL textures and trace elements recorded similar hydrothermal conditions over a large area during the formation of these deposits.

Besides carbonates and quartz, it has been shown that apatite records hydrothermal fluid interactions in Carlin-type deposits, and exhibits different distinguishable textures under CL according to their provenance (Barker et al., 2009). In altered and unmineralized rocks, apatite shows a single overgrowth generation, whereas in mineralized samples of apatite, several generations are observed. Therefore, it is possible to detect mineralization by examining detrital apatite in samples including residual (sedimentary) deposits.

6.4.3.3. Volcanic Massive Sulfides (VMS) and Sedimentary-Exhalative (Sedex) Deposits

The CL of quartz has been applied to the analysis of VMS and Sedex deposits. At Noranda, Ben Nevis, and Matagami VMS districts, Abitibi Greenstone Belt, Canada, quartz occurs in the main ore, stockwork, veins, and fillings between lava pillows (Ioannou et al., 2003). The CL characteristics of quartz, which are consistent over 250 km, indicate a hydrothermal environment. The common occurrence of growth zoning provides evidence that quartz has not recrystallized, which means that the fluids entrapped in inclusions are representative of VMS fluids. Chronological relations between quartz deposition and economic mineralization can be established as well as a common trend between growth zoning development and the proximity with sea-floor surface. The CL may have a strong transient character, especially at Ben Nevis, where a short-lived (<20 s) yellow emission is interpreted as typical of fast crystallization in a hydrothermal environment (Götze et al., 2015).

At Neves Corvo (Portugal), quartz associated with the VMS sulfide ore luminesces yellow and occasionally exhibits sector zoning, which are preserved primary features (Moura et al., 2003; Götze et al., 2015). Therefore, no recrystallization has occurred and fluid inclusions can reliably be studied for as a proxy of mineralizing fluids. In the study of Red Dog Zn-Pb-Ag Sedex deposit, Alaska, USA, the CL of quartz also helped to better constrain fluid inclusion data (Leach et al., 2004), notably by relating CL characteristics to specific mineralization stages.

Sphalerite in VMS and Sedex deposits luminesces bright yellow-orange, as in Kuruko-type VMS deposits in Turkey and in the Franklin and Ogdensburg Sedex deposits. The CL activation is due to Mn^{2+} and is quenched by Fe^{2+} (Karakus et al., 2008).

Scheelite luminesces blue, with complex growth and oscillatory zoning, as shown in massive sulfide deposits from the Britannia Mine, British Columbia, Canada (Poulin et al., 2016). Scheelite has been used as an ore guide, since this mineral is an alteration product developed during sulfide deposition (Bell et al., 1980 in Hagni, 1984).

6.4.3.4. Mississippi-Valley Type (MVT) and Low-T Fluorite Deposits

The large-scale fluid flow processes leading to MVT deposits are accompanied with regional epidiagenetic mineralizing events. Large-scale zonal correlations of mineral signatures are thus possible and help isolate areas with high mineralization potential. CL petrography constitutes an extremely valuable exploration tool since many minerals in MVT deposits are luminescent, such as calcite, dolomite, sphalerite, barite, and fluorite (e.g., Ebers & Kopp, 1979; Kopp et al., 1990; Montanez, 1996; Leach, 1997; Wei et al., 2015). Quartz in MVT deposits is commonly late and commonly euhedral in form. CL textures in quartz from several southeast US, MVT deposits show abundant well-developed euhedral growth zones. Sector zonation is extremely common, and chalcedonic textures are common as well. Variations in CL intensity are correlated with fluctuations in Al and Li concentrations in roughly the same range as are observed in epithermal quartz. MVT and epithermal quartz have the highest Al concentrations of all hydrothermal quartz, but the range is wide from a few ppm Al up to ~4000 ppm Al (Rusk et al., 2008).

In the MVT mineralization of the Mascot-Jefferson City district, Tennessee, USA, CL distinguished six different zones in gangue dolomite that luminesces bright to dull red (Ebers & Kopp, 1979). These characteristics can be used for microstratigraphic correlation through the district and for exploration as three of the recognized dolomite zones are associated with sphalerite mineralization. In the mineralized Upper Knox carbonates, Southern Appalachians, USA, diagenetic cements that are associated with trace amounts of MVT ores can be correlated zonally using their luminescence colors (Montanez, 1996). In the Maoping Pb-Zn deposit, Yunnan, China, five dolomitic stages and a final calcite stage were distinguished based on CL (Wei et al., 2015). Sulfide mineralization is typically associated with zoned dolomite, which can be used as an aid to exploration. Similarly, a sparry dolomite showing four different zones characterized by a distinct luminescence and associated with MVT mineralization can be traced over tens (and maybe hundreds) of kilometers in the Reelfoot Rift Complex, Ozark region, USA (Leach, 1997).

The CL of sphalerite was also used for a similar purpose in MVT ores from the Drau Range, Slovenia-Austria (Kuhlemann et al., 2001). Seven generations of sphalerite were recognized based on CL color and texture, which can be correlated throughout the district.

Fluorite mineralization is common in MVT systems and CL is advantageously used to distinguish the different stages of deposition and replacement involving fluorite and carbonates (e.g., Gigoux et al., 2015; Mondillo et al., 2016). The CL of fluorite is dominated by a bright blue emission due to Eu^{2+} activation (e.g., Marshall, 1988). This prominent blue emission often overwhelms the CL from other elements substituting for Ca^{2+} in the fluorite lattice, typically REE^{3+}. This can be mitigated by digital processing of the CL images or spectral imaging (Fig. 6.13 & 6.14; Baele et al., 2012).

Besides growth zoning, strong sector zoning affects many fluorite crystals from south Belgium (Baele et al., 2012), which formed in a low-T environment (Baele, 1999). The investigated samples exhibit mixed crystal morphologies ranging from the cube to the dodecahedron end-members. Other crystal morphologies such as the octahedron are very rare, which is consistent with crystallization at low-T (Kyrianova & Glikin, 1999). The growth of the cube and dodecahedron crystal faces produced |100| and |110| sectors, respectively (Fig. 6.13). These sectors are distinguished under CL in some samples. |100| sectors have a bluish-green to bluish-grey CL, which turns to deep blue in less than 10 s under common excitation conditions (15 kV, 5–10 μA/mm^2). Prolonged electron irradiation also induces a semipermanent purple color in this sector only. This makes the internal crystal textures of fluorite visible with a conventional optical microscope (e.g., Dickson, 1980), which is useful for further in-situ micro-analysis. |110| sectors may exhibit a stable pinkish to violet CL. Spectral imaging revealed that sector zoning actually occurs in all the investigated fluorite samples, even those with a homogeneous blue CL. Heavy-REE activation is enhanced in |100| relative to |110| sectors, which in turn are more intensely activated by light REE. LA-ICP-MS analyses have confirmed that enhanced activation corresponds to actual enrichment in REE with a |110|/|100| partitioning coefficient ranging from 20 to 0.7 from La through Lu (Baele et al., 2012). This means that

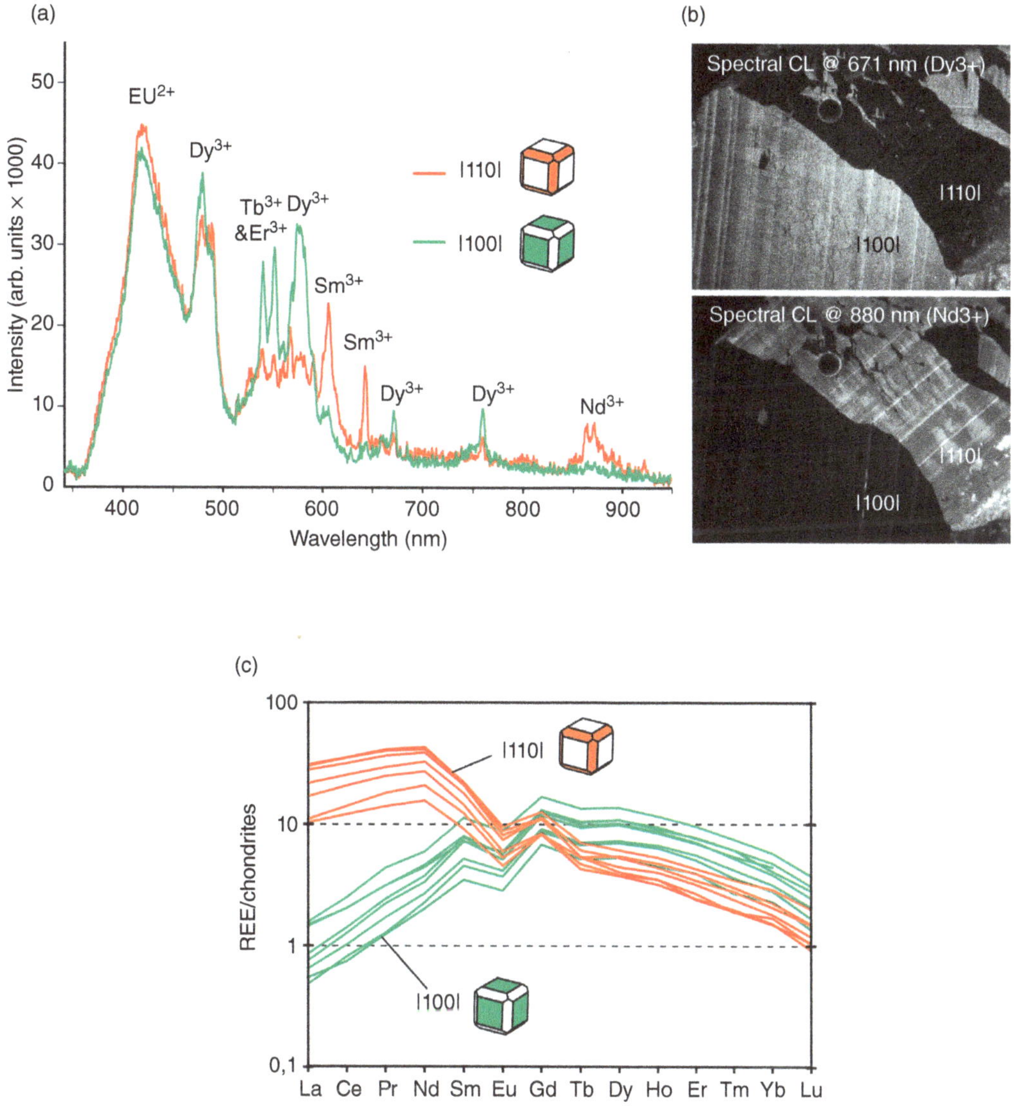

Figure 6.13 Cathodoluminescence characteristics and REE patterns in a single fluorite crystal from Ardenne, South Belgium. This example illustrates that fluorite can be affected by strong compositional heterogeneity due to partitioning of trace-elements in the different crystallographic sectors, which may not be observable in color images but are present in spectral CL imaging. (a) CL spectra recorded in |100| and |110| sectors (cube and dodecahedron, respectively). Note the difference in emission intensity for the light (Sm, Nd) compared to the heavy (Dy, Tb/Er) REE. (b) CL spectral imaging at 671 and 880 nm, which isolates the emission of Dy^{3+} and Nd^{3+}, respectively. As expected from the spectra in (a), both spectral images are the reverse of each other. (c) Chondrite-normalized REE patterns of the fluorite sample in (b) confirming the observed partitioning, with light-REE enrichment in |110| sector and slight heavy-REE enrichment in |100| sector (LA-ICP-MS data from Baele et al., 2012). A laser ablation crater is visible in (b).

there is a 20-fold La enrichment in the |110| sector relative to its time-equivalent |100| sector in a single crystal.

Thus, crystal growth may exert a dramatic influence on REE incorporation in fluorite. The effect may be so strong that REE patterns from a single crystal can look very different, except for Ce and Eu anomalies, which seem unaffected. Such compositional heterogeneity should be taken into consideration when conducting bulk or in-situ analysis of trace elements in fluorite, as well as when interpreting the resulting REE data. As an example, the partitioning of trace elements into the crystal sectors of fluorite described here may change the Tb/La ratio by up to two orders of magnitude in the Tb/Ca-Tb/La plot of Möller et al. (1976).

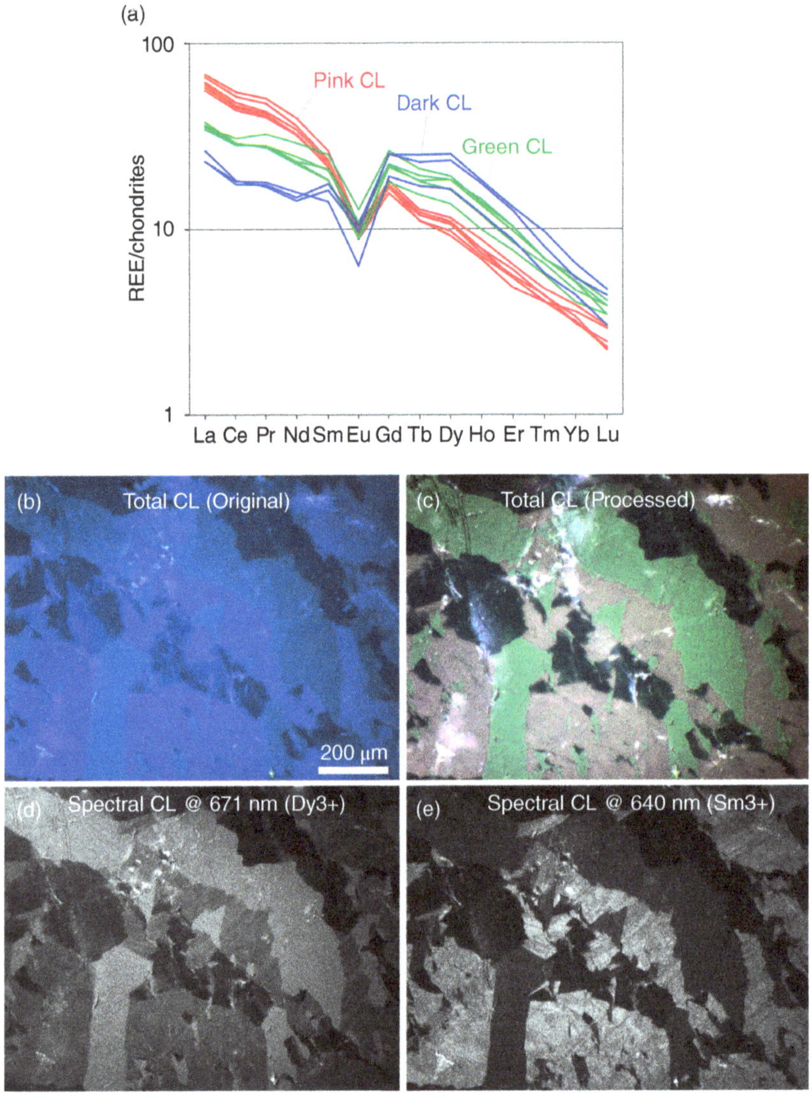

Figure 6.14 REE patterns and cathodoluminescence of a typical fluorite ore from Fontsante mine, south France. (a) Chondrite-normalized REE patterns showing an overall consistent shape but significant differences in heavy- and light-REE enrichment as a function of the CL (chondrite values after Taylor & McLennan, 1985). (b) Total (color) CL image of the fluorite sample. A texture is visible but it is masked by a strong blue component due to the ubiquitous Eu^{2+} emission. (c) Digital enhancement (contrast maximization of each RGB color channel) of the image in (b) shows three different CL colours : dark, pink and green. (d) and (e): Spectral CL of the Dy^{3+} (d) and Sm^{3+} (e) emissions. Texture relations (growth zoning, zig zag boundaries) indicate the three different CL emissions are related to crystal sectorization. The sectors cannot be crystallographically assigned due to the lack of freely-grown crystal faces on the sample. However, the (blue-)green-luminescence could correspond to the cube (|100|), and the (blue-)pink, to either the dodecahedron (|110|) or octahedron (|111|). There is a good consistency between the relative spectral CL intensity and REE content for these two sectors but not for the heavy REE in the dark-luminescing sector.

A similar REE partitioning into crystal sectors was observed in fluorite from Nordvik, Russia (Smolyanskii et al., 2009), fluorite from Long Lake and Bingham, USA (Bosze & Rakovan, 2002), and apatite from Llallagua, Bolivia, and Minas Gerais, Brazil (Rakovan & Reeder, 1994). These studies point to the role of crystal growth mechanisms, surface structure of the symmetrically nonequivalent crystal faces and variation of the characteristics (ionic radius) of the REE from La to Lu. Investigation of a typical fluorite ore from Fontsante, France, yielded similar sector zoning, although the effect is less pronounced and a third, unidentified sector was observed (Fig. 6.14). The factors controlling the partitioning of trace elements into the different crystal sectors are still unclear and cathodoluminescence is a valuable tool in this research topic.

6.4.4. Phosphates in Sedimentary Environment

The CL of apatite in phosphorites and other sedimentary phosphates is commonly, but not always, weak, and activation by Mn^{2+} and REE has been established (Waychunas, 2002).

Apatite associated with iron ore in a sedimentary context has been fairly well investigated. For instance, most of the carbonate-fluorapatite encountered in Clinton-type iron ore luminesces blue to violet, with a few bluish-white grains (Hagni, 1984). In the Lorraine ironstone, carbonate-fluorapatite grains, which exhibit more than ten morphological forms, show various CL colors that are dull-grey, bluish, or yellow-green due to different concentrations in Mn^{2+} and REE (Karakus et al., 2001). Cathodoluminescence is also an efficient complementary tool for sourcing ironstone based on the CL of apatite, which is a ubiquitous accessory mineral in these ores. A range of CL colors, textures, and replacement/depositional sequences is indeed observed in Paleozoic oolitic ironstones in Belgium, which can be used as a fingerprint in addition to other mineralogical and petrographical signatures (Baele et al., 2016).

In the sedimentary phosphate deposits of Florida, which are hosted in dolomite, collophane exhibits dull purple-brown and bright blue luminescence. CL facilitates the distinction of phosphate-bearing grains from the red-luminescent dolomite in flotation concentrates during the processing of the ore (Hagni, 1984).

Diagenetic monazite (so-called grey or dark monazite) forms during late diagenesis to low-grade metamorphism of shale sequences (Donnot et al., 1973; Rosenblum & Mosier, 1983; Reed et al., 1987; Burnotte et al., 1989; Copjakova et al., 2011). Monazite-bearing shales can feed placer systems, which may be of economic interest (Cobert et al., 2015). Thus, the delimitation of potential source rocks is essential for early exploration strategies. CL imaging enables the rapid evaluation of the content and mineralogy of common REE minerals in shale. First, luminescent radiation-induced halos in quartz can be used to locate REE-bearing minerals, which usually contain sufficient actinides for damaging the lattice of adjacent quartz over geological timescales. The only limitation of this method is when the shale is pure argilite without any silt grains, which is not common. Subsequently, the CL characteristics will positively identify the mineral. Diagenetic monazite is dark to blue luminescent (with a strong out-of-focus effect) and bright under spectral CL centered on the emission wavelength of Nd^{3+} (Figs. 6.2 & 6.15). It very often exhibits a simple concentric zoning consisting of a darker, sometimes

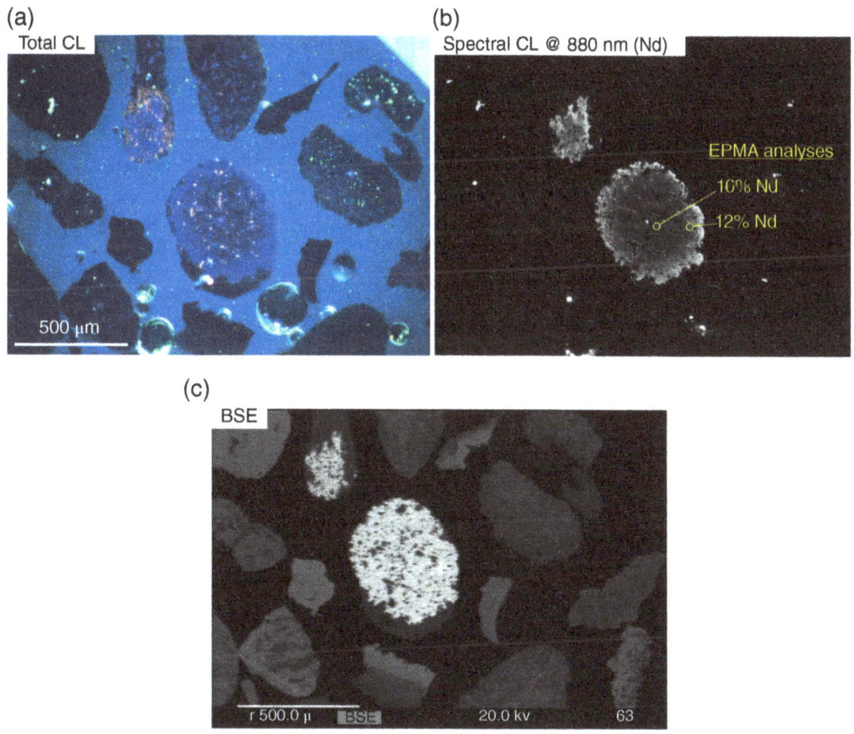

Figure 6.15 Cathodoluminescence (CL) and backscattered electrons (BSE) images of panned stream sediments from Ardenne, Belgium. (a) Grey monazite nodules are barely distinguished from other grains under total (color) CL because their luminescence is very weak (the embedding epoxy luminesces blue). (b) The monazite nodules are revealed under spectral CL (Nd^{3+} emission), which also shows their typical concentric zoning. In fact, the dark, inner core of the nodules has a higher concentration in Nd than the bright outer rim (EPMA analyses in the bright outermost rim in other nodules not shown here yielded < 10% Nd), which could indicate self-quenching. (c) The zoning in diagenetic monazite is not visible under backscattered electrons.

slightly more reddish inner core and a brighter outer rim. The zoning is best observed under spectral CL at ~880 nm and could relate to self-quenching as dark-CL zones are more enriched in Nd than bright-CL zones (Fig. 6.11). A bottle-green, usually homogeneous CL (Richter et al., 2006), and no Nd^{3+} spectral emission but a Dy^{3+} emission instead, indicate xenotime. Many zircons luminesce blue to bright blue, but occasional green luminescent zircon can be distinguished from xenotime by its fine growth zoning. If needed, a detailed examination of the emission spectrum also helps (Richter et al., 2006). All of the allanite we have observed is nonluminescent, likely because this mineral incorporates Fe^{2+}. In addition, allanite does not show a Nd^{3+} emission in the NIR spectrum. Apatite usually does not contain sufficient amounts of actinides to induce noticeable luminescence halos in adjacent quartz. However, apatite in shale will still be unambiguously distinguished from monazite or xenotime by its bright CL of varying color, with activation by Dy^{3+} and Nd^{3+} simultaneously and/or by Mn^{2+}.

6.5. SUMMARY AND PROSPECTS

CL expands the capability of conventional observation techniques (optical and electron microscope) and provides an important link between petrography and geochemistry. This review has reported a range of applications of CL to ore geology, which can be summarized as follows:

The CL characteristics of minerals provide signatures (fingerprints) that can be applied for:

1. evaluating mineralization potential;
2. determining the provenance of stream, shore, and soil sediments;
3. distinguishing and tracing mineralizing episodes in multistage deposits;
4. drawing geochemical, alteration or paleogeographical trends associated with temperature gradients, flow patterns, and so on at the scale of the deposit to that of the district;
5. quickly and easily visualizing luminescent minerals in fundamental and applied (beneficiation) problems.

Furthermore, CL textures in minerals and their associated geochemical/structural heterogeneities are a unique record, which can be used for reconstructing mineralization processes, relating mineralizing events to specific physiochemical conditions, and guiding in-situ micro-analysis for geochemical, geothermometrical, geochronological, and fluid inclusion studies, and interpreting the results. All of this contributes to a better understanding of ore deposits.

Quartz and apatite appear as the most versatile and promising minerals in CL studies applied to ore deposits (Table 6.1) as they are ubiquitous in many geological environments and can record their formation conditions with great sensitivity. Quartz is especially abundant in felsic to intermediate magmatic and hydrothermal settings while apatite also occurs in alkaline magmatic and sedimentary systems. The CL of other minerals such as fluorite, calcite/dolomite, zircon, feldspars, scheelite, cassiterite, and sphalerite, has great potential in ore geology as well, but either in a more restricted range of environments or for more specific applications.

While progress is being made, our rather poor understanding of CL phenomena in natural minerals still limits our capability of interpreting CL data in terms of temperature, pressure, composition, and so on, of the formation environment. This is particularly evident for

Table 6.1 Summary of cathodoluminescent minerals that have been used in the study of various ore deposits reported in this chapter (except diamond and strontianite).

Environment	Deposit type	anh	ap	cal/dol	cst	fl	fsp	mnz	qtz	sch	sp	xtm	zrn
Magmatic alkaline	Alkaline and carbonatite complexes		X	X		X		X					X
	IOCG	X	X				X						
Magmatic-hydrothermal	Cu-Mo-W Porphyry and epithermal	X	X						X	X			
	Granite (Sn-W and U)		X		X	X			X	X			X
Hydrothermal	Orogenic gold								X	X			X
	Carlin gold		X	X					X				
	VMS and sedex								X	X	X		
	MVT and low-T fluorite			X		X					X		
Sedimentary	Sedimentary phosphates (and iron)		X					X	X			X	
Diverse	Uranium								X				

Note: See text for references. anh: anhydrite; ap: apatite; cal/dol: calcite/dolomite; fl: fluorite; fsp: feldspar; mnz: monazite; qtz: quartz; sch: scheelite; sp: sphalerite; xtm: xenotime; zrn: zircon.

quartz, whose CL is mainly generated by a variety of crystal defects rather than substitutional impurities (Steven-Kalceff et al., 2000). Elucidating its CL properties then necessitates the use of advanced analytical techniques as demonstrated by Götze et al. (2015), who recently unraveled the yellow CL of hydrothermal quartz.

ACKNOWLEDGMENTS

We would like to acknowledge Sci-Tech at Western Washington University for access to and assistance with the instrumentation. The Royal Museum for Central Africa is warmly thanked for providing the study samples coming from Burundi. The first version of the manuscript benefited from the comments of Ian Coulson and an anonymous reviewer. We are also grateful to Laurence Robb, who handled the manuscript.

REFERENCES

Agangi, A., Hofmann, A., & Przybyłowicz, W. (2014). Trace element zoning of sulfides and quartz at Sheba and Fairview gold mines: Clues to Mesoarchean mineralisation in the Barberton Greenstone Belt, South Africa. *Ore Geology Reviews*, *56*, 94–114. https://doi.org/10.1016/j.oregeorev.2013.08.016.

Assadzadeh, G. E., Samson, I. M., & Gagnon, J. E. (2017). The trace element chemistry and cathodoluminescence characteristics of fluorite in the Mount Pleasant Sn-W-Mo deposits: Insights into fluid character and implications for exploration. *Journal of Geochemical Exploration*, *172*, 1–19. https://doi.org/10.1016/j.gexplo.2016.09.010.

Baele, J. M. (1999). Reliques silicifiées et minéralisées de paléokarsts post-varisques sur le Dévonien en Belgique méridionale. *Annales de la Société Géologique du Nord*, *6*, 127–133.

Baele, J. M. (2014). Application of cathodoluminescence total and spectral imaging to ore geology. Paper presented at 24th Earth Sciences Meeting (RST2014), Pau, France.

Baele, J. M., Dreesen, R., & Dusar, M. (2016). Assessing apatite cathodoluminescence as a tool for sourcing oolitic ironstones. In C. Billard, D. Bosquet, R. Dreesen, E. Goemaere, C. Hamon, Y. Jadin, H. Salomon, & X. Savary (Eds.), *About Hematite. Procurement and transformation during Recent Prehistory, Anthropologica et Praehistorica* (Vol. *126*/2015(2), pp. 57–67). Brussels: Studia Praehistorica Belgica.

Baele, J. M., Monin, L., Navez, J., & André, L. (2012). Systematic REE partitioning in cubo-dodecahedral fluorite from Belgium revealed by cathodoluminescence spectral imaging and laser ablation-ICP-MS. In M. A. T. M. Broekmans (Ed.), *Proceedings of the 10th International Congress for Applied Mineralogy (ICAM)* (pp. 23–30). Berlin Heidelberg: Springer-Verlag. https://doi.org/10.1007/978-3-642-27682-8_4.

Ballouard, C., Poujol, M., Mercadier, J., Deloule, E., Boulvais, P., & Baele, J. M., et al. (2017). Uranium metallogenesis of the peraluminous leucogranite from the Pontivy-Rostrenen magmatic complex (French Armorican Variscan belt): The result of long-term oxidized hydrothermal alteration during strike-slip deformation. *Mineralium Deposita*. https://doi.org/10.1007/s00126-017-0761-5.

Bao, Z., Sun, W., Li, C., & Zhao, Z. (2014). U-Pb dating of hydrothermal zircon from the Dongping gold deposit in North China: Constraints on the mineralization processes. *Ore Geology Reviews*, *61*, 107–119. https://doi.org/10.1016/j.oregeorev.2014.02.006

Barbarand, J., & Pagel, M. (2001). Cathodoluminescence study of apatite crystals. *American Mineralogist*, *86*(4), 473–484. https://doi.org/10.2138/am-2001-0411.

Barker, S. L., Hickey, K. A., Dipple, G. M., & Layne, G. (2009). Apatite as a paleohydrothermal fluid recorder in Carlin-type gold deposits. Paper presented at AGU Joint Assembly 2009, Toronto, Canada.

Beaton, N. B. (2015). Diagenetic Controls on Hydrothermal Fluid Flow in the Osiris, Isis, and Isis East Carlin-type Au showings, Nadaleen Trend, Yukon. Master's Thesis, Department of Earth and Atmospheric Sciences, University of Alberta. https://doi.org/10.7939/R3B853R92.

Bell III, H., Carpenter, R. H., & Feiss, P. G. (1980). Volcanogenic ore deposits of the Carolina slate belt. *Excursions in Southeastern Geology*, *1*, 149–178.

Belousova, E., Griffin, W. L., O'Reilly, S. Y., & Fisher, N. L. (2002). Igneous zircon: Trace element composition as an indicator of source rock type. *Contributions to Mineralogy and Petrology*, *143*(5), 602–622. https://doi.org/10.1007/s00410-002-0364-7.

Belousova, E. A., Griffin, W. L., & Pearson, N. J. (1998). Trace element composition and cathodoluminescence properties of southern African kimberlitic zircons. *Mineralogical Magazine*, *62*(3), 355–366. https://doi.org/10.1180/002646198547747.

Ben Abdallah, R., Medhioub, M., Hatira, N., Chaftar, R., Baele, J. M., Yans, J., & Jamoussi, F. (2013). Hydrothermal and meteoric alteration of Triassic materials in the Oued Belif structure (Nefza, Northern Tunisia). *European Journal of Scientific Research*, *98*(4), 470–480.

Blanc, P., Baumer, A., Cesbron, F., Ohnenstetter, D., Panczer, G. & Rémond, G. (2000). Systematic cathodoluminescence spectral analysis of synthetic doped minerals : anhydrite, apatite, calcite, fluorite, scheelite and zircon. In M. Pagel, V. Barbin, P. Blanc, & D. Ohnenstetter (Eds.), *Cathodoluminescence in Geosciences* (pp. 127–160). Heidelberg, New York: Springer Verlag.

Boggs, S., & Krinsley, D. (2006). *Application of Cathodoluminescence Imaging to the Study of Sedimentary Rocks*. New York: Cambridge University Press.

Bosze, S., & Rakovan, J. (2002). Surface-structure-controlled sectoral zoning of the rare earth elements in fluorite from Long Lake, New York, and Bingham, New Mexico, USA. *Geochimica et Cosmochimica Acta*, *66*, 997–1009. https://doi.org/10.1016/S0016-7037(01)00822-5

Botis, S., Nokhrin, S.M., Pan, Y., Xu, Y. & Bonli, T. (2005). Natural radiation-induced damage in quartz. I. Correlations between cathodoluminescence colors and paramagnetic defects. *The Canadian Mineralogist*, *43*, 1565–1580. https://doi.org/10.2113/gscanmin.43.5.1565

Botis, S., Pan, Y., Bonli, T., Xu, Y., Zhang, A., Nokhrin, S., & Sopuck, V. (2006). Natural radiation-induced damage in quartz. II. Distribution and implications for uranium mineralization in the Athabasca basin, Saskatchewan, Canada. *The Canadian Mineralogist*, *44*(6), 1387–1402. https://doi.org/10.2113/gscanmin.44.6.1387

Bouzari, F., Hart, C. J., Bissig, T., & Barker, S. (2016). Hydrothermal alteration revealed by apatite luminescence and chemistry: a potential indicator mineral for exploring covered porphyry copper deposits. *Economic Geology*, *111*(6), 1397–1410. https://doi.org/10.2113/econgeo.111.6.1397

Breiter, K., Korbelová, Z., Chládek, Š., Uher, P., Knesl, I., Rambousek, P., et al. (2017). Diversity of Ti-Sn-W-Nb-Ta oxide minerals in the classic granite-related magmatic-hydrothermal Cínovec/Zinnwald Sn-W-Li deposit (Czech Republic). *European Journal of Mineralogy*, *29*(4), 727–738. https://doi.org/10.1127/ejm/2017/0029-2650

Brokus, S. A., Silletti, D. K., Lunderberg, J. M., DeYoung, P. A., Peaslee, G. F., Carpenter, D. E., & Buscaglia, J. (2015). Cathodoluminescence dependence on feldspar mineral structure and implications for forensic geology. *American Mineralogist*, *100*(2–3), 451–458. https://doi.org/ 10.2138/am-2015-4661.

Burnotte, E., Pirard, E., & Michel, G. (1989). Genesis of gray monazites; evidence from the Paleozoic of Belgium. *Economic Geology*, *84*, 1417–1429. https://doi.org/10.2113/gsecongeo.84.5.1417.

Campbell, L. S., & Henderson, P. (1997). Apatite paragenesis in the Bayan Obo REE-Nb-Fe ore deposit, Inner Mongolia, China. *Lithos*, *42*(1–2), 89–103. https://doi.org/10.1016/s0024-4937(97)00038-8.

Cobert, C., Baele, J. M., Boulvais, P., Decrée, S., Dupont, N., & Spagna, P. (2015). Grey Monazite Paleoplacers in Lower Cretaceous Continental Formations in the Mons Basin, Belgium. *Proceedings of the 13th Biennial SGA Meeting*, *2*, 703–706.

Copjakova, R., Novak. M., & Francu, E. (2011). Formation of authigenic monazite-(Ce) to monazite-(Nd) from Upper Carboniferous graywackes of the Drahany Upland: Roles of the chemical composition of host rock and burial temperature. *Lithos*, *127*, 373–385. https://doi.org/10.1016/j.lithos.2011.08.001.

Coppola, V., Boni, M., Gilg, H. A., Balassone, G., & Dejonghe, L. (2008). The «calamine» nonsulfide Zn-Pb deposits of Belgium : Petrographical, mineralogical and geochemical characterization. *Ore Geology Reviews*, *33*, 187–210. https://doi.org/10.1016/j.oregeorev.2006.03.005.

Corriveau, L., Montreuil, J. F., & Potter, E. G. (2016). Alteration facies linkages among iron oxide copper-gold, iron oxide-apatite, and affiliated deposits in the Great Bear magmatic zone, Northwest Territories, Canada. *Economic Geology*, *111*(8), 2045–2072. https://doi.org/10.2113/econgeo.111.8.2045.

Corriveau, L., Ootes, L., Mumin, H., Jackson, V., Bennett, V., Cremer, J. F., et al. (2007). Alteration vectoring to IOCG (U) deposits in frontier volcano-plutonic terrains, Canada. *Proceedings of Exploration*, *7*, 1171–1177.

Corriveau, L., Williams, P. J., & Mumin, A. H. (2010). Alteration vectors to IOCG mineralization: from uncharted terranes to deposits. *Geological Association of Canada Short Course Notes*, *20*, 89–110.

Coulson, I. M. (Ed.) (2014). Cathodoluminescence and its application to geoscience. *Mineralogical Association of Canada Short Series Course*, *45*.

Decrée, S., Marignac, C., De Putter, T., Yans, J., Clauer, N., Dermech, M., et al. (2013). The Oued Belif Hematite-Rich Breccia: A Miocene Iron Oxide Cu-Au-(U-REE) Deposit in the Nefza Mining District, Tunisia. *Economic Geology*, *108*(6), 1425–1457. https://doi.org/10.2113/econgeo.108.6.1425.

Decrée, S., Boulvais, P., Cobert, C., Baele, J. M., Midende, G., Gardien, V., et al. (2015). Structurally controlled hydrothermal alteration in the syntectonic Neoproterozoic Upper Ruvubu Alkaline Plutonic Complex (Burundi): Implications for REE and HFSE mobilities. *Precambrian Research*, *269*, 281–295. https://doi.org/10.1016/j.precamres.2015.08.016

Decrée, S., Boulvais, P., Tack, L., André, L., & Baele, J. M. (2016). Fluorapatite in carbonatite-related phosphate deposits: the case of the Matongo carbonatite (Burundi). *Mineralium Deposita*, *51*(4), 453–466. https://doi.org/10.1007/s00126-015-0620-1.

De St. Jorre, L. D. S., & Smith, D. G. (1988). Cathodoluminescent gallium-enriched feldspars from the Thore Lake rare metal deposits, Northwest Territories. *The Canadian Mineralogist*, *26*(2), 301–308.

De Toledo, M. C. M., Lenharo, S. L., Ferrari, V. C., Fontan, F., De Parseval, P., & Leroy, G. (2004). The compositional evolution of apatite in the weathering profile of the Catalão I alkaline-carbonatitic complex, Goiás, Brazil. *The Canadian Mineralogist*, *42*(4), 1139–1158.

Donovan, J. J., Lowers, H. A.,& Rusk, B. G. (2011). Improved electron probe microanalysis of trace elements in quartz. *American Mineralogist*, *96*, 274–282. https://doi.org/10.2138/am.2011.3631.

Dickson, J. A. D. (1980). Artificial coloration of fluorite by electron bombardment. *Mineralogical Magazine*, *43*, 820–822.

Dill, H. G., & Weber, B. (2010). Accessory minerals of fluorite and their implication regarding the environment of formation (Nabburg-Wölsendorf fluorite district, SE Germany), with special reference to fetid fluorite ("Stinkspat"). *Ore Geology Reviews*, *37*, 65–86. https://doi.org/10.1016/j.oregeorev.2010.01.004

Donnot, M., Guigues, J., Lulzac, Y., Magnien, A., Parfenoff, A., & Picot, P. (1973). Un nouveau type de gisement d'europium : La monazite grise à europium en nodules dans les schistes paléozoïques de Bretagne. *Mineralium Deposita*, *8*, 7–18. https://doi.org/10.1007/BF00203346.

Ebers, M. L., & Kopp, O. C. (1979). Cathodoluminescent microstratigraphy in gangue dolomite, the Mascot-Jefferson City district, Tennessee. *Economic Geology*, *74*(4), 908–918. https://doi.org/10.2113/gsecongeo.74.4.908.

Emsbo, P., Hofstra, A. H., Lauha, E. A., Griffin, G. L., & Hutchinson, R. W. (2003). Origin of high-grade gold ore, source of ore fluid components, and genesis of the Meikle and neighboring Carlin-type deposits, northern Carlin trend, Nevada. *Economic Geology*, *98*(6), 1069–1105. https://doi.org/10.2113/gsecongeo.98.6.1069

Erickson, S. D. (2011). Cathodoluminescence of sodic calcic alteration, White Rock Stock contact aureole, Elk Mountains, Colorado. Paper presented at 2011 GSA Annual Meeting, Minneapolis, USA.

Farmer, C. B., Searl, A., & Halls, C. (1991). Cathodoluminescence and growth of cassiterite in the composite lodes at South Crofty Mine, Cornwall, England. *Mineralogical Magazine*, *55*(380), 447–458. https://doi.org/10.1180/minmag.1991.055.380.14.

Frelinger, S. N., Ledvina, M. D., Kyle, J. R., & Zhao, D. (2015). Scanning electron microscopy cathodoluminescence of quartz: Principles, techniques and applications in ore geology. *Ore Geology Reviews*, *65*, 840–852. https://doi.org/10.1016/j.oregeorev.2014.10.008

Gaft, M., Panczer, G., Uspensky, E., & Reisfeld, R. (1999). Laser-induced time-resolved luminescence of rare-earth elements in scheelite. *Mineralogical Magazine*, *63*(2), 199–210. https://doi.org/10.1180/002646199548439.

Gigoux, M., Delpech, G., Guerrot, C., Pagel, M., Augé, T., Négrel, P., & Brigaud, B. (2015). Evidence for an Early Cretaceous mineralizing event above the basement/sediment unconformity in the intracratonic Paris Basin: paragenetic sequence and Sm-Nd dating of the world-class Pierre-Perthuis stratabound fluorite deposit. *Mineralium Deposita*, *50*(4), 455–463. https://doi.org/10.1007/s00126-015-0592-1.

Götze, J. (2010). Chemistry, textures and physical properties of quartz: Geological interpretation and technical application. *Mineralogical Magazine*, *73*(4), 645–671. https://doi.org/10.1180/minmag.2009.073.4.645.

Götze, J., Habermann, D., Neuser, R. D., & Richter, D. K. (1999). High-resolution spectrometric analysis of rare earth elements-activated cathodoluminescence in feldspar minerals. *Chemical Geology*, *153*(1), 81–91. https://doi.org/10.1016/S0009-2541(98)00153-3.

Götze, J., Plötze, M., Graupner, T., Hallbauer, D. K., & Bray, C. J. (2004). Trace element incorporation into quartz: a combined study by ICP-MS, electron spin resonance, cathodoluminescence, capillary ion analysis, and gas chromatography. *Geochimica et Cosmochimica Acta*, *68*(18), 3741–3759. https://doi.org/10.1016/j.gca.2004.01.003.

Götze, J., Plötze, M., & Trautmann, T. (2005). Structure and luminescence characteristics of quartz from pegmatites. *American Mineralogist*, *90*, 13–21. https://doi.org/10.2138/am.2005.1582.

Götze, J., & Kempe, U. (2008). A comparison of optical microscope (OM) and scanning electron microscope (SEM) based cathodoluminescence (CL) imaging and spectroscopy applied to geosciences. *Mineralogical Magazine*, *72*, 909–924. https://doi.org/10.1180/minmag.2008.072.4.909.

Götze, J. (2012). Application of cathodoluminescence microscopy and spectroscopy in geosciences. *Microscopy and Microanalysis*, *18*(6), 1270–1284. https://doi.org/10.1017/S1431927612001122.

Götze, J., Pan, Y., Stevens-Kalceff, M., Kempe, U., & Müller, A. (2015). Origin and significance of the yellow cathodoluminescence (CL) of quartz. *American Mineralogist*, *100*(7), 1469–1482. https://doi.org/10.2138/am-2015-5072

Graupner, T., Niedermann, S., Rhede, D., Kempe, U., Seltmann, R., Williams, C. T., & Klemd, R. (2010). Multiple sources for mineralizing fluids in the Charmitan gold (-tungsten) mineralization (Uzbekistan). *Mineralium Deposita*, *45*(7), 667–682. https://doi.org/10.1007/s00126-010-0299-2.

Graupner, T., Mühlbach, C., Schwarz-Schampera, U., Henjes-Kunst, F., Melcher, F., & Terblanche, H. (2015). Mineralogy of high-field-strength elements (Y, Nb, REE) in the world-class Vergenoeg fluorite deposit, South Africa. *Ore Geology Reviews*, *64*, 583–601. https://doi.org/10.1016/j.oregeorev.2014.02.012.

Groves, D. I., & Vielreicher, N. M. (2001). The Phalabowra (Palabora) carbonatite-hosted magnetite–copper sulfide deposit, South Africa: an end-member of the iron-oxide copper-gold-rare earth element deposit group? *Mineralium deposita*, *36*(2), 189–194. https://doi.org/10.1007/s001260050298.

Grzela, D. (2017). *Chemical composition of indicator minerals from orogenic gold deposits and glacial sediments of the Val-d'Or district (Québec, Canada)*. Master's thesis (https://explomin.ggl.ulaval.ca/fileadmin/aem/documents/Documents/DGrzela.pdf). Université de Laval, Québec.

Haberman, D., Neuser, R., & Richter, D. K. (2000). Quantitative high resolution spectral analysis of Mn^{2+} in sedimentary calcite. In M. Pagel, V. Barbin, P. Blanc, D. Ohnenstetter (Eds.), *Cathodoluminescence in Geosciences* (pp. 331–358). Heidelberg, New York: Springer Verlag.

Hagni, R. D. (1984). Cathodoluminescence microscopy applied to mineral exploration and beneficiation. In W. C. Park., D. M. Hausen, & R. D. Hagni (Eds.), *Proceedings of the Second International Congress on Applied Mineralogy in the Minerals Industry* (pp. 41–66). New York: AIME.

Hagni, R. D., & Shivdasan, P. R. (2001). Application of cathodoluminescence microscopy to the study of carbonatite-related fluorspar ores, host rock carbonatites and beneficiation products at Okorusu, Namibia. *Minerals & Metallurgical Processing*, *18*(4), 203–208.

Hagni, R. D. (2012) The replacement of carbonatite and marble by fluorspar ores and its relationship to beneficiation problems at Okorusu, Namibia. In M. A. T. M. Broekmans (Ed.), *Proceedings of the 10th International Congress for Applied Mineralogy (ICAM)* (pp. 297–305). Berlin, Heidelberg: Springer-Verlag. https://doi/org/10.1007/978-3-642-27682-8_36.

Hayward, C. L., & Jones, A. P. (1991). Cathodoluminescence petrography of middle Proterozoic extrusive carbonatite from Qasiarsuk, South Greenland. *Mineralogical Magazine*, *55*(381), 591–603.

Hu, B., Pan, Y., Botis, S., Rogers, B., Kotzer, T., & Yeo, G. (2008). Radiation-induced defects in drusy quartz, Athabasca Basin, Canada: A new aid to exploration of uranium deposits. *Economic Geology*, *103*(7), 1571–1580. https://doi.org/10.2113/gsecongeo.103.7.1571.

Huang, R. & Audetat, A. (2012). The titanium-in-quartz (TitaniQ) thermobarometer: A critical examination and re-calibration. *Geochimica et Cosmochimica Acta*, *84*, 75–89. https://doi.org/10.1016/j.gca.2012.01.009.

Ioannou, S. E., Götze, J., Weiershäuser, L., Zubowski, S. M., & Spooner, E. T. C. (2003). Cathodoluminescence characteristics of Archean volcanogenic massive sulfide related quartz: Noranda, Ben Nevis and Matagami districts, Abitibi Subprovince, Canada. *Geochemistry Geophysics Geosystems*, *5*(2). https://doi.org/10.1029/2003GC000613

Karakus, M., Hagni, R. D., & Spreng, A. C. (2001). Cathodoluminescence petrography and chemistry of the phosphate grains in the lower Jurassic (Aalenian) ironstones of Lorraine, France. In R. D. Hagni, (Ed.), *Studies on Ore Deposits, Mineral Economics, and Applied Mineralogy: With Emphasis on Mississippi Valley-Type Base Metals and Carbonate-Related Ore Deposits* (pp. 335–363). University of Missouri-Rolla.

Karakus, M., Hagni, R. D., Koenig, A. E., & Ciftc, E. (2008). Cathodoluminescence, Laser Ablation Inductively Coupled Plasma Mass Spectrometry, Electron Probe Microanalysis and Electron Paramagnetic Resonance Analyses of Natural Sphalerite. In *Proceedings of the 9th International Congress for Applied Mineralogy, ICAM 2008* (pp. 113–124). Carlton Victoria: Australasian Institute of Mining and Metallurgy Publication Series.

Kayama, M., Nishido, H., & Ninagawa, K. (2009). Cathodoluminescence of feldspar and the assignment of luminescence centers. *Advances in ESR Applications*, *25*, 13–18.

Kempe, U., Gruner, T., Nasdala, L. & Wolf, D. (2000). Relevance of cathodoluminescence for the interpretation of U-Pb zircon ages, with an example of an application to a study of zircons from the Saxonian Granulite Complex, Germany. In M. Pagel, V. Barbin, P. Blanc, & D. Ohnenstetter (Eds.), *Cathodoluminescence in Geosciences* (pp. 425–456). Heidelberg, New York: Springer Verlag.

Kempe, U., & Götze, J. (2002). Cathodoluminescence (CL) behaviour and crystal chemistry of apatite from rare-metal deposits. *Mineralogical Magazine*, *66*(1), 151–172. https://doi.org/10.1180/0026461026610019

Kiryanova, E. V. & Glikin, A. E. (1999). The laws of fluorite and calcite habit formation in terms of the morphogenetic structural-chemical concepts. *Journal of Crystal Growth*, *198*/199, 697–703. https://doi.org/10.1016/S0022-0248(98)01148-8.

Kopp, O. C., Reeves, D. K., Rivers, M. L., & Smith, J. V. (1990). Synchrotron X-ray fluorescence analysis of zoned carbonate gangue in Mississippi Valley-type deposits (USA). *Chemical Geology*, *81*(4), 337–347. https://doi.org/10.1016/0009-2541(90)90054-B.

Kuhlemann, J., Vennemann, T., Herlec, U., Zeeh, S., & Bechstädt, T. (2001). Variations of sulfur isotopes, trace element compositions, and cathodoluminescence of Mississippi Valley-type Pb-Zn ores from the Drau Range, eastern Alps (Slovenia-Austria): Implications for ore deposition on a regional versus microscale. *Economic Geology*, *96*(8), 1931–1941. https://doi.org/10.2113/gsecongeo.96.8.1931.

Landtwing, M. R., & Pettke, T. (2005). Relationships between SEM-cathodoluminescence response and trace-element composition of hydrothermal vein quartz. *American Mineralogist*, *90*, 122–131. https://doi.org/10.2138/am.2005.1548.

Leach, D. L. (1997). Evidence for hot Mississippi Valley-type brines in the Reelfoot rift complex, south-central United States, in Late Pennsylvanian-Early Permian. *USGS Geological Survey Professional Paper* (Vol. *1577*). Washington, DC: US Government Printing Office.

Leach, D. L., Marsh, E., Emsbo, P., Rombach, C. S., Kelley, K. D., & Anthony, M. (2004). Nature of hydrothermal fluids at the shale-hosted red dog Zn-Pb-Ag deposits, Brooks Range, Alaska. *Economic Geology*, *99*(7), 1449–1480. https://doi.org/10.2113/gsecongeo.99.7.1449.

Leeman, W. P., MacRae, C. M., Wilson, N. C., Torpy, A., Lee, C. T. A., Student, J. J., et al. (2012). A study of cathodoluminescence and trace element compositional zoning in natural quartz from volcanic rocks: Mapping titanium content in quartz. *Microscopy and Microanalysis*, *18*(6), 1322–1341. https://doi.org/10.1017/S1431927612013426.

Lubben, J. D., Cline, J. S., & Barker, S. L. (2012). Ore fluid properties and sources from quartz-associated gold at the Betze-Post Carlin-type gold deposit, Nevada, United States. *Economic Geology*, *107*(7), 1351–1385. https://doi.org/10.2113/econgeo.107.7.1351.

Lypaczewski, P., Normandeau, P. X., Paquette, J., & McMartin, I. (2013). Petrographic and cathodoluminescence characterization of apatite from the Sue-Dianne and Brooke IOCG mineralization systems, Great Bear magmatic zone, Northwest Territories, Canada. *Geol Surv Can, Open File 7319*, https://doi.org/10.4095/292369.

Machel, H. G. (1985). Cathodoluminescence in calcite and dolomite and its chemical interpretation. *Geosciences Canada*, *12*, 139–147.

Machel, H. G., (2000). Application of cathodoluminescence to carbonate diagenesis. In M. Pagel, V. Barbin, P. Blanc, & D. Ohnenstetter (Eds.), *Cathodoluminescence in Geosciences* (pp. 271–301). Heidelberg, New York: Springer Verlag.

Maki, S., Ohgo, S., & Nishido, H. (2016). Cathodoluminescence characterization of feldspar minerals from granite-syenite rocks in Iwagijima Island, Ehime Prefecture, Japan. *Naturalistae*, *20*, 13–18.

Mao, W., Rusk, B., Yang, F., & Zhang, M. (2017). Physical and chemical evolution of the dabaoshan porphyry mo deposit, South China: Insights from fluid inclusions, cathodoluminescence, and trace elements in quartz. *Economic Geology*, *112*(4), 889–918. https://doi.org/10.2113/econgeo.112.4.889.

Marshall, D. J. (1988). *Cathodoluminescence of Geological Materials*. Boston: Unwin Hyman.

Mariano, A. N., & Ring, P. J. (1975). Europium-activated cathodoluminescence in minerals. *Geochimica et Cosmochimica Acta*, *39*(5), 649–660. https://doi.org/10.1016/0016-7037(75)90008-3.

Mariano, A. N. (1979). Report of the geology and economic potential of the Prairie Lake carbonatite-alkalic complex (Consultants report to Nuinsco Resources Limited). Retrieved from http://www.geologyontario.mndm.gov.on.ca/mndmfiles/afri/data/imaging/42E02SE0013/42E02SE0013.Pdf.

Maydagán, L., Franchini, M., Rusk, B., Lentz, D. R., McFarlane, C., Impiccini, A., et al. (2015). Porphyry to Epithermal Transition in the Altar Cu-(Au-Mo) Deposit, Argentina, Studied by Cathodoluminescence, LA-ICP-MS, and Fluid Inclusion Analysis. *Economic Geology*, *110*(4), 889–923. https://doi.org/10.2113/econgeo.110.4.889.

Mitchell, R. H. (2014). Cathodoluminescence of apatite. *Mineralogical Association of Canada Short Course*, *45*, 143–167.

Möller, P., Parekh, P., & Schneider, H. J. (1976). The application of Tb/Ca-Tb/La abundance ratios to problems of fluorspar genesis. *Mineralium Deposita*, *11*, 111–116. https://doi.org/10.1007/BF00203098.

Mondillo, N., Boni, M., Balassone, G., Spoleto, S., Stellato, F., Marino, et al. (2016). Rare earth elements (REE): Minerals in the Silius fluorite vein system (Sardinia, Italy). *Ore Geology Reviews*, *74*, 211–224. https://doi.org/10.1016/j.oregeorev.2015.11.016.

Montanez, I. P. (1996). Application of Cathodoluminescent Cement Stratigraphy for Delineating Regional Diagenetic and Fluid Migration Events Associated with Mississippi Valley-Type Mineralization in the Southern Appalachians. In D. F. Sangster (Ed.), *Carbonate-Hosted Lead-Zinc Deposits: 75th Anniversary Volume* (Vol. 4, pp. 432–447). Littleton, CO, Special Publication of the Society of Economic Geologists.

Monteiro, L. V., Xavier, R. P., de Carvalho, E. R., Hitzman, M. W., Johnson, C. A., de Souza Filho, C. R., & Torresi, I. (2008). Spatial and temporal zoning of hydrothermal alteration and mineralization in the Sossego iron oxide-copper-gold deposit, Carajás Mineral Province, Brazil: paragenesis and stable isotope constraints. *Mineralium Deposita*, *43*(2), 129–159. https://doi.org/10.1007/s00126-006-0121-3.

Moreto, C. P. N., Monteiro, L. V. S., Xavier, R. P., Creaser, R. A., DuFrane, S. A., Tassinari, et al. (2015). Neoarchean and Paleoproterozoic iron oxide-copper-gold events at the Sossego

Deposit, Carajás Province, Brazil: Re-Os and U-Pb geochronological evidence. *Economic Geology*, *110*(3), 809–835. https://doi.org/10.2113/econgeo.110.3.809.

Moura, A., Gotze, J., & Sodré Borges, F. (2003). Preliminary results on the cathodoluminescence of quartz from the Neves Corvo VMS deposit (Portugal). *Boletín de la Sociedad Española de Mineralogía*, (*26-A*), 175–176.

Müller, A., Seltmann, R., Halls, C., Siebel, W., Dulski, P., Jeffries, et al. (2006). The magmatic evolution of the Land's End pluton, Cornwall, and associated pre-enrichment of metals. *Ore Geology Reviews*, *28*(3), 329–367. https://doi.org/10.1016/j.oregeorev.2005.05.002.

Müller, A., Herrington, R., Armstrong, R., Seltmann, R., Kirwin, D. J., Stenina, N. G., & Kronz, A. (2010). Trace elements and cathodoluminescence of quartz in stockwork veins of Mongolian porphyry-style deposits. *Mineralium Deposita*, *45*(7), 707–727. https://doi.org/10.1007/s00126-010-0302-y.

Nasdala, L., Zhang, M., Kempe, U., Panczer, G., Gaft, M., Andrut, M., & Plötze, M. (2003). Spectroscopic methods applied to zircon. *Reviews in Mineralogy and Geochemistry*, *53*(1), 427–467.

Nasdala, L., Grambole, D., Götze, J., Kempe, U., & Váczi, T. (2010). Helium irradiation study on zircon. *Contributions to Mineralogy and Petrology*, *161*, 777–789. https://doi.org/10.1007/s00410-010-0562-7.

Neuser, R. D., Bruhn, F., Götze, J., Habermann, D. & Richter, D. K. (1995). Kathodolumineszenz: Methodik und Anwendung. *Zentralbl Geol Paläontol Teil*, *I*(H 1/2), 287–306.

Ni, P., Pan, J. Y., Wang, G. G., Chi, Z., Qin, H., Ding, J. Y., & Chen, H. (2017). A CO_2-rich porphyry ore-forming fluid system constrained from a combined cathodoluminescence imaging and fluid inclusion studies of quartz veins from the Tongcun Mo deposit, South China. *Ore Geology Reviews*, *81*, 856–870. https://doi.org/10.1016/j.oregeorev.2016.07.007

Ogola, J. S., Behr, H. J., & van den Kerkhof, A. M. (1994). Fluid inclusion and cathodoluminescence studies on fluorite from the Kerio Valley, Kenya. *Journal of African Earth Sciences*, *18*(4), 309–323. https://doi.org/10.1016/0899-5362(94)90070-1.

Penniston-Dorland, S. C. (2001). Illumination of vein quartz textures in a porphyry copper ore deposit using scanned cathodoluminescence: Grasberg Igneous Complex, Irian Jaya, Indonesia. *American Mineralogist*, *86*(5–6), 652–666. https://doi.org/10.2138/am-2001-5-606.

Poulin, R. S., McDonald, A. M., Kontak, D. J., & McClenaghan, M. B. (2016). On the Relationship Between Cathodoluminescence and the Chemical Composition of Scheelite From Geologically Diverse Ore-Deposit Environments. *The Canadian Mineralogist*, *54*(5), 1147–1173. https://doi.org/10.3749/canmin.1500023.

Pudack, C., Halter, W. E., Heinrich, C. A., & Pettke, T. (2009). Evolution of magmatic vapor to gold-rich epithermal liquid: the porphyry to epithermal transition at Nevados de Famatina, northwest Argentina. *Economic Geology*, *104*(4), 449–477. https://doi.org/10.2113/gsecongeo.104.4.449.

Rae, D. A., Coulson, I. M., & Chambers, A. D. (1996). Metasomatism in the North Qôroq centre, South Greenland: apatite chemistry and rare-earth element transport. *Mineralogical Magazine*, *60*(1), 207–220. https://doi.org/10.1180/minmag.1996.060.398.14.

Rakovan, J., & Reeder, R. J. (1994). Differentiel incorporation of trace elements and dissymmetrization in apatite: the role of surface structure during growth. *American Mineralogist*, *79*, 892–903. https://doi.org/10.1007/978-94-015-9179-9_6.

Read, D., Cooper, D. C., & McArthur, J. M. (1987). The composition and distribution of nodular monazite in the Lower Palaeozoic rocks of Great Britain. *Mineralogical Magazine*, *51*, 271–280. https://doi.org/10.1180/minmag.1987.051.360.09.

Rice, C. M., Mark, D. F., Selby, D., Neilson, J. E., & Davidheiser-Kroll, B. (2016). Age and Geologic Setting of Quartz Vein-Hosted Gold Mineralization at Curraghinalt, Northern Ireland: Implications for Genesis and Classification. *Economic Geology*, *111*(1), 127–150. https://doi.org/10.2113/econgeo.111.1.127.

Richter, D. K., Krampitz, H., Görgen, P., Götte, T. & Neuser, R., (2006). Xenotime in the Lower Buntsandstein of Central Europe: evidence from cathodoluminescence investigation. *Sedimentary Geology*, *183*, 261–268. https://doi.org/10.1016/j.sedgeo.2005.09.017.

Richter, D. K., Görgen, P., & Götte, T. (2008). Monazite cathodoluminescence: A new tool for heavy mineral analysis of siliciclastic sedimentary rocks. *Sedimentary Geology*, *209*(1), 36–41. https://doi.org/10.1016/j.sedgeo.2008.06.008.

Roeder, P. L., MacArthur, D., Ma, X. P., Palmer, G. R., & Mariano, A. N. (1987). Cathodoluminescence and microprobe study of rare-earth elements in apatite. *American Mineralogist*, *72*(7–8), 801–811.

Rosenblum, S., & Mosier, E. L. (1983). Mineralogy and occurrence of europium-rich dark monazite. *USGS Professional Paper* (Vol. *1181*). Washington, DC: US Government Printing Office.

Rusk, B. (2012). Cathodoluminescent textures and trace elements in hydrothermal quartz. In J. Götze, & R. Möckel (Eds.), *Quartz: Deposits, Mineralogy and Analytics* (pp. 307–329). Berlin, Heidelberg: Springer-Verlag.

Rusk, B., & Reed, M. (2002). Scanning electron microscope: Cathodoluminescence analysis of quartz reveals complex growth histories in veins from the Butte porphyry copper deposit, Montana. *Geology*, *30*(8), 727–730.

Rusk, B. G., Reed, M. H., Dilles, J. H., & Kent, A. J. (2006). Intensity of quartz cathodoluminescence and trace-element content in quartz from the porphyry copper deposit at Butte, Montana. *American Mineralogist*, *91*(8–9), 1300–1312. https://doi.org/10.2138/am.2006.1984.

Rusk, B. G., Lowers, H. A. & Reed, M. H. (2008). Trace elements in hydrothermal quartz: relationships to cathodoluminescent textures and insights into vein formation. *Geology*, *36*(7), 547–550. https://doi.org/10.1130/G24580A.1.

Rusk, B., Koenig, A., & Lowers, H. (2011). Visualizing trace element distribution in quartz using cathodoluminescence, electron microprobe, and laser ablation-inductively coupled plasma-mass spectrometry. *American Mineralogist*, *96*(5–6), 703–708. https://doi.org/10.2138/am.2011.3701.

Scholonek, C., & Augustsson, C. (2016). Can cathodoluminescence of feldspar be used as provenance indicator?. *Sedimentary Geology*, *336*, 36–45. https://doi.org/10.1016/j.sedgeo.2015.11.005.

Schönenberger, J., Köhler, J., & Markl, G. (2008). REE systematics of fluorides, calcite and siderite in peralkaline plutonic rocks from the Gardar Province, South Greenland. *Chemical Geology*, *247*(1), 16–35. https://doi.org/10.1016/j.chemgeo.2007.10.002.

Singh, R. K., & Tiwari, R. N. (2010). Sectoral zoning in natural fluorites from carbonatite rocks of Ambadongar, Gujarat. *Journal of the Geological Society of India*, *76*(3), 244–246. https://doi.org/10.1007/s12594-010-0096-7.

Smolyanskii, P. L., Proskurnin, V. F., & Bogomomlov, E. S. (2009). Growth dissymmetrization of rare earth element distribution and the Sm-Nd isotope system in fluorite crystals of the Nordvik deposit (Taymyr). *Doklady Earth Sciences*, *424*, 86–89. https://doi.org/10.1134/S1028334X09010188.

Steven-Kalceff, M. A., Phillips, M. R., Moon, A. R., & Kalceff, W. (2000). Cathodoluminescence micro-caracterisation of silicon dioxide polymorphs. In M. Pagel, V. Barbin, P. Blanc, & D. Ohnenstetter (Eds.), *Cathodoluminescence in Geosciences* (pp. 193–224). Heidelberg, New York: Springer Verlag.

Stefanova, E., Driesner, T., Zajacz, Z., Heinrich, C. A., Petrov, P., & Vasilev, Z. (2014). Melt and fluid inclusions in hydrothermal veins: The magmatic to hydrothermal evolution of the Elatsite porphyry Cu-Au deposit, Bulgaria. *Economic Geology*, *109*(5), 1359–1381. https://doi.org/10.2113/econgeo.109.5.1359.

Takahashi, R., Müller, A., Matsueda, H., Okrugin, V. M., Ono, S., Van Den Kerkhof, A., et al. (2008). Cathodoluminescence and trace elements in quartz: clues to metal precipitation mechanisms at the Asachinskoe gold deposit in Kamchatka. In H. Okada, S. F. Mawatari, N. Suzuki, P. Gautam (Eds.), *Origin and Evolution of Natural Diversity, Proceedings of International Symposium "The Origin and Evolution of Natural Diversity"* (pp. 175–184). Hokkaido University.

Taylor, S. R., & McClennan, S. M. (1985). *The Continental Crust: Its Composition and Evolution*. Oxford: Blackwell.

Thomas, J. B., Watson, E. B., Spear, F. S., Shemella, P. T., Nayak, S. K. & Lanzirotti, A. (2010). TitaniQ under pressure: the effect of pressure and temperature on the solubility of Ti in quartz. *Contrib. Mineral Petrol.*, *152*, 743–754. https://doi.org/10.1007/s00410-006-0123-1.

Tichomirowa, M., Grosche, G., Götze, J., Belyatsky, B. V., Savva, E. V., Keller, J., & Todt, W. (2006). The mineral isotope composition of two Precambrian carbonatite complexes from the Kola Alkaline Province: Alteration versus primary magmatic signatures. *Lithos*, *91*(1), 229–249. https://doi.org/10.1016/j.lithos.2006.03.019.

Tichomirowa, M., Whitehouse, M. J., Gerdes, A., Götze, J., Schulz, B., & Belyatsky, B. V. (2013). Different zircon recrystallization types in carbonatites caused by magma mixing: evidence from U-Pb dating, trace-element and isotope composition (Hf and O) of zircons from two Precambrian carbonatites from Fennoscandia. *Chemical Geology*, *353*, 173–198. https://doi.org/10.1016/j.chemgeo.2012.11.004.

Townsend, P. D., & Rowlands, A. P. (2000). Information encoded in cathodoluminescence emission spectra. In M. Pagel, V. Barbin, P. Blanc, & D. Ohnenstetter (Eds.), *Cathodoluminescence in Geosciences* (pp. 41–57). Heidelberg New York: Springer Verlag.

Tremblay, J., Bédard, L. P., & Matton, G. (2017). Columbitization of fluorcalciopyrochlore by hydrothermalism at the Saint-Honoré alkaline complex, Québec (Canada): New insights on halite in carbonatites. *Ore Geology Reviews*, *91*, 695–707. https://doi.org/10.1016/j.oregeorev.2017.08.027.

Valley, P. M., Hanchar, J. M., & Whitehouse, M. J. (2011). New insights on the evolution of the Lyon Mountain Granite and associated Kiruna-type magnetite-apatite deposits, Adirondack Mountains, New York State. *Geosphere*, *7*(2), 357–389. https://doi.org/10.1130/GES00624.1.

Vasyukova, O. V., Goemann, K., Kamenetsky, V. S., MacRae, C. M., & Wilson, N. C. (2013). Cathodoluminescence properties of quartz eyes from porphyry-type deposits: Implications for the origin of quartz. *American Mineralogist*, *98*(1), 98–109. https://doi.org/10.2138/am.2013.4018.

Vaughan, J. R., Hickey, K. A., & Barker, S. L. (2016). Isotopic, chemical, and textural evidence for pervasive calcite dissolution and precipitation accompanying hydrothermal fluid flow in low-temperature, carbonate-hosted, Gold systems. *Economic Geology*, *111*(5), 1127–1157. https://doi.org/10.2113/econgeo.111.5.1127.

Wang, L. X., Marks, M. A., Wenzel, T., Von Der Handt, A., Keller, J., Teiber, H., & Markl, G. (2014). Apatites from the Kaiserstuhl Volcanic Complex, Germany: new constraints on the relationship between carbonatite and associated silicate rocks. *European Journal of Mineralogy*, *26*(3), 397–414. https://doi.org/10.1127/0935-1221/2014/0026-2377.

Wang, X., Chen, J., & Ren, M. (2016). Hydrothermal zircon geochronology: Age constraint on Nanling Range tungsten mineralization (Southeast China). *Ore Geology Reviews*, *74*, 63–75. https://doi.org/10.1016/j.oregeorev.2015.10.034.

Wang, Y. Y., van den Kerkhof, A., Xiao, Y., Sun, H., Yang, X., Lai, J., & Wang, Y. (2017). Geochemistry and fluid inclusions of scheelite-mineralized granodiorite porphyries from southern Anhui Province, China. *Ore Geology Reviews*, *89*, 988–1005. https://doi.org/10.1016/j.oregeorev.2017.08.004.

Wark, D. A., & Watson, E. B. (2006). TitaniQ: a titanium-in-quartz geothermometer. *Contributions to Mineralogy and Petrology*, *152*, 743–754. https://doi.org/10.1007/s00410-006-0132-1.

Waychunas, G. A. (2002). Apatite luminescence. *Reviews in Mineralogy and Geochemistry*, *48*(1), 701–742. https://doi.org/10.2138/rmg.2002.48.19.

Wei, A., Xue, C., Xiang, K., Li, J., Liao, C., & Akhter, Q. J. (2015). The ore-forming process of the Maoping Pb–Zn deposit, northeastern Yunnan, China: Constraints from cathodoluminescence (CL) petrography of hydrothermal dolomite. *Ore Geology Reviews*, *70*, 562–577. https://doi.org/10.1016/j.oregeorev.2015.02.007.

Westhues, A., Hanchar, J. M., Whitehouse, M. J., & Martinsson, O. (2016). New constraints on the timing of host-rock emplacement, hydrothermal alteration, and iron oxide-apatite mineralization in the Kiruna district, Norrbotten, Sweden. *Economic Geology*, *111*(7), 1595–1618. https://doi.org/10.2113/econgeo.111.7.1595.

Williams, P. J., Barton, M. D., Johnson, D. A., Fontboté, L., De Haller, A., Mark, G., et al. (2005). Iron oxide copper-gold deposits: Geology, space-time distribution, and possible modes of origin. In J. W. Hedenquist, J. F. H. Thompson, R. J. Goldfarb, J. P. Richards (Eds.), *100th Anniversary Volume 1905–2005, Economic Geology* (pp. 371–405). Littleton CO, Special Publication of the Society of Economic Geologists.

Yacobi, B. G., & Holt, D. B. (1990). *Cathodoluminescence Microscopy of Inorganic Solids*. New York: Plenum Press.

Yang, Z., Chang, Z., Paquette, J., White, N. C., Hou, Z., & Ge, L. (2015). Magmatic Au mineralization at the Bilihe Au deposit, China. *Economic Geology*, *110*(7), 1661–1668. https://doi.org/10.2113/econgeo.110.7.1661.

Zeng, L. P., Zhao, X. F., Li, X. C., Hu, H., & McFarlane C. (2016). In situ elemental and isotopic analysis of fluorapatite from the Taocun magnetite-apatite deposit, Eastern China: Constraints on fluid metasomatism. *American Mineralogist*, *101*, 2468–2483. https://doi.org/10.2138/am-2016-5743.

Zhang, R., Lu, J., Lehmann, B., Li, C., Li, G., Zhang, L., et al. (2017). Combined zircon and cassiterite U-Pb dating of the Piaotang granite-related tungsten–tin deposit, southern Jiangxi tungsten district, China. *Ore Geology Reviews*, *82*, 268–284. https://doi.org/10.1016/j.oregeorev.2016.10.039.

Zinkernagel, U. (1978). Cathodoluminescence of quartz and its application to sandstone petrology. *Contributions to Sedimentology*, *8*, Stuttgart, E. Schweizerbart'sche Verlagsbuchhandlung.

7

Transition Metal Isotopes Applied to Exploration Geochemistry: Insights from Fe, Cu, and Zn

Ryan Mathur[1] and Da Wang[2]

ABSTRACT/INTRODUCTION

Transition metal isotope geochemistry has been applied to many different aspects of economic concentrations of metal on the surface and in the subsurface of the Earth. Here we focus on Fe, Cu, and Zn isotope fractionation measured in ores, waters, rocks, soils, and other materials in an attempt to display the range of potential applications these relatively new techniques can provide (Fig. 7.1).

From an exploration geological standpoint, two aspects of the isotopic data can be used. First, the large fractionation found to occur at the surface of the Earth permits the use of waters, weathered rocks, soils, and plants as vectors to ore in the subsurface (Mathur et al., 2009; Asael et al., 2012; Kimball et al., 2009; Li et al., 2016, and others). Equally interesting is the fact that the high temperature ores also show a large enough separation in isotopic space to allow for the potential vector. Second is the genetic and source information for the process that led to the ores. This chapter will focus on these two aspects of how these transition metal isotopes can aid in exploration geology by starting with the deepest magmatic systems and working upward to the supergene systems.

7.1. ANALYTICAL ASPECTS

7.1.1. Copper and Zinc Isotopes

Copper and zinc isotope data in the literature spans a large range of chemical matrices. Two approaches have been attempted to measure copper isotope compositions in materials, a solution entry into the mass spectrometer and laser ablation techniques. For zinc, the majority of the data is presented as solution. From the solution side, aside from native copper, chalcopyrite ($CuFeS_2$), chalcocite (CuS_2), copper oxides, and sphaerlite (ZnS) anion exchange chromatography has been used to purify copper and zinc from complex chemical matrices. Most studies use the BioRad MP-1 anion exchange resin with several studies using AG1X8 with similar yields (Balistrieri et al., 2008; Bermin et al., 2006; Borrok et al., 2007; Chapman et al., 2006; Li et al., 2009; Marechal & Albarede, 2002; Marechal et al., 1999; Mason et al., 2004; Mathur et al., 2012; Mathur et al., 2009; Moynier et al., 2007; Peel et al., 2008; Petit et al., 2008; Pokrovsky et al., 2008; Pribil et al., 2009; Sossi et al., 2015; Vance et al., 2008; Zhu et al., 2002). The key aspects pointed to these studies are that the resin can cause fractionation copper within the sample if not all of the copper is recovered, over saturation of the resin can lead to the stripping of copper earlier in the elution sequence, and multiple column treatments are necessary if an abundance of cations are present. Only a few studies have attempted to measure copper isotopes through laser ablation (Graham et al., 2004; Ikehata & Hirata, 2013; Ikehata et al., 2008; Ikehata et al., 2011; Jackson & Günther, 2003; Kuhn et al., 2007) and have found that matrix matching is imperative to generate meaningful isotopic data.

[1] *Department of Geology, Juniata College, Huntingdon, Pennsylvania, USA*

[2] *State Key Laboratory of Geological Processes and Mineral Resources, School of Earth Sciences and Resources, China University of Geosciences, Beijing, China*

Ore Deposits: Origin, Exploration, and Exploitation, Geophysical Monograph 242, First Edition.
Edited by Sophie Decrée and Laurence Robb.

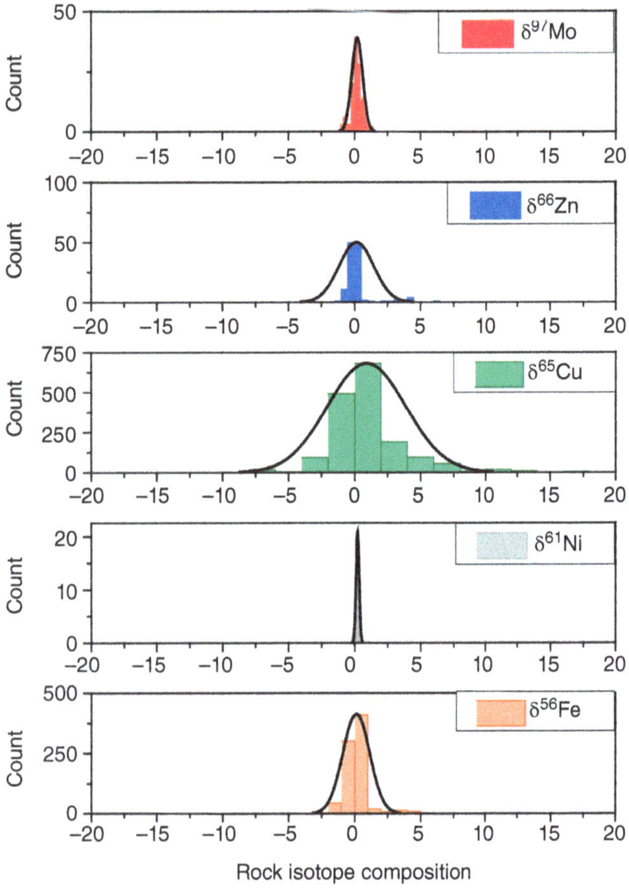

Figure 7.1 Variations of transition metal isotope values in rocks and minerals. Data taken from the literature (Mathur & Fantle, 2015) and all data are compared to the same international isotope standard.

The presentation of the data is in the traditional per mil format where: $\delta^{65}Cu$ (‰) = [(^{65}Cu /^{63}Cu)$_{sample}$/(^{65}Cu /^{63}Cu)$_{NIST\,976}$ – 1] × 1000, although the first published articles by Zhu et al. (2002, 2000) reported ε values with changes in the forth decimal place because no one had documented the large isotopic range that is now known to exist. The majority of copper isotope data in the current literature use the NIST 976 isotopic standard. This standard can no longer be purchased and several papers have developed new standards relative to this standard so that all of the isotope data can be compared. Up to now, there is no unified international Zn standard. There are several general Zn standards that include JMC-Lyon, IRMM-3702, and AA-ETH (Archer et al., 2017). The conversion relation is $\delta^{66}Zn_{AA\text{-}ETH}$ = $\delta^{66}Zn_{IRMM\text{-}3702}$ = $\delta^{66}Zn_{JMC\ 3\text{-}0749\ L}$ – 0.28‰ (Archer et al., 2017). All the data in this paper are conversed to $\delta^{66}Zn$ values in JMC-Lyon standard.

$$\delta^{66}Zn(‰)=\left[\left({}^{66}Zn/{}^{64}Zn\right)_{Sample}/\left({}^{66}Zn/{}^{64}Zn\right)_{Standard}-1\right]\times 1000$$

$$\delta^{68}Zn(‰)=\left[\left({}^{68}Zn/{}^{64}Zn\right)_{Sample}/\left({}^{68}Zn/{}^{64}Zn\right)_{Standard}-1\right]\times 1000$$

The mass spectrometry for the measurement of the copper and zinc isotope ratios vary for the three major brands of multicollector mass spectrometers (Neptune, Micromass Isoprobe, and Nu). The similarities during measurement of copper isotope ratios are as follows: the mass spectrometers are in low-resolution mode; background copper must be subtracted for each measurement on the half mass or on peak; the standard and samples should match intensities within 50%; no auxiliary gases are used for the plasma; and at least 25 ratios generate data with low enough error. Sample uptake rates, concentration of copper, HT settings, types of sample introduction, and a many other variables have been reported in different settings for all instruments. There have not been reports of differences in values measured using wet versus dry plasma.

Mass bias corrections for Cu are important as there is no internal check for mass dependent fractionation because there are only two copper isotopes. Two elements have been used to correct for mass bias in copper isotope measurements, which are nickel and zinc. Both elements have produced highly accurate and precise measurements with 2σ errors around 0.04‰ (Chapman et al., 2006; Chapman et al., 2004; Li, 2015; Liu et al., 2015; Peel et al., 2008; Sossi et al., 2015). The studies all used the exponential law to correct for mass bias in the samples and standards; the corrected values were then used for traditional standard-sample-standard bracketing. It has also been shown that not using nickel or zinc produces the accurate values; however the 2σ error is a little larger on the order of 0.12‰. With regard to zinc, copper isotopes have been used to correct for mass bias and mass dependent fractionation can be monitored as there are more than two isotopes of zinc.

Isobars are a common problem for all isotopic analysis and, for copper, on all types of mass spectrometers used. The common element interferences are caused by excessive Ti, Na, and Mg. These all form molecules that interfere at 63 or 65. It has also been noted that transition metals present at greater than 10:1 (transition metal: copper) have caused issues during measurement.

7.1.2. Iron Isotopes

Fe is the highest abundant valence-variable element in the Earth; the valence states are usually 0, +2, +3. The Fe-bearing minerals mainly include pyrite, magnetite, hematite, siderite, chlorite, goethite, and iron chlorite. Fe has four stable isotopes, including ^{54}Fe (5.84%),

^{56}Fe (91.76%), ^{57}Fe (2.12%), and ^{58}Fe (0.28%; Wang & Zhu, 2012). The δ formulas are as follows:

$$\delta^{56}\text{Fe}(‰)=\left[\left({}^{56}\text{Fe}/{}^{54}\text{Fe}\right)_{\text{sample}}/\left({}^{56}\text{Fe}/{}^{54}\text{Fe}\right)_{\text{standard}}-1\right]\times1000$$

$$\delta^{57}\text{Fe}(‰)=\left[\left({}^{57}\text{Fe}/{}^{54}\text{Fe}\right)_{\text{sample}}/\left({}^{57}\text{Fe}/{}^{54}\text{Fe}\right)_{\text{standard}}-1\right]\times1000$$

$$\Delta^{x}\text{Fe}_{A-B}=\delta^{x}\text{Fe}_{A}-\delta^{x}\text{Fe}_{B}\ \text{and}\ \Delta^{56}\text{Fe}=0.678\Delta^{57}\text{Fe}\ (\text{Mass fractionation}).$$

Fe isotope purification also uses the AG-X and MP-1 anion exchange resins from BioRad and there are several slightly different protocols for elution of Fe from the resin. The key aspect of the chromatography is that the chromium is removed from the matrix as this is a direct interference with ^{54}Fe. The measurement of Fe isotope ratios must be conducted in high-resolution mode on ICP-MS multicollectors to resolve the large argide interference at mass 56. Solutions must be at greater concentration, usually around 3–5 ppm to conduct successful measurements. All studies measure on the right shoulder of peak with mass bias corrected for with bracketing or use of elements like Cu and Ni to correct for mass bias with the exponential law. There are two international standards for Fe isotope: (1) the average value of 15 Earth igneous rock and 5 lunar high-titanic basalt; and (2) the IRMM-014 supplied by Council of Europe (Taylor et al., 1992). The conversion relation is $\delta^{56}\text{Fe}_{\text{Igneous Rock}}=\delta^{56}\text{Fe}_{\text{IRMM-014}}-0.09‰$, $\delta^{57}\text{Fe}_{\text{Igneous Rock}}=\delta^{57}\text{Fe}_{\text{IRMM-014}}-0.11‰$ (Johnson et al., 2004). All the data in this paper are converted to δ^{56}Fe values relative to IRMM-014 standard.

7.2. DEPOSIT TYPES

7.2.1. High Temperature Ore-Forming Systems

7.2.1.1. Layered Mafic Intrusions, Porphyry Copper Deposits, Skarns-Cu Isotopes

Interpretation of copper isotope (and Zn and Fe) values from high-temperature mineralization and magmatic systems have suffered from the fact that few experimental studies to this point in the literature have constrained the mechanisms that could impact the copper isotope values. Thus, most papers attempt to empirically constrain the mechanisms for fractionation by comparing the copper isotope values in concert with other concentration and isotopic data. The fact that the studies have been conducted in systems where known geochemical reactions have been modeled, allows for further constraint of what the isotope values may indicate. Nonetheless, there are important aspects of the data that can be used by exploration geologists that relate to sources of the metals and vectoring potential.

The highest temperature systems concerning sulfide mineralization are in layered mafic intrusions and related to mafic-type deposits. This suggests these systems have multiple steps where copper isotopes could fractionate. Ikehata and Hirata (2012) showed that native copper found in these mafic rocks shows little fractionation of copper where no significant amounts of sulfide exist. This suggests that without sulfide segregation, no fractionation of the copper isotope occurs. In contrast, three papers about Ni-Cu PGE deposits by Malitch et al. (2014), Ripley et al. (2015), and Zhao et al. (2017) provide data from Norilsk, midcontinental USA, and Talgren, China, in systems where measurable copper isotope fractionation exists. Malitch et al. (2014) show a progression from lower copper isotope values in distal mineralization (Kharaelakh deposit (−2.3 to −0.9‰) and the Talnakh occurrences (−1.1 to 0.0‰) to the higher values proximal to the Noril'sk-1 intrusion (−0.1 to 0.6‰). These authors assert that the differences between the deposits in the Noril'sk region may be due to different mechanisms of isotope fractionation during magmatic fractionation of Cu isotopes, to differences in the Cu isotope ratio of primary parental magmas, or to assimilation of Cu from external sources. Ripley et al. (2015) recorded a distinct difference between sulfides in disseminated sheet style versus conduit style mineralization. They analyzed surrounding sedimentary rocks to assess the potential that Cu could be derived during assimilation. The Fe isotope compositions of the ores could not be related to local stratigraphy through geochemical mixing models. This led the authors to conclude that Fe isotope fractionation in ores could be related to different mantle sourced magmas or complex magmatic processes.

Zhao et al. (2017) have direct exploration focus and point out that the distal parts of the system have lower isotope values than the proximal samples. Zhao et al. (2017) point to redox reactions known to hypothetically occur during differentiation of the mafic intrusion as a cause for the systematic variation (Fig. 7.2).

Porphyry copper deposits have received the most attention with regard to copper isotope studies. Zhu et al. (2000) first reported measurable fractionation in museum specimens of chalcopyrite. Larson et al. (2003) were the next to provide a more detailed analysis of samples from Bagdad and other PCDs in the USA. This study pointed out that there are variations in higher temperature systems and that there could be systematic changes in copper isotope values related to the physiochemical properties of the hydrothermal system. Graham et al. (2004) provided the first detailed study of copper isotope fractionation within a deposit and did so using laser ablation of multiple copper phases in various parts of the hydrothermal

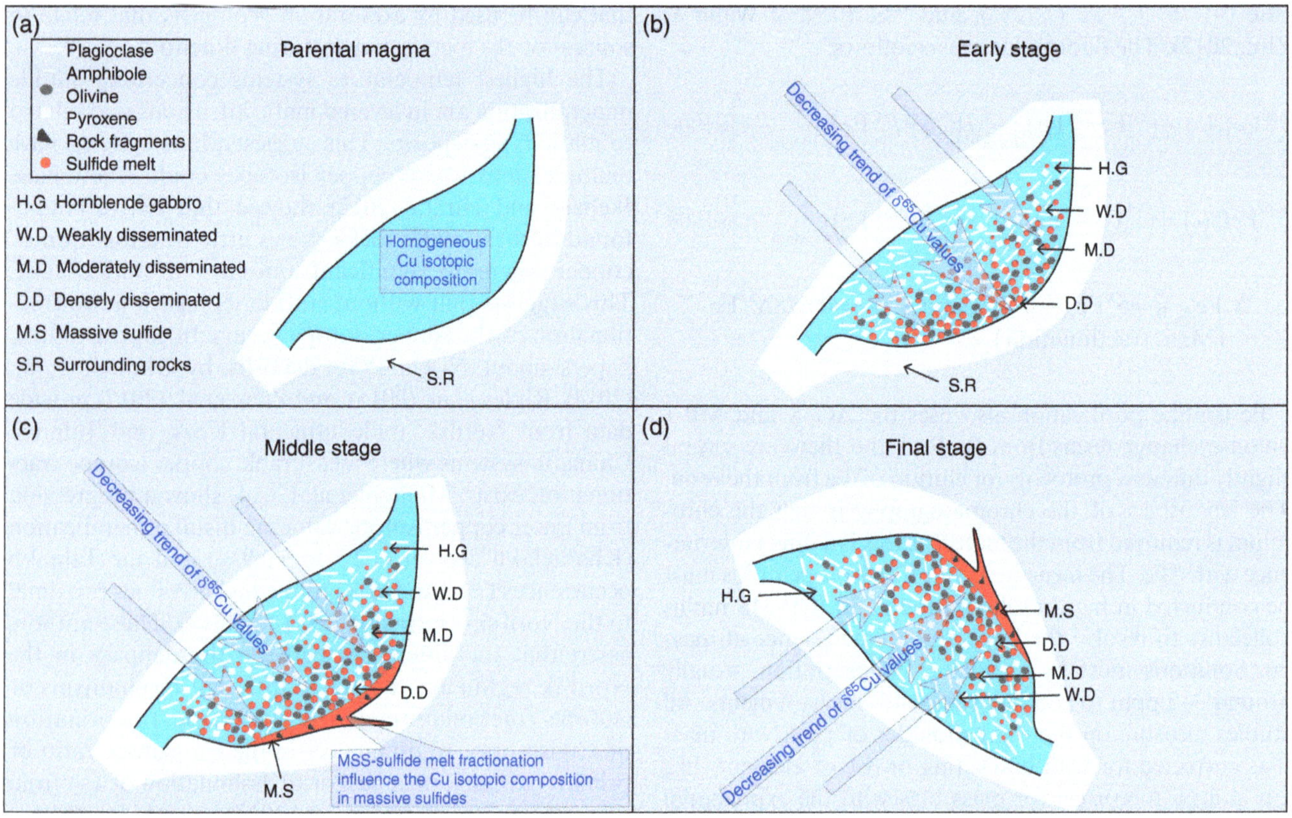

Figure 7.2 Zhao et al. (2017) depicts the isotopic evolution and structural evolution of the Tulgren massive sulfide deposit. Copper isotope variations are tied to magmatic differentiation that is induced by redox reactions of copper into different phases of the cooling intrusion.

system. They found the Grasberg samples had a successive variation of heavier values with later intrusions and that there are variations in the skarns, and surrounding pyrite shell in the ore deposit indicate different sources of crustal metals for these subsequent mineralization events. Li et al. (2010) examined the North Parkes PCD and focused on chalcopyrite in the core of the system and noticed that a systematic variation of the copper isotope values were lower in the core of the system and became progressively larger in the more distal parts. Dendas (2011), Mathur et al. (2013) (Fig. 7.3), and Song et al. (2016) noticed the same pattern in both chalcopyrite and bornite on a larger scale from the potassic to phyllic alteration at Bingham, Pebble, and Dexing respectively. Duan et al. (2016) noticed a similar pattern in this Tibean deposit and predicted that there are missing parts to the system given the lack of the heavier values on the periphery of the system.

A combination of all these observations point to two important aspects that exploration geologists could use while searching for the deposits. First, given the systematic change in isotope composition from the proximal to distal location in the hydrothermal system, copper isotope values in ores could be used to vector to the deposit by using sulfide found in the drill core (Fig. 7.3). Second, all of these studies note that the lowest values recorded in all copper phases are found at the core of the system. Therefore, the copper isotope values found in the deepest parts of the systems are probably representative of the source of the metal from which there is a systematic fractionation that occurs when the fluids are exsolved from these magmas. Given that most nonaltered mantle-like rocks and the bulk Earth value is approximately 0 per mil (Sossi et al., 2015) and the similarities across multiple PCDs, the lower values probably indicate a mantle source for these metals. As more data are amassed about these systems, mixing models could be used to indicate various amounts of crustal and mantle metal in these systems, which could relate to the overall size of the deposit.

Other types of mineralization associated with the formation of PCDs have not received as much attention. Maher and Larson (2007) show systematic variations in the copper isotope values found in skarns (Fig. 7.3), which are inverse to those found in chalcopyrite in PCD mineralization. They found lighter values are found in more distal parts of the systems. Later, Maher et al.

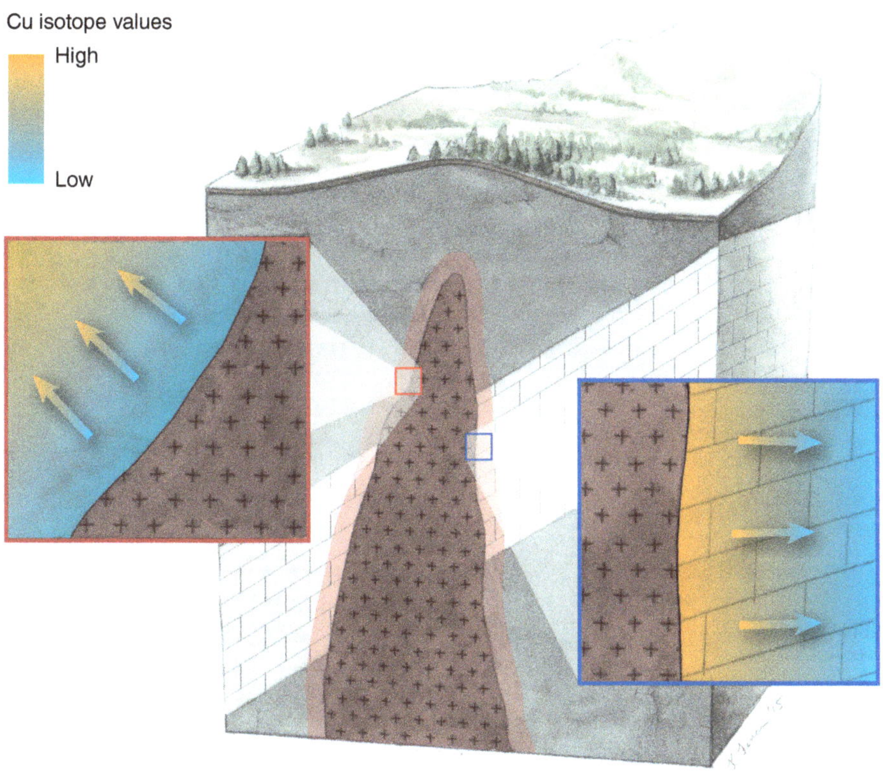

Figure 7.3 Patterns of copper isotope values in chalcopyrite surrounding PCD and skarn deposits. The data can be used to vector and understand the shape of the hydrothermal system.

(2011) provided experimental work to show liquid-vapor transitions might cause the fractionation measured in hydrothermal systems. Yao et al. (2016) further augmented this experimental data with deriving empirical evidence (from Dauhtang skarn) of fluid inclusions, other transition metal, and lighter isotope data that liquid-vapor transitions cause the variations measured in the system. Epithermal mineralization that may or may not be related to PCDs has shown that there are lower values found in both native Au and chalcopyrite. Saunders et al. (2015) related this to nanoparticle transport of metal from the mantle into the shallow portions of the crust. Markl et al. (2006) examined hydrothermal veined chalcopyrite in felsic rock hosts that have multiple metals present in the large-scale veined system. In this setting, the copper isotopes did not show systematic variations and had values that clustered around 0 per mil and slightly more negative.

7.2.1.2. Magmatic, Magmatic Hydrothermal, and Skarn Fe Deposits: Fe Isotopes

These studies demonstrate the potential of Fe isotopes to trace the metal source and provide insights into ore-forming evolution. The studies focus on the ores and their relationships with associated igneous rocks. No studies have found direct evidence for using the isotope values to vector, however the data show promise, in that minerals that formed later in the system do have different isotope values than earlier formed minerals.

Within the highest temperature magmatic Fe related deposit, Bayan Obo Fe-rare earth element (REE) deposit in China, studied by Sun et al. (2013), the Fe isotopes show no fractionation (with values clustering near 0‰; Fig. 7.4). No relationships exist between concentration and isotope values for these ores. Zhu et al. (2016) have researched the Fe isotopic variation in the Han-Xing skarn Fe deposit in China, diorite samples have marked variation in Fe isotopic compositions, combining with the limited variation in immobile elements (e.g., Si, Ti, Al, and P) and negative correlations between LOI (loss on ignition) and TFe_2O_3. The large variation of $\delta^{56}Fe$ in the diorites cannot be the result of early magmatic processes and late secondary alteration (due to lack of minerals associated with these reactions); the skarn-type alteration should be the main cause of Fe isotopic fractionation. The Fe isotopic similarities between ore minerals and diorites confirmed the magmatic hydrothermal origin instead of recycling preexisting mineralization or altering associated igneous rocks. Bilenker et al. (2016) examined iron oxide apatite deposits in Chile, which have a debated magmatic or low-temperature origin. They found similarities in the Fe isotope values of the ores and

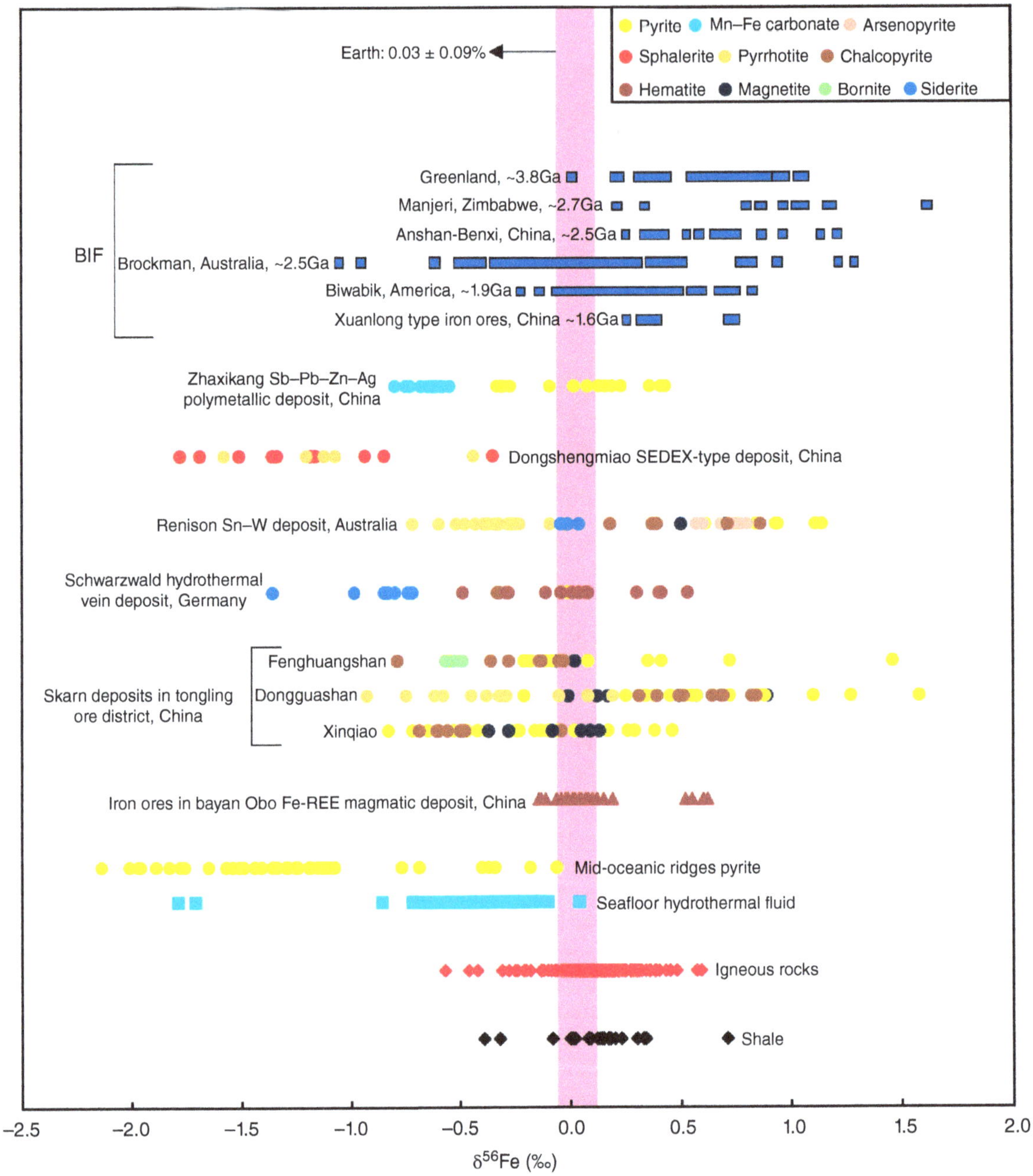

Figure 7.4 Fe isotopic compositions of the Bulk Silicate Earth (Wang & Zhu, 2012), shale (Beard et al., 2003a; Rouxel et al., 2005), igneous rocks (Beard & Johnson, 1999; Zhu et al., 2002; Beard et al., 2003b; Poitrasson et al., 2004; Poitrasson & Freydier, 2005; Dauphas er al., 2004, 2010; Heimann et al., 2008; Schuessler et al., 2009; Wang et al., 2011, 2015; Telus et al., 2012; Foden et al., 2015; Wawryk & Foden, 2015), seafloor hydrothermal fluids (Sharma et al., 2001; Beard et al., 2003a; Severmann et al., 2004; Rouxel et al., 2008; Bennett et al., 2009), midoceanic ridges pyrite (Rouxel et al., 2004; Rouxel et al., 2008), the Bayan Obo Fe-REE magmatic-type deposit in China (Sun et al., 2013), the skarn-type deposits in Tongling ore district in China (Wang et al., 2011), the Schwarzwald hydrothermal vein deposit in Germany (Markl et al., 2006), the Renison Sn-W deposit in Australia (Wawryk & Foden, 2015), the Dongshengmiao SEDEX-type deposit in China (Gao et al., 2017), and the Zhaxikang Sb-Pb-Zn-Ag polymetallic deposit in China (Wang et al., 2017).

the associated igneous rocks for several deposits, from which they interpret that high temperature magmatic processes caused the consistent Fe isotope signatures.

Wang et al. (2011, 2015) studied the Fe isotopes of the Xinqiao skarn Cu-S-Fe-Au, the Dongguashan skarn Cu-Au, and the Fenghuangshan skarn Cu-Fe-Au deposits in Tongling Ore District. The results excluded the possibility that Fe originated from sedimentary strata. These Fe isotopes matched the igneous source rocks and mineralization. The δ^{56}Fe values of sulfides gradually increased both from early to late stages and away from the ore-related igneous rocks (Fig. 7.5). These studies from Wang et al. (2011, 2015) also revealed that the Fe isotopes fractionate during fluid exsolution and the ore-forming fluid is enriched in light isotopes relative to ore-related igneous rocks. To the contrary, Wawryk and Foden (2015) investigated the Fe-isotope fractionation in the Renison magmatic-hydrothermal Sn-W (300 ~ 400°C) deposit in Australia and discovered that Fe isotopic compositions of pyrite (0.61‰ ~ 1.14‰), chalcopyrite (0.18‰ ~ 0.71‰), and magnetite (0.50‰ ~ 0.70‰) are isotopically heavier than Renison granite (0.27‰ ~ 0.45‰; Fig. 7.4). They hypothesized that a magmatic-hydrothermal fluid exsolved from an isotopically heavy reduced magma, which resulted in the deposition of isotopically heavy ore minerals. This would dictate through mass balance that the oxidized magmas crystallize magmatic magnetite, melt, and fluid that is isotopically lighter.

7.2.1.3. VHMS (VMS) and Irish-Type Zn Deposits-Cu and Zn Isotopes

Copper isotope studies of volcanogenic massive sulfide deposits (VMS) have shown the largest range of copper isotope values in high temperature systems. Rouxel et al. (2004) present recent copper isotope values from an active VMS on the seafloor of the mid-Atlantic Ridge and report up to 3 per mil variation, which is attributed to reworking and oxidation of sulphides on the chimney after initial precipitation. Berkenbosch et al. (2015) examined sulphides from VMS forming in the South Pacific and noted up to 1.5 per mil variations, which were attributed to liquid vapor fractionation as these vapors have been noted to be metal rich. In contrast, in ancient systems, not much fractionation is seen. Mason et al. (2005) document a very limited range (about 0.5 per mil variation in several textures of samples, n = 27) for a Devonian aged VMS in the Ural Mountains. Ikehata et al. (2011) found an identical spread for samples from two Beshi-type deposits in Japan using laser ablation. The difference between ancient and recent systems could be related to the metamorphic events and fluid migration that occurs during the suturing of these systems onto continents.

Previous studies (Marechal et al., 2002; Archer et al., 2004; John et al. 2008; Wang & Zhu, 2010) have proposed a Rayleigh distillation model to explain an increasing trend in δ^{66}Zn values within precipitates over time for the hydrothermal fluid. This would suggest that Zn could be used as a vector to ore using sphalerite. The sulfides precipitate from the ore-forming fluid, and the ^{64}Zn-enriched sulfides form during the early phase resulting from the kinetics of the chemical reactions, which cause the heavier and heavier δ^{66}Zn values of residual fluids and later precipitates. This Rayleigh distillation model is also supported by the continuous batch experimental research of Borrok et al. (2008), which shows the fluids preferentially enriched in heavier Zn isotopes compared with related rocks during the solution-solid reaction. In actual deposit research, the δ^{66}Zn values from the core to the edge of the Alexandrinka volcanic hosted massive sulphide (VHMS) deposit in Russia (Mason et al., 2005) and from early to late stages of Midlands Irish-type deposit in Ireland (Wikinson et al., 2005; Gagnevin et al., 2012) both gradually increased (Fig. 7.6), which can be the example verification of this Rayleigh distillation model.

7.2.2. Intermediate to Low-Temperature Mineralization Systems

7.2.2.1. Sedimentary Copper Deposits-Cu Isotopes

Asael et al. (2007) published the first data from a traditional sedimentary copper deposit in Israel and pointed out the potential of using the copper isotopes to trace redox reactions and understand mass transfer in these systems. Later, this group enhanced the chemical modeling of phase transitions accompanying different ions in solutions through expanded work into the Kuperschiefer in Germany (Asael et al., 2012; Asael et al., 2009). This work also pointed to using the copper isotope data to trace sources along with understanding the redox geochemistry of how these deposits form at intermediate (100–200°C) temperatures.

The obvious application for tracing metals and fluids pathways in these deposits is the sedimentary copper deposits in Africa. Scant data currently exists with only one deposit-scale study reported in the literature by Haest et al. (2009). The Dikulushi deposit experienced multiple episodes of reduction and oxidative fluids, the mineralization deposited during these times was subsequently reworked by supergene weathering. They found a systematic variation where the copper isotope composition of chalcocite becomes heavier in the proximal part of the system. The data were interpreted as indicating changes in the reducing brine at higher temperatures in the mineralization events or a supergene overprint. The only other published data from this vast expanse of copper mineralization are reported in Wilson et al. (2016) where native copper and chalcopyrite from the Trident deposit in Zambia have values > 6 per mil.

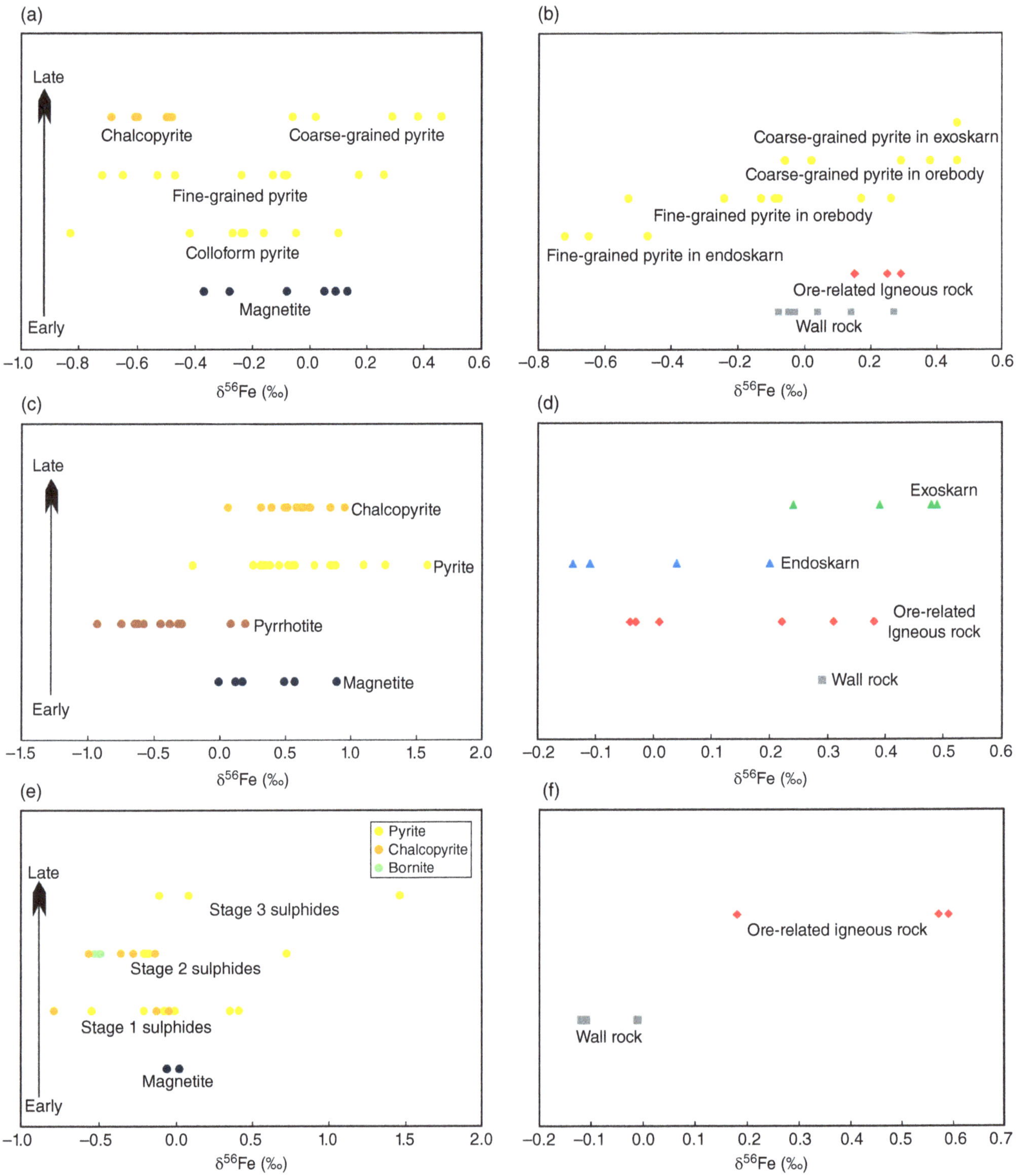

Figure 7.5 (a) The temporal Fe isotopic variation in the Xinqiao skarn Cu-S-Fe-Au deposit; (b) the spatial Fe isotopic variation in the Xinqiao skarn Cu-S-Fe-Au deposit; (c) the temporal Fe isotopic variation in the Dongguashan skarn Cu-Au deposit; (d) the spatial Fe isotopic variation in the Dongguashan skarn Cu-Au deposit; (e) the temporal Fe isotopic variation in the Fenghuangshan skarn Cu-Fe-Au deposit; (f) the spatial Fe isotopic variation in the Fenghuangshan skarn Cu-Fe-Au deposit (modified from Wang et al., 2011, 2015).

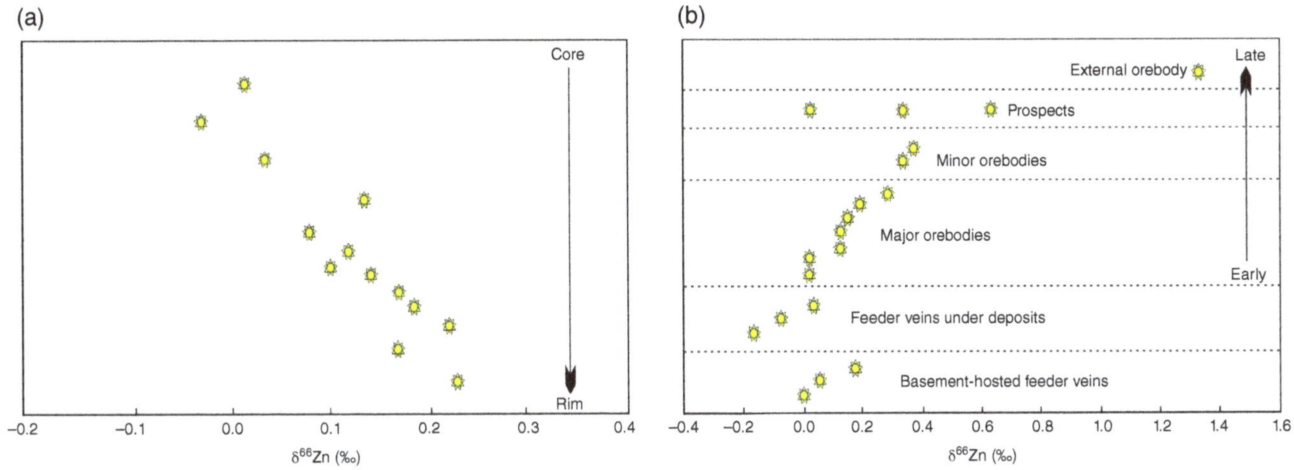

Figure 7.6 (a) The $\delta^{66}Zn$ values gradually increase from core to rim in the Alexandrinka VHMS type deposit in Russia (modified from Mason et al., 2005); (b) the $\delta^{66}Zn$ values gradually increase from early to late stages in the Midlands Irish-type deposit in Ireland (modified from Wilkinson et al., 2005).

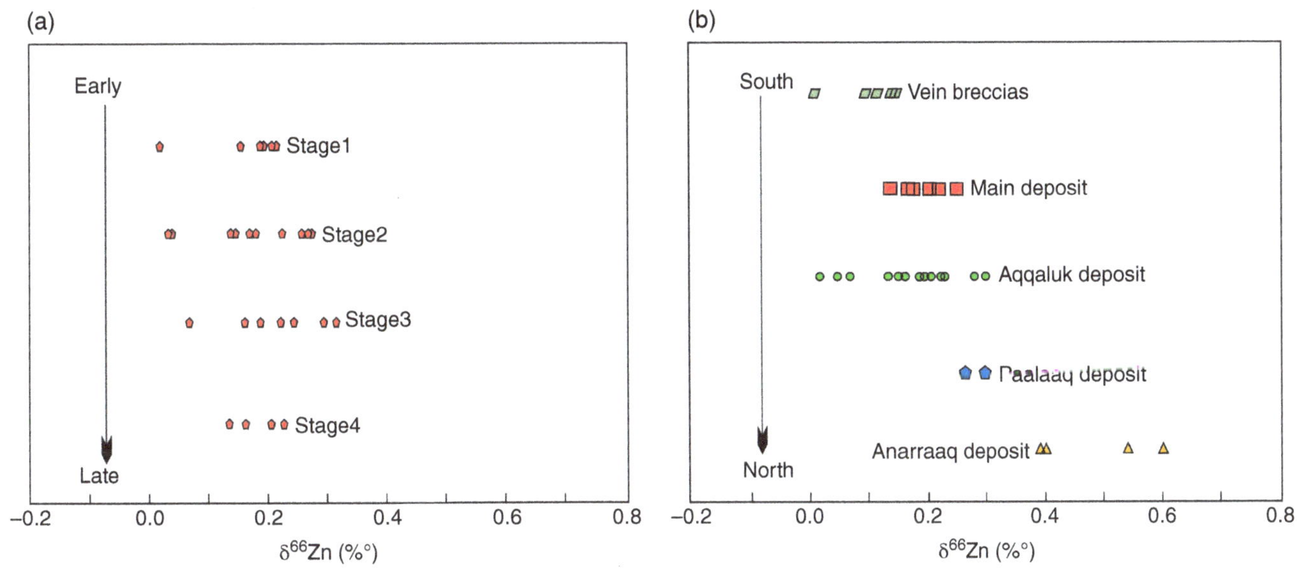

Figure 7.7 (a) The $\delta^{66}Zn$ values gradually increase from early to late stages in the Red Dog ore district in Alaska; (b) The $\delta^{66}Zn$ values gradually increase from south to north in the Red Dog ore district in Alaska (modified from Kelley et al., 2009).

This deposit type deserves more study as there are multiple changing redox conditions associated with the low- to medium-grade metamorphic events associated with mineralization.

7.2.2.2. SEDEX Deposits: Zn and Fe Isotopes

Based on the theory that the $\delta^{66}Zn$ values of minerals precipitating from the same hydrothermal fluids become heavier over time, (Marechal et al.,2002; Archer et al., 2004; John et al., 2008), Gao et al. (2017) constrain the evolution of the hydrothermal system of the Dongshengmiao SEDEX deposit in Inner Mongolia by the lateral trend of increasing $\delta^{66}Zn$ and $\delta^{56}Fe$ values from southwest to northeast within the main ore body combining with the homogeneous Pb-isotope composition of ore sulphides. Kelley et al. (2009) used the homogeneous $\delta^{66}Zn$ values and the gradually increasing trend from early to late stages and from south to north within the Red Dog ore district in Alaska to inverse the temporal and spatial ore-forming fluid evolution and constrain the single Zn source with SEDEX genesis (Fig. 7.7). In the meantime, the Fe/Mn ratios have a negative correlation with the increasing $\delta^{66}Zn$ values in sphalerite from the Red Dog Ore District, which is caused by the conservative characteristics and rapid precipitation of Fe with hydrothermal fluids mixing and cooling (Seewald & Seyfried, 1990; Metz & Trefry, 2000). However, no similar correlation has been found between Cu concentration and $\delta^{66}Zn$ values due to the lack of Cu in this ore district.

Wang et al. (2017) determined the Fe-Zn isotopic variations and compositions of pyrite, sphalerite, Mn-Fe carbonate, and slate from the Zhaxikang deposit (Zheng et al., 2012; SEDEX overprinted by later hydrothermal fluid) in southern Tibet. The temporally increasing $\delta^{56}Fe$ and decreasing $\delta^{66}Zn$ values recorded in the deposit coincide with an increase in alteration provide evidence for two pulses of mineralization (Fig. 7.8). The heavier $\delta^{56}Fe$ values of stage 3 pyrite also traced the magmatic hydrothermal fluid origin of the second pulse of mineralization, and the $\delta^{56}Fe$ values of the first pulse of ore-forming fluid (–0.54‰ to –0.34‰) have been calculated using theoretical equations.

7.2.2.3. BIF Fe Deposits-Fe Isotopes

Multiple contributions (Johnson et al., 2003, 2008; Dauphas et al., 2004, 2007; Rouxel et al., 2005; Dauphas & Rouxel, 2006; Johnson & Beard, 2006; Anbar & Rouxel, 2007; Frost et al., 2007; Whitehouse & Fedo, 2007; Li et al., 2008a, b, 2012; Steinhoefel et al., 2009, 2010; Planavsky et al., 2009, 2011; Czaja et al., 2010; Tsiko et al., 2010; Heimann et al., 2010; Yan et al., 2010; Craddock & Dauphas, 2011; Halverson et al., 2011; Hou et al., 2014) have studied the Fe isotopic compositions of Archeozoic and Proterozoic BIF deposits all over the world. The $\delta^{56}Fe$ values ranged from −2.05‰ to 3.15‰, and the heaviest and lightest values appeared in magnetite and siderite samples, respectively. All the deposits have a similar Fe isotopic composition with the largest variation range and relatively heavier $\delta^{56}Fe$ values compared with other kinds of deposits (Fig. 7.4). Zhu et al. (2008) considered that during the ore-forming process, with the precipitation of magnetite and hematite, the residual $\delta^{56}Fe$ values of Fe^{2+} ions became lighter and lighter, and then these Fe^{2+} ions combined with CO_3^{2-} and S^{2-} to form the sulfides and carbonates. This hypothesis has also been proved by other research (Yamaguchi et al., 2005; Archer & Vance, 2006; Severmann et al., 2006; Fehr et al., 2008; Von Blanckenburg et al., 2008; Hofmann et al., 2009; Bekker et al., 2010) that shows the sedimentary pyrite (especially before 2.3Ga) and Early Cambrian BIF siderite has relatively lighter $\delta^{56}Fe$ values of −3.60‰ ~ 1.17‰ and −2.04‰ ~ 1.04‰, respectively.

7.2.2.4. Carbonate-Hosted Pb-Zn Deposits-Zn Isotopes

Zhou et al. (2014) analyzed the Zn isotopes of sphalerite (–0.26‰ ~ 0.71‰) and related rocks (–0.24‰ ~ 0.44‰) from the Tainqiao and Bangbangqiao Pb-Zn deposits in the Sichuan-Yunnan-Guizhou Pb-Zn metallogenic province and compared them with other types of Pb-Zn deposits (Fig. 7.9; including the Red Dog SEDEX-type ore district in Alaska [0‰ ~ 0.60‰; Kelley et al., 2009], the Alexandrinka volcanic hosted massive sulphide [VHMS]-type deposit in Russia [–0.43‰ ~ 0.25‰; Mason et al., 2005], Irish [–0.17‰ ~ 1.33‰; Wikinson et al., 2005], the Cévennes Mississippi Valley-type [MVT] deposit in France [–0.06‰ ~ 0.47‰; Albarède, 2004], and the Gorno and Raibl magmatic-type deposit in Italy [0‰ ~ 0.50‰; Marechal et al., 1999] deposits). They identified the unique carbonate-hosted genesis of these deposits combined with S-Pb isotopic data. The Paleozoic carbonate host rocks and Precambrian basements are considered to be the origin of metals, and these rocks have lighter $\delta^{66}Zn$ values (–0.52‰ to 0.16‰) than the sphalerite from the Tianqiao (–0.54‰ to 0.30‰) and Bangbangqiao (–0.21‰ to 0.43‰) deposits. This also supports the theory that the fluids become preferentially enriched in heavier Zn isotopes than related rocks during the solution-solid reaction (Fernandez & Borrok, 2009).

7.2.3. Supergene Systems

7.2.3.1. Copper Deposits: Cu Isotopes

The largest range of copper isotope fractionation has been recorded in these systems, which display total variation of up to 20 per mil. There are various aspects of the system that have been studied both experimentally and in the field. This discussion will first focus on how the experimental work constrains mechanisms for fractionation and then show how the fractionation in nature found in rocks, soils, waters, and plants can be used for exploration geochemistry.

Multiple studies have clearly demonstrated that the loss or gain of electrons in different reactions involving copper leads up to 3 per mil fractionation of the two copper isotopes. Oxidation is known to favor the heavier metal isotope and reduction favors the lighter isotope. Marechal et al. (1999) first showed this with the changing bond environment with stripping copper from the MP-1 resin. Zhu et al. (2002) and Ehrlich et al. (2004) conducted reduction reactions with different copper minerals and recorded 3 per mil variations. Mathur et al. (2005) showed that the oxidation of chalcocite produces slightly heavier copper in solution than the oxidation of chalcopyrite.

Continued experimental work by Wall et al. (2011a) and Wall et al. (2011b) used real-time synchrotron XRD data linked to the oxidation of chalcocite and bornite to study the cause behind the shifts reported by weathering of different phases. Their work clearly shows that the kinetics of the reactions impact how the phases respond with regard to the copper isotope composition of the fluid. It was also noted that chalcocite produced the largest

Figure 7.8 (a) Relationship between increasing alteration of samples and heavier δ^{56}Fe values of pyrite; (b) relationship between increasing alteration of samples and lighter δ^{66}Zn values of sphalerite; (c) relationship between increasing alteration of samples and lighter δ^{66}Zn values of Mn-Fe carbonate; (d) relationship between increasing alteration of samples and lighter δ^{66}Zn values of slate. Abbreviations are as follows: Mcar 1 = stage 1 fine-grained Mn-Fe carbonate; Mcar 2 = stage 2 coarse-grained Mn-Fe carbonate; Py 2 = stage 2 pyrite; Sp 2 = stage 2 sphalerite; Py 3 = stage 3 pyrite; Sp 3 = stage 3 sphalerite; Gn 3 = stage 3 galena; Qtz 3 = stage 3 quartz; Blr 4 = stage 4 boulangerite; Qtz 4 = stage 4 quartz; Sti 5 = stage 5 stibnite; Qtz 6 = stage 6 quartz (from Wang et al., 2017).

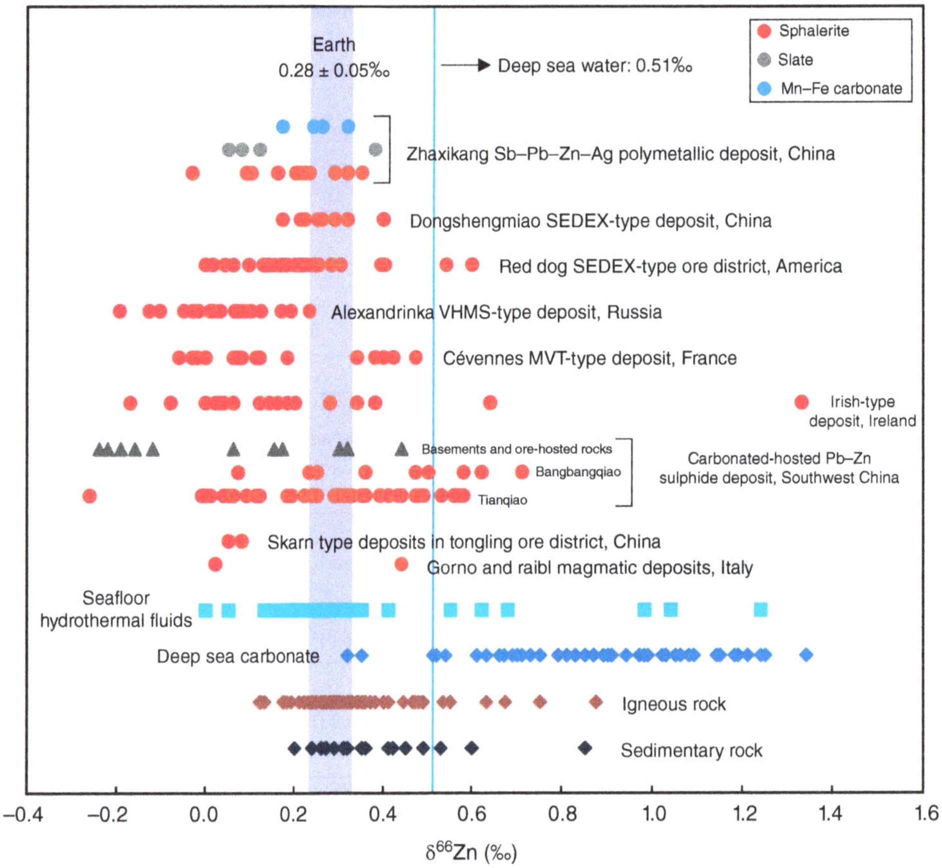

Figure 7.9 Zn isotopic compositions of the Bulk Silicate Earth (Chen et al., 2013), sedimentary rocks (Marechal et al., 2000; Weiss et al., 2007; Bentahila et al., 2008), igneous rocks (Veirs et al., 2007; Bentahila et al., 2008; Toutain et al., 2008; Herzog et al., 2009; Telus et al., 2012; Chen et al., 2013), deep-sea carbonates (Pichat et al., 2003), seafloor hydrothermal fluids (John et al., 2008), deap sea water (John & Conway, 2014; Little et al., 2014; Zhao et al., 2014); and other deposits with different geneses: the Gorno and Raibl magmatic-type deposit in Italy (Marechal et al., 1999), the skarn-type deposits in the Tongling ore district in China (Wang & Zhu, 2010), the Tianqiao and Bangbangqiao carbonated-hosted Pb-Zn sulphide deposits in China (Zhou et al., 2014), the Irish-type deposit in Ireland (Wilkinson et al., 2005), the Cévennes MVT deposit in France (Albarède, 2004), the Alexandrinka VHMS-type deposit in Russia (Mason et al., 2005), the Red Dog SEDEX-type ore district in Alaska (Kelley et al., 2009), Dongshengmiao SEDEX-type deposit in China (Gao et al., 2017), and the Zhaxikang Sb-Pb-Zn-Ag polymetallic deposit in China (Wang et al., 2017).

range of fractionation. In summary, the oxidation of copper within the phases as weathering occurs most likely relates to the changing isotopic composition of the fluids liberating copper.

Biological processes could potentially change the Cu isotopic composition of solutions and solids involved. Biological activities from microbe to plant scale plays a role in fractionating Cu isotopically (Fujii et al., 2013; Kimball et al., 2009; Navarrete et al., 2011a; Navarrete et al., 2011b). Biological material (i.e., cells) in most of the studies appears to incorporate the light nuclide preferentially, though the degree of fractionation varies from experiment to experiment. The importance here is that most of the copper in supergene system is not housed in the bacteria in comparison to the amount of copper in solution. Thus, abiotic reaction most likely provides a much larger mass of copper into these systems in comparison with the minor amounts of copper housed in organic materials.

The experimental data can be applied to the natural system. Within supergene systems, distinct differences in δ^{65}Cu between the three reservoirs that compose such systems (i.e., the leach cap, enrichment blanket, and non-weathered ore; Fig. 7.10). Leach caps have lower δ^{65}Cu values than both the enrichment and proto ore, while the enrichment blanket has the highest δ^{65}Cu (Fig. 7.10). Copper mobilization, transport, and precipitation as secondary minerals generate soils and leach cap Fe-oxides

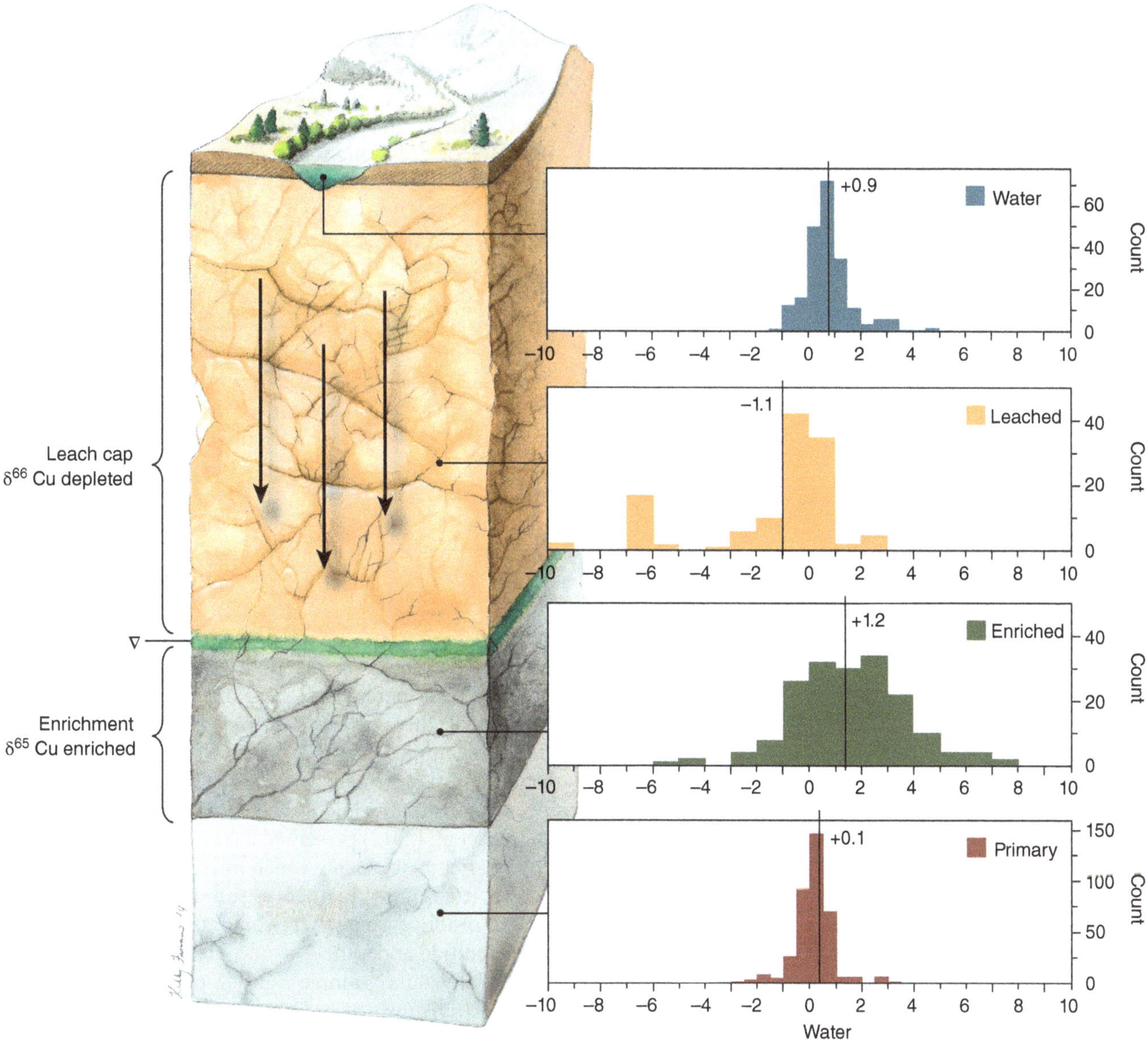

Figure 7.10 Distinct copper isotope compositions in the three main reservoirs in the supergene environment.

that have relatively low δ^{65}Cu values, yet over quite a large range (−9‰ to +2‰; Braxton & Mathur, 2011; Mathur et al., 2010; Mathur et al., 2013; Mathur & Schlitt, 2010; Mirnejad et al., 2010).

From an exploration standpoint, there are several avenues where this fractionation can be used. The first was pointed out in Mathur et al. (2005), where solutions that possess heavier copper could be indicative of weathering of copper sulfide. This relationship was recorded in nature by the following authors in different ore deposit types: Borrok et al. (2008), Kimball et al. (2009), Mathur et al. (2013), and Mathur et al. (2014). Specifically, Kimball et al. (2009) shows that the copper isotope value in one AMD drainage changes from heavier values proximal to the source and lighter values in more distal parts. Mathur et al. (2013) amplified this idea on a large scale and showed that heavier water existed above the known porphyry copper deposit in Pebble, Alaska, in area that spans 10s of Kms. They argued the cause for the heavier values was due to oxidative weathering of copper minerals. Mathur et al. (2014) took this idea a step further and showed that weathering of copper deposits with different copper minerals will produce copper isotope signatures in fluids, which relate to the mineralogy of the copper phase being oxidized (Fig. 7.11). Thus, copper isotope values in water could be used as an inference for the mineral being weathered if the initial copper isotope value of the sulfide can be hypothesized. Importantly, the

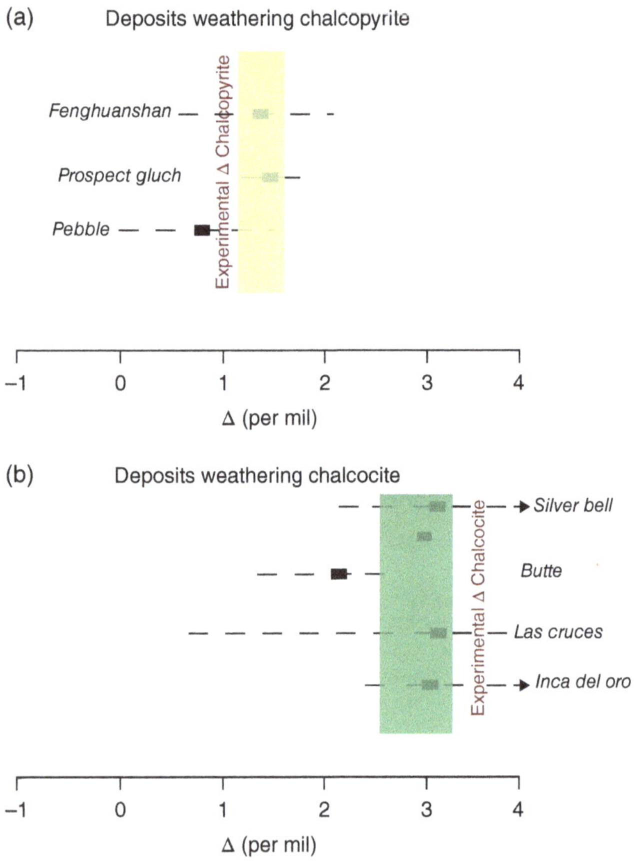

Figure 7.11 Fractionation factors between the solution and the economic copper mineral of interest used to identify copper mineral being weathered. Shaded areas indicate the experimental ranges for weathering of chalcopyrite and chalcocite.

document shows that the mineralogy, not deposit type, relates to the copper isotope signature in the waters.

The residual products of the weathering process, like soils or Fe-oxides in leach caps, reflect this process and can be used in exploration. Braxton and Mathur (2011) and Mirnejad et al. (2010) show that Fe oxides are isotopically lighter than enrichment chalcocite and the starting proto-ore. They proposed that leach cap where chalcocite was weathered (reworked) would have heavier copper isotopes in comparison with areas where no reworking of copper occurred. At both locations (Iran and the Philippines), the leach cap was isotopically lighter but they were unable to compare with places where no mineralization occurred. Mathur et al. (2016) presented data from Queaveco, Peru, where Fe-oxides above enrichment at depth show consistently heavier values in comparison with areas that lacked mineralization at depth.

Cu-oxides and sulfides from laterally transported copper, which form exotic deposits, could be used to vector to the source of ores. Braxton and Mathur (2011) show a pattern in chalcocite that has migrated laterally from the deposit where the heaviest values are found closest to the source of the PCD. Palacios et al. (2011) also document the migration and cycling of copper found in oxide phases of weathered deposits in the Atacama. All of these studies model or suggest that the higher amounts of fractionation seen in the copper isotopes indicates that greater degrees of cycling of copper have occurred in the geologic past. Mathur et al. (2010) pointed to this also in the supergene samples in PCDs in Arizona. They pointed out that larger degrees of cycling may lead to the reduction of total copper preserved in supergene mineralization by pointing to larger differences in the copper isotope composition of starting chalcopyrites compared with supergene chalcocite.

The copper isotope compositions of soils are complicated and have not been used directly in ore deposit studies. A majority of the studies have an environmental monitoring slant, which either ties sources of copper isotope values in ores to soils or traces the migration of copper with the fractionation found in soils (Babcsányi et al., 2016; Bigalke et al., 2010; Bigalke et al., 2009; Bigalke et al., 2011; Li et al., 2016; Pokrovsky et al. (in press); Pokrovsky et al., 2008; Song et al., 2016). In a general sense, the copper isotope composition of the soils represents the parent rock at some level. If the parent material of the soils possesses elevated copper isotope values in comparison with rocks devoid of copper sulfide derived from supergene process, there should be differences in soils that would indicate mineralization at depth. Mathur et al. (2012) presented a study of a weathering of a pyrite-rich black shale, which mirrors the fractionation seen in leach caps. In theory, sulfide-rich parent rocks should produce soils that would indicate the starting copper sulfide at depth.

Natural waters "sample" sulfide phases as they travel through the subsurface. For exploration geologists, the waters are probes into the subsurface and the dissolution of sulfide minerals leads to the liberation of ^{65}Cu in greater amounts than ^{63}Cu. The ultimate composition of the fluids that contained dissolved copper derived from sulfide will directly relate to the starting copper isotope composition found in the ores. Importantly, the average copper isotope composition of seawater and natural waters has been determined to be approximately +0.6 per mil, which is heavier than the average bulk earth value (Bermin et al., 2006; Little et al., 2014; Vance et al., 2008). The elevated values found in these reservoirs could be related to weathering of sulphides (Mathur & Fantle, 2015).

Two studies directly explore how the isotope values could be used in waters. Mathur et al. (2013) present water data from seeps that are actively weathering the Pebble deposit in Alaska. The samples were waters from glacial tills, which lie at least 100 m above the deposit.

They found that waters lying directly above the deposit have an average of +1.47 per mil and samples lying outside the defined mineralization average -0.27 per mil. The data contour above the deposit and demonstrate the vectoring potential of using the waters. They also pointed out that the Cu isotope values indicated weathering of sulfides in waters that contained 5 ppb Cu. Mathur et al. (2014) expanded the study to areas that possess different types of copper deposits and found that the heavier isotope values consistently occurred in areas weathering copper sulfide. Deposit type did not impact the copper isotope values measured. However, they did point out that copper phases could change the copper isotope values in waters.

7.2.3.2. Iron Deposits-Fe Isotopes

Cheng et al. (2015) studied the Fe isotopic composition and variation in Gaosong Sn-polymetallic deposit in Yunnan province (−0.33‰ ~ 0.20‰). They found that the δ^{56}Fe values and oxidation degree have a positive correlation (Fig. 7.12). The δ^{56}Fe values of primary sulfide ores are lower than the oxidized ores, while the δ^{56}Fe values of gossan samples with highest oxidation degree, are heavier than the oxidized ore samples. This indicates that Fe isotopes experienced fractionation during supergene weathering process, and the more oxidized samples are preferentially enriched in heavier Fe isotopes. The gradually increasing δ^{56}Fe values from igneous rocks to ground surface also trace the metal source from igneous rocks rather than country/ host rock. The enrichment of light Fe isotopes in primary sulfide ores suggests that Fe isotope fractionation occurred during fluid exsolution from parental magma and the Fe ores have a magmatic source. Likewise, Markl et al. (2006) have measured the δ^{56}Fe values of the pyrite, hematite, goethite, siderite, Fe-bearing arsenates, and sedimentary iron ores from the Schwarzward hydrothermal-vein deposit. The results show that the product of high-temperature alteration has heavier δ^{56}Fe values than the original mineral; low-temperature alteration almost does not modify the Fe isotopic composition of the original mineral (Fig. 7.4). These two studies demonstrate that the Fe isotopes have the potential to trace supergene geochemical cycling and ore-deposit formation/evolution processes.

7.3. SUMMARY AND FUTURE DIRECTIONS

7.3.1. Copper Isotopes

The above summary shows the current breadth of how copper isotopes have been used in several different types of geological media. The applications for exploration geologists can range from brown to greenfield-type projects. The scale of the projects ranges from several meters to kilometers. Given the analytical capabilities to measure in the nanogram range in both solution and laser ablation techniques and the many different aspects of copper isotope geochemistry, this metal provides exploration geologists multiple avenues for applications.

Future copper isotope studies need to examine nonproducing ore-forming systems to determine how they relate to the economic deposits. Also, a need for greater studies of soils and plants will add to the ability to use the copper isotope system to identify mineralization at depth.

7.3.2. Iron and Zinc Isotopes

As for Fe-Zn isotopes, the previous research demonstrates that both of the Fe-Zn isotopes fractionate during

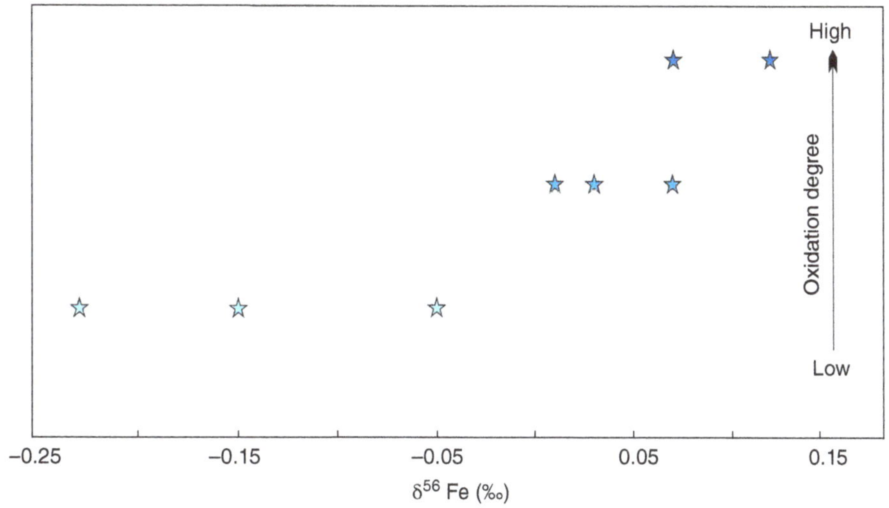

Figure 7.12 Fe isotopic compositions of the oxidized ores with various oxidation degrees from the Gaosong deposit (modified from Cheng et al., 2015).

igneous processes, fluid exsolution, mineral precipitation, and supergene process. The different types of deposits have the different Fe-Zn isotopic characteristics and variation history. Therefore, the Fe-Zn isotopes have the potential to trace the metal source and provide insights into ore-forming processes. The future work on Fe-Zn isotopic application in ore deposits should deliberately sample with exploration geologic intention.

REFERENCES

Albarède, F. (2004). The stable isotope geochemistry of copper and zinc. *Reviews in Mineralogy and Geochemistry*, *55*(1), 409–427.

Anbar, A. D., & Rouxel, O. (2007). Metal stable isotopes in paleoceanography. *Annual Review Earth Planetary Science*, *35*, 717–746.

Archer, C., Vance, D., & Butler, I. (2004). Abiotic Zn isotope fractionations associated with ZnS precipitation. 14th Annual V M Goldschmidt Conference. *Geochimica et Cosmochimica Acta*, *68*(11), A325.

Archer, C., & Vance, D. (2006). Coupled Fe and S isotope evidence for Archean microbial Fe (III) and sulfate reduction. *Geology*, *34*(3), 153–156.

Archer, C., Andersen, M. B., Cloquet, C., Conway, T. M., Dong, S., Ellwood, M., et al. (2017). Inter-calibration of a proposed new primary reference standard AA-ETH Zn for zinc isotopic analysis. *Journal of Analytical Atomic Spectrometry*, *32*(2), 415–419.

Asael, D., Matthews, A., Bar-Matthews, M., & Halicz, L. (2007). Copper isotope fractionation in sedimentary copper mineralization (Timna Valley, Israel). *Chemical Geology*, *243*(3–4), 238–254.

Asael, D., Matthews, A., Bar-Matthews, M., Harlavan, Y., & Segal, I. (2012). Tracking redox controls and sources of sedimentary mineralization using copper and lead isotopes. *Chemical Geology*, *310–311*, 23–35.

Asael, D., Matthews, A., Oszczepalski, S., Bar-Matthews, M., & Halicz, L. (2009). Fluid speciation controls of low temperature copper isotope fractionation applied to the Kupferschiefer and Timna ore deposits. *Chemical Geology*, *262*(3–4), 147–158.

Babcsányi, I., Chabaux, F., Granet, M., Meite, F., Payraudeau, S., Duplay, J., & Imfeld, G. (2016). Copper in soil fractions and runoff in a vineyard catchment: Insights from copper stable isotopes. *Science of the Total Environment*, *557–558*, 154–162.

Balistrieri, L. S., Borrok, D. M., Wanty, R. B., & Ridley, W. I. (2008). Fractionation of Cu and Zn isotopes during adsorption onto amorphous Fe(III) oxyhydroxide: Experimental mixing of acid rock drainage and ambient river water. *Geochimica et Cosmochimica Acta*, *72*(2), 311–328.

Beard, B. L., and Johnson, C. M. (1999). High precision iron isotope measurements of terrestrial and lunar materials. *Geochimica et Cosmochimica Acta*, *63*(11), 1653–1660.

Beard, B. L., Johnson, C. M., Skulan, J. L., Nealson, K. H., Cox, L., and Sun, H. (2003a). Application of Fe isotopes to tracing the geochemical and biological cycling of Fe. *Chemical Geology*, *195*(1), 87–117.

Beard, B. L., Johnson, C. M., Von Damm, K. L., Poulson, R. L. (2003b). Iron isotope constraints on Fe cycling and mass balance in oxygenated earth oceans. *Geology*, *31*(7), 629–632.

Bekker, A., Slack, J. F., Planavsky, N., Krapež, B., Hofmann, A., Konhauser, K. O., & Rouxel, O. J. (2010). Iron formation: the sedimentary product of a complex interplay among mantle, tectonic, oceanic, and biospheric processes. *Economic Geology*, *105*(3), 467–508.

Bennett, S. A., Rouxel, O., Schmidt, K., Garbe-Schoznberg, D., Statham, P. J., German, C. R. (2009). Iron isotope fractionation in a buoyant hydrothermal plume, 5°S Mid-Atlantic Ridge. *Geochimica et Cosmochimica Acta*, *73*(19), 5619–5634.

Bentahila, Y., Othman, D. B., & Luck, J. M. (2008). Strontium, lead and zinc isotopes in marine cores as tracers of sedimentary provenance: A case study around Taiwan orogen. *Chemical Geology*, *248*(1), 62–82.

Berkenbosch, H. A., de Ronde, C. E. J., Paul, B. T., & Gemmell, J. B. (2015). Characteristics of Cu isotopes from chalcopyrite-rich black smoker chimneys at Brothers volcano, Kermadec arc, and Niuatahi volcano, Lau basin. *Mineralium Deposita*, *50*(7), 811–824.

Bermin, J., Vance, D., Archer, C., & Statham, P. J. (2006). The determination of the isotopic composition of Cu and Zn in sea water. *Chemical Geology*, *226*(3–4), 280–297.

Bigalke, M., Weyer, S., & Wilcke, W. (2009). Stable copper isotopes: A novel tool to trace copper behavior in hydromorphic soils. *Soil Sci. Soc. Am. J.*, *74*(1), 60–73.

Bigalke, M., Weyer, S., & Wilcke, W. (2011). Stable Cu isotope fractionation in soils during oxic weathering and podzolization. *Geochimica et Cosmochimica Acta*, *75*(11), 3119–3134.

Bigalke, M., Weyer, S., Kobza, J., & Wilcke, W. (2010). Stable Cu and Zn isotope ratios as tracers of sources and transport of Cu and Zn in contaminated soil. *Geochimica et Cosmochimica Acta*, *74*(23), 6801–6813.

Bilenker, L. D., Simon, A. C., Reich, M., Lundstrom, C. C., Gajos, N., Bindeman, I., & Munizaga, R. (2016). Fe-O stable isotope pairs elucidate a high-temperature origin of Chilean iron oxide-apatite deposits. *Geochimica et Cosmochimica Acta*, *177*, 94–104.

Borrok, D. M., Wanty, R. B., Ridley, W. I., Wolf, R., Lamothe, P. J., & Adams, M. (2007). Separation of copper, iron, and zinc from complex aqueous solutions for isotopic measurement. *Chemical Geology*, *242*(3–4), 400–414.

Borrok, D. M., Nimick, D. A., Wanty, R. B., & Ridley, W. I. (2008). Isotopic variations of dissolved copper and zinc in stream waters affected by historical mining. *Geochimica et Cosmochimica Acta*, *72*(2), 329–344.

Braxton, D., & Mathur, R. (2011). Exploration applications of copper isotopes in the supergene environment: A case study of the Bayugo porphyry copper-gold deposit, Southern Philippines. *Economic Geology*, *106*(8), 1447–1463.

Chapman, J. B., et al. (2004). An adapted column chemistry procedure for separation of Fe, Cu and Zn from geological matrices, and natural Zn isotopic variations in geological standard reference materials BCR-027, BCR-030 and NOD-P-1. *Abstracts with Programs–Geological Society of America*, *36*(5), 448.

Chapman, J. B., Mason, T. F. D., Weiss, D. J., Coles, B. J., & Wilkinson, J. J. (2006). Chemical separation and isotopic variations of Cu and Zn from five geological reference materials. *Geostandards and Geoanalytical Research*, *30*(1), 5–16.

Chen, H., Savage, P. S., Teng, F. Z., Helz, R. T., and Moynier, F. (2013). Zinc isotope fractionation during magmatic differentiation and the isotopic composition of the bulk Earth. *Earth and Planetary Science Letters*, *369*, 34–42.

Cheng, Y., Mao, J., Zhu, X., & Wang, Y. (2015). Iron isotope fractionation during supergene weathering process and its application to constrain ore genesis in Gaosong deposit, Gejiu district, SW China. *Gondwana Research*, *27*(3), 1283–1291.

Craddock, P. R., & Dauphas, N. (2011). Iron and carbon isotope evidence for microbial iron respiration throughout the Archean. *Earth and Planetary Science Letters*, *303*(1), 121–132.

Czaja, A. D., Johnson, C. M., Beard, B. L., Eigenbrode, J. L., Freeman, K. H., & Yamaguchi, K. E. (2010). Iron and carbon isotope evidence for ecosystem and environmental diversity in the ~2.7 to 2.5 Ga Hamersley Province, Western Australia. *Earth and Planetary Science Letters*, *292*(1), 170–180.

Dauphas, N., Janney, P. E., Mendybaev, R. A., Wadhwa, M., Richter, F. M., Davis, A. M., Zuilen, M. V., Hines, R., and Foley, C. N. (2004). Chromatographic separation and multicollection-ICPMS analysis of iron. Investigating mass-dependent and-independent isotope effects. *Analytical Chemistry*, *76*(19), 5855–5863.

Dauphas, N., & Rouxel, O. (2006). Mass spectrometry and natural variations of iron isotopes. *Mass Spectrometry Reviews*, *25*(4), 515–550.

Dauphas, N., Van Zuilen, M., Busigny, V., Lepland, A., Wadhwa, M., & Janney, P. E. (2007). Iron isotope, major and trace element characterization of early Archean supracrustal rocks from SW Greenland: protolith identification and metamorphic overprint. *Geochimica et Cosmochimica Acta*, *71*(19), 4745–4770.

Dauphas, N., Teng, F. Z., & Arndt, N. T. (2010). Magnesium and iron isotopes in 2.7 Ga Alexo komatiites: mantle signatures, no evidence for Soret diffusion, and identification of diffusive transport in zoned olivine. *Geochimica et Cosmochimica Acta*, *74*(11), 3274–3291.

Dendas, M. (2011). Copper isotopes tracers of fluid flow at Bingham Canyon, Utah, University of Arizona, Tuscon.

Duan, J., Tang, J., Li, Y., Liu, S. A., Wang, Q., Yang, C., & Wang, Y. (2016). Copper isotopic signature of the Tiegelongnan high-sulfidation copper deposit, Tibet: Implications for its origin and mineral exploration. *Mineralium Deposita*, *51*(5), 591–602.

Ehrlich, S., Butler, I., Halicz, L., Rickard, D., Oldroyd, A., & Matthews, A. (2004). Experimental study of the copper isotope fractionation between aqueous Cu (II) and covellite, CuS. *Chemical Geology*, *209*(3–4), 259–269.

Fehr, M. A., Andersson, P. S., Hålenius, U., & Mörth, C. M. (2008). Iron isotope variations in Holocene sediments of the Gotland Deep, Baltic Sea. *Geochimica et Cosmochimica Acta*, *72*(3), 807–826.

Fernandez, A., and Borrok, D. M. (2009). Fractionation of Cu, Fe, and Zn isotopes during the oxidative weathering of sulfide-rich rocks. *Chemical Geology*, *264*(1), 1–12.

Foden, J., Sossi, P. A., and Wawryk, C. M. (2015). Fe isotopes and the contrasting petrogenesis of A-, I-and S-type granite. *Lithos*, *212*, 32–44.

Frost, C. D., Von Blanckenburg, F., Schoenberg, R., Frost, B. R., & Swapp, S. M. (2007). Preservation of Fe isotope heterogeneities during diagenesis and metamorphism of banded iron formation. *Contributions to Mineralogy and Petrology*, *153*(2), 211.

Fujii, T., Moynier, F., Abe, M., Nemoto, K., & Albarède, F. (2013). Copper isotope fractionation between aqueous compounds relevant to low temperature geochemistry and biology. *Geochimica et Cosmochimica Acta*, *110*(0), 29–44.

Gagnevin, D., Boyce, A. J., Barrie, C. D., Menuge, J. F., Blakeman, R. J. (2012). Zn, Fe and S isotope fractionation in a large hydrothermal system. *Geochimica et Cosmochimica Acta*, *88*, 183–198.

Gao, Z., Zhu, X. K, Sun, J., Luo, Z. H, Bao, C., Tang, C., and Ma, J. X. (2017). Spatial evolution of Zn-Fe-Pb isotopes of sphalerite within a single ore body: A case study from the Dongshengmiao ore deposit, Inner Mongolia, *China. Mineralium Deposita*, 1–11.

Graham, S., Pearson, N., Jackson, S., Griffin, W., & O'Reilly, S. Y. (2004). Tracing Cu and Fe from source to porphyry; in situ determination of Cu and Fe isotope ratios in sulfides from the Grasberg Cu-Au deposit. *Chemical Geology*, *207*(3–4), 147–169.

Haest, M., Muchez, P., Petit, J. C. J., & Vanhaecke, F. (2009). Cu isotope ratio variations in the Dikulushi Cu-Ag deposit, DRC: Of primary origin or induced by supergene reworking. *Economic Geology*, *104*(7), 1055–1064.

Halverson, G. P., Poitrasson, F., Hoffman, P. F., Nédélec, A., Montel, J. M., & Kirby, J. (2011). Fe isotope and trace element geochemistry of the Neoproterozoic syn-glacial Rapitan iron formation. *Earth and Planetary Science Letters*, *309*(1), 100–112.

Heimann, A., Beard, B. L., and Johnson, C. M. (2008). The role of volatile exsolution and sub-solidus fluid/rock interactions in producing high 56Fe/54Fe ratios in siliceous igneous rocks. *Geochimica et Cosmochimica Acta*, *72*(17), 4379–4396.

Heimann, A., Johnson, C. M., Beard, B. L., Valley, J. W., Roden, E. E., Spicuzza, M. J., & Beukes, N. J. (2010). Fe, C, and O isotope compositions of banded iron formation carbonates demonstrate a major role for dissimilatory iron reduction in ~2.5 Ga marine environments. *Earth and Planetary Science Letters*, *294*(1), 8–18.

Herzog, G. F., Moynier, F., Albarède, F., Berezhnoy, A. A. (2009). Isotopic and elemental abundances of copper and zinc in lunar samples, zagami, pele's hairs, and a terrestrial basalt. *Geochimica et Cosmochimica Acta*, *73*(19), 5884–5904.

Hofmann, A., Bekker, A., Rouxel, O., Rumble, D., & Master, S. (2009). Multiple sulphur and iron isotope composition of detrital pyrite in Archaean sedimentary rocks: a new tool for provenance analysis. *Earth and Planetary Science Letters*, *286*(3), 436–445.

Hou, K., Li, Y., Gao, J., Liu, F., & Qin, Y. (2014). Geochemistry and Si–O–Fe isotope constraints on the origin of banded iron formations of the Yuanjiacun Formation, Lvliang Group, Shanxi, China. *Ore Geology Reviews*, *57*, 288–298.

Ikehata, K., & Hirata, T. (2012). Copper isotope characteristics of copper-rich minerals from the Horoman Peridotite Complex, Hokkaido, northern Japan. *Economic Geology*, *107*(7), 1489–1497.

Ikehata, K., & Hirata, T. (2013). Evaluation of UV-fs-LA-MC-ICP-MS for pprecise in situ copper isotopic microanalysis of cubanite. *Analytical Sciences*, *29*(12), 1213–1217.

Ikehata, K., Notsu, K., & Hirata, T. (2011). Copper isotope characteristics of copper-rich minerals from Besshi-type volcanogenic massive sulfide deposits, Japan, determined using a femtosecond LA-MC-ICM-MS. *Economic Geology*, *106*(2), 307–316.

Ikehata, K., Notsu, K., & Hirata, T. (2008). In situ determination of Cu isotope ratios in copper-rich materials by NIR femtosecond LA-MC-ICP-MS. *Journal of Analytical Atomic Spectrometry*, *23*(7), 1003–1008.

Jackson, S. E., & Günther, D. (2003). The nature and sources of laser induced isotopic fractionation in laser ablation-multicollector-inductively coupled plasma-mass spectrometry. *Journal of Analytical Atomic Spectrometry*, *18*(3), 205–212.

John, S. G., Rouxel, O. J., Craddock, P. R., Engwall, A. M., Boyle, E. A. (2008). Zinc stable isotopes in seafloor hydrothermal vent fluids and chimneys. *Earth and Planetary Science Letters*, *269*(1), 17–28.

John, S. G., and Conway, T. M. (2014). A role for scavenging in the marine biogeochemical cycling of zinc and zinc isotopes. *Earth and Planetary Science Letters*, *394*, 159–167.

Johnson, C. M., Beard, B. L., Beukes, N. J., Klein, C., & O'Leary, J. M. (2003). Ancient geochemical cycling in the Earth as inferred from Fe isotope studies of banded iron formations from the Transvaal Craton. *Contributions to Mineralogy and Petrology*, *144*(5), 523–547.

Johnson, C. M., Beard, B. L., Roden, E. E., Newman, D. K., & Nealson, K. H. (2004). Isotopic constraints on biogeochemical cycling of Fe. *Reviews in mineralogy and geochemistry*, *55*(1), 359–408.

Johnson, C. M., & Beard, B. L. (2006). Fe isotopes: an emerging technique for understanding modern and ancient biogeochemical cycles. *GSA TODAY*, *16*(11), 4.

Johnson, C. M., Beard, B. L., Klein, C., Beukes, N. J., & Roden, E. E. (2008). Iron isotopes constrain biologic and abiologic processes in banded iron formation genesis. *Geochimica et Cosmochimica Acta*, *72*(1), 151–169.

Kelley, K. D., Wilkinson, J. J., Chapman, J. B., Crowther, H. L., Weiss, D. J. (2009). Zinc isotopes in sphalerite from base metal deposits in the red dog district, northern Alaska. *Economic Geology*, *104*(6), 767–773.

Kimball, B. E., Mathur, R., Dohnalkova, A. C., Wall, A. J., Runkel, R. L., & Brantley, S. L. (2009). Copper isotope fractionation in acid mine drainage. *Geochimica et Cosmochimica Acta*, *73*(5), 1247–1263.

Kuhn, H.-R., Pearson, N., & Jackson, S. (2007). The influence of the laser ablation process on isotopic fractionation of copper in LA-MC-ICP-MS. *Journal of Analytical Atomic Spectrometry*, *22*(5), 547–552.

Larson, P. B., Maher, K., Ramos, F. C., Chang, Z., Gaspar, M., & Meinert, L. D. (2003). Copper isotope ratios in magmatic and hydrothermal ore-forming environments. *Chemical Geology*, *201*(3–4), 337–350.

Li, S.-Z., Zhu, X.-K., Wu, L.-H., & Luo, Y.-M. (2016). Cu isotopic compositions in Elsholtzia splendens, Influence of soil condition and growth period on Cu isotopic fractionation in plant tissue. *Chemical Geology*, *444*, 49–58.

Li, W., Jackson, S. E., Pearson, N. J., Alard, O., & Chappell, B.W. (2009). The Cu isotopic signature of granites from the Lachlan Fold Belt, SE Australia. *Chemical Geology*, *258*(1–2), 38–49.

Li, W., Jackson, S. E., Pearson, N. J., & Graham, S. (2010). Copper isotopic zonation in the Northparkes porphyry Cu-Au deposit, SE Australia. *Geochimica et Cosmochimica Acta*, *74*(14), 4078–4096.

Li, Z. H., Zhu, X. K., & Tang, S. H. (2008a). Characters of Fe isotopes and rare earth elements of banded iron formations from Anshan–Benxi area: implications for Fe source. *Acta Petrologica et Mineralogica*, *27*(4), 285–290 (In Chinese with English abstract).

Li, Z. H., Zhu, X. K., Tang, S. H., & Li, Y. H. (2008b). Fe isotope fractionation between magnetite and pyrite during green schist-lower amphibolites facies metamorphism. *Acta Petrologica et Mineralogica*, *27*(4), 291–297 (In Chinese with English abstract).

Li, Z. H., and Zhu, X. K. (2012). Geochemical features of Xuanlong type iron ore deposit in Hebei Province and their geological significances. *Acta Petrologica Sinica*, *28*(9), 2903–2911 (In Chinese with English abstract).

Little, S. H., Vance, D., Walker-Brown, C., & Landing, W. M. (2014). The oceanic mass balance of copper and zinc isotopes, investigated by analysis of their inputs, and outputs to ferromanganese oxide sediments. *Geochimica et Cosmochimica Acta*, *125*(0), 673–693.

Liu, S.-A., Huang, J., Liu, J., Wörner, G., Yang, W., Tang, Y. J., & Li, S. (2015). Copper isotopic composition of the silicate Earth. *Earth and Planetary Science Letters*, *427*, 95–103.

Maher, K. C., & Larson, P. B. (2007). Variation in copper isotope ratios and controls on fractionation in hypogene skarn mineralization at Coroccohuayco and Tintaya, Peru. *Economic Geology*, *102*(2), 225–237.

Maher, K. C., Jackson, S., & Mountain, B. (2011). Experimental evaluation of the fluid-mineral fractionation of Cu isotopes at 250°C and 300°C. *Chemical Geology*, *286*(3–4), 229–239.

Malitch, K. N., Latypov, R. M., Badanina, I. Y., & Sluzhenikin, S. F. (2014). Insights into ore genesis of Ni-Cu-PGE sulfide deposits of the Noril'sk Province (Russia): Evidence from copper and sulfur isotopes. *Lithos*, *204*, 172–187.

Marechal, C., & Albarede, F. (2002). Ion-exchange fractionation of copper and zinc isotopes. *Geochimica et Cosmochimica Acta*, *66*(9), 1499–1509.

Maréchal, C. N., Emmanuel, N., Chantal, D., Francis, A. (2000). Abundance of zinc isotopes as a marine biogeochemical tracer. *Geochemistry, Geophysics, Geosystems*, *1*(5).

Marechal, C. N., Telouk, P., & Albarede, F. (1999). Precise analysis of copper and zinc isotopic compositions by plasma-source mass spectrometry. *Chemical Geology*, *156*(1–4), 251–273.

Markl, G., Lahaye, Y., & Schwinn, G. (2006). Copper isotopes as monitors of redox processes in hydrothermal mineralization. *Geochimica et Cosmochimica Acta*, *70*(16), 4215–4228.

Mason, T. F. D., Weiss, D. J., Horstwood, M., Parrish, R. R., Russell, S. S., Mullane, E., & Coles, B. J. (2004). High-precision Cu and Zn isotope analysis by plasma source mass spectrometry Part 1. Spectral interferences and their correction. *Journal of Analytical Atomic Spectrometry*, *19*(2), 209–217.

Mason, T. F. D., Weiss, D. J., Chapman, J. B., Wilkinson, J. J., Tessalina, S. G., Spiro, B., & Coles, B. J. (2005). Zn and Cu isotopic variability in the Alexandrinka volcanic-hosted massive sulphide (VHMS) ore deposit, Urals, Russia. *Chemical Geology*, *221*(3–4), 170–187.

Mathur, R., & Fantle, M. S. (2015). Copper isotopic perspectives on supergene processes: implications for the global cu cycle. *Elements*, *11*(5), 323–329.

Mathur, R., & Schlitt, W. J. (2010). Identification of the dominant Cu ore minerals providing soluble copper at Cañariaco, Peru through Cu isotope analyses of batch leach experiments. *Hydrometallurgy*, *101*(1–2), 15–19.

Mathur, R., Ruiz, J., Titley, S., Liermann, L., Buss, H., & Brantley, S. (2005). Cu isotopic fractionation in the supergene environment with and without bacteria. *Geochimica et Cosmochimica Acta*, *69*(22), 5233–5246.

Mathur, R., Titley, S., Barra, F., Brantley, S., Wilson, M., Phillips, A., & Hart, G. (2009). Exploration potential of Cu isotope fractionation in porphyry copper deposits. *Journal of GeoChemical Exploration*, *102*(1), 1–6.

Mathur, R., Jin, L., Prush, V., Paul, J., Ebersole, C., Fornadel, A., & Brantley, S. (2012). Cu isotopes and concentrations during weathering of black shale of the Marcellus Formation, Huntingdon County, Pennsylvania (USA). *Chemical Geology*, *304–305*(0), 175–184.

Mathur, R., Munk, L., Nguyen, M., Gregory, M., Annell, H., & Lang, J. (2013). Modern and paleofluid pathways revealed by cu isotope compositions in surface waters and ores of the Pebble Porphyry Cu-Au-Mo Deposit, Alaska. *Economic Geology*, *108*(3), 529–541.

Mathur, R., Munk, L. A., Townley, B., Gou, K. Y., Miguélez, N. G., Titley, S., & Ruiz, J. (2014). Tracing low-temperature aqueous metal migration in mineralized watersheds with Cu isotope fractionation. *Applied Geochemistry*, *51*(0), 109–115.

Mathur, R., Dendas, M., Titley, S., & Phillips, A. (2010). Patterns in the copper isotope composition. of minerals in porphyry copper deposits in southwestern United States. *Economic Geology*, *105*(8), 1457–1467.

Metz, S., & Trefry, J. H. (2000). Chemical and mineralogical influences on concentrations of trace metals in hydrothermal fluids. *Geochimica et Cosmochimica Acta*, *64*(13), 2267–2279.

Mirnejad, H., Mathur, R., Einali, M., Dendas, M., & Alirezaei, S. (2010). A comparative copper isotope study of porphyry copper deposits in Iran. *Geochemistry: Exploration, Environment, Analysis*, *10*(4), 413–418.

Moynier, F., Blichert-Toft, J., Telouk, P., Luck, J.-M., & Albarede, F. (2007). Comparative stable isotope geochemistry of Ni, Cu, Zn, and Fe in chondrites and iron meteorites. *Geochimica et Cosmochimica Acta*, *71*(17), 4365–4379.

Navarrete, J. U., Borrok, D. M., Viveros, & M., Ellzey, J. T. (2011a). Copper isotope fractionation during surface adsorption and intracellular incorporation by bacteria. *Geochimica et Cosmochimica Acta*, *75*(3), 784–799.

Navarrete, J. U., Viveros, M., Ellzey, J. T., & Borrok, D. M. (2011b). Copper isotope fractionation by desert shrubs. *Applied Geochemistry*, *26*, Supplement(0), S319–S321.

Palacios, C., Rouxel, O., Reich, M., Cameron, E. M., & Leybourne, M. I. (2011). Pleistocene recycling of copper at a porphyry system, Atacama Desert, Chile; Cu isotope evidence. *Mineralium Deposita*, *46*(1), 1–7.

Peel, K., Weiss, D., Chapman, J., Arnold, T., & Coles, B. (2008). A simple combined sample-standard bracketing and inter-element correction procedure for accurate mass bias correction and precise Zn and Cu isotope ratio measurements. *Journal of Analytical Atomic Spectrometry*, *23*(1), 103–110.

Petit, J. C., De Jong, J., Chou, L., & Mattielli, N. (2008). Development of Cu and Zn isotope MC-ICP-MS measurements: Application to suspended particulate matter and sediments from the Scheldt Estuary. *Geostandards and Geoanalytical Research*, *32*(2), 149–166.

Pichat, S., Douchet, C., Albarède, F. (2003). Zinc isotope variations in deep-sea carbonates from the eastern equatorial Pacific over the last 175 ka. *Earth Planetary Science Letters*, *210*(1), 167–178.

Planavsky, N., Rouxel, O., Bekker, A., Shapiro, R., Fralick, P., & Knudsen, A. (2009). Iron-oxidizing microbial ecosystems thrived in late Paleoproterozoic redox-stratified oceans. *Earth and Planetary Science Letters*, *286*(1), 230–242.

Planavsky, N., Rouxel, O. J., Bekker, A., Hofmann, A., Little, C. T., & Lyons, T. W. (2011). Iron isotope composition of some Archean and Proterozoic iron formations. *Geochimica et Cosmochimica Acta*, *80*, 158–169.

Poitrasson, F., Halliday, A. N., Lee, D. C., Levasseur, S., and Teutsch, N. (2004). Iron isotope differences between Earth, Moon, Mars and Vesta as possible records of contrasted accretion mechanisms. *Earth and Planetary Science Letters*, *223*(3), 253–266.

Poitrasson, F., and Freydier, R. (2005). Heavy iron isotope composition of granites determined by high resolution MC-ICP-MS. *Chemical Geology*, *222*(1), 132–147.

Pokrovsky, O. S., Viers, J., Emnova, E. E., Kompantseva, E. I., & Freydier, R. (in press). Copper isotope fractionation during its interaction with soil and aquatic microorganisms and metal oxy(hydr)oxides: Possible structural control. *Geochimica et Cosmochimica Acta*, In Press, Corrected Proof.

Pokrovsky, O. S., Viers, J., Emnova, E. E., Kompantseva, E. I., & Freydier, R. (2008). Copper isotope fractionation during its interaction with soil and aquatic microorganisms and metal oxy(hydr)oxides; possible structural control. *Geochimica et Cosmochimica Acta*, *72*(7), 1742–1757.

Pribil, M. J., Wanty, R. B., Ridley, W. I., & Borrok, D. M. (2009). Influence of sulfur-bearing polyatomic species on high precision measurements of Cu isotopic composition. *Chemical Geology*, *272*(1–4), 49–54.

Ripley, E. M., Dong, S., Li, C., & Wasylenki, L. E. (2015). Cu isotope variations between conduit and sheet-style Ni-Cu-PGE sulfide mineralization in the Midcontinent Rift System, North America. *Chemical Geology*, *414*, 59–68.

Rouxel, O., Fouquet, Y., & Ludden, J. N. (2004). Copper isotope systematics of the Lucky Strike, Rainbow, and Logatchev sea-floor hydrothermal fields on the Mid-Atlantic Ridge. *Economic Geology*, *99*(3), 585–600.

Rouxel, O. J., Bekker, A., and Edwards, K. J. (2005). Iron isotope constraints on the Archean and Paleoproterozoic ocean redox state. *Science*, *307*(5712), 1088–1091.

Rouxel, O., Iii, W. C. S., Bach, W., Edwards, K. J. (2008). Integrated Fe- and S-isotope study of seafloor hydrothermal vents at East Pacific Rise 9–10°N. *Chemical Geology*, *252*(3), 214–227.

Saunders, J. A., Mathur, R., Kamenov, G. D., Shimizu, T., Brueseke, M. E. (2015). New isotopic evidence bearing on bonanza (Au-Ag) epithermal ore-forming processes. *Mineralium Deposita*, *51*(1), 1–11.

Schuessler, J. A., Schoenberg, R., and Sigmarsson, O. (2009). Iron and lithium isotope systematics of the Hekla volcano, Iceland—evidence for Fe isotope fractionation during magma differentiation. *Chemical Geology*, *258*(1), 78–91.

Seewald, J. S., & Seyfried, W. E. (1990). The effect of temperature on metal mobility in subseafloor hydrothermal systems: constraints from basalt alteration experiments. *Earth and Planetary Science Letters*, *101*(2–4), 388–403.

Severmann, S., Johnson, C. M., Beard, B. L., German, C. R., Edmonds, H. N., Chiba, H. and Green, D. R. H. (2004). The effect of plume processes on the Fe isotope composition of hydrothermally derived Fe in the deep ocean as inferred from the rainbow vent site, Mid-Atlantic ridge, 36°14'N. *Earth and Planetary Science Letters*, *225*(1), 63–76.

Severmann, S., Johnson, C. M., Beard, B. L., & McManus, J. (2006). The effect of early diagenesis on the Fe isotope compositions of porewaters and authigenic minerals in continental margin sediments. *Geochimica et Cosmochimica Acta*, *70*(8).

Sharam, M., Polizzotto, M., and Anbar, A. D. (2001). Iron isotopes in hot springs along the Juan de Fuca Ridge. *Earth and Planetary Science Letters*, *194*(1), 39–51.

Song, S., Mathur, R., Ruiz, J., Chen, D., Allin, N., Guo, K., & Kang, W. (2016). Fingerprinting two metal contaminants in streams with Cu isotopes near the Dexing Mine, China. *Science of The Total Environment*, *544*, 677–685.

Sossi, P. A., Halverson, G. P., Nebel, O., & Eggins, S. M. (2015). Combined Separation of Cu, Fe and Zn from rock matrices and improved analytical protocols for stable isotope determination. *Geostandards and Geoanalytical Research*, *39*(2), 129–149.

Steinhoefel, G., Horn, I., & von Blanckenburg, F. (2009). Micro-scale tracing of Fe and Si isotope signatures in banded iron formation using femtosecond laser ablation. *Geochimica et Cosmochimica Acta*, *73*(18), 5343–5360.

Steinhoefel, G., von Blanckenburg, F., Horn, I., Konhauser, K. O., Beukes, N. J., & Gutzmer, J. (2010). Deciphering formation processes of banded iron formations from the Transvaal and the Hamersley successions by combined Si and Fe isotope analysis using UV femtosecond laser ablation. *Geochimica et Cosmochimica Acta*, *74*(9), 2677–2696.

Sun, J., Zhu, X. K., Chen, Y. L., and Fang, N. (2013). Iron isotopic constraints on the genesis of Bayan Obo ore deposit, Inner Mongolia, China. *Precambrian Research*, *235*, 88–106.

Taylor, P. D. P., Maeck, R., & De Bievre, P. (1992). Determination of the absolute isotopic composition and atomic weight of a reference sample of natural iron. *International Journal of Mass Spectrometry and Ion Processes*, *121*(1–2), 111–125.

Telus, M., Dauphas, N., Moynier, F., Tissot, F. L., Teng, F. Z., Nabelek, P. I., Craddock, P. R., and Groat, L. A. (2012). Iron, zinc, magnesium and uranium isotopic fractionation during continental crust differentiation: The tale from migmatites, granitoids, and pegmatites. *Geochimica et Cosmochimica Acta*, *97*, 247–265.

Toutain, J. P., Sonke, J., Munoz, M., Nonell, A., Polvé, M., Viers, J., & Sumarti, S. (2008). Evidence for Zn isotopic fractionation at Merapi volcano. *Chemical Geology*, *253*(1), 74–82.

Tsikos, H., Matthews, A., Erel, Y., & Moore, J. M. (2010). Iron isotopes constrain biogeochemical redox cycling of iron and manganese in a Palaeoproterozoic stratified basin. *Earth and Planetary Science Letters*, *298*(1), 125–134.

Vance, D., Archer, C., Bermin, J., Perkins, J., Statham, P. J., Lohan, M. C., & Mills, R. A. (2008). The copper isotope geochemistry of rivers and the oceans. *Earth and Planetary Science Letters*, *274*(1–2), 204–213.

Wall, A. J., et al. (2011a). A flow-through reaction cell that couples time-resolved x-ray diffraction with stable isotope analysis. *Journal of Applied Crystallography*, *44*(2), 429–432.

Wall, A. J., Mathur, R., Post, J. E., & Heaney, P. J. (2011b). Cu isotope fractionation during bornite dissolution: An in situ x-ray diffraction analysis. *Ore Geology Reviews*, *42*(1), 62–70.

Wang, D., Sun, X., Zheng, Y. Y, Wu, S., Xia, S. L, Chang, H. F., and Yu, M. (2017). Two pulses of mineralization and genesis of the Zhaxikang Sb–Pb–Zn–Ag deposit in southern Tibet: Constraints from Fe–Zn isotopes. *Ore Geology Reviews*, *84*, 347–363.

Wang, Y., and Zhu, X. K. (2010). Application of Zn Isotopes to study of Mineral Deposits: A review. *Mineral Deposits*, *29*(5), 843–852 (In Chinese with English abstract).

Wang, Y., and Zhu, X. K. (2012). Fe isotope systematics and its implications in ore deposit geology. *Acta Petrologica Sinica*, *28*(11), 3638–3654 (In Chinese with English abstract).

Wang, Y., Zhu, X. K., Mao, J. W., Li, Z. H., and Cheng, Y. B. (2011). Iron isotope fractionation during skarn-type metallogeny: a case study of Xinqiao Cu–S–Fe–Au deposit in the Middle–Lower Yangtze valley. *Ore Geology Review*, *43*(1), 194–202.

Wang, Y., Zhu, X. K., and Cheng, Y. (2015). Fe isotope behaviours during sulfide-dominated skarn-type mineralisation. *Journal of Asian Earth Sciences*, *103*, 374–392.

Wawryk, C. M., and Foden, J. D. (2015). Fe-isotope fractionation in magmatic-hydrothermal mineral deposits: a case study from the Renison Sn–W deposit, Tasmania. *Geochimica et Cosmochimica Acta*, *150*, 285–298.

Weiss, D. J., Rausch, N., Mason, T. F. D., Coles, B. J., Wilkinson, J. J., Ukonmaanaho, L., Arnold, T. and Nieminen, T. M. (2007). Atmospheric deposition and isotope biogeochemistry of zinc in ombrotrophic peat. *Geochimica et Cosmochimica Acta*, *71*(14), 3498–3517.

Whitehouse, M. J., & Fedo, C. M. (2007). Microscale heterogeneity of Fe isotopes in > 3.71 Ga banded iron formation from the Isua Greenstone Belt, southwest Greenland. *Geology*, *35*(8), 719–722.

Wilkinson, J. J., Weiss, D. J., Mason, T. F. D., Coles, B. J. (2005). Zinc isotope variation in hydrothermal systems: preliminary evidence from the Irish Midlands ore field.

Wilson, M., Wilson, D. L., & Mathur, R. (2016). Tracing the source of native copper mineral specimens with copper isotope values. *Rocks & Minerals*, *91*(4), 352–356.

Yamaguchi, K. E., Johnson, C. M., Beard, B. L., & Ohmoto, H. (2005). Biogeochemical cycling of iron in the Archean–Paleoproterozoic Earth: constraints from iron isotope variations in sedimentary rocks from the Kaapvaal and Pilbara Cratons. *Chemical Geology*, *218*(1), 135–169.

Yan, B., Zhu, X. K., Zhang, H. F., & Tang, S. H. (2010). Fe isotopic characteristics of the Neoproterozoic BIF in Guangxi province and its implications. *Acta Geologica Sinica*, *84*(7), 1080–1086 (In Chinese with English abstract).

Yao, J., Mathur, R., Sun, W., Song, W., Chen, H., Mutti, L., & Luo, X. (2016). Fractionation of Cu and Mo isotopes caused by vapor-liquid partitioning, evidence from the Dahutang W-Cu-Mo ore field. *Geochemistry, Geophysics, Geosystems*, *17*(5), 1725–1739.

Zhao, Y., Vance, D., Abouchami, W., and De Baar, H. J. W. (2014). Biogeochemical cycling of zinc and its isotopes in the Southern Ocean. *Geochimica et Cosmochimica Acta*, *125*, 653–672.

Zhao, Y., Xue, C., Liu, S. A., Symons, D. T., Zhao, X., Yang, Y., & Ke, J. (2017). Copper isotope fractionation during sulfide-magma differentiation in the Tulaergen magmatic Ni-Cu deposit, NW China. *Lithos*, *286*, 206–215.

Zheng, Y. Y., Liu, M. Y., Sun, X., Yuan, E. H., Tian, L. M., Zheng, H. T., Zhang, G. Y., Zhang, L. H. (2012). Type, discovery process and significance of Zhaxikang antimony polymetallic ore deposit, Tibet. *Earth Science*, *37*(5), 1004–1014 (In Chinese with English abstract).

Zhou, J. X., Huang, Z. L., Zhou, M. F., Zhu, X. K., Muchez, P. (2014). Zinc, sulfur and lead isotopic variations in carbonate-hosted Pb–Zn sulfide deposits, southwest China. *Ore Geology Review*, *58*, 41–54.

Zhu, B., Zhang, H. F., Zhao, X. M., & He, Y. S. (2016). Iron isotope fractionation during skarn-type alteration: implications for metal source in the Han-Xing iron skarn deposit. *Ore Geology Review*, *74*, 139–150.

Zhu, X. K., Li, Z. H., Tang, S. H., & Li, Y. H. (2008). Fe isotope characteristics of early Precambrian pyrite deposits and their geological significance: examples from Shandong and Hebei Provinces. *Acta Petrologica et Mineralogica*, *27*(5), 429–434 (In Chinese with English abstract).

Zhu, X. K., Guo, Y., Williams, R. J. P., O'nions, R. K., Matthews, A., Belshaw, N. S., & Salvato, B. (2002). Mass fractionation processes of transition metal isotopes. *Earth and Planetary Science Letters*, *200*(1–2), 47–62.

Zhu, X. K., O'Nions, R. K., Guo, Y., Belshaw, N. S., & Rickard, D. (2000). Determination of natural Cu-isotope variation by plasma-source mass spectrometry; implications for use as geochemical tracers. *Chemical Geology*, *163*(1–4), 139–149.

8

Exploring for Carbonate-Hosted Ore Deposits Using Carbon and Oxygen Isotopes

Shaun L. L. Barker[1,2,3] and Gregory M. Dipple[2]

ABSTRACT

Carbonate-hosted ore deposits often have very limited mineralogical and lithogeochemical alteration halos, as the highly reactive carbonate host rocks neutralize acidic hydrothermal fluids, limiting the ability of those fluids to cause hydrothermal alteration or transport pathfinder elements into the surrounding rocks. However, carbon and oxygen stable isotope ratios in the rocks surrounding carbonate-hosted ore deposits often record large alteration halos (on the order of hundreds of meters to kilometers) which can be used to identify and vector toward ore bodies. In this contribution we review the theory of carbon and oxygen isotope alteration during hydrothermal fluid flow, and present various case studies carried out over the last 50 years which demonstrate stable isotope alteration in carbonate-hosted ore deposits. In particular, it is clear that world-class ore deposits, such as the Mount Isa Cu ore bodies, and the Carlin Trend gold deposits, are surrounded by very large (3–8 kilometre) oxygen isotope alteration halos. We discuss advances in analytical technology that make stable isotope analysis a practical tool for mineral exploration, and highlight potential future advances. Finally, practical details of sampling, and factors to consider before carrying out a stable isotope study are outlined.

Plain Language Summary

Carbonate rocks such as limestones and dolostones often host large hydrothermal ore deposits, of gold, copper, lead, zinc, and other metals. By measuring the carbon and oxygen isotope composition of the carbonate rocks and veins, which surround ore deposits, valuable information, including potential tools to help discover deposits, can be obtained. This paper is recommended to any academic researcher or mineral explorer interested in carbonate-hosted mineral deposits.

[1] School of Science, University of Waikato, Hamilton, New Zealand

[2] Mineral Deposit Research Unit, University of British Columbia, Vancouver, BC, Canada

[3] Centre for Ore Deposit and Earth Sciences, University of Tasmania, Hobart, Tasmania

8.1. INTRODUCTION

A number of important types of mineral deposits are found in carbonate-rich host rocks. Examples include skarn and other carbonate-replacement deposits, Mississippi-Valley Type deposits, sedimentary copper, and Carlin-type gold deposits (see reviews and compilations of Megaw et al., 1988; Kesler et al., 1995; Hofstra & Cline, 2000; Cline et al., 2005; Sangster, 1997; Leach et al., 2005; Meinert et al., 2005). Carbonate-hosted ore deposits are found within sedimentary rocks (limestone, dolostone, sandstone, or shale), and often are associated with the intrusion of felsic dykes or stocks (Morris, 1986). Due to their highly reactive nature, carbonate-hosted mineral deposits tend to have relatively limited mineralogical alteration halos, wherein carbonate minerals effectively neutralize acid present in the hydrothermal fluid. Indeed, mineralogical alteration is often limited to the region proximal to the igneous body (where carbonate minerals have been replaced by calcsilicate skarn

Ore Deposits: Origin, Exploration, and Exploitation, Geophysical Monograph 242, First Edition.
Edited by Sophie Decrée and Laurence Robb.

minerals such as epidote, amphibole, garnet, pyroxene, and iron-oxide minerals), or in areas where ore is present due to intense fluid-rock interaction and precipitation of jasperoid has occurred (e.g., Carlin-type gold deposits). Megaw et al. (1988), in their paper on manto and chimney deposits, note that "razor-sharp contacts between mineralization and megascopically unaltered carbonate wall rocks are common in these deposits" highlighting the challenge of recognizing alteration around carbonate-hosted ore deposits (cf. Hitzman et al., 2002).

Due to the often limited presence of mineralogical and lithogeochemical alteration halos, carbonate-hosted ore deposits tend to be challenging mineral exploration targets. However, the carbon and (particularly) oxygen isotope composition of carbonate rocks can be used as an extremely robust indicator of whether a carbonate rock has been altered by a hydrothermal fluid. A early stable isotope study paper published during the 1950s documented carbon and oxygen isotope composition of carbonate rocks in the Leadville District, Colorado, and noted that variations in oxygen isotope composition are likely due to variations in temperature and fluid abundance (Engel et al., 1958). Oxygen isotope alteration in carbonate rocks has been detected at distances ranging from meters to kilometers from hydrothermal ore deposits (Frimmel, 1992), at greater distances than any mineralogical or lithogeochemical indicator of hydrothermal alteration (Kesler et al., 1995; Escalante, 2008; Barker et al., 2013). In addition, the carbonate mineral component of rocks is particularly easy to analyze for its isotope composition, as most carbonate minerals can easily be converted to CO_2 suitable for isotopic analysis by gas source isotope ratio mass spectrometry (IRMS) or infrared spectroscopy.

In the following sections, we review the theory for isotopic alteration of carbonate rocks during hydrothermal fluid flow; the variation in the carbon and oxygen isotope composition of rocks through geological time; case studies of carbon and oxygen isotopes from various carbonate-hosted ore deposits; and practical considerations of sampling, analysis, and the way to integrate stable isotope analysis as a standard tool in the exploration workflow for carbonate-hosted ore deposits. We note that this review is in no way intended to provide deposit descriptions or comprehensive reviews of carbonate deposit genesis (except where relevant). For readers interested in specifics about particular deposits, or stable isotope alteration patterns, we refer them to the cited papers that focus on individual deposits, and references contained within those papers.

8.2. INTRODUCTION TO CARBON AND OXYGEN ISOTOPES

There are two stable carbon isotopes, ^{12}C representing 98.9% of carbon, and ^{13}C, representing 1.1% of carbon. There are three stable oxygen isotopes: ^{16}O, representing 99.76% of oxygen; ^{17}O, representing 0.04% of oxygen; and ^{18}O, representing 0.2% oxygen, of which only ^{16}O and ^{18}O are measured routinely. Isotopic abundances are reported as ratios relative to the isotopic ratios within a defined standard. For carbon, the standard is the Vienna PeeDee Belemnite (VPDB) where: $\delta^{13}C_{VPDB} = ((^{13}C/^{12}C_{sample})/(^{13}C/^{12}C_{VPDB})-1)*1000$, and values are reported as per mil, ‰, or parts per thousand.

For oxygen isotope ratios the standards are either VPDB or Vienna Standard Mean Ocean Water (VSMOW). $\delta^{18}O_{VSMOW} = ((^{18}O/^{16}O_{sample})/(^{18}O/^{16}O_{VSMOW})-1)*1000$, and values are reported as per mil, ‰, or parts per thousand.

Typically, in economic geology, the VSMOW scale is used for reporting oxygen isotope data. In this article all oxygen isotope data are presented relative to the VSMOW standard. The VPDB and VSMOW scales can be related to one another by $\delta^{18}O_{VSMOW} = \delta^{18}O_{VPDB}*1.03091 + 30.91‰$ (Coplen, 1988).

The isotopic composition of sedimentary carbonate rocks reflects several factors including the isotopic composition of the water (which will typically have been ocean water) from which the carbonate mineral grains (either chemical or biological) making up the rock precipitated, the temperature of the seawater at which carbonate precipitation occurred, and any subsequent changes to the isotopic composition of the rock, which might involve diagenetic, metamorphic, or hydrothermal processes. When considering the hydrothermal isotopic alteration of carbonate rocks, it is critical to understand the underlying variation that occurs in the isotopic composition of sedimentary carbonate rocks through geological time, which reflects changes in the isotopic composition of the ocean over time, along with variations in seawater temperature (these variations are reflected in the data shown in Fig. 8.1). Veizer et al. (1999) and Shields and Veizer (2002) published compilations of stable carbon and oxygen isotope data from ~3800 Ma to the present day (Fig. 8.1). There are significantly more data available from ~500 Ma onward, and the carbonate isotopic compositions are better constrained for this time period. These compilations provide an initial isotopic "background" for carbonate rocks of a particular age, against which values from a particular area can be compared. Many studies of isotopic alteration of carbonate rocks that host ore deposits have found that rocks most distal to mineralization have carbon and oxygen isotope compositions that are consistent with the global background values for that time period (e.g., Kesler et al., 1995; Stenger et al., 1998; Vazquez et al., 1998; Vaughan, 2013; Barker et al., 2013). Here we use the term *background* to mean the oxygen and carbon isotope composition of rocks that have been unaffected by hydrothermal activity, and thus represent the background isotope value to compare against hydrothermally altered rock values. Thus, the review papers of Veizer (1999) and Shields and Veizer (2002) are an excellent

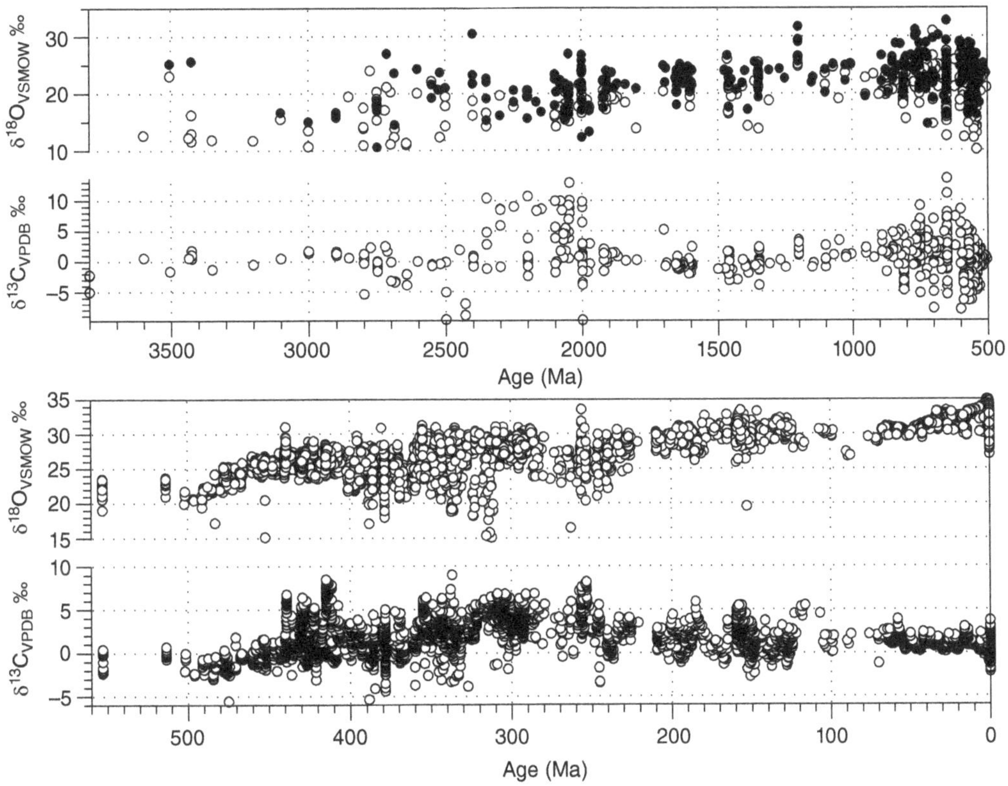

Figure 8.1 Age versus $\delta^{13}C$ and $\delta^{18}O$ for (top) the Precambrian (data from Shields and Veizer., 2002) and (bottom) the Phanerozoic (data from Veizer et al., 1999). Note that for the Precambrian data, the mean isotopic compositions are shown, with the open circles for $\delta^{18}O$ from calcite and the filled circles $\delta^{18}O$ from dolomite. All Phanerozoic data from calcite only.

starting point to establish the likely background isotope composition of the carbonate host rocks, which host a mineral deposit of interest. Local background should always try to be established wherever practicable to include regional variations due to diagenesis and metamorphism, via sampling the freshest rocks distal to metamorphism, igneous intrusions, or visible hydrothermal alteration (e.g., veins). Should isotopic values be measured that depart significantly from global isotope compositions for that time period in what are considered to be unaltered rocks, then more effort may be required to determine what is representative of "local" background, and what represents an anomaly. Alternatively, a significant excursion of isotopic data away from expected values could reflect poor control on the age of the host rocks, which is more common in Precambrian carbonate rocks. Indeed, stable isotopes can be used as a tool for regional stratigraphic correlation in Precambrian rocks (Kah et al., 1999; Shields & Veizer, 2002).

It is important to note that when we discuss here the isotopic composition of carbonate rocks, we are discussing only the isotopic composition of the carbonate minerals contained within that rock. The analytical results described here are all based on the analysis of carbon dioxide, which is produced by the reaction of high concentration (~85–103%) phosphoric acid with the carbonate minerals. This reaction is not capable of liberating oxygen or carbon from other mineral phases.

8.3. THEORY OF ISOTOPIC EXCHANGE AND ISOTOPIC ALTERATION DURING FORMATION OF CARBONATE-HOSTED ORE DEPOSITS

8.3.1. Temperature Controls on Fractionation

Carbon dioxide in a mineral has a different isotopic composition from the CO_2 in the fluid from which the mineral precipitated. This is because the heavier and lighter isotopes are differently proportioned between the gas, liquid, and solid phases in which CO_2 occurs. This differential proportionation, isotopic fractionation, is due to slight differences in the bond strength of isotopes of different mass, with heavier masses forming stronger bonds and typically having slower reaction rates. The difference in isotope distribution between different phases, is termed *isotopic fractionation*, or the *fractionation factor*. Isotopic fractionation is the partitioning of different isotope ratios between various components present within a chemical system (Hoefs, 2013). Changes in temperature will change the fractionation factor between water and mineral, and thus the isotopic composition of the mineral in equilibrium with a fluid of fixed composition. Fractionation factors are largest at low temperatures. However, at higher temperatures (~400 °C and above) the change in fractionation factor with increasing temperature becomes negligible (O'Neil et al., 1969). The significant variation in oxygen isotope ratios in altered carbonate-bearing rocks, which occur around many carbonate-hosted hydrothermal deposits (often exceeding 10 to 15‰), exceeds any likely variation induced by variation in changes in temperature alone (Cathles, 1993, 1997). For example, cooling a water-only hydrothermal fluid from 300 °C to 200 °C would change the measured oxygen isotope value in calcite precipitated from the same hydrothermal fluid by ~ + 4‰ (O'Neil et al., 1969).

8.3.2. Changes in Carbonate Mineral Isotopic Composition Due to Exchange with Other Rock Components

The isotopic composition of the carbonate fraction of a rock sample can change during exchange with the other components in the rock (e.g., O in silicates, C in organics) and during mass transfer between the rock sample and its surroundings. Even in closed systems and fixed whole rock isotopic values, the isotopic composition of the constituent minerals will shift in response to the changing mineral-mineral fractionation factors during temperature change. For example, in Fig. 8.2, the isotopic values of quartz and calcite shift to values closer together and nearer to the bulk rock composition at high temperatures, reflecting the lower values of mineral-mineral fractionation at elevated temperatures. Shifts in oxygen isotopic composition due to exchange between rock components are generally small (at most a few per mil) while carbon isotopic composition can be shifted more significantly during metamorphic devolatilization reactions (Baumgartner & Valley, 2001). Oxidation/reduction reactions with organic carbon can also cause more significant variations in carbon isotope composition. Isotopic exchange and mass transfer between the rock and its surroundings can occur by diffusion and advection, but the small length-scales of diffusion under crustal conditions and timescales generally limit alteration halos to meters or less (e.g., Bowman et al., 1994). Thus, large shifts in isotopic ratios over hundreds of meters or more can be reliably attributed to isotopic exchange during fluid infiltration except in extreme circumstances.

A final consideration is the mechanism by which isotopes are exchanged between the solid and fluid phases. Surface reaction mechanisms such as dissolution and reprecipitation are several orders of magnitude faster

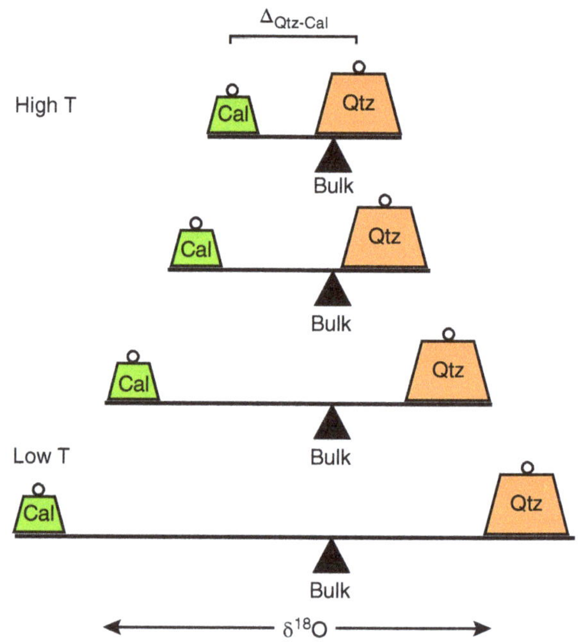

Figure 8.2 Conceptual representation of the variation in mineral isotopic composition due to temperature- dependent mineral-mineral fractionation in a closed system.

than solid-state diffusion at low to moderate temperatures (Cole & Ohmoto, 1986). A difference in isotopic composition of minerals and fluids alone does not provide a sufficient energetic drive for mineral reactions to proceed at a significant rate. Therefore, the kinetics of fluid-rock isotope exchange at these temperatures are controlled by mineral alteration and dissolution-reprecipitation reactions, and isotopic alteration in carbonate minerals will reflect carbonate mineral reactivity and solubility in the presence of hydrothermal fluids (Northrop & Clayton, 1966; Sheppard & Schwarcz, 1970; Lassey & Blattner, 1988; Chacko et al., 1991; Bowman et al., 1994; Pokrovsky et al., 2009).

8.3.3. Other Causes of Isotopic Alteration in Carbonate Rocks

In the absence of fluid infiltration, there are several alternative mechanisms that can drive significant isotopic alteration in rocks via various isotopic exchange processes. One example is metamorphic processes, during which carbon and oxygen can be lost from a package of rocks via the release of volatile species (e.g., H_2O, CO_2). During prograde metamorphism, partitioning of ^{18}O and ^{13}C during devolatilization of H_2O and CO_2 will alter bulk rock isotopic composition. As O and C are lost from the rock, equilibrium partitioning of ^{18}O and ^{13}C between H_2O-CO_2 fluids and most crustal minerals drives rock

$\delta^{18}O$ and $\delta^{13}C$ to lower values. The amount of depletion depends on how much of the elemental reservoir is volatilized and on the mechanism of fluid release, and is reviewed by Baumgartner and Valley (2001). In the most extreme cases, complete dehydration and decarbonation of marly sediments releases at most 40% of the O and as much as 100% of the C originally present. Thus the isotopic shifts resulting from devolatilzition are much larger for C than for O. The resulting patterns of strong ^{13}C depletions and modest ^{18}O depletions (up to 5 per mil) are distinct from the alteration patterns resulting from hydrothermal fluid flow (Baumgartner & Valley, 2001).

Redox processes can also drive strong ^{13}C alteration, but will also generally preserve the ^{18}O composition of rocks. Biological fractionation imparts a distinctive strongly negative $\delta^{13}C$ value to organic carbon. Exchange of carbon between the organic and inorganic pools can thus drive $\delta^{13}C$ depletions of 10s per mil. Within the fluid phase, the rates of isotopic equilibration between CH_4 and CO_2 at temperatures of $<300\,°C$ exceed 1,000,000 years (Ohmoto & Goldhaber, 1997), meaning that for the majority of intrusion-related hydrothermal systems isotopic equilibrium is likely never reached between CH_4 and CO_2 species (Cathles et al., 1997). In lower-temperature hydrothermal systems, isotopic equilibrium is likely never reached between CH_4 and CO_2 due to the extremely slow kinetics of carbon isotope exchange between these species at low temperatures. This means that in order to incorporate carbon from organic carbon species (such as CH_4) into carbonate minerals, oxidation of organic species is required. Thus, significantly depleted $\delta^{13}C$ values in low-temperature hydrothermal systems are most likely indicative of oxidation of organic carbon during hydrothermal fluid flow.

8.3.4. Reactive Transport Theory Applied to Isotope Halos Around Ore Deposits

The conventional framework for interpreting the alteration patterns resulting from fluid flow and reaction is reactive transport theory. Excellent reviews on its application to stable isotope alteration are presented elsewhere (e.g., Lassey & Blattner, 1988; Frimmel, 1992; Bowman et al., 1994; Baumgartner & Valley, 2001). Here we focus on a few highlights relevant to interpreting large-scale alteration halos in mineralizing hydrothermal systems. Reactive transport theory is predicated on mass balance considerations. As reactive fluid passes through rock, reaction progress is recorded by changes in both fluid and rock composition. Mass conservation dictates that isotopic alteration experienced by the rock is matched by compensating isotopic shifts in the fluid. In most hydrothermal systems, the fluid that produced alteration is rarely preserved, and reaction processes are gleaned from the alteration pattern left in the rock.

Directional fluid percolation imposes a pattern on isotopic alteration (Korzhinsky, 1959). As fluid infiltrates rock with which it is not in isotopic equilibrium, stable isotope fronts, akin to chromatographic fronts, will propagate in the direction of fluid flow (Figs. 8.3, 8.4, 8.5, and 8.6). The stable isotope ratios (e.g., $^{18}O/^{16}O$) change abruptly at the isotopic alteration front. In the "upstream zone," rock isotope ratios are altered toward equilibrium with the infiltrating fluid. In the "downstream zone," rock isotope compositions are unchanged and fluid isotope ratios approach equilibrium with the rock due to fluid-rock exchange in the reaction zone. Although the volume of infiltrated fluid remains constant along the flow path, the system shifts from a fluid-dominated region near the inlet where the composition of both fluid and rock reflect the composition of the incoming fluid (fluid dominated), to a rock-dominated region downstream in which both rock and fluid composition reflect that of the unaltered rock. The shift reflects the progressive exchange of the fluid along the flow path until the reactive capacity of the fluid has been exceeded and it approaches a value in isotopic equilibrium with the unaltered host rock. The shape and position of the isotope alteration front depends on the amount of fluid flow, the mechanism of isotope transport, the isotope exchange kinetics, and the initial composition of the fluid and the rock (Blattner, 1985; Lassey & Blattner, 1988; Frimmel, 1992; Bowman et al., 1994; Gerdes et al., 1995).

The distance traveled by a reaction front is proportional to the amount of fluid flow (as time-integrated fluid flux) and the partitioning of the tracer species (such as the $^{18}O/^{16}O$ isotope ratio) between rock and fluid (Bickle & McKenzie, 1987, Figs. 8.3, 8.4, 8.5, and 8.6). Chemical species that are abundant in fluid and rare in rock (such as hydrogen) will cause more significant alteration of a volume of rock during a given volume of hydrothermal fluid flow. This corresponds to a larger altered rock volume and farther distance traveled by the reaction front. In comparison, chemical species that are in low abundance in fluid compared with rock (such as Zr) can only impose very minor amounts of reaction on the rock, and the corresponding reaction front will be located closer to the fluid source or inlet. A single fluid-flow event therefore produces multiple reaction fronts for the various species in the fluid (Fig. 8.4). Each front travels a different distance from the fluid source, depending on its relative abundance in the fluid compared with the rock. Alteration reactions of species that are abundant in the fluid travel farther than those of less soluble species. For example, in Fig. 8.5b, it is shown that a fluid with a higher CO_2 content will have a ^{13}C reaction front that extends considerably farther than the ^{13}C reaction front for a fluid with a lower abundance of CO_2.

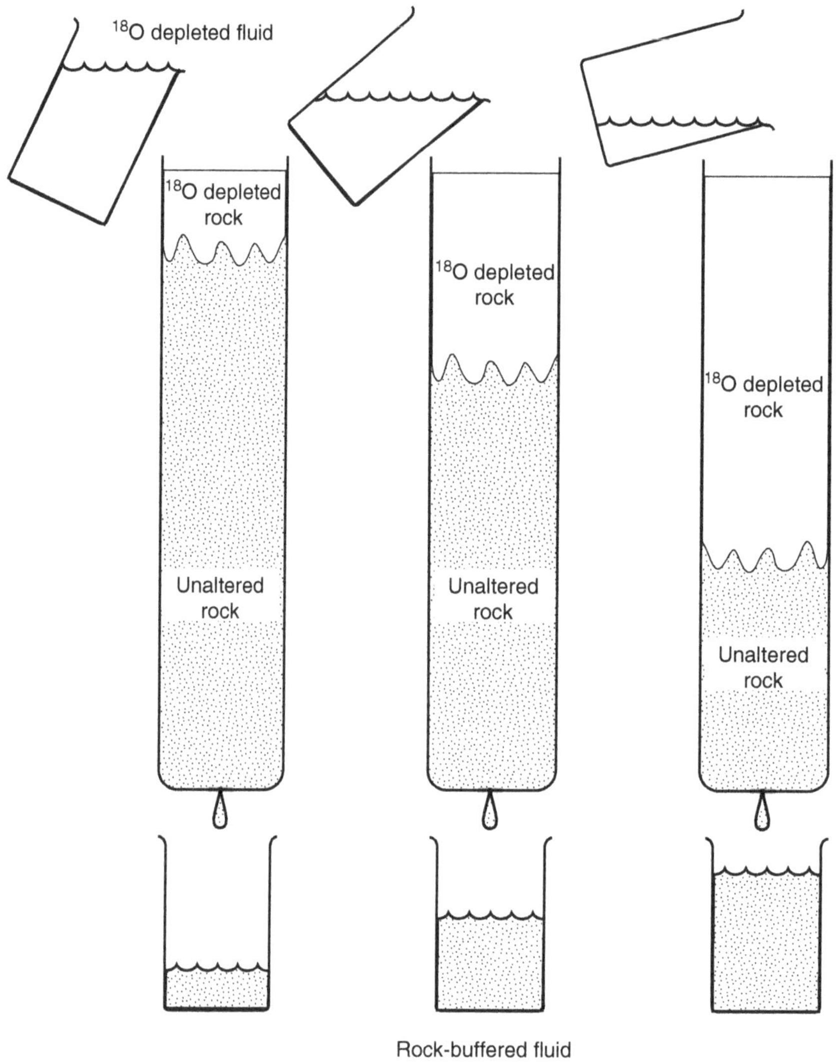

Figure 8.3 Conceptual diagram showing the progressive propagation of an oxygen isotope alteration front with increasing amounts of fluid infiltration.

As reaction fronts migrate away from the fluid source, the slower traveling alteration reactions overprint the fastest traveling reactions in a concentric alteration pattern centered on the fluid source. This has direct import to the zonation in alteration observed in hydrothermal mineral deposits, and can be useful in vectoring to the core of mineralized systems (Kelley et al., 2006). For example, relatively soluble pathfinder elements such as arsenic and thallium are present in elevated abundance on the periphery of hydrothermal systems, while visible mineral alteration and concentrations of ore are concentrated closer to the cores, nearer to the fluid inlets (e.g. Chaffee, 1976; Ford, 1978; Halley et al., 2015). This pattern can be interpreted to reflect the relative mobility of alteration fronts from fluid-partitioned (e.g., pathfinder elements) and rock-partitioned (e.g., metals) tracers. Isotopic alteration in carbonate-hosted systems often extends to the limits of pathfinder elemental halos and beyond. The relative velocity of ^{18}O and ^{13}C fronts depends on the CO_2 concentration in the fluid phase as both O and C are relatively abundant in carbonate rock. In CO_2-rich fluids, the velocity of the ^{13}C alteration front exceeds that of the ^{18}O alteration front, but at more modest CO_2 concentrations, typical of crustal aqueous fluids, the O front travels farther (Fig. 8.5b and 8.7).

Vigorous hydrothermal systems will generate larger isotopic alteration footprints because the distance traveled by a reaction front is proportional to the amount of fluid flow (Bickle & McKenzie, 1987). The intensity of hydrothermal fluid flow is tracked using the time-integrated fluid flux (e.g., Baumgartner & Ferry, 1991). If isotopic equilibrium between rock and fluid is maintained, the ^{18}O front propagation distance is 12 cm per mole H_2O/cm^2 rock in silicate and carbonate rocks

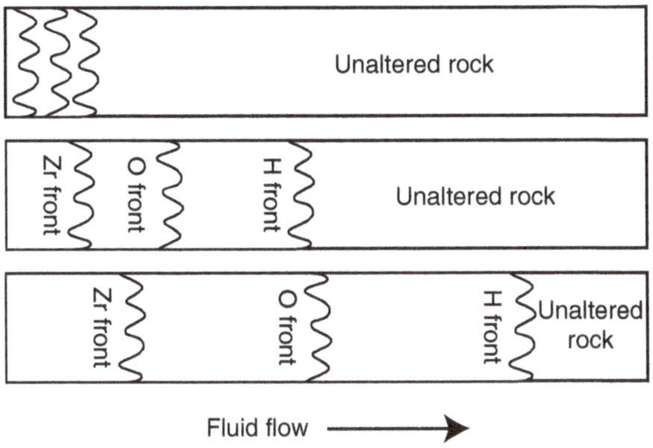

Figure 8.4 Conceptual diagram showing the progressive propagation of multiple alteration fronts (represented by different chemical tracers) with increasing amounts of fluid infiltration. Differential rock-fluid partitioning for different chemical tracers results in different propagation velocities for each reaction front.

(Dipple & Ferry, 1992). On this basis, time-integrated fluid fluxes of 1000 moles H_2O/cm^2 will produce isotopic alteration over a distance of 120 meters, and 10^4 moles H_2O/cm^2 will produce a km-scale ^{18}O halo (Fig. 8.5c).

8.3.5. Controls on the Development of Reaction Fronts

A number of chemical and physical processes operate to broaden the sharp reaction fronts depicted in Fig. 8.5. Chemical diffusion within pore fluid contributes to broadening reaction fronts, but for large-scale fluid infiltration and isotopic alteration, the predominant processes are hydrodynamic dispersion and kinetic dispersion (Cathles, 1993; Bowman et al., 1994; Cathles, 1997). Hydrodynamic dispersion reflects the inherent heterogeneity of fluid flow through porous media (Bear, 1972; Zheng, 1999). Dissimilar path lengths of fluid flow

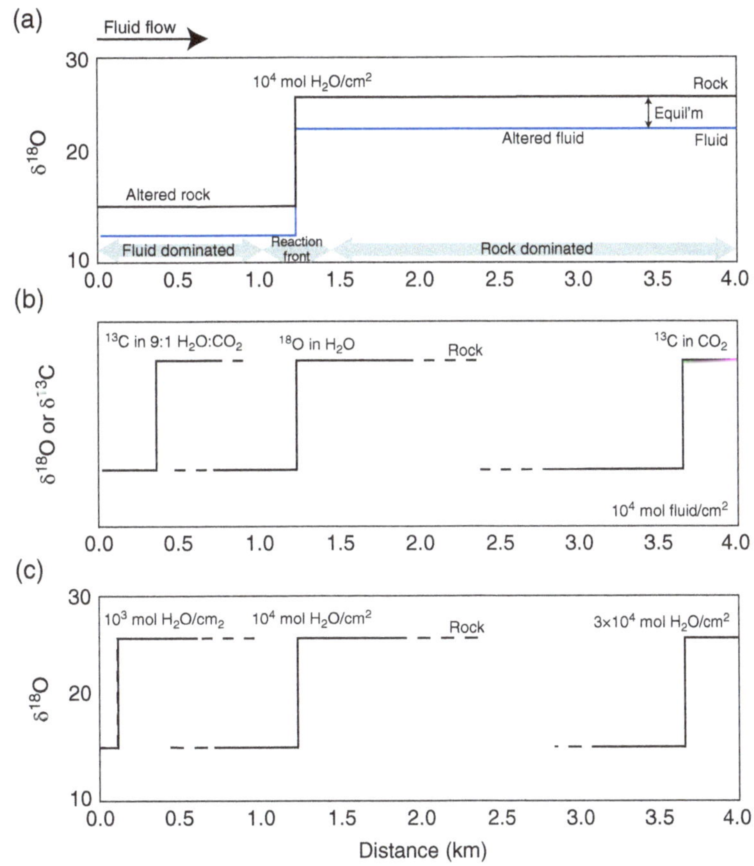

Figure 8.5 Stable isotope alteration fronts illustrated for an idealized one-dimensional system. Isotopic composition of rock and fluid, in which local isotopic equilibrium is maintained, are plotted versus distance along the path of fluid flow. The isotopic composition of coexisting rock and fluid differ by an amount equal to the equilibrium fractionation factor, which is temperature dependent. (a) ^{18}O composition of rock and fluid with time integrated fluid flux of 10^4 moles/cm^2. (b) Rock isotopic composition illustrating reaction fronts at fixed fluid flux for ^{13}C in pure CO_2 fluid, ^{13}C in H^2O-rich, CO_2 bearing fluid, and ^{18}O in pure H^2O fluid. (c) Relative positions of ^{18}O alteration fronts in rock for different time-integrated fluid fluxes.

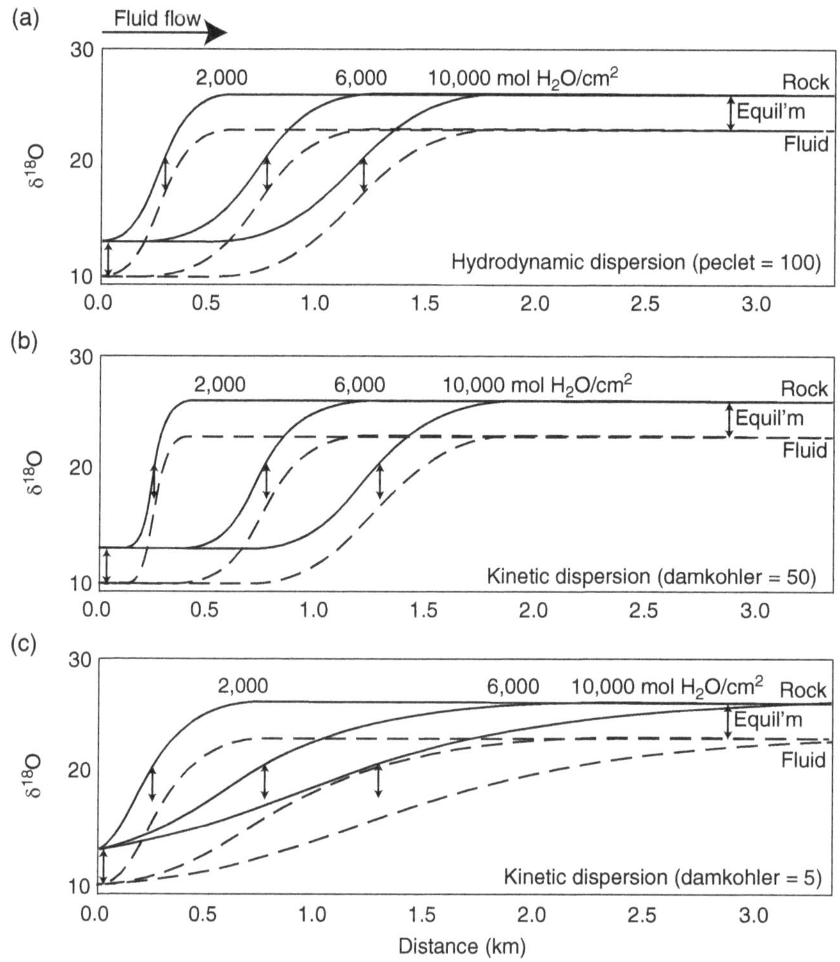

Figure 8.6 Stable isotope alteration fronts for one dimensional systems with hydrodynamic (a) and kinetic (b, c) dispersion after Bowman et al. (1994). Isotopic composition of rock and fluid is plotted versus distance along the flow path. Dispersion broadens the idealized sharp reaction fronts in Fig. 8.2. Hydrodynamic dispersion is tracked with the dimensionless Peclet number (see Bowman et al.., 1994 for details), and reflects mixing of fluid particles due to tortuous fluid flow paths. Fluid and rock maintain local isotopic equilibrium in Fig. 8.3a. Isotopic disequilibrium between rock and fluid result from sluggish isotope exchange reactions and are tracked using the dimensionless Damkohler number (Bowman et al., 1994). Lower Damkohler numbers correspond to slower reaction rates relative to the rate of fluid flow. In Fig. 8.6b,c, there is no hydrodynamic dispersion (Peclet = infinity).

through tortuous or irregularly fractured porous media result in fluid mixing. The result is a broadening of the reaction front even as fluid-rock isotopic equilibrium is maintained (represented by the constant difference between the fluid and rock isotope compositions shown in Fig. 8.6a). As noted above, isotopic exchange between minerals and fluid does not have a strong thermodynamic driving force, and, as a result, the attainment of isotopic equilibrium usually requires the intervention of mineral alteration or dissolution-reprecipitation processes. Sluggish exchange of stable isotopes between fluid and rock may fail to keep pace with the transport of stable isotopes by fluid flow, resulting in significant disequilibrium between the isotopic composition of the fluid and the rock. Disequilibrium likewise imparts dispersion to reaction fronts, but these systems are typified by differences in the isotopic composition of rock and fluid that depart from equilibrium fractionation (Fig. 8.6b, c). In exhumed hydrothermal systems, this may be reflected in vein isotopic compositions out of equilibrium with altered wall rocks. Sluggish isotope exchange results in incomplete equilibration of rock and fluid because isotopic reactions in the wallrock may proceed at a different rate to those involving precipitation of minerals directly into a fluid-filled crack, leaving the fluid more reactive as it continues along the flow path. Thus, in systems where local isotopic equilibrium between rock and fluid are not maintained, including most hydrothermal mineralization systems, the shifts in isotopic ratios within rock are less than predicted by equilibrium models but their lateral extent is greater, thus producing larger isotopic footprints (e.g., Fig. 8.6c).

The idealized one-dimensional profiles in Figs. 8.5 and 8.6 can be reconciled with patterns of isotopic alteration in hydrothermal systems in which samples (of whole carbonate rock) are collected along the primary fluid flow path such as faults and bedding controlled hydrothermal "aquifers" (e.g., Bowman et al., 1994). More commonly, however, the controls on permeability are less obvious, and heterogeneous and tortuous fluid flow produces isotopic alteration patterns that are more difficult to interpret. Irregular isotopic alteration patterns in heterogeneous two-dimensional fluid flow are illustrated following Gerdes et al. (1995) to extend the concepts of dispersion and disequilibrium to alteration patterns to multiple spatial dimensions. Figure 8.7a, b illustrates isotopic alteration in such a system in which fluid-mineral isotopic exchange is close to equilibrium. More permeable zones experience more fluid flow, and more widespread isotopic alteration (Fig. 8.7a). As in the one-dimensional systems, fluid-partitioned chemical tracers have more extensive alteration footprints (Fig. 8.7a) than rock-partitioned tracers for the same amount of fluid flow (Fig. 8.7b). In exhumed systems, alteration patterns in high-permeability zones can faithfully image the fluid flow pathways, and concentric zoning of tracers within them can permit vectoring toward the core of mineralized hydrothermal systems.

Heterogeneous fluid flow results in mixing of fluid particles with different rock-exchange histories, thus producing hydrodynamic dispersion. Sharp isotopic alteration fronts are lacking, especially in the zones of highest permeability (Fig. 8.7a). Sluggish isotope exchange and fluid-rock isotope disequilibrium compounds the dispersion processes, resulting in even more widespread alteration and broader isotope alteration fronts. The sharpest gradients in isotopic composition develop across strong permeability contrasts and are parallel to the dominant fluid flow direction (Fig. 8.7d). Thus these "fronts" are effectively stationary during fluid flow and form at the margins of fluid flow paths. Yardley and Lloyd (1995) referred to these features as alteration "sides". In exhumed systems, sides can be distinguished from fronts because strong gradients in fluid-partitioned

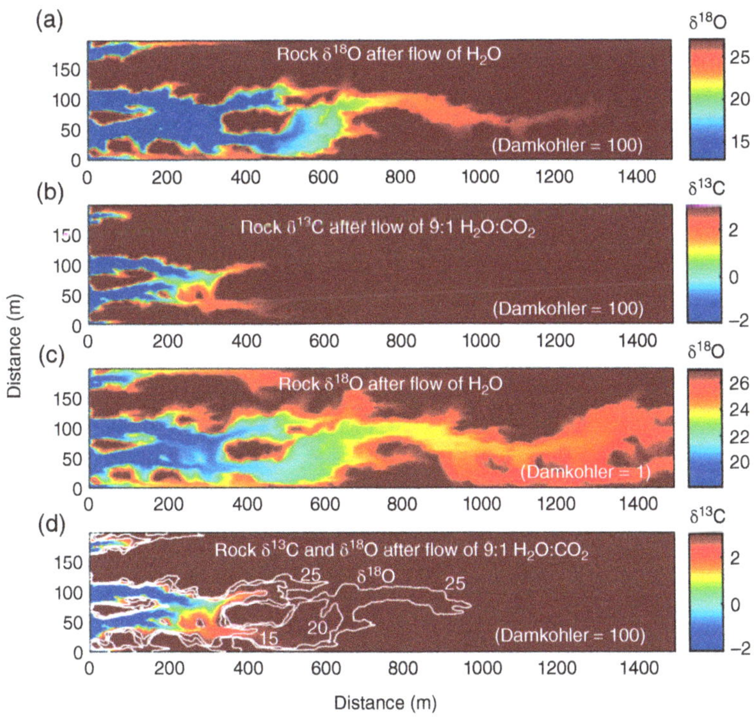

Figure 8.7 Two dimensional plots (200 m by 1500 m) of rock stable isotopic composition resulting from fluid percolation through a heterogeneous permeability field after Gerdes et al. (1995). Initial rock composition is $\delta^{18}O=27$ per mil and $\delta^{13}C = 3$ per mil, and rock equilibrated with infiltrated fluid has $\delta^{18}O=13$ per mil and $\delta^{13}C = -2$ per mil. Time-integrated fluid flux is the same for all panels. (a) $\delta^{18}O$ with Damkohler number = 100. (b) $\delta^{13}C$ for fluid with 10 mole percent CO_2 (cf., Fig 5b). Damkohler number = 100. (c) $\delta\ ^{18}O$ with Damkohler number = 1. (d) Overlay of $\delta\ ^{18}O$ (white contours) and $\delta\ ^{13}C$ (color map) from panels (a) and (b). Note that gradients in $\delta\ ^{13}C$ and $\delta\ ^{18}O$ coexist at high permeability contrasts, corresponding to alteration sides.

and rock-partitioned tracers will co-exist at sides while at fronts they will be separated by large distances. Coincidence of multiple alteration profiles has been documented in skarn systems. According to one-dimensional reactive transport theory, oxygen isotope and wollastonite skarn reaction fronts should be widely separated (Dipple & Gerdes, 1998) whereas at Elkhorn, Montana (Bowman et al., 1985), and Valentine, New York (Gerdes & Valley, 1994), they coincide within centimeters. In this interpretation, fluid flow at these localities was dominantly parallel to the alteration gradients (e.g., Fig. 8.7d). Mapping of multiple alteration fronts in hydrothermal systems can therefore serve to identify sharp paleo-permeability contrasts and infer patterns of hydrothermal fluid flow.

8.4. CASE STUDIES OF CARBON AND OXYGEN ISOTOPES AROUND CARBONATE-HOSTED ORE DEPOSITS

8.4.1. Skarn and Manto Deposits

The stable carbon and oxygen isotope composition of carbonate rocks surrounding skarn deposits (and closely related manto deposits) has been documented in numerous studies (Shimazaki et al., 1986; Shimazaki & Kusakabe, 1990; Naito et al., 1995; Vazquez et al., 1998; Escalante, 2008) and was comprehensively reviewed in the paper by Bowman (1998).

In general, skarn deposits do not show smooth, systematic variations in carbon and oxygen isotope ratios away from ore bodies, likely due to the tortuous pathways, which hydrothermal fluids will take due to heterogeneous permeability and hydrodynamic dispersion (Fig. 8.7). In general, samples more proximal to ore bodies have more depleted ^{18}O values. As an example, Vazquez et al. (1998) collected samples around the El Mochito skarn sulfide-oxide limestone deposit in Honduras (Fig. 8.8). The deposit is hosted in the Atima Limestone, which consists of fossilferous to cherty and shaly, thin-bedded limestones, shaley limestone, and limey shales, as well as thinly bedded light grey limestones. Mineralization is hosted both in chimneys (vertical structures) as well as mantos (horizontal structures). In detailed sampling traverses around ore bodies at the El Mochito deposit, Vazquez et al. (1998) demonstrated that in all cases oxygen isotope alteration halos extended for more than 30 to 40 meters around specific ore bodies, and at a property scale oxygen isotope alteration halos extended up to 4 kilometers from mineralization at the El Mochito deposit. In contrast, carbon isotope halos are smaller, up to ~30 meters around ore bodies, and often back to regional limestone (i.e., background) compositions within a few meters of skarn and ore (Vazquez et al., 1998).

At the Antamina Cu-Zn skarn deposit in the western Peruvian Andes, copper and zinc skarn ore bodies are developed around a series of Miocene quartz-monzonite porphyries and dykes (Escalante, 2008). The igneous

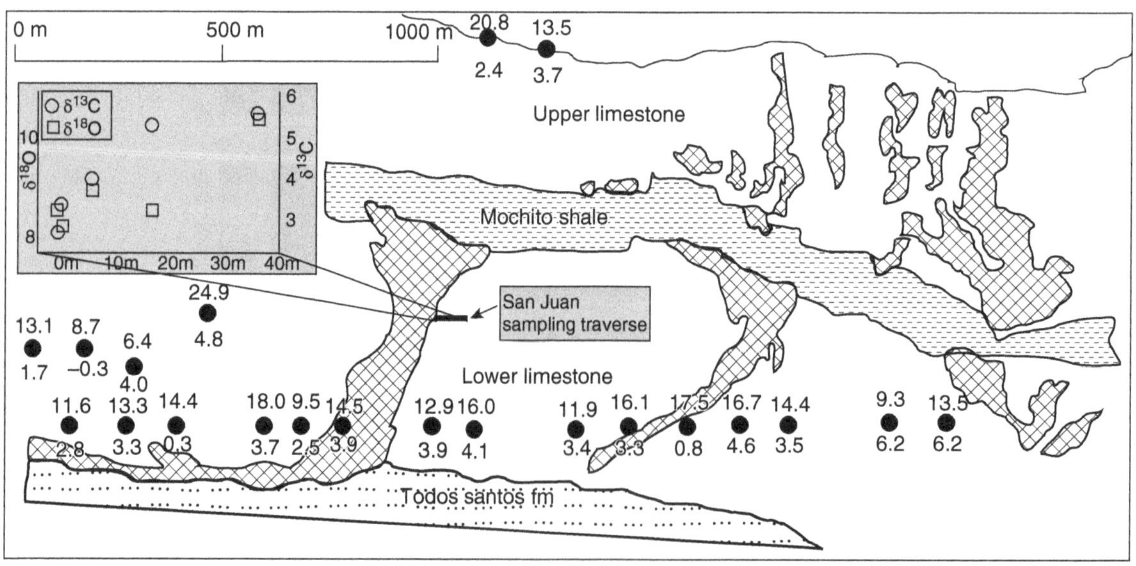

Figure 8.8 Longitudinal section from the El Mochito skarn deposit, Honduras (modified from Vazquez et al, 1998) showing the distribution of carbonate host rocks, and skarn-manto style mineralization (diagonal hatched symbol). Black dots mark the location of whole rock carbon and oxygen isotope samples with accompanying $\delta^{18}O$ (number above sampling dot) and $\delta^{13}C$ composition (number below dot) compositions. Inset is a more detailed sampling traverse around the San Juan ore body.

rocks intrude into Upper Cretaceous grey, thick bedded relatively pure limestones interbedded with thin layers of marly limestone and marls, which grade upward into thinly bedded marls, marly limestones, and shales. A carbon and oxygen isotope study by Escalante (2008) on carbonate rocks surrounding the Antamina ore body demonstrated that hydrothermal fluids exploited more permeable lithologies and faults, leading to isotopic alteration being measured along "fluid escape" structures surrounding the main skarn body (Escalante, 2008). At Antamina, Escalante (2008) identified that isotopic halos on the order of 20 to 150 meters surround inferred fluid escape structures over scales of hundreds of meters above the deposit (Fig. 8.9).

Manto (and closely related chimney) deposits are high-temperature, polymetallic deposits closely associated with skarns and related intrusive bodies, which form slightly more distally to the main intrusive body. Manto deposits are believed to form as high-temperature, magmatic hydrothermal fluids are channeled along higher permeability fluid-flow pathways such as preexisting faults and lithological layering (Morris, 1986). Manto deposits are likely part of a continuum of deposit types, which include skarn, manto, chimney, and Ag-Pb-Zn vein deposits. Manto deposits commonly have a very limited mineralogical alteration halo of recrystallized and bleached marble, and occasional silicification (Engel et al., 1958; Megaw et al., 1988; Megaw, 1990).

Megaw (1990, his Fig. 109) demonstrated that oxygen isotope halos on the orders of tens to hundreds of meters occurred around manto and chimney deposits in the Buena Tierra, Santa Ria, and San Antonio mines in the Santa Eulalia mining district, Northern Mexico. Megaw also noted that stable isotope alteration was influenced by fractures. Limestones collected several kilometers from mineralization yield oxygen isotope values of ~21–22‰ (Megaw, 1990), lower than would be expected for pristine limestones of lower Cretaceous age (Veizer et al., 1999), indicating that a subtle, larger scale isotopic alteration halo may be present. The potential presence of such large but relatively subtle halos highlights the importance or defining the "background" isotopic composition of equivalent host carbonates well away from any recognized hydrothermal alteration. It should be noted that even outside the immediate 30 m isotope alteration halo defined by both carbon and oxygen isotopes at El Mochito by Vazquez et al (1998), oxygen isotope ratios in limestones remain 7 to 15‰ lower than the regional "background" host rocks. Thus, oxygen isotopes around skarn ore bodies can be used both to define large-scale alteration halos and provide more direct vectors to ore bodies within a larger hydrothermal alteration cell.

In comparison with the examples described above, a study of oxygen and carbon isotope ratios from the Superior district, Arizona, reveal little or no hydrothermal alteration accompanying the formation of those carbonate-replacement deposits (Friehauf & Pareja, 1998). This was interpreted to be partly due to diagenetic dolomites, which had altered the primary composition of host limestones to lower values (effectively altering the "background" composition of host carbonate rocks). In addition, while the kinetics of oxygen isotope exchange between fluid and mineral for dolomite and calcite are similar, it appears likely that the key process causing isotopic alteration in many of these systems is not equilibrium exchange of isotopes, but dissolution-precipitation processes (see section 8.3.5). Dolomite is less susceptible to dissolution than calcite at similar pH conditions (Benezeth et al., 2013), and so is less likely to undergo dissolution-precipitation reactions with a hydrothermal fluid of moderate pH, and therefore does not record as significant a stable isotope exchange as calcite at equivalent conditions.

Kesler et al. (1995) presented stable isotope data from surface rocks above a series of chimney-manto deposits, including the El Mochito deposit (Honduras), Naica deposit (northern Mexico), and Tsumeb deposit (Tanzania). At El Mochito, only one sample (7 km from the known ore body), had an oxygen isotope ratio consistent with oxygen isotope ratios for unaltered limestones, indicating that a large isotopic alteration halo existed at Mochito (cf. Vazquez et al., 1998). For Naica, the oxygen isotope anomaly extended for over 3 km from the deposit, with only the closest sample to the deposit (a marble) showing any macroscopic evidence of alteration. All samples within 2 km of the Naica deposit had a $\delta^{18}O$ value of < 15‰, and thus were significantly isotopically altered below the background $\delta^{18}O$ of ~25‰ (Kesler et al., 1995; see summary of results in Fig. 8.10). In contrast, the Tsumeb deposit in Tanzania, predominantly hosted in dolostones, had very limited oxygen isotope alteration. The lack of an oxygen isotope alteration halo at Tsumeb was attributed to the dolostone host rocks (similar to the Superior district, outlined above), and potentially to the lower temperature nature of the Tsumeb deposit compared with the Naica and Mochito deposits (Kesler et al., 1995).

8.4.2. Sedimentary-Hosted Copper Deposits

Numerous stable isotope studies have been carried out on carbonate host rocks and hydrothermal carbonate minerals associated with sedimentary copper mineralization. The majority of those studies have been designed to determine hydrothermal processes that may have been controlling mineralization. Sediment-hosted copper deposits may include those formed via basin-scale fluid flow processes (such as the Kuperscheifer and Zambia;

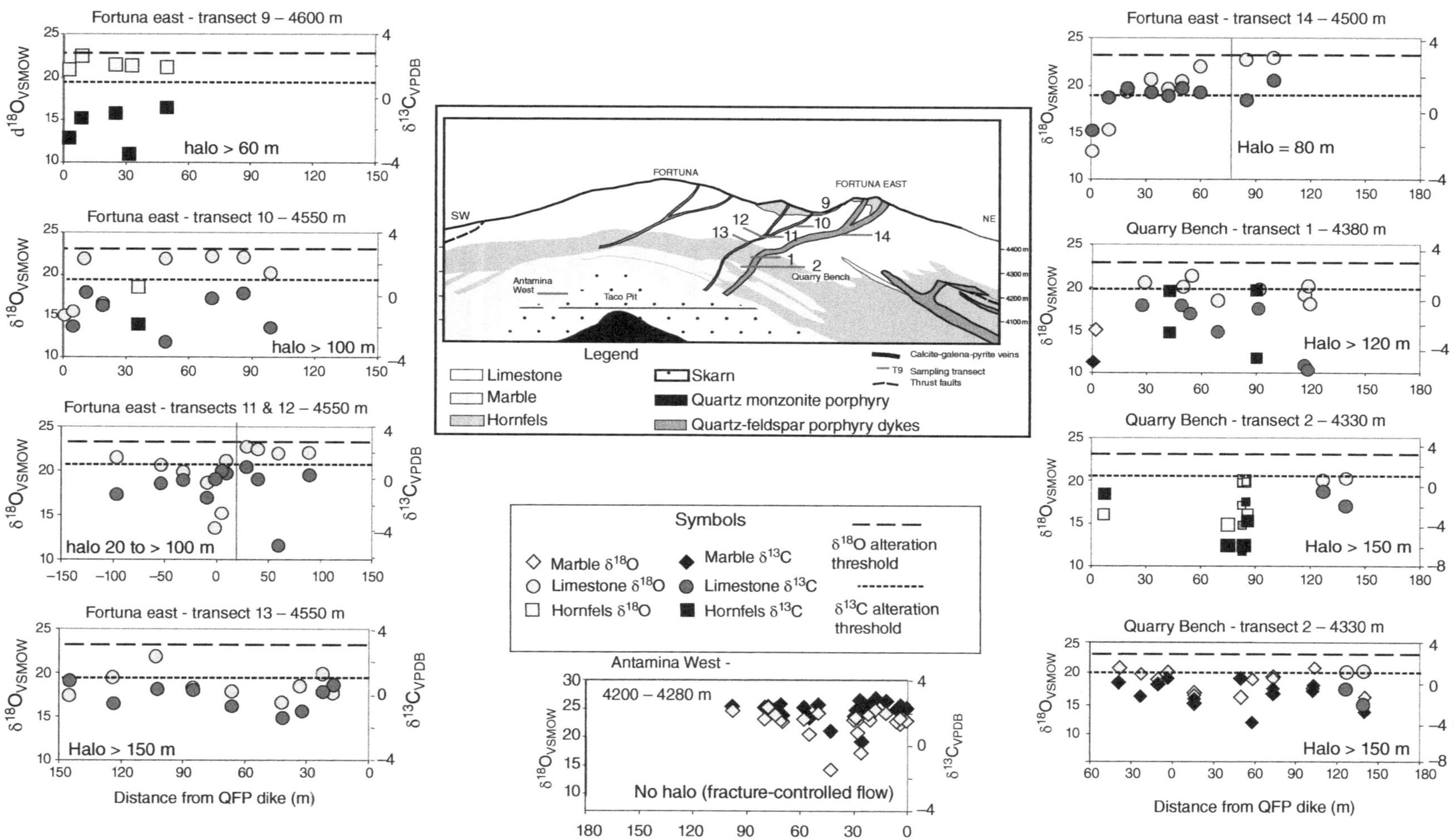

Figure 8.9 Schematic cross section from the Antamina-Fortuna area, showing the location of sampling traverses for stable isotope analyses. Numbers show the location of each stable isotope traverse, with distance "0" representing the locality of the quartz porphyry dyke from which each traverse began. Each traverse has the position of samples with corresponding $\delta^{13}C$ and $\delta^{18}O$ values separated by host rock type (marble, hornfels, or limestone). Plotted on each traverse are the background thresholds for both $\delta^{13}C$ and $\delta^{18}O$, representing values that would be considered equivalent to those founds in rocks unaffected by hydrothermal fluids. Note that isotope alteration halos are generally larger than the scale of the sampling traverses (greater than ~ 100–150 m in size), and appear to become narrower higher above the deposit (modified from Escalante, 2008).

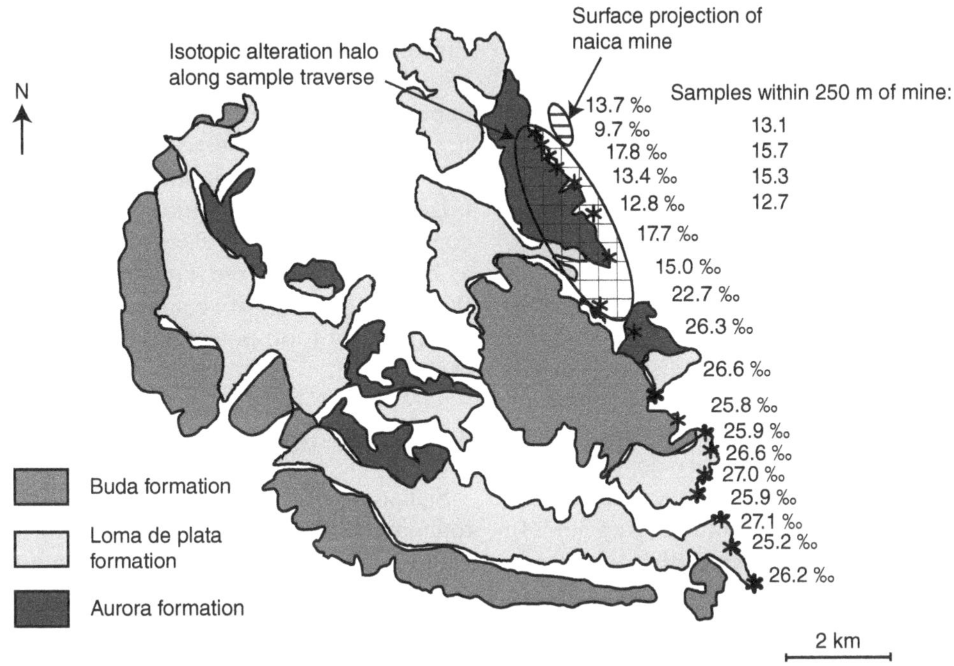

Figure 8.10 Sketch map of the limestones that host the Naito carbonate replacement deposit, northern Mexico. Shown is the location of the deposit (projected to surface), along with oxygen isotope values from limestones collected as the surface along a traverse leading southeast away from the deposit. Note that there is an oxygen isotope alteration halo of approximately 4 km length highlighted by the analyses, which increases in intensity toward the deposit (i.e., lowest oxygen isotopes values are adjacent to the deposit). Modified from Kesler et al., 1995.

Hitzman et al., 2005), as well as those formed during higher-temperature metamorphism (e.g., Mount Isa, Heinrich et al., 1989).

In the Zambian copper belt, carbon and oxygen isotopes reveal that unaltered sedimentary carbonate rocks have isotope values typical of Neoproterozoic marine carbonates ($\delta^{13}C$ ~5 to +10‰ and $\delta^{18}O$ +19 to +26‰). In contrast, altered, ore-bearing horizons have isotopically lighter compositions with $\delta^{13}C$ between −26 to −5‰ and $\delta^{18}O$ ~5 to 23‰ (Selley et al. 2005). The variations in carbon isotopes are thought to represent interaction with organic-rich fluids (Selley et al., 2005). The oxygen and carbon isotope changes measured between unaltered sedimentary carbonates and altered and mineralized carbonates suggest that a stable isotopic vector toward ore bodies may exist in the Zambian copper belt (and other) sedimentary hosted copper deposits. For example, it may be that changes in $^{13}C/^{12}C$ will reflect mixing zones bordering ore bodies. Case studies collecting systematic stable carbon and oxygen data across Zambian deposits need to be carried out to test for potential vectors to ore bodies.

In contrast with the sediment-hosted, lower temperature mineralization of the Zambian Copper Belt, Mount Isa-type copper deposits are believed to form during metamorphism, with examples including Mount Isa and the Nifty deposit (Waring et al., 1998; Hitzman et al., 2005). At Mount Isa, extensive stable carbon and oxygen isotope studies have been carried out over a period of more than 30 years, with over 4000 analyses carried out over various academic studies, and as part of internal Mount Isa Mine studies (Smith et al., 1978; Valenta, 1988, Heinrich et al., 1989; Waring, 1991; Waring et al., 1998; Painter, 2003; Chapman, 2004;). The bulk of the data was collected by the Mount Isa Mines mineral exploration group. The most extensive publicly available stable isotope study at Mount Isa is Waring et al. (1998), which documents the size of the stable oxygen isotope alteration halo produced during emplacement of the Mount Isa copper ore bodies. Dolomite within the silicate-dolomite alteration at Mount Isa has a narrow range of $\delta^{18}O$ values of 10–12‰, and $\delta^{13}C$ of −8 to −3‰. These values are significantly depleted in ^{18}O compared with texturally unaltered background dolostones, which have $\delta^{18}O$ of 20–22‰ and $\delta^{13}C$ of −2 to 0‰. Thus, there is a significant difference of ~8–12‰ in $\delta^{18}O$ between unaltered wall rock and ore zones, and a more subtle shift of ~3 to 6‰ in $\delta^{13}C$. The studies of Waring (1991) and Waring et al. (1998) demonstrated that dolomitic shales up to 1 km from ore, which show no evidence of proximity to Cu, were still depleted in ^{18}O. At distances 500 m to 1000 m from ore bodies, dolomitic veins are typically ~1.5‰

more depleted in ^{18}O than adjacent wall rock, giving an additional potential vector toward ore.

Further work at Mount Isa, based on a further ~800 stable isotope analyses, were sufficient to allow isotopic alteration contours to be generated on a series of cross sections and on a surface plan (Waring et al., 1998). This work demonstrated that (a) the oxygen isotope alteration halo at Mount Isa is enormous, around 9 km long by 2 km wide, and extends at least 1 km deep; (b) dolomite veins have isotopic compositions approximately equal to host rock proximal to ore, but become depleted in ^{18}O compared to wall rock by ~1 to 1.5‰ at distances of 500 to 1000 meters from ore likely due to the disequilibrium processes described above (section 8.3.5); and (c) wall rock isotopic compositions become very depleted adjacent to some fault zones, interpreted to be due to some faults acting as feeder structures for mineralizing fluid. In addition, Waring et al. (1998) suggested that the change in oxygen isotope composition with distance could be used to predict distance to mineralization. This is consistent with the isotopic front concept discussed above, and highlighted graphically in Figures 8.3, 8.4, 8.5, 8.6 and 8.7.

In addition to the potential to use oxygen isotopes as a vector toward ore in Mount Isa-type copper deposits, Waring et al. (1998) suggested that the overall size of an oxygen isotope alteration halo be used to predict the amount of contained copper. This relies on the assumption that copper mineralization is limited by the solubility of copper in the mineralizing fluid, and that the migration of more fluid will lead to a larger copper deposit (representing a larger amount of mass transfer). This being the case, then the size of the isotopic alteration halo will reflect the total volume of fluid transferred, and therefore the potential size of the contained ore body. The 10 km scale halo documented by Waring at Mount Isa implies a time-integrated fluid flux of at least 10^5 moles/cm^2 (Fig. 8.5). A smaller halo elsewhere would indicate less potential for a large ore body.

The Nifty copper deposit is a syndeformational., sediment-replacement Cu deposit hosted in subgreenschist-grade rocks of the Neoproterozoic Paterson orogeny in Western Australia, approximately 450 km southeast of Port Hedland (Anderson et al., 2001). The Nifty deposit is also hosted by dolomitic wall rocks, similar to Mount Isa. For the Nifty deposit, a relatively narrow range of stable isotope compositions was determined for dolomite in unaltered wall rocks ($\delta^{18}O$ ~19–23‰, $\delta^{13}C$ ~0 to 5‰) while dolomite within ore zones and altered wall rocks ranges from $\delta^{18}O$ ~14 to 20‰, and $\delta^{13}C$~−10 to 0‰ (Anderson et al., 2001). Stable isotope compositions in and surrounding ore zones are thus significantly depleted, which has been attributed to metamorphic decarbonation changing ^{13}C compositions, while ^{18}O alteration was attributed to fluid-rock reaction (Anderson et al., 2001).

The examples above highlight that there are clear stable carbon and oxygen isotope alteration signals associated with both low-temperature sedimentary-hosted copper deposits as well as higher temperature sedimentary-hosted copper deposits associated with metamorphism. The studies by Waring and coauthors document a very large stable isotope alteration halo around the Mount Isa copper deposits, which clearly offers significant potential as both a vector to ore, but also as a potential guide to the total volume of contained copper (and thus a prospectivity tool, potentially useable at early stages of exploration).

8.4.3. Carlin-Type Gold Deposits

Signatures of oxygen isotope alteration within and surrounding Carlin-type gold deposits are well established, with a variety of studies demonstrating that Carlin-type gold mineralization is almost ubiquitously associated with oxygen (and in some cases carbon) isotope alteration (Radtke et al., 1980; Stenger et al., 1998; Arehart & Donelick, 2006; Lepore, 2012; Barker et al., 2013; Vaughan, 2013). The study of Stenger et al. (1998) revealed that oxygen isotope values varied between ~24‰ (representing host carbonate rocks of unaltered isotopic composition) and ~10‰, while carbon isotope values ranged between ~0 (representing unaltered host rocks) to −10‰.'Stenger et al. (1998) demonstrated that ore zones had more depleted oxygen isotope compositions, which graded outward to less depleted oxygen isotope compositions in surrounding host rocks, which therefore defined an oxygen isotope halo several hundred meters wide around ore bodies. Arehart and Donelick (2006) published stable isotope results from the Pipeline Carlin-type gold deposit, and found evidence for an isotopic alteration halo on the order of several kilometers. Arehart and Donelick also make the important point that stable isotope data display significant variation over short distance scales, likely due to fracture-controlled permeability, and that therefore larger data sets (50 samples or more) will enable isotopic alteration halos around Carlin-type gold deposits to be recognized with a greater degree of confidence.

Lepore (2012) reported carbon and oxygen isotope data from approximately 2500 samples from limestones and dolostones, which host the Long Canyon Carlin-type gold deposit in northeastern Nevada. Lepore noted that dolostones had very little evidence of isotopic alteration, which was attributed to the stability of dolomite in ore-stage hydrothermal fluids. This was evidenced by breccias, where clasts of limestone were well rounded, yet coexisting clasts of dolostone retained angular shapes, which Lepore interpreted as evidence of dolomite not reacting with ore-stage fluids. Limestones above ore shoots controlled by faults between dolostone boudins at Long

Canyon displayed oxygen isotope alteration ~100 meters vertically above the 100 ppb gold anomaly defined from gold assay results. In addition, significant isotopic alteration could be detected in drill pulps, as well as in individual hand samples ~300 meters laterally from mineralization.

Barker et al. (2013) and Vaughan (2013) reported data from ~2000 stable carbon and oxygen isotope analyses from the northern Carlin Trend, collected from two cross sections lying to the west and north of the giant Betze-Post Carlin-type gold deposits (~50 MOz of contained gold). The samples analyzed were predominantly assay pulps collected on 1.5 m, 3 m, and 6 m intervals produced for gold assay during mineral exploration drilling. They identified an isotopic alteration halo at least 3 km wide to the west of the deposits. The halo was characterized by oxygen isotope alteration (very little or no carbon isotope alteration was distinguished), with the western most drill holes sampled displaying the least isotopic alteration, with isotopic alteration becoming progressively greater, and more variable with increasing proximity to gold mineralization. Isotopic anomalies were commonly centered on faults previously identified by geological reconstructions carried out by Barrick geologists, emphasizing the role of fault-controlled fluid flow within Carlin-type gold deposits. These studies emphasize that Carlin-type gold deposits are almost certain to have large stable oxygen (and potentially carbon) isotopic anomalies.

8.4.4. Assessing Regional Flow Systems Using Carbon and Oxygen Isotopes

Carbonate alteration and veining are a common feature of many ore-forming hydrothermal systems. Distinguishing the carbonate signal associated with regional scale hydrothermal systems (capable of forming hydrothermal ore deposits) versus short-distance carbonate redistribution from local carbonate-bearing rocks can be difficult to determine from field relationships or trace element studies. However, stable carbon and oxygen isotopes can be used to constrain fluid sources and recognize whether carbonate cements and alteration are locally sourced or part of a much larger hydrothermal alteration system. For example, Cartwright et al. (1992) used stable isotopes of carbonate veins and rocks to demonstrate local, restricted mass transfer near a major thrust fault. In contrast, Oliver et al. (1993) and Marshall et al. (2006) documented major fluxes of deep-seated fluids (and variations in carbon and oxygen isotope ratios) in veins and alteration associated with iron oxide-copper-gold (IOCG) mineralization in the eastern Mount Isa block, Queensland, Australia. In this case, despite a large reservoir of (metamorphosed) marine carbonates in the sequences containing the vein systems and ore deposits, veins and ores show isotope signals indicating large fluxes of ^{18}O- and ^{13}C-depleted fluids, requiring major input of CO_2 derived from magmas. Thus, stable isotope studies carried out at regional scale may be used to evaluate prospectivity of IOCG and other districts where the recognition of large-scale fluid volumes is desirable for identifying where maximum fluid transfer has occurred.

8.4.5. Conclusions from Case Studies

The case studies discussed above demonstrate that:

1. Changes in the stable oxygen isotope ratios of carbonate rocks, which host hydrothermal systems, almost ubiquitously accompany hydrothermal mineralization processes, and define alteration halos of varying geometry and size around mineral deposits.
2. Changes in stable carbon isotope ratios also occur, although not to the same spatial extent or change in ratios as for oxygen isotope ratios. Changes in carbon isotope ratios appear to be restricted more proximal to some ore deposit types (skarn and manto).
3. The size of oxygen isotope alteration halos likely reflects the total volume of fluid that has migrated through the carbonate rock, which may be a proxy for total metal endowment (assuming that mineralization is limited by the ability of the fluid to transport metal).
4. The geometry and pattern of oxygen isotope alteration will reflect the permeability structure of the host rocks, which will reflect both primary permeability (e.g., lithological variations), as well as secondary permeability induced by faulting and associated fracturing.

While isotopic alteration patterns can be used independently of geological information, they can be far more robustly interpreted when used in conjunction with geological, structural, and geochemical information. Stable carbon and oxygen isotopes are not a "silver bullet," but they can add significant and important information when exploring for carbonate-hosted ore deposits.

8.5. PRACTICAL CONSIDERATIONS AND FUTURE APPLICATIONS

8.5.1. Instrumentation

Despite the case studies (and others) outlined above, which clearly demonstrate that large stable isotope alteration halos surround many carbonate-hosted mineral deposits, stable isotope studies are not routinely utilized during mineral exploration for carbonate-hosted ore deposits. This likely reflects a variety of factors, including a lack of industry (and academic) knowledge about the potential utility and best practice of incorporating stable isotope analyses into mineral exploration, costs traditionally associated with collecting stable isotope data, but perhaps most important, the inaccessibility of stable isotope data on timescales relevant to mineral exploration.

Many academic and government labs around the world are capable of measuring the carbon and oxygen isotope composition of carbonate minerals. However, most of these labs do not have the capacity to deal with the hundreds to thousands of analyses that could be submitted simultaneously as part of a stable isotope survey for mineral exploration. The technology traditionally used for stable isotope analysis, gas source isotope ratio mass spectrometry (IRMS), is expensive, highly sensitive equipment, which needs high purity He and other gases, and has high maintenance demands requiring skilled technical staff. Most IRMS instruments can analyze ~40–50 carbonate samples per day.

Over the past 10 years or so, instrumentation based on cavity enhanced infrared absorption spectroscopy has become commercially available, which allows for stable isotope ratios of gases such as CH_4, CO_2 and H_2O to be measured cheaply and easily. These instruments typically can make stable isotope measurements more rapidly, and at far lower cost than IRMS instruments, with comparable precision and accuracy. The studies highlighted in Lepore (2012), Barker et al. (2013), and Vaughan (2013) utilized a modified Los Gatos Research CCIA analyzer (utilizing off-axis integrated cavity output spectroscopy), operating at a wavelength of ~1.6 μm to measure the carbon and oxygen isotope ratio of CO_2 liberated from carbonate minerals (Barker et al., 2011). The method developed by Barker et al. (2011), and further developed in a second generation analyzer by Beinlich et al. (2017), allows carbon and oxygen isotope analyses to be made on carbonate minerals at a fraction of the time and cost of traditional IRMS analysis. This is because the analyses are more rapid (~5 minutes per sample); there are very few consumables; and any person with basic laboratory and chemical handling skills can be taught to generate data with accuracy of ±0.5‰ in both $\delta^{13}C$ and $\delta^{18}O$ with approximately 2–3 hours of training. This paradigm shift in analytical technology enabled the large number of samples analyzed in those studies to be collected by undergraduate and graduate students over time frames of a few weeks. The value of having large numbers of stable isotope analyses to help identify the spatial scale and variations in stable isotope ratios is clearly demonstrated by the case studies discussed above.

Further development of infrared spectrometry instruments and developments in autosampler technology will allow stable carbon and oxygen isotope data to be collected more rapidly. Currently, there are several commercial infrared absorption instruments available capable of measuring both ^{13}C and ^{18}O isotopes in CO_2, including the Los Gatos Research carbon dioxide isotope analysers; Thermo Delta Ray™; and Aerodyne Carbon Dioxide Isotope Monitor.

8.5.2. Clumped Isotopes

Over the last 10 years, "clumped" stable isotope geochemistry has emerged as a new research field. Clumped isotope measurements focus on measuring the distribution of rare isotopes in natural materials. In this instance, it has been established that the distribution of $^{13}C - {}^{18}O$ bonds (the rare isotopes of carbon and oxygen) in carbonate minerals reflect the temperature at which the carbonate mineral formed, independent of the bulk isotopic composition of the solution from which the carbonate mineral formed (Eiler, 2007). Clumped stable isotope analyses have been used recently to constrain temperatures of fault-hosted carbonate cements (Swanson et al., 2012). Thus, clumped stable isotope analysis could potentially be used to constrain what temperature fluid a carbonate vein formed, as well as the isotopic composition of the fluid from which the vein formed (Mering et al., 2018). This could have significant implications for the application of stable isotope analysis of carbonate minerals when searching for carbonate-hosted ore deposits. This approach could be used to discern high-temperature veins (perhaps associated with magmatic fluids) from moderate or low-temperature fluids (perhaps associated with diagenetic, tectonic or metamorphic processes). Alternatively, veins collected over an area could be analyzed by clumped isotope techniques to seek steeper temperature gradients that might lead to the hotter part of the source hydrothermal system, where ore is more likely to be hosted. In addition, recently published work has demonstrated that carbonates that recrystallize in sedimentary basins retain the temperature of recrystallization (Shenton et al., 2015). The demonstration that the formation temperature of carbonates that are recrystallized can be retained in the clumped isotope record has important implications for carbonate rocks that host ore deposits. If recrystallization temperatures are preserved, then this implies that during the dissolution-precipitation process, altered host rocks would likely retain the temperature at which dissolution-precipitation occurred. This means that isotopic anomalies related to low-temperature fluids could be identified, allowing "false" isotopic anomalies, unrelated to high-temperature hydrothermal systems, to be ignored.

Currently, clumped stable isotope analyses are relatively slow and require specially designed IRMS instruments, which are only available in relatively few laboratories. However, infrared absorption instruments have shown promise for allowing clumped isotope measurements to be made, with the recent development of midinfrared analyzers offering very strong CO_2 absorptions, which will allow the rare isotopologues of CO_2 to be quantified relatively rapidly and with sufficient accuracy to allow paleotemperature information to be determined (Nelson et al., 2015).

8.5.3. Sampling Scale

An important consideration when designing any stable isotope survey for mineral exploration is determining the purpose of the study, and the sampling strategy best suited for that purpose. The case studies outlined above suggest that large mineral deposits are surrounded by isotopic alteration halos on the order of several kilometers (Kesler et al., 1995; Vazquez et al., 1998; Waring et al., 1998; Arehart & Donelick, 2006; Barker et al., 2013). However, the size and geometry of halos can vary (e.g., Stenger et al., 1998; Escalante, 2008), as will the distribution of isotopic alteration, which would be dependent on the permeability structure of the host rocks. Large carbonate-hosted mineral deposits are likely to have significant oxygen isotope anomalies, which likely extend for more than one kilometer (as demonstrated by the studies carried out at Mount Isa by Waring et al., 1998, and the northern Carlin Trend by Barker et al., 2013). Therefore, a regional survey of whole rock carbon and oxygen isotope compositions (plus analysis of any hydrothermal carbonate veins identified in the field) carried out with a grid spacing of ~1 km would be an appropriate scale to identify regions of significant potential hydrothermal alteration, and areas for follow-up prospecting. Kesler et al. (1995) pointed out that northern Mexico has ~750,000 km^2 of carbonate terrain potentially suitable for the formation of skarn and manto deposits, and other large areas of carbonate host rocks are found in Peru, Bolivia, and Argentina. In another example, Oliver et al. (1993, 2004) and Marshall et al. (2006) analyzed carbon and oxygen isotopes in carbonate minerals in unaltered wall rocks, altered rocks, and veins over the Mount Isa Inlier, Australia, an area approximately 150 x 30 km. This area is synonymous with iron oxide-copper-gold (IOCG) mineralization, and the broad scale of isotope sampling allowed fluid pathways, fluid sources, and alteration footprints to be constrained. Together, these studies point to the value of broad-scale assessments of isotopic variations.

Once anomalous isotopic anomalies have been detected at a regional scale, isotope surveys could be carried out at the ~250 to 500 meter scale to delineate the size of isotopic alteration halos, identify the largest halos, and identify the most prospective regions, in conjunction with other mineral exploration techniques, including lithogeochemistry (indeed, we would recommend that any sample submitted for isotope analysis should also have four-acid ICPMS analysis carried out to correlate any isotopic anomalies to pathfinder element anomalies, as well as potentially identify the carbonate mineral species by examining Ca:Mg and Ca:Fe ratios).

Once a prospective area and/or metal anomalies/mineralization are identified, then stable isotope data can be used to vector toward ore bodies by identifying stable isotope gradients, which should provide a direction toward regions of enhanced fluid-rock reaction and fluid flow pathways (see section 8.3.5), which are more likely to host mineralization (Waring et al., 1998; Barker et al., 2013). At this point, the sampling density will likely be dependent on drill hole coverage, complexity of lithology and structure, and available budget.

8.5.4. Stable Isotope Sampling

Once the sampling scale has been chosen (e.g., regional, district, or deposit), then the sampling medium and scale must be chosen. If sampling at a regional scale, with carbonate outcrops being sampled, then a representative sample must be chosen (in the same way that a representative lithogeochemical sample would be collected). For example, if collecting a sample from a fossiliferous limestone, a sufficiently large sample should be chosen to avoid undue influence from a single fossil fragment. Typically, a fist-sized sample would be considered representative for stable isotope sampling. Samples collected and crushed to a suitable grain size for lithogeochemical analysis will be suitable for isotopic analysis, and a split of the powder prepared for lithogeochemical analysis can be used. If the sampled outcrop contains carbonate veins (of known or unknown significance), then the wall rock should be sampled without the influence of macroscopic veins. Microscopic veins, unless they compose more than ~10% of the rock by mass (at which point they should be visible), are unlikely to significantly influence the overall isotopic composition of the rock (Fig. 8.11). Sampling of veins (if thought to be related to hydrothermal mineralization), and adjacent wall rock, may also be desirable, as wallrock-vein pairs have been suggested to provide useful indicators of proximity to ore (Oliver et al., 1993; Waring et al., 1998). However, the presence of multiple vein generations will complicate interpretation of wallrock-vein pairs, as it may not be obvious which veins are associated with the ore-forming hydrothermal system, and significant paragenetic work may be needed to establish what overprinting relationships exist.

If samples are being collected from drill holes, then several types of samples are available. These include analyzing composited intervals of core (potentially the same intervals submitted for lithogeochemistry/assay, and consequently crushed and homogenized to allow isotope data to be directly correlated with geochemical data); analyzing selected hand specimens on the basis of logging; or using a subsampling technique, such as a Dremel® tool fitted with diamond drill bit to sample a particular lithology or vein (Fig. 8.11). The choice of sampling scale/medium will be dependent upon the geological complexity, as well as the question(s) that the stable isotope data are being used to solve. However, composite samples may yield robust and meaningful data for many geological situations. Waring et al. (1998) used a mixture of composite samples (20 meter composites), hand samples, and individually sampled veins to constrain the isotopic alteration around the Mount Isa copper ore bodies. Waring (1991) reported that composite samples yield isotopic samples representative of several whole rock samples from the same composite interval, which might yield a "noisy" result. However, composite samples may act to mask subtle isotopic alteration on the edges of alteration systems (Fig. 8.11). Lepore (2012) and Barker et al. (2013) compared results of hand sampling (~10 cm core intervals), versus 1.5 m drill composites from the Long Canyon Carlin-type gold deposit. In general, there was a consistent relationship between composite and hand samples, with isotopically altered composite intervals almost invariably having altered hand specimens. In contrast, composite intervals indistinguishable from background near the fringes of the isotopic alteration

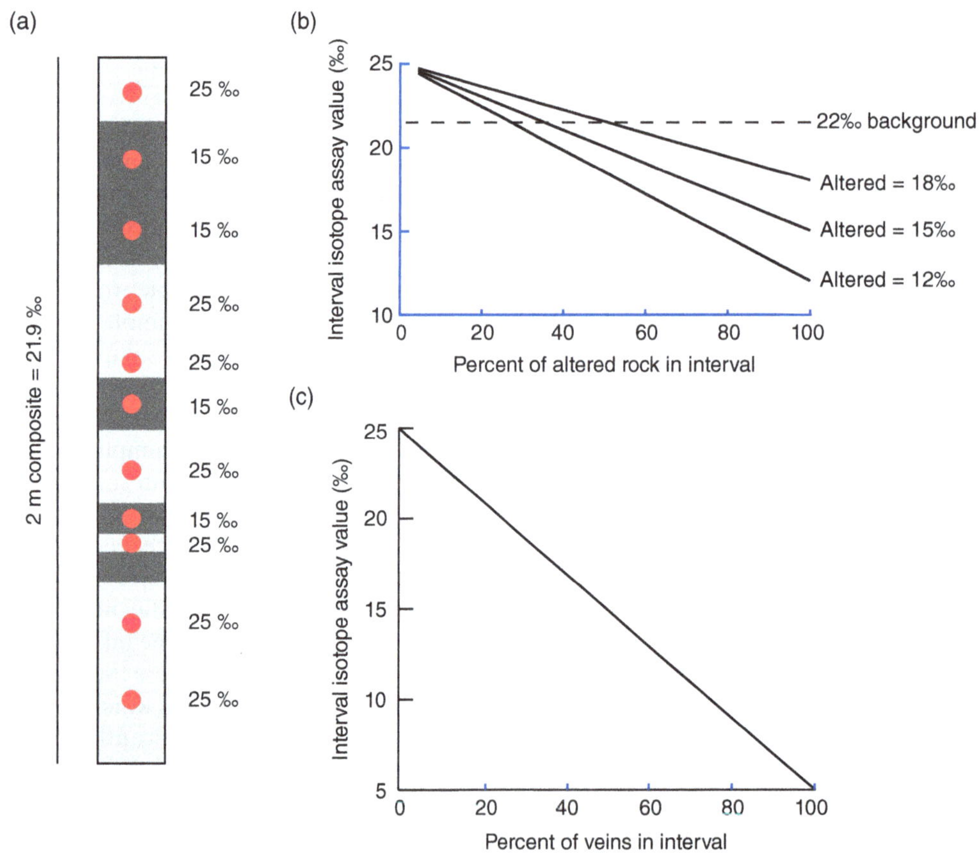

Figure 8.11 (a) Schematic diagram showing a length of drill core containing discrete intervals of hydrothermally altered rock with a $\delta^{18}O$ = 15‰ contained within unaltered rock with $\delta^{18}O$ = 25‰, with individual sampling points (red dots) that might be taken using a hand sampling or microdrilling approach. The overall composite would yield an oxygen isotope assay value of $\delta^{18}O$ = 21.9‰ as the altered and unaltered rocks are mixed together. (b) Graph showing the influence of mixing unaltered rock (25‰ background value) with varying percentages of hydrothermally altered rock of different oxygen isotope values (18‰, 15‰, and 12‰). If a background oxygen isotope cutoff of 22‰ were established, then varying amounts of the composite would need to be altered before the assay showed up as anomalous. This highlights the potential utility of finer scale sampling in areas where fluid flow is heterogeneously distributed. (c) Graph showing the resulting oxygen isotope assay value of an interval as the proportion of veins ($\delta^{18}O$ = 5‰) is increased within the interval (modified after Vaughan, 2013).

halo could contain hand samples that were isotopically altered. Lepore (2012) identified these hand specimens to have slightly different lithological make up for surrounding material, which were likely slightly more permeable to hydrothermal fluid flow and consequently susceptible to isotopic alteration. Vazquez et al. (1998) evaluated the difference between analyzing whole rock samples versus microdrilling carbonate from the same samples. In that case study, they found little difference between the two sampling scales when analyzing samples of whole rock.

We suggest that in many cases, composite samples are likely to yield stable isotope results that are representative and reliable for discerning hydrothermal alteration, particularly proximal to mineralizing systems. More distally, there is a chance that composited intervals will mask isotopic alteration, where more careful hand sampling might enable hydrothermally altered fractures, fault zones, or lithologies to be identified. One advantage of using composited intervals is that, similarly to lithogeochemical or assay results, they are less likely to suffer from a "nugget" effect than small hand samples or microdrilling. In addition, the composited intervals can also be used for lithogeochemical and assay analysis, enabling the data sets to be directly compared. Figure 8.11 shows the distribution of isotopically altered lithological layers within an interval of otherwise isotopically unaltered rock down a drill hole (Fig. 8.11a), and the influence of changing the proportion of isotopically altered rock on the overall isotopic assay from that interval (Fig. 8.11b). Despite the presence of several layers containing clearly hydrothermally altered rock of $\delta^{18}O$ ~15‰, the overall interval has a value of $\delta^{18}O = 21.9$‰, compared with the background value of 25‰. This scenario highlights the potential value of analysing selected lithological layers rather than composite samples.

If composited intervals are being analyzed, the influence of veins can be evaluated by utilizing core photographs (or logs of vein density) to estimate the number of veins over the interval of core and converting this into a volume. For example, if an interval of otherwise unaltered rock with an isotopic value of $\delta^{18}O = 25$‰ contained 20 % veins with a value of $\delta^{18}O = 0$‰ completely unrelated to the ore-forming hydrothermal system, then the composite value would change by approximately 5‰, to $\delta^{18}O = 20$‰. Depending on your threshold of isotopic alteration, this might falsely highlight an area of weak hydrothermal alteration. However, inspection of the core photos or vein density logs should quickly highlight that the weak anomaly is associated with a dense interval of veining. The influence of percentage of veining (assuming veins have a $\delta^{18}O = 5$‰) on the isotopic assay value of an interval is shown in Figure 8.11c. At a vein proportion of ~15%, the isotopic value of the interval has lowered to $\delta^{18}O = 22$‰. However, 15% veining should be readily identifiable via visual or automated logging technologies. As with any analytical technique, good geological logging and core photographs will enable far more robust interpretations of isotopic data.

8.5.5. Factors to Consider Before Planning an Isotopic Study

The studies of Friehauf and Pareja (1998) and Lepore (2012) highlighted that dolomite-rich host rocks (dolostones) may respond differently to host rocks dominated by calcite (limestones), likely due to the differential response of calcite and dolomite to dissolution in hydrothermal fluids at the same pH. However, the Mount Isa studies of Waring (1991) and Waring et al. (1998) clearly demonstrate that dolostones can record large and significant stable isotope alteration halos. Thus, mineralogy should clearly be considered when undertaking an isotopic survey, and a good understanding of carbonate mineralogy will help immensely with interpretation of stable isotope data. However, dolomite-dominant carbonate mineralogy should not dissuade a stable isotope survey from being undertaken.

A commonly raised concern we have encountered when discussing stable isotope analyses with the mineral exploration community is the influence of diagenesis and/or metamorphism on lowering background host rock compositions below accepted background values and thus creating false isotope anomalies. Thermal metamorphism drives devolatilization reactions that can nearly or completely decarbonate marly sediments and lower $^{13}C/^{12}C$ ratios (see section 8.3.2, and also Baumgartner & Valley, 2001). Even in the most extreme cases, however, ^{18}O depletions are less than about 5 per mil. The resulting coupled strong depletion in ^{13}C and modest depletion in ^{18}O is distinct from the alteration patterns that result from hydrothermal fluid flow (Baumgartner & Valley, 2001). There is no appreciable pressure dependence on stable isotope fractionation over the range of lithostatic pressure in the crust.

Diagenesis can influence both carbon and oxygen isotope compositions, depending on the diagenetic history of the carbonate rock, and is often on the order of 1–2‰ for both carbon and oxygen for meteoric diagenesis (Allan & Matthews, 1982). The best way of understanding and controlling for diagenesis will be to utilize regional stratigraphy that is thought to be unaffected by hydrothermal fluids (but presumably has the same diagenetic history). Having relevant, regional control on isotopic background compositions will allow greater confidence of data interpretation than the published values for carbon and oxygen isotope compositions of carbonates over geological time (Veizer et al., 1999).

Another common concern is the influence of multiple hydrothermal systems on isotopic values, some of which may have little or no associated mineralization, as is the case when interpreting other geochemical and geophysical data sets. This is often raised as an issue, most likely due to the fact that mineralizing hydrothermal systems are expected to create chemical changes to the rock that can be measured (e.g., pathfinder elements), whereas isotopic anomalies can be induced by any fluid. While this is certainly a valid concern, there are multiple case studies, which we detailed above, where the isotopic anomalies identified appear to be focused around the major ore deposits (Waring, 1991; Kesler et al., 1995; Naito et al., 1995; Stenger et al., 1998; Waring et al., 1998; Barker et al., 2013; Vaughan et al., 2016), and even in areas with evidence for multiple hydrothermal fluid flow and igneous intrusion events such as the northern Carlin Trend (Barker et al., 2013; Vaughan et al., 2016), and multiple hydrothermal events and metamorphism (Mount Isa, Waring et al., 1998). The isotopic alteration halo preserved seems to reflect (at least on the basis of geometry) the major mineralizing event(s). This may be because these events are larger in scale and fluid volume than other, nonmineralizing processes.

8.6. CONCLUSIONS

The case studies and theory outlined above clearly demonstrate that stable isotope alteration accompanies the formation of most carbonate-hosted ore deposits. In our opinion, mineral exploration for carbonate-hosted ore deposits should utilize stable isotope data to help inform exploration activities, and should be used routinely at various stages of mineral exploration to identify prospective areas, and vector toward ore bodies using the strong theoretical understanding and numerous case studies already available as a guide.

ACKNOWLEDGMENTS

Thanks to Ken Hickey, Jeremy Vaughan, Will Lepore, Ben Andrew, John Mering, Andreas Beinlich, Craig Hart, Abraham Escalante, Paul Dobak, Francois Robert, Lucas Marshall, John Thompson, Sarah Rice, Matt Leybourne, Mary Doherty, Moira Smith, Peter Megaw, Lyle Hansen, and many other individuals who have supported research, development, and application of our work on stable isotope alteration of carbonate-hosted ore deposits. The authors acknowledge support from ALS Minerals, Barrick, Teck, Newmont, and MAG Silver for financial support. Parts of this research were supported by NSERC CRD. The authors thank Nick Oliver, who provided an extremely valuable review, which substantially clarified and improved the manuscript.

REFERENCES

Allan, J. R., & Matthews, R. K. (1982). Isotope signatures associated with early meteoric diagenesis. *Sedimentology*, *29*, 797–817.

Anderson, B. R., Gemmell, J. B., & Berry, R. F. (2001). The geology of the Nifty copper deposit, Throssell Group, Western Australia: Implications for ore genesis. *Economic Geology*, *96*(7), 1535–1565.

Arehart, G. B., & Donelick, R. A. (2006). Thermal and isotopic profiling of the Pipeline hydrothermal system: Application to exploration for Carlin-type gold deposits. *Journal of Geochemical Exploration*, *91*(1–3), 27–40. doi:10.1016/j.gexplo.2005.12.005.

Barker, S. L. L., Dipple, G. M., Dong, F., & Baer, D. S. (2011). Use of laser spectroscopy to measure the 13C/ 12C and 18O/ 16O compositions of carbonate minerals. *Anal. Chem.*, *83*(6), 2220–2226. doi:10.1021/ac103111y.

Barker, S. L. L., Dipple, G. M., Hickey, K. A., Lepore, W. A., & Vaughan, J. R. (2013). Applying stable isotopes to mineral exploration: teaching an old dog new tricks. *Economic Geology*, *108*, 1–9.

Baumgartner, L. P., & Ferry, J. M. (1991). A model for coupled fluid-flow and mixed-volatile mineral reactions with applications to regional metamorphism. *Contributions to Mineralogy and Petrology*, *106*(3), 273–285. doi:10.1007/BF00324557.

Baumgartner, L. P., & Valley, J. W. (2001). Stable isotope transport and contact metamorphic fluid flow. *Reviews in Mineralogy and Geochemistry*, *43*, 415–468.

Bear, J. (1972). *Dynamics of fluids in porous media.* Elsevier.

Beinlich, A., Barker, S. L. L., Gupta, M., & Baer, D. (2017). Stable isotope (δ13C, δ18O) analysis of sulfide-bearing carbonate samples using laser absorption spectrometry. *Economic Geology*, *112*, 693–700.

Benezeth, P., Stefansson, A., Gautier, Q., & Schott, J. (2013). Mineral solubility and aqueous speciation under hydrothermal conditions to 300 C, the carbonate system as an example. *Reviews in Mineralogy and Geochemistry*, *76*(1), 81–133. doi:10.2138/rmg.2013.76.4.

Bickle, M. J., & McKenzie, D. (1987). The transport of heat and matter by fluids during metamorphism. *Contributions to Mineralogy and Petrology*, *95*, 384–392.

Blattner, P. (1985). Isotope shift data and the natural evolution of geothermal systems. *Chemical Geology*, *49*(1), 187–203.

Bowman, J. R. (1998). Stable-isotope systematics of skarns, in D. Lentz (Ed.), *Mineralized Intrusion-Related Skarn Systems*, vol. *26* (pp. 99–145). Mineralogical Association of Canada, Quebec.

Bowman, J. R., O'Neil, J. R., & Essene, E. J. (1985). Contact skarn formation at Elkhorn, Montana; II, Origin and evolution of COH skarn fluids. *American Journal of Science*, *285*, 621–660.

Bowman, J. R., Willett, S. D., & Cook, S. J. (1994). Oxygen isotopic transport and exchange during fluid flow: One-dimensional models and applications. *American Journal of Science*, *294*, 1–55.

Cartwright, I., Power, W. L., Oliver, N. H. S., Valenta, R. K., & McLatchie, G. S. (1994). Fluid migration and vein formation during deformation and greenschist facies metamorphism at Ormiston Gorge, central Australia. *Journal of Metamorphic Geology*, *12*, 373–386.

Cathles, L. M. (1993). Oxygen isotope alteration in the Noranda mining district, Abitibi greenstone belt, Quebec. *Economic Geology*, *88*(6), 1483–1511.

Cathles, L. M. (1997). Thermal aspects of hydrothermal ore deposits, in *Geochemistry of hydrothermal ore deposits* (pp. 191–227). John Wiley & Sons, New York.

Cathles, L. M., Erendi, H. J., & Barrie, T. (1997), How long can a hydrothermal system be sustained by a single intrusive event? *Economic Geology*, *92*, 766–771.

Chacko, T., Mayeda, T. K., Clayton, R. N., & Goldsmith, J. R. (1991). Oxygen and carbon isotope fractionations between CO^2 and calcite. *Geochimica et Cosmochimica Acta*, *55*(10), 2867–2882. doi:10.1016/0016-7037(91)90452-B.

Chaffee, M. A. (1976). The zonal distribution of selected elements above the Kalamazoo porphyry copper deposit, San Manuel district, Pinal County, Arizona. *Journal of Geochemical Exploration*, *5*(1), 145–165. doi:10.1016/0375-6742(76)90042-X.

Chapman, L. (2004). Geology and genesis of the George-Fisher Zn-Pb-Ag deposit, Mount Isa, Australia. James Cook University, unpublished PhD thesis. doi: 10.2113/gsecongeo.99.2.233.

Cline, J. S., Hofstra, A. H., Muntean, J. L., Tosdal, R. M., & Hickey, K. A. (2005). Carlin-type gold deposits in Nevada: Critical geologic characteristics and viable models. *Economic Geology 100th Anniversary Volume*, 451–484.

Cole, D. R., & Ohmoto, H. (1986). Kinetics of isotopic exchange at elevated temperatures and pressures. *Reviews in Mineralogy and Geochemistry*, *16*, 41–90.

Coplen, T. B. (1988). Normalization of oxygen and hydrogen isotope data. *Chemical Geology: Isotope Geoscience section. 72*, 293–297

Dipple, G. M., & Ferry, J. M. (1992). Metasomatism and fluid flow in ductile fault zones. *Contributions to Mineralogy and Petrology*, *112*, 149–164.

Dipple, G. M., & Gerdes, M. L. (1998). Reaction-infiltration feedback and hydrodynamics at the skarn front, in *Mineralized Intrusion-related skarn systems* (pp. 71–97). Mineralogical Association of Canada.

Eiler, J. M. (2007). "Clumped-isotope" geochemistry: The study of naturally occurring, multiply substituted isotopologues. *Earth and Planetary Science Letters*, *262*(3–4), 309–327. doi:10.1016/j.epsl.2007.08.020.

Engel, A., Clayton, R. N., & Epstein, S. (1958). Variations in isotopic composition of oxygen and carbon in Leadville limestone (Mississippian, Colorado) and in its hydrothermal and metamorphic phases. *The Journal of Geology*, *66*, 374–393.

Escalante, A. (2008). Patterns of distal alteration zonation around Antamina Cu-Zn skarn and Uchucchacua Ag-base metal vein deposits, Peru: Mineralogical, chemical and isotopic evidence for fluid composition, and infiltration, and implications for mineral exploration. University of British Columbia, 24 June.

Ford, J. H. (1978). A chemical study of alteration at the Panguna porphyry copper deposit, Bougainville, Papua, New Guinea. *Economic Geology*, *73*(5), 703–720. doi:10.2113/gsecongeo.73.5.703.

Friehauf, K. C., & Pareja, G. A. (1998). Can oxygen isotope halos be produced around high-temperature dolostone-hosted ore deposits? Evidence from the Superior District, Arizona. *Economic Geology*, *93*, 639–650.

Frimmel, H. E. (1992). Isotopic fronts in hydrothermally mineralized carbonate rocks. *Mineralium Deposita*, *27*(4), 257–267. doi:10.1007/BF00193396.

Gerdes, M. L., & Valley, J. W. (1994). Fluid flow and mass transport at the Valentine wollastonite deposit, Adirondack Mountains, New York State. *Journal of Metamorphic Geology*, *12*, 589–608.

Gerdes, M. L., Baumgartner, L. P., Person, M., & Rumble, D. (1995). One-and two-dimensional models of fluid flow and stable isotope exchange at an outcrop in the Adamello contact aureole, Southern Alps, Italy. *American Mineralogist*, *80*, 1004–1019.

Halley, S., Dilles, J. H., & Tosdal, R. M. (2015). *Footprints: hydrothermal alteration and geochemical dispersion around porphyry copper deposits*. Society of Economic Geologists Newsletter.

Heinrich, C. A., Andrew, A. S., Wilkins, R. W. T., & Patterson, D. J. (1989). A fluid inclusion and stable isotope study of synmetamorphic copper ore formation at Mount Isa, Australia. *Economic Geology*, *84*(3), 529–550. doi:10.2113/gsecongeo.84.3.529.

Hitzman, M., Kirkham, R., Broughton, D., & Thorson, J. (2005). The sediment-hosted stratiform copper ore system, in J. W. Hedenquist, J. F. H. Thompson, R. J. Goldfarb, & J. P. Richards (Eds.), *Economic Geology 100th Anniversary Volume* (pp. 609–642). Society of Economic Geologists.

Hitzman, M., Redmond, P. B., & Beaty, D. W. (2002). The carbonate-hosted Lisheen Zn-Pb-Ag deposit, Country Tipperary, Ireland. *Economic Geology*, *97*, 1627–1655.

Hoefs, J. (2013). *Stable Isotope Geochemistry*. Springer Science & Business Media.

Hofstra, A. H., & Cline, J. S. (2000). Characteristics and models for Carlin-type gold deposits. *Reviews in Economic Geology*, *13*, 163–220.

Kah, L., Sherman, A., Narbonne, G., Knoll, A. & Kaufman, A. (1999) $\delta^{13}C$ stratigraphy of the Proterozoic Bylot Supergroup, Baffin Island, Canada: Implications for regional lithostratigraphic correlations. *Canadian Journal of Earth Science*, *36*, 313–332.

Kelley, D. L., Kelley, K. D., Coker, W. B., Caughlin, B., & Doherty, M. E. (2006). Beyond the obvious limits of ore deposits: The use of mineralogical, geochemical, and biological features for the remote detection of mineralization. *Economic Geology*, *101*(4), 729–752.

Kesler, S. E., Vennemann, T. W., Vazquez, R., & Stenger, D. P. (1995). Application of large-scale oxygen isotope halos to exploration for chimney-manto Pb-Zn-Cu-Au deposits. *Geology and Ore Deposits of the America Cordillera Proceedings Volume III*, 1383–1396.

Korzhinsky, D. S. (1959). The advancing wave of acidic components in ascending solutions and hydrothermal acid-base differentiation. *Geochimica et Cosmochimica Acta*.

Lassey, K. R., & Blattner, P. (1988). Kinetically controlled oxygen isotope exchange between fluid and rock in one-dimensional advective flow. *Geochimica et Cosmochimica Acta*, *52*(8), 2169–2175. doi:10.1016/0016-7037(88)90197-4.

Leach, D. L., Sangster, D. F., Kelley, K. D., Large, R., Carven, G., & Allen, C. (2005). Sediment-hosted lead-zinc deposits: a global perspective, in *Economic Geology One Hundredth Anniversary Volume* (pp. 561–608). Society of Economic Geologists, Littleton, Colorado.

Lepore, W. A. (2012). Petrophysical and physicochemical controlling parameters on stable isotope depletion patterns in carbonate rocks from auriferous hydrothermal fluid infiltration at the Long Canyon sediment-hosted gold deposit: NE Nevada. University of British Columbia, unpublished MSc thesis. doi: 10.14288/1.0074052.

Marshall, L. J., Oliver, N. H. S., & Davidson, G. J. (2006). Carbon and oxygen isotope constraints on fluid sources and fluid-wallrock interaction in regional alteration and iron-oxide–copper–gold mineralisation, eastern Mt Isa Block, Australia. *Mineralium Deposita. 41*, 429–452

Megaw, P. (1990). Geology and geochemistry of the Santa Eulalia mining district, Chihuahua, Mexico. University of Arizona, unpublished PhD thesis.

Megaw, P., Ruiz, J., & Titley, S. R. (1988). High-temperature, carbonate-hosted Ag-Pb-Zn (Cu) deposits of northern Mexico. *Economic Geology*, *83*, 1856–1885.

Meinert, L. D., Dipple, G. M., & Nicolescu, S. (2005). World skarn deposits, in *Economic Geology One Hundredth Anniversary Volume* (pp. 299–337). Society of Economic Geologists, Littleton, Colorado.

Mering, J., Barker, S. L. L., Huntington, K. W., Simmons, S., Dipple, G., Andrew, B. & Schauer, A. (2018). Carbonate clumped isotope thermometry in hydrothermal ore deposits. *Economic Geology*, *113*, DOI:10.5382/econgeo.2018.4608.

Morris, H. T. (1986). *Descriptive Model of Polymetallic Replacement Deposits*. United States Geological Survey.

Naito, K., Fukahori, Y., Peiming, H., Sakurai, W., Shimazaki, H., & Matsuhisa, Y. (1995). Oxygen and carbon isotope zonations of wall rocks around the Kamioka Pb-Zn skarn deposits, central Japan: Application to prospecting. *Journal of Geochemical Exploration*, *54*(3), 199–211.

Nelson, D., McManus, B., & Herndon, S. (2015). Recent progress in development of a laser based, ultra-high precision isotope monitor for carbon dioxide. *EGU General Assembly*, *17*.

Northrop, D. A., & Clayton, R. N. (1966). Oxygen-isotope fractionations in systems containing dolomite. *The Journal of Geology*, *74*(2), 174–196

Oliver, N. H. S., Cartwright, I., Wall, V. & Golding, S. 1993. The stable isotope signature of kilometre-scale fracture dominated metamorphic fluid pathways, Mary Kathleen, Australia. *Journal of metamorphic Geology*, *11*, 705–720.

Oliver, N. H. S., Cleverley, J., Mark, G., Pollard, P.., Fu, B., Marshall, L., Rubenach, M., Williams, P., & Baker, T. (2004). Geochemistry and geochemical modelling of fluid-rock interaction in the eastern Mt. Isa Block, Australia: The role of sodic alteration in the genesis of iron oxide-copper-gold deposits. *Economic Geology*, *99*, 1145–1176.

Ohmoto, H., & Goldhaber, M. B. (1997). Sulfur and carbon isotopes, in H. L. Barnes (Ed.), *Geochemistry of Hydrothermal Ore Deposits* (pp. 517–611). John Wiley & Sons, New York.

O'Neil, J., Clayton, R., & Mayeda, T. 1969. Oxygen isotope fractionation in divalent metal carbonates. *The Journal of Chemical Physics*, *51*, 5547.

Painter, M. G. M. (2003). The geochemical and mineralogical halos around the Mount Isa base metal orebodies. The University of Queensland, unpublished PhD thesis.

Pokrovsky, O. S., Golubev, S. V., Schott, J., & Castillo, A. (2009). Calcite, dolomite and magnesite dissolution kinetics in aqueous solutions at acid to circumneutral pH, 25 to 150 °C and 1 to 55 atm pCO^2: New constraints on CO^2 sequestration in sedimentary basins. *Chemical Geology*, *265*(1–2), 20–32. doi:10.1016/j.chemgeo.2009.01.013.

Radtke, A. S., Rye, R. O., & Dickson, F. W. (1980). Geology and stable isotope studies of the Carlin gold deposit, Nevada. *Economic Geology*, *75*(5), 641–672.

Sangster, D. F. 1997. Carbonate-hosted lead-zinc deposits. Society of Economic Geologists Special Publication No. 4. Society of Economic Geologists, Littleton, CO.

Selley, D., Broughton, D., Scott, R. J., Hitzman, M., Bull, S. W., Large, R. R., et al. (2005). A new look at the geology of the Zambian Copperbelt, in J. W. Hedenquist, J. F. H. Thompson, R. J. Goldfarb, & J. P. Richards (Eds.), *Economic Geology 100th Anniversary Volume* (pp. 965–1000). Society of Economic Geologists, Littleton, CO.

Shenton, B. J., Grossman, E. L., Passey, B. H., Henkes, G. A., Becker, T. P., Laya, J. C., et al. (2015). Clumped isotope thermometry in deeply buried sedimentary carbonates: The effects of bond reordering and recrystallization. *Geological Society of America Bulletin*, *127*, 1036–1051, doi:10.1130/B31169.1.

Sheppard, S. M. F., & Schwarcz, H. P. (1970). Fractionation of carbon and oxygen isotopes and magnesium between coexisting metamorphic calcite and dolomite. *Contributions to Mineralogy and Petrology*, *26*(3), 161–198. doi:10.1007/BF00373200.

Shields, G., & Veizer, J. (2002). Precambrian marine carbonate isotope database: Version 1.1. *Geochem. Geophys. Geosyst.*. doi:10.1029/2001GC000266.

Shimazaki, H., & Kusakabe, M. (1990). Oxygen isotope study of the Kamioka Zn-Pb skarn deposits, Central Japan. *Mineralium Deposita*, *25*, 221–229. doi:10.1007/BF00190385.

Shimazaki, H., Shimizu, M., & Nakano, T. (1986). Carbon and oxygen isotopes of calcites from Japanese skarn deposits. *Geochemical Journal*.

Smith, J. W., Burns, M. S., & Croxford, N. J. W. (1978). Stable isotope studies of the origins of mineralization at Mount Isa. *I. Mineralium Deposita*, *13*(3), 369–381. doi:10.1007/BF00206570.

Stenger, D. P., Kesler, S. E., & Vennemann, T. (1998). Carbon and oxygen isotope zoning around Carlin-type gold deposits: A reconnaissance survey at Twin Creeks, NV. *Journal of Geochemical Exploration*, *63*(2), 105–121.

Swanson, E. M., Wernicke, B. P., & Eiler, J. M. (2012). Temperatures and fluids on faults based on carbonate clumped-isotope thermometry. *American Journal of Science*, *312*, 1–21.

Valenta, R. K. (1988). Deformation, Fluid Flow, and Mineralization in the Hilton Area, Mount Isa, Australia. Monash University, unpublished PhD thesis.

Vaughan, J. R. (2013). Tracing hydrothermal fluid flow in the rock record: Geochemical and isotopic constraints on fluid flow in Carlin-type gold systems, University of British Columbia, unpublished PhD thesis. doi: 10.14288/1.0074127.

Vaughan, J., Hickey, K. A., & Barker, S. L. L. (2016). Isotopic, Chemical, and Textural Evidence for Pervasive Calcite Dissolution and Precipitation Accompanying Hydrothermal Fluid Flow in Low-Temperature, Carbonate-Hosted, Gold Systems. *Economic Geology*, *111*, 1127–1157.

Vazquez, R., Vennemann, T. W., & Kesler, S. E. (1998). Carbon and oxygen isotope halos in the host limestone, El Mochito Zn-Pb-(Ag) skarn massive sulfide-oxide deposit, Honduras. *Economic Geology*, *93*, 15–31.

Veizer, J., Ala, D., Azmy, K., Bruckschen, P., Buhl, D, & Bruhn, F. (1999). $^{87}Sr/^{86}Sr$, $\delta^{13}C$ and $\delta^{18}O$ evolution of Panerozoic seawater. *Chemical Geology*, *161*, 59–88.

Waring, C. L. (1991). Carbonate Isotopic and Major Element Geochemical Exploration Techniques. Report to Mount Isa Mines Exploration Pty Ltd.

Waring, C. L., Andrew, A. S., & Ewers, G. R. (1998). Use of O, C, and S stable isotopes in regional mineral exploration. *AGSO Journal of Australian Geology and Geophysics*, *17*, 301–313.

Yardley, B. W. D., & Lloyd, G. E. (1995). Why metasomatic fronts are really metasomatic sides. *Geology*, *23*(1), 53–56.

Zheng, Y.-F. (1999). Oxygen isotope fractionation in carbonate and sulfate minerals. *Geochemical Journal*, *33*, 109–126.

9

The Importance of Large Scale Geophysical Investigations for Mineral Exploration

Susan J. Webb[1], Stephanie E. Scheiber-Enslin[1], and Janine Cole[1,2]

ABSTRACT

Exploration for mineral resources using geophysical methods generally focuses on depths of less than 2 km, because the cost of exploring and mining below that depth is prohibitive. However, this strategy neglects the added value of a broader context, possibly resulting in missed opportunities for the discovery of significant deposits and their large-scale extensions. Larger-scale geophysical methodologies, such as considering the entire crust and upper mantle in the conceptual model, offer great potential for the recognition and exploitation of new, potentially large, ore deposits and their extensions. The Bushveld Complex from South Africa provides a rich lesson in how changing concepts resulted in new exploration strategies.

9.1. INTRODUCTION

Exploration for and delineation of mineral deposits has traditionally focused on follow-up of the near surface expression of orebodies discovered through geological, geochemical or geophysical mapping and confirmed by drilling. The larger scale structural model and its extension at depth are often delineated after the initial discovery. This small scale approach can fail to identify the full lateral extent of an orebody, or prospective region, as the initial model becomes the entrenched idea. The Bushveld Complex provides a rich lesson in how large scale conceptual models can change exploration strategies.

9.2. THE BUSHVELD COMPLEX – INTERFERING ANOMALIES AND LONG WAVELENGTH CONTRIBUTIONS

The Bushveld Complex in South Africa is the world's largest layered mafic intrusion, and due to its large areal extent (~400 × 400 km) and thickness (6–9 km) it also qualifies as a Large Igneous Province (LIP; Ernst, 2014). It hosts a rich array of mineral deposits, most notably Platinum Group Metals (PGMs), vanadium, chromite, and dimension stone, which have been comprehensively summarized in Lee (1996). Given the growing body of evidence for the connectivity between the exposed Eastern and Western Limbs of the complex from geophysical and petrological observations and measurements (Cawthorn & Webb, 2001), as well as geophysical modeling (Webb et al., 2004; Cole et al., 2014), it is very likely that substantial ore deposits may be present beneath the cover rocks in the central portions of the Bushveld Complex, covering an area of around 150,000 km^2 (Fig. 9.1).

The Bushveld Complex is clearly delineated in the regional scale gravity and magnetic maps published by the Council for Geoscience (1997) (Fig. 9.2). The mafic lithologies of the Rustenburg Layered Suite have an average density that is higher than the average crustal density (Ashwal et al., 2005; Maré & Tabane, 2004). The largest gravity values are always found down dip of the outcrop, indicating the thickness increases inward. The upper zone hosts extensive layers of magnetite with large susceptibilities, remanence, and a strong component of demagnetization, resulting in magnetic anomalies that are challenging to model with available measured remanent

[1] School of Geosciences, University of the Witwatersrand, Johannesburg, South Africa

[2] Geophysics and Remote Sensing Unit, Council for Geoscience, Silverton, Pretoria, South Africa

Ore Deposits: Origin, Exploration, and Exploitation, Geophysical Monograph 242, First Edition.
Edited by Sophie Decrée and Laurence Robb.

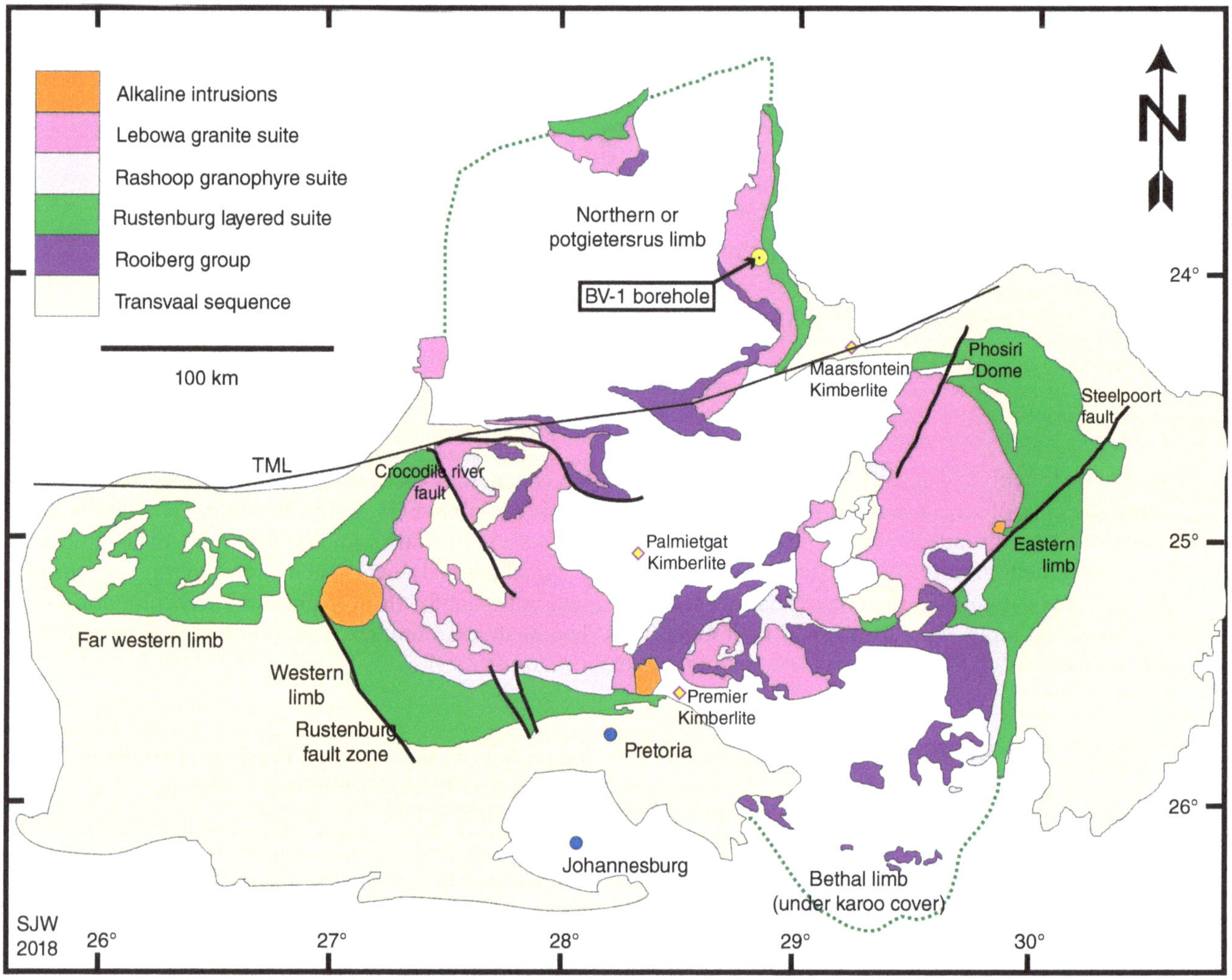

Figure 9.1 Simplified geological map of the Bushveld Complex. The Crocodile River Fragment is to the west of the Crocodile River Fault. The BV-1 research borehole added critical knowledge of physical properties, mineralogical and petrological information. The Palmietgat kimberlite, located close to the center, helped establish the connectivity between the Western and Eastern Limbs.

parameters and physical properties (Ashwal et al., 2005; Maré & Tabane, 2004; Cole, 2018).

The Bushveld Complex provides an ideal lesson on the importance of incorporating large-scale geophysical modeling to the study of a laterally extensive ore body, in which the mineralization is stratigraphically controlled. Studies have shown the importance of including the long-wavelength gravity signal to deduce the connectivity between the Western and Eastern Limbs (Cawthorn et al., 1998). A connected Bushveld Complex means the volume is much larger than previously thought, perhaps as much as 1–2 x 10^6 km^3. While most of this proposed extension in the buried central part is likely too deep to mine with current technology, it represents an extensive brown fields target that is completely underexplored. Here we review the evolving ideas about the three-dimensional geometry of the Bushveld Complex that brought us to the current model, coupled with the growing body of geophysical and geological data, to outline the importance of incorporating crustal- and upper mantle-scale information into ore body modeling.

9.2.1. Early Conceptual Models

The geologic, stratigraphic, and petrologic similarities between the Western and Eastern Limbs of the Bushveld Complex were recognized by early workers, and interpreted in terms of a large, lopolithic intrusion (Molengraaff, 1901; Hall, 1932; Du Toit, 1954) (Fig. 9.3). These early models highlight an important enduring problem in geological mapping: how do we portray uncertainty in geological maps and drawings? These sketches are based only on outcrops: the interior geometry is purely speculative. However, the lopolith interpretation provided a working conceptual model for ~50 years of Bushveld study. Exploration at this time was largely confined to delineating

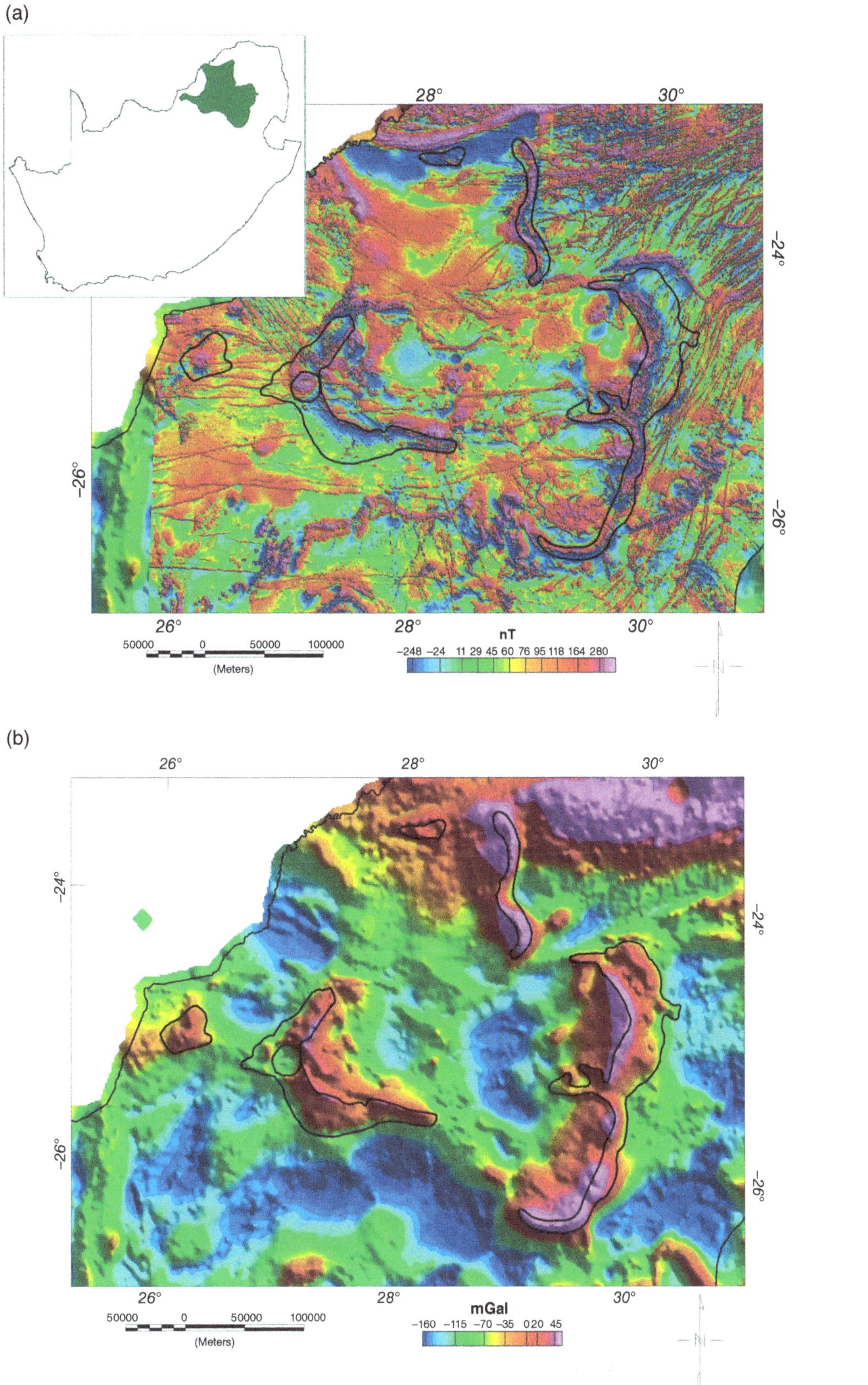

Figure 9.2 The locality of the Bushveld Complex in north-eastern South Africa (inset – green fill) and the magnetic and gravity data available for South Africa. (a) IGRF corrected and merged aeromagnetic data over the Bushveld Complex (Stettler et al., 1999; Stettler et al., 2000). (b) The simple Bouguer gravity regional anomaly map (Venter et al., 1999). The Bushveld Complex is readily apparent in both maps due to complex magnetization and high density of the Bushveld Complex rocks. The outcrop outline is the thin black line in (a) and (b).

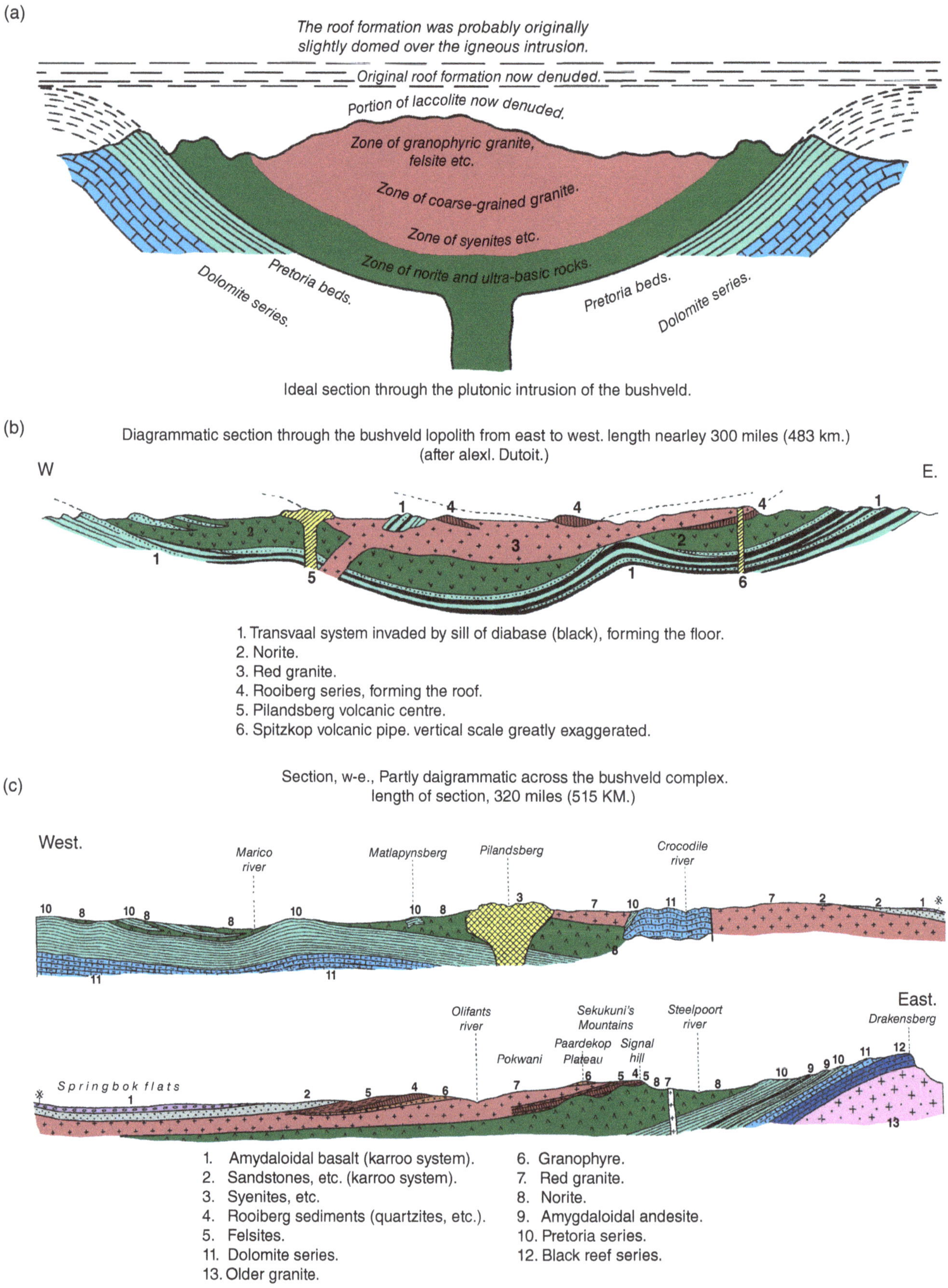

Figure 9.3 Three early conceptual models of the Bushveld Complex: (a) Molengraaff (1901) was the first to describe the Bushveld and proposed a large plutonic laccolite; (b) du Toit (1954) suggested a continuous lopolith; (c) Hall (1932) did not illustrate the deeper structure, but implied the outcropping similarities suggested connection (modified from Cole et al., 2014). Published with the permission of the Journal of African Earth Science (Elsevier).

near surface outcrops; the details of the interior of the model did not influence exploration. Feeder zones were drawn near the center of the intrusion, based on speculation. This conceptual model almost completely dominated exploration efforts, which focused on mapping of field relationships and ultramafic rocks (Wagner, 1929).

9.2.2. Cousins: Early Gravity Data Refutes Lopolith

By 1959, the Geological Survey of South Africa had collected sufficient gravity data for the physicist Cedric Cousins to test the lopolith geometry by forward modeling (Cousins, 1959). His analysis of the data did not support the lopolith model, which he rejected in favor of two distinctly separated, synformal intrusive bodies, one each in the Western and Eastern Limbs. Cousins centered the intrusive bodies beneath the ~50–100 mGal Bouguer gravity anomaly and described them as "dyke-like bodies having a structure closely resembling that calculated by Weiss (1940) for the Great Dyke of Southern Rhodesia." Thin dyke feeders were proposed to extend to the mantle source below (Fig. 9.4). In the same publication, Cousins

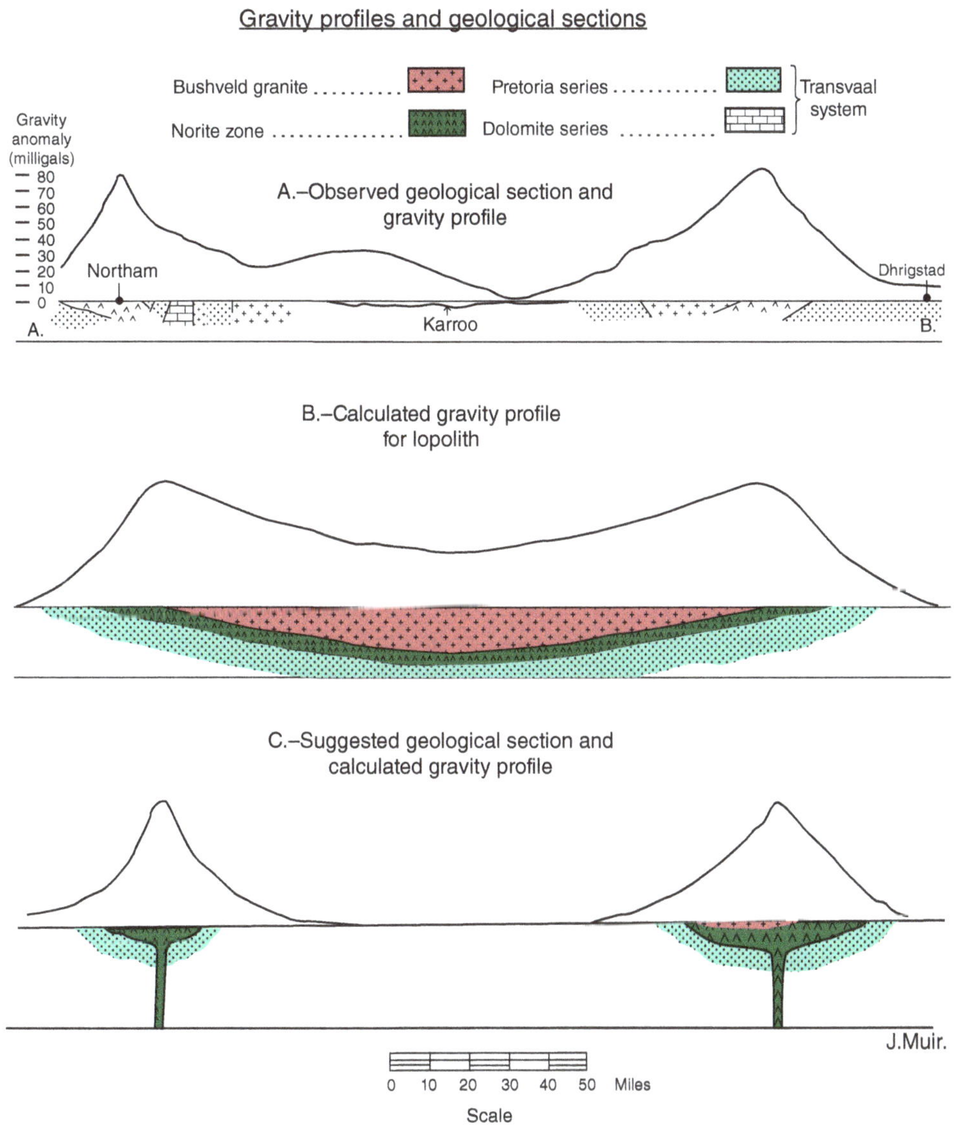

Figure 9.4 The gravity data (top panel) and models (middle and lower panels) used by Cousins to argue for separated intrusions in the Bushveld Complex (Cousins, 1959). The calculated gravity profile over a large connected lopolith model does not fit the observed gravity profile due to large gravity values in the center, while the calculated profile over two separate intrusions does. He used the term *synformal intrusive bodies,* which extend north and south as dykes. We use the same term to avoid confusion between the larger connected lopolith model and the separated "synformal intrusions" of Cousins, which look like small lopoliths (modified from Cousins, 1959). Published with the permission of the South African Journal of Geology.

postulated that there was a "Bushveld Chain of Intrusives" extending from the Great Dyke all the way to the Trompsburg (Cousins, 1959), suggesting a fracture extending deep into the Earth's crust. The 11 pages of comments and replies to this manuscript document the passionate debate about his new model. In overturning the lopolith model, Cousins points out that the original lopolith model was "a structural hypothesis which appeared to be universally accepted." Thus he acknowledges the ensuing resistance to his new conceptual model, which slowly gained acceptance. Cousins's new model largely ignored the geological similarities between the Western and Eastern Lobes, but suggests a rich exploration target. The feeder zones are located near the edge in this model: Could they be associated with enriched mineralization?

Debate of Cousin's model was still taking place in the 1960s as evidenced by a comment in the publication of the South African gravity data in 1962. This insightful comment by Hales and Gough (1962, p. 365) about the modeling of the Bushveld Complex gravity, states: "the increased load of the gabbroic rocks of the Bushveld would [similarly] have caused mantle failure and downwarping." Although this idea was never tested with modeling or developed into a publication, it hints that debates were still taking place. Earlier in this publication, Hales and Gough state, "it is reasonable to conclude that no considerable mass of gabbro, or other rock denser than granite, can underlie the central part of the Bushveld Igneous Complex" (Hales & Gough, 1962, p. 365), suggesting Cousins's model was now generally accepted.

As more gravity data were obtained, and more efficient gravity field calculation algorithms were developed, Cousins's (1959) model of separated intrusive bodies was tested by calculating gravity anomalies for profiles at a number of new locations around the Bushveld Complex (Van der Merwe, 1976; Molyneux & Klinkert, 1978; Hattingh, 1980). In these trough-shaped, synformal models, the thickest extent of the intrusive body lies beneath the maximum Bouguer anomaly. The largest Bouguer anomalies were suggested to overlie feeder zones (Sharpe & Snyman, 1980; Sharpe et al., 1981; Viljoen, 1999). Kinloch (1982) postulated that feeders were associated with volatiles, enhanced mineralization, and increased numbers of potholes. This led several companies to focus exploration efforts on locating feeder zones, which were assumed to be more richly mineralized, despite a paucity of robust evidence that this is the case.

9.2.3. Dipping Sheets and New Geophysical Constraints

Du Plessis and Kleywegt (1987), however, noted that a model of two widely separated intrusive bodies with synformal geometry is contradicted by outcrops at the northwestern end of the Crocodile River fragment (Fig. 9.1). Here the dip of the mafic lithologies is in the opposite direction than that predicted for the inner limb of a synformal structure. This led him them to propose a dipping sheet model, where the sheet tapers with depth to thin, unresolvable feeder dykes, up to 100 kilometers inward of outcropping lithologies. A similar model was supported by the ultradeep resistivity soundings of the CSIR (Meyer & de Beer, 1987). Using current electrode spacings of 40 km (up to 100 in places), they mapped the geoelectric stratigraphy identifying two distinct less -resistive layers: the magnetite bearing Upper Zone (~400 ohm-m) and a graphitic shale (<10 ohm-m) in the underlying Silverton Formation of the Transvaal sediments (Fig. 9.5). Based on these large-scale resistivity soundings across the Bushveld Complex, they argued that mafic rocks and even the Transvaal lithologies are absent from the center concluding that an inward dipping sheet model satisfies both the gravity and resistivity data better than the Cousins's synformal intrusive model. However, an alternative model (Webb, 2009) suggests that a connected Bushveld Complex also fits the resistivity data if realistic thicknesses are used and the connection is deeper.

While the conceptual change from a separated synformal intrusive model to one involving dipping sheets that thin toward the center appears to be minor, it impacted exploration. The feeders of the dipping sheet model would be too thin and deep to mine. The dipping sheet model, summarized by Eales et al. (1993), dominated for a decade, although a number of papers continued to suggest locations for Cousins's elusive feeders for the synformal intrusive model (Viljoen, 1999; Sharpe, et al. 1980; Sharpe et al., 1981).

From 1959 until 1998, all the Bushveld Complex conceptual models based on gravity data prohibited the presence of mafic rocks in the center. However, all gravity modeling is inherently ambiguous, meaning that there are many variations of models that can fit the data equally well. This is well illustrated by the fact that the gravity values calculated for the synformal intrusive model of Cousins match the gravity data values as effectively as the calculations for the dipping sheet model of Meyer and de Beer (1987), and DuPlessis and Kleyvegt (1987). In fact, the dipping sheet models were inspired by considering new data provided by outcrops in the case of Du Plessis and Keyvegt, and deep resistivity soundings in the case of Meyer and de Beer. Thus, gravity models can only be used to reject specific models; they never prove a particular model. The dipping sheet model is an early example of the importance of incorporating additional constraints to gravity modeling, emphasizing the requirement that all data, including geological constraints, must be equally satisfied.

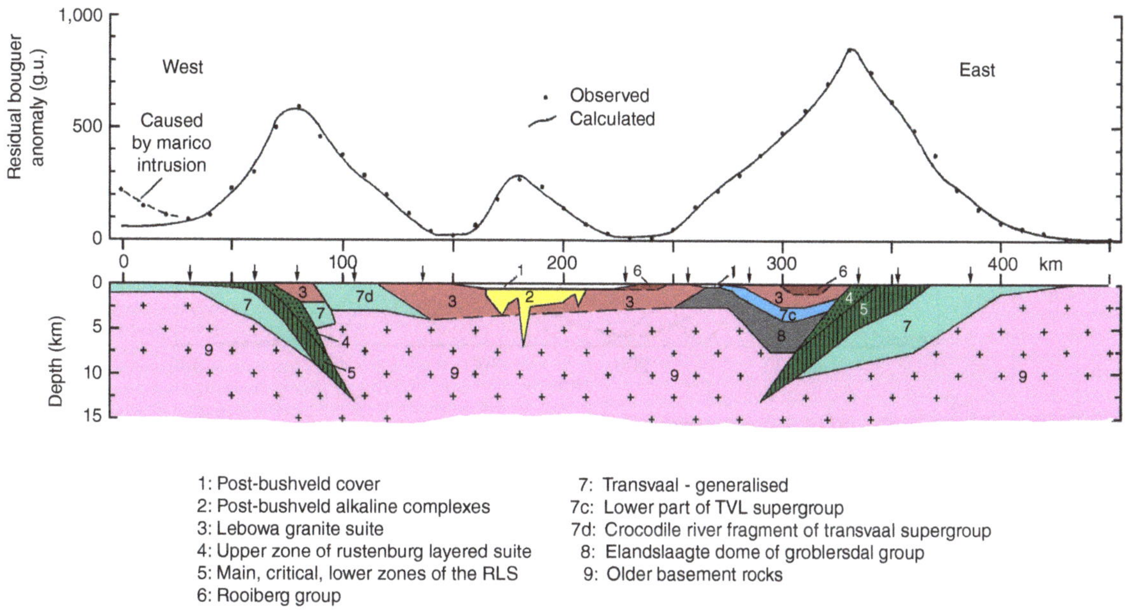

Figure 9.5 The dipping sheet conceptual model as proposed by Meyer and de Beer (1987), similar to that proposed by Du Plessis and Kleywegt (1987). This conceptual model moves the feeders back to the center as thin, unresolvable dykes that would extend vertically downward from the tips of the sheets. It ignores the geological similarities of the outcrops and moves the feeder zones away from the outcrop regions (from Meyer and de Beer, 1987). Published with the permission of Nature (Springer).

9.2.4. The Lopolith Once Again: Flexure and Regional Scale Deep Geophysical Data

In 1998, the lopolith model was reexamined by Cawthorn et al. (1998) in a proof of concept paper that incorporated crustal flexure at the Moho due to the weight of the intruded Bushveld Complex and overturned the accepted dipping sheet model (Fig. 9.6). They demonstrated that a lopolith model could fit the gravity data as well as all other proposed models, by allowing the Moho to flex. This creates interfering gravity anomalies. The high gravity values at the edges are generated by the dense, outcropping mafic lithologies. The lopolith connecting the Western and Eastern Limbs produces a high gravity value in the center, as noticed first by Cousins (center panel Fig. 9.4). However the flexing of the Moho directly counteracts this, as it produces a gravity anomaly with low values of similar amplitude and wavelength. In this preliminary model, the amount of flexure was estimated with simple isostatic calculations and was inspired by the work of Watts (2001) pointing out the necessity of incorporating flexure for studies of sources larger than 50 km in extent. By adding the high gravity values produced by the lopolith to the low values produced by flexing the Moho, the resulting gravity model fits the data as well as all previous models.

This lopolith with flexed Moho idea was further developed by incorporating geological evidence for connection combined with more detailed gravity modeling based on outcrop constraints and essential physical property data from the BV-1 borehole (Ashwal et al., 2005) in Cawthorn and Webb (2001). Both of these gravity models propose a lopolith conceptual model and predict a ~6 km thicker crust in the center of the Bushveld Complex.

The prediction of a thicker crust due to a deeper Moho is a testable prediction and the Southern African Seismic Experiment (SASE) of the Kaapvaal Project provided the data to test this model (James et al., 2001; Nguuri et al., 2001). The Kaapvaal Project deployed seismic stations across southern Africa from April 1997 to July 1999 and these data continue to be incorporated into global, regional, and local studies. The first crustal thickness results (Nguuri et al., 2001) and crustal velocity models for P and S waves (Nguuri, 2002) were incorporated into a gravity model by Webb et al. (2004). This model confirmed that the lopolith model with a flexed Moho satisfies the gravity data, crustal thickness values, and crustal velocity constraints. The crustal thickness determined from the seismic data corresponds to that originally predicted in Cawthorn et al. (1998) (Fig. 9.7). These results also show thickened crust beneath the Northern Limb, implying that this limb extends much farther west from the mapped outcrop. The slight fold modeled in Rustenburg Layered Suite introduced in this model accounts for the elevated gravity values in the center. While this idea is poorly constrained, the implication is clear; this could be a target worth exploring.

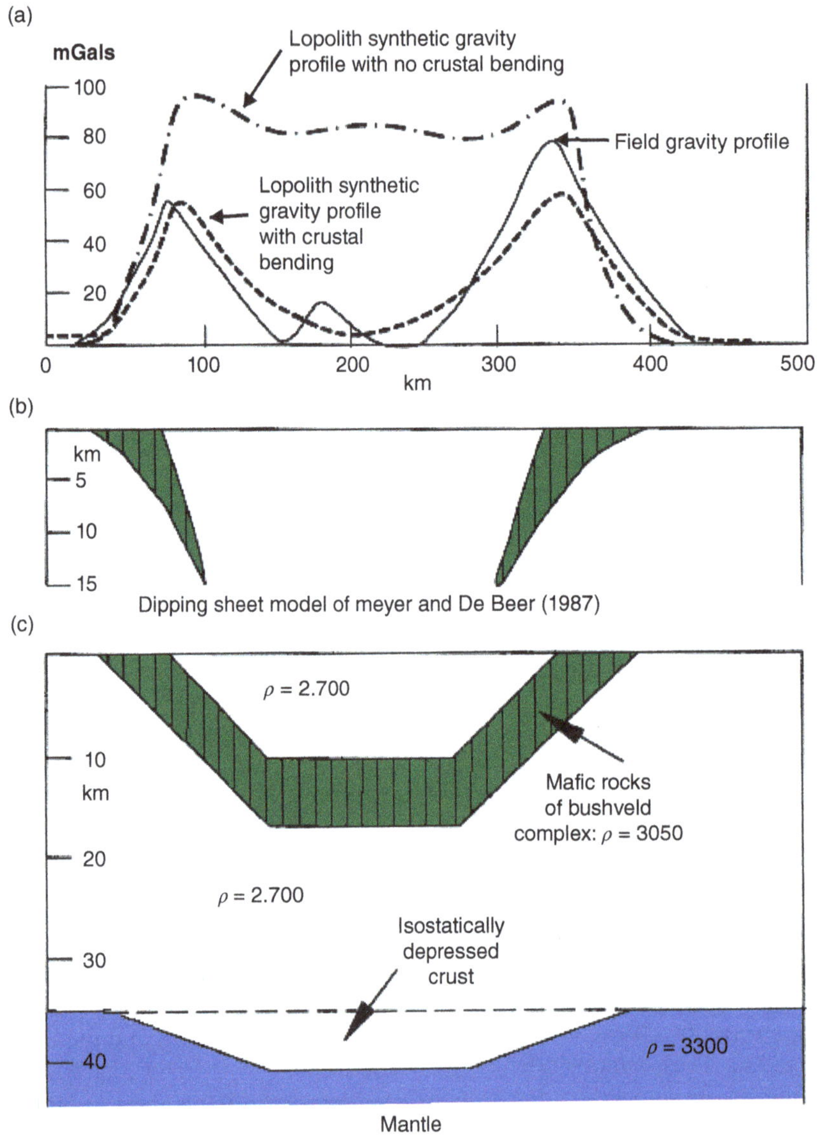

Figure 9.6 The proof of concept model demonstrating a lopolith can satisfy the gravity data if the crust is allowed to flex into the mantle at the Moho. (a) The observed gravity data are the solid line; the dot-dash line is the model calculated from a lopolith only that reproduces Cousins (1959) calculations. The dashed line is the calculated gravity values for the model combining lopolith calculations with crustal flexure calculations. (b) the dipping sheet model of Meyer and Beer (1987) and Du Plessis and Levitt (1987) also produces gravity values shown by the dashed line in (a). In (c) we illustrate a lopolith such as Cousins modeled, which produces the dot-dash line in (a), which clearly does not fit the observed gravity data. However adding the gravity anomaly produced by the isostatically depressed crust to the lopolith model results in a good fit to the data (dashed line in top panel) (from Cawthorn et al, 1998). Published with the permission of the South African Journal of Geology.

The novel suggestion of looking for Bushveld Complex crustal xenoliths in the Palmeitgat kimberlite as a test of the connectivity was provided by Edgar Stettler in a manuscript review. This kimberlite cluster consists of six pipes and is located in the center of the Bushveld Complex (Fig. 9.1). The pipes were emplaced in the late Cretaceous (~100 Ma) through the Karoo Supergroup (180 Ma) and are mined for diamonds. During a research visit to the site, five small (~4 cm across) pyroxenitic rocks were recovered. These samples were found in the high-density concentrate waste discarded by the Sortex machine during diamond recovery. Using textures, petrographic analysis, and mineral compositions, these samples were determined to be chromite-bearing feldspathic pyroxenites of the Upper Critical Zone of the Bushveld Complex (Webb et al., 2011). This discovery

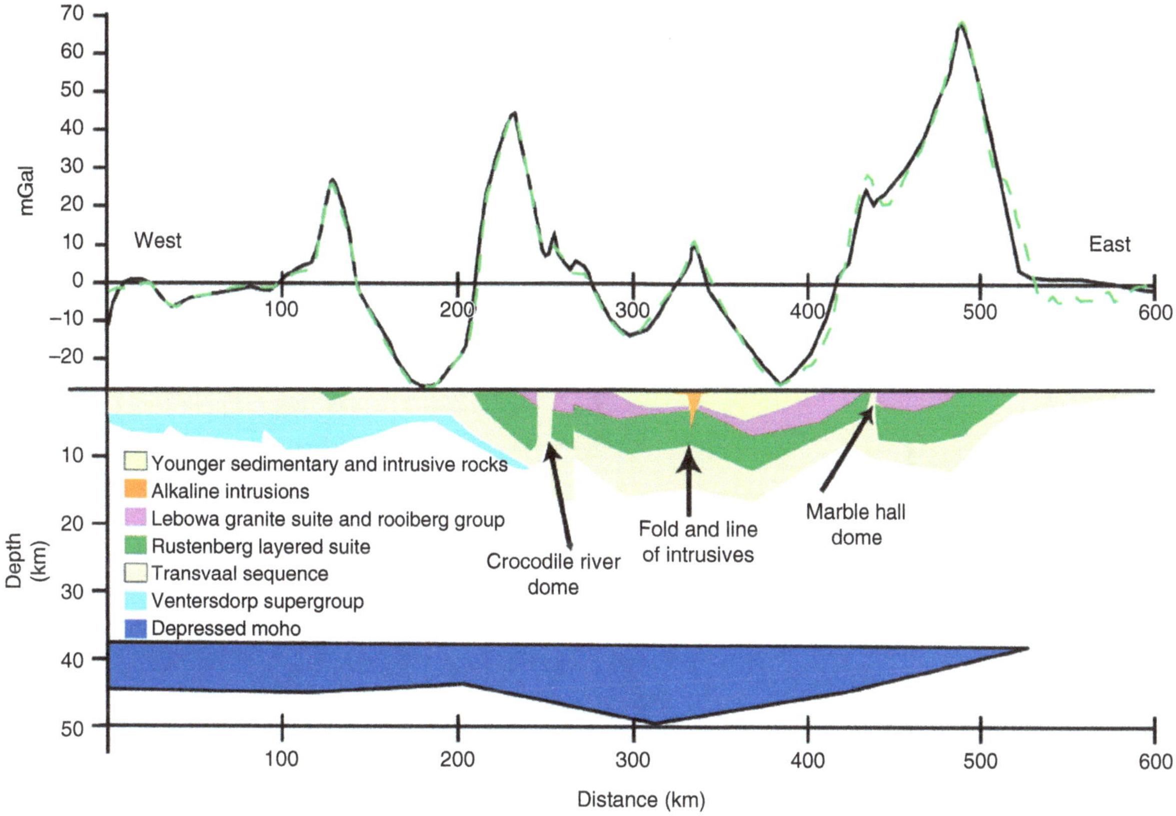

Figure 9.7 The 2.5D gravity model of the Bushveld Complex that incorporates measured crustal thickness and constrains from P and S crustal velocity models. In this model, an alkaline intrusion in the center of the Bushveld Complex is modeled, which aligns with other known alkaline complexes. This does not account for the lateral extent of the elevated gravity values in the center, which are modeled as a slight fold in the Rustenburg Layered Suite. The depth of this fold is poorly constrained by gravity modeling and it suggests a promising exploration target because it brings potentially mineralized rocks closer to the surface (from Webb et al., 2004). Published with the permission of the South African Journal of Geology.

provides convincing evidence for connectivity between the Western and Eastern Limbs at the stratigraphic level of the well-mineralized Upper Critical Zone and generated some exploration activity.

Since the publication of the first crustal thickness model derived from teleseimic data (Nguuri et al., 2001), other crustal thickness models have been produced from these data using different methods as algorithms were improved (for example Nair et al., 2006; Kgaswane et al., 2009, 2012; Youssof et al. 2013; Delph & Porter, 2015). All of the crustal thickness models determined from seismic data show thickened crust underneath the central and northern parts of the Bushveld Complex (Fig. 9.8). This robust result provides strong support for the lopolith model, and requires all future models to incorporate the contribution of thickened crust in their gravity calculation. The fact that models for the Northern Limb also demonstrate thickened crust suggests that this limb extends far to the west of its outcrop area and is a viable exploration target.

9.3. THE BUSHVELD COMPLEX MODEL IN 3D: CONSISTENT GEOLOGY AND GEOPHYSICS

All previous gravity models of the Bushveld Complex have been limited to 2D or 2.5D, which is a simplification for visualization and to shorten computing time. However, this processing methodology is not strictly appropriate in the case of the Bushveld Complex and impacts the calculated gravity field with incorrect amplitudes due to the assumed 2D structure. With the development of faster computers and new algorithms, full 3D modeling is becoming standard (Cole, 2017). The first 3D gravity models of the Bushveld Complex investigated the effect of a range of published crustal thickness models. These broad-scale regional models reveal that different crustal thickness models will affect the resulting thickness and geometry of the Bushveld Complex (Fig. 9.9) (Cole et al., 2014; Cole, 2018). This important result emphasizes the value of crustal scale studies, and points to regions that warrant consideration as exploration targets.

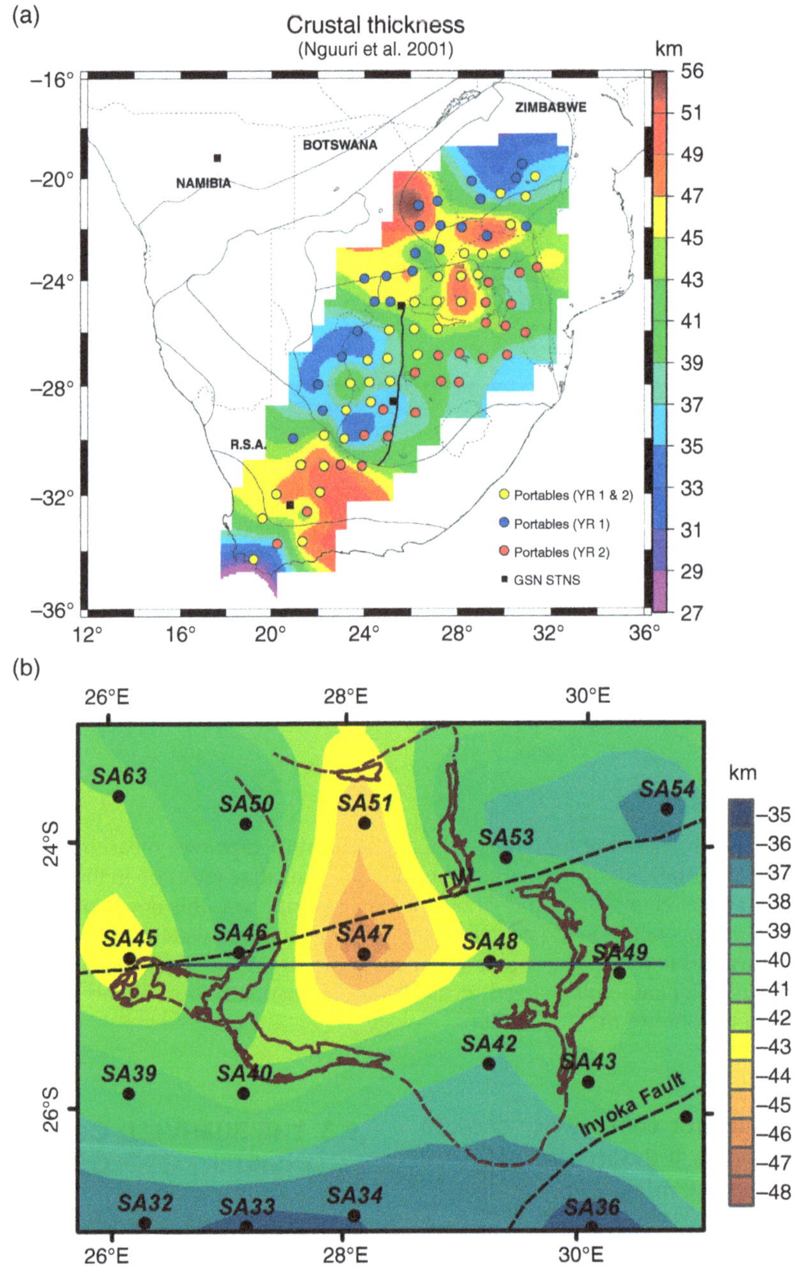

Figure 9.8 Crustal thickness models determined using teleseismic data from the Kaapvaal project. (a) The initial crustal thickness results produced by Nguuri et al. (2001) for the whole of South Africa (from Webb (2009)). Tectonic provinces are outlined. (b) Average of five crustal thickness models (Nguuri et al., 2001; Nair et al., 2006; Kgaswane et al., 2009, 2012; Youssof et al., 2013; Delph & Porter, 2015) for the Bushveld Complex and the surrounding areas (from Cole (2018)). The outline of the complex is marked and dashed lines indicate faults.

A detailed 3D model of the Northern Limb of the Bushveld Complex that includes crustal thickness reveals ~4 km thickness of flat-lying layered mafic rocks, under ~2 km of Waterberg Group sediments (Fig. 9.1) (Finn et al., 2015). Magnetotelluric data from the Southern African Magnetotelluric Experiment (SAMTEX) were used as an additional constraint to the modeling (Jones et al., 2009; Khoza et al., 2013). This integrated geophysical model also demonstrated the continuation of the Transvaal Supergroup under most of the Northern Limb. These results imply that exciting exploration targets might exist across the Northern Limb, including the extension of the mineralized Platreef ~200 km westward, and a richly mineralized proposed feeder zone west of Polokwane (Figs. 9.1 and 9.10).

More complex 3D models of the central and south-eastern sections of the Bushveld Complex have been created

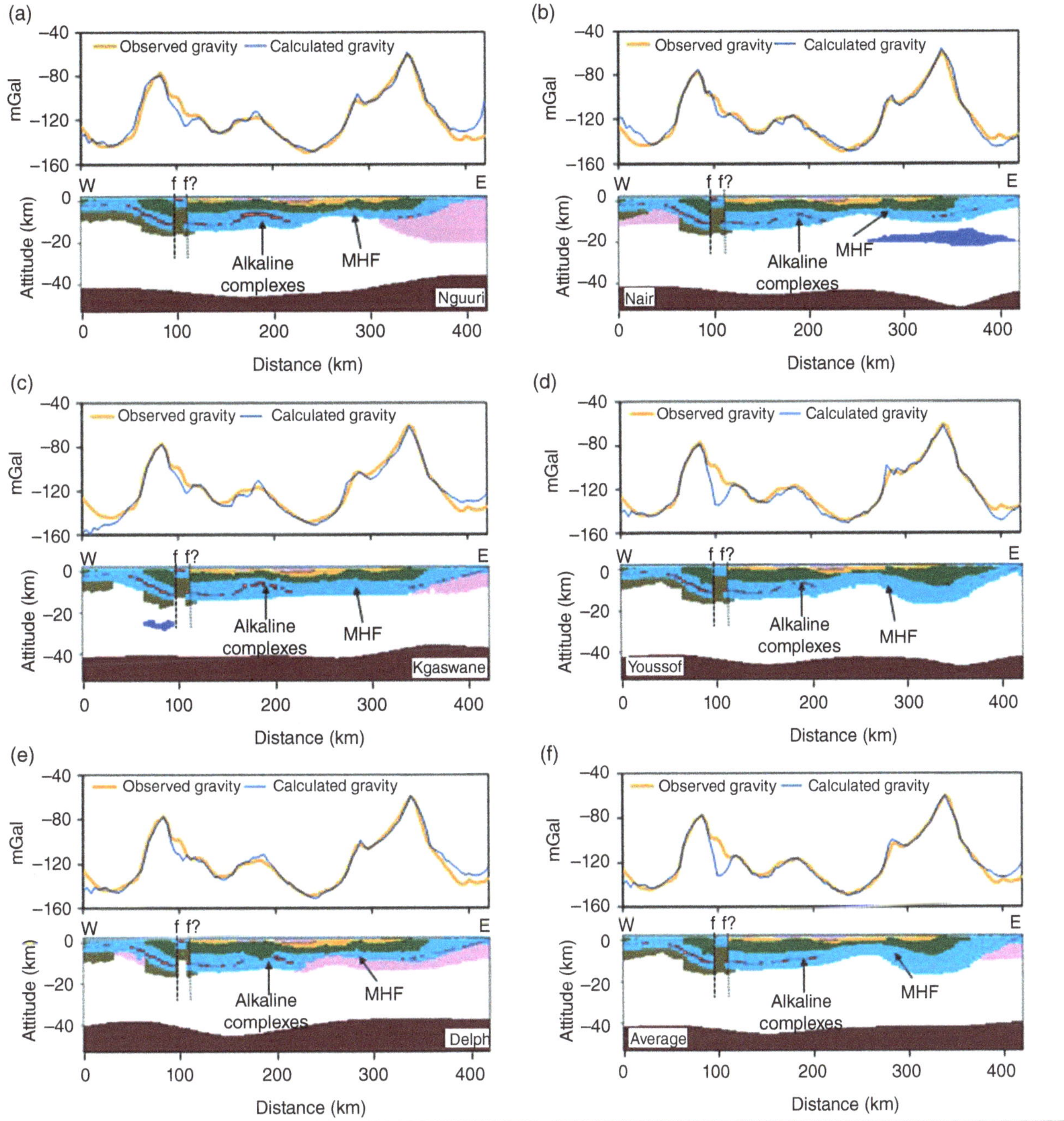

	Lithology	Density (kg/m^3)
	Karoo basalt	2767
	Karoo sediments	2506
	Rooiberg group felsite	2725
	Granite/granophyre	2620
	Rustenbutg layered suite	3009

	Lithology	Density (kg/m^3)
	Transvaal supergroup – BIF (penge Fm)	3178
	Transvaal supergroup	2737
	Witwatersrand/ventersdorp supergroup	2744
	Dense material	3100
	Lower density archaean granite	2580
	Background	2700
	Upper mantle	3375

Figure 9.9 Models extracted from simple 3D gravity models along the East-West profile shown in Figure 9.8b (from Cole (2018)). The models incorporate crustal thickness from several studies: (a) Nguuri et al., (2001), (b) Nair et al. (2006), (c) Kgaswane et al. (2009, 2012); (d) Youssof et al. (2013); (e) Delph and Porter (2015), and (f) the average crustal thickness model. They demonstrate that different models of crustal thickness affect Rustenburg layered suite (RLS) thickness, although the overall geometry remains relatively similar. MHF – Marble Hall Fragment.

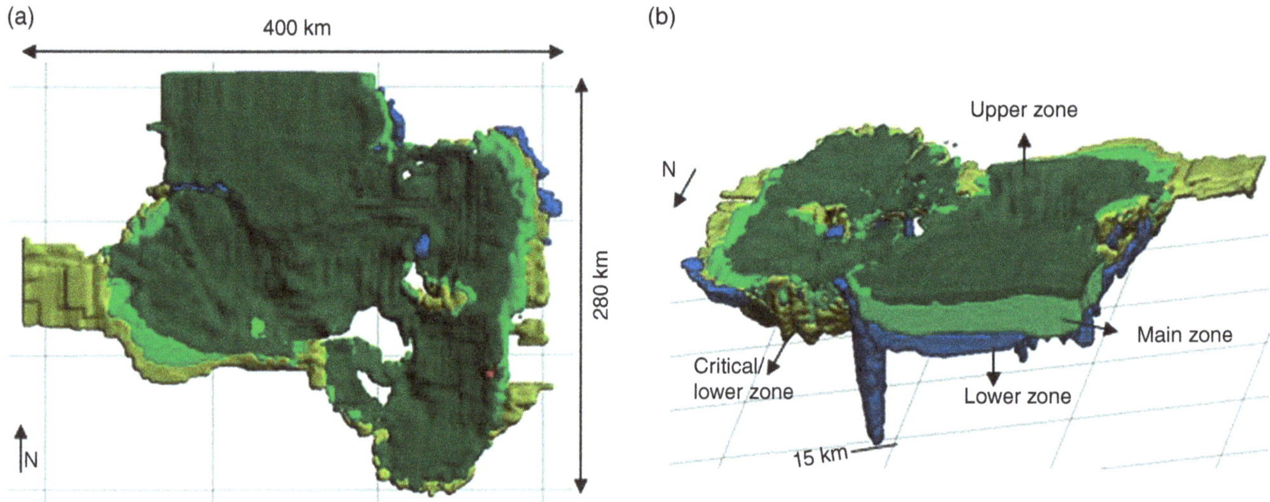

Figure 9.10 Full 3D models of the Bushveld Complex based on joint forward modeling of gravity and magnetic data. (a) Top view of the 3D model of the central Bushveld Complex (only RLS shown); (b) view of the 3D model from the north showing the feeder zone west of Polokwane.

by joint forward modeling of regional gravity and magnetic data (Fig. 9.10) (Cole, 2018). These models are constrained using published crustal thickness, density and susceptibility data, geologic mapping, boreholes, and limited seismic reflection data (Campbell, 1990, 2011; Odgers, 1998; Odgers et al., 1993; Sargeant, 2011). Such models reveal complex geometries, with 5–8 km thick, continuous RLS between the Western and Eastern Limbs and in the southeastern lobe (Cole, 2018). Bushveld granites and younger sediments form a 4–7 km thick cover over the central RLS. The modeling also reveals updomed basement material (underlying Transvaal sediments) in the Eastern Lobe, implying prospective regions on the dome flanks.

9.4. THE PHOSIRI PROJECT AND FURTHER EXPLORATION PROSPECTS

Due to the easily mapped lateral and down dip extentions of the UG2 and Merensky reefs, there has been little need for exploration of the Bushveld Complex interior. However, the model of diapirism proposed by Uken and Watkeys (1997) suggests that these mineralized reefs may be dragged close to surface on the limbs of uplifted Transvaal domes. A few junior exploration companies, which have little access to prospects in the main mining regions, have pursued this idea (Scoon, 2002; Campbell, 2011). Scoon (2002) details successful targeting of the Merensky Reef on the southeastern flank of the Phosiri dome (also called the Zaaikloof dome, Fig. 9.1) in the western regions of the northeastern Bushveld Complex, using a conceptual model of structurally controlled compartments. These different compartments are a result of floor highs. These floor highs affected lateral magma flow, effectively forming sectors or compartments with some lithostratigraphic differences in the Lower and Lower Critical Zones (Clarke et al., 2009; Scoon and Teigler, 1995; Cameron, 1980). He argues that each separate lobe has a distinct stratigraphy; this is similar to Cousins's model, and implies no connection in the central Bushveld Complex. Scoon (2002) documents a complicated sequence of emplacement and diapirism, and although he discusses a magnetic interpretation to locate marker horizons, his model is only constrained by borehole intersections, without satisfying gravity, crustal thickness, or aeromagnetic data.

Campbell (2011, 2007) describe a successful integrated geophysical exploration approach on the northeastern side of the Phosiri dome (a continuation of the Zaaikloof dome). Here he integrated gravity, aeromagnetic, and 2D seismic data to guide the exploration drilling strategy. Using borehole control that was fed back into the 3D seismic and gravity models, a 3D map of the Merensky Reef at depth was produced (Campbell, 2007).

Both of these successful exploration projects implicitly depend on the lopolith model involving crustal flexure. They demonstrate that the majority of the Bushveld Complex stratigraphy continues down dip from outcropping regions, and can be brought to near-surface on the edges of Transvaal diapirs. This suggests that the collection of detailed gravity and seismic data could help delineate further regions for exploration, especially near known Transvaal fragments.

9.5. FINAL SUMMARY

Currently the use of geophysical techniques for mineral exploration is commonplace, but they are often poorly integrated with regional geophysical and geological data sets not set in a tectonic framework. In many geological settings, long-wavelength geophysical signatures mask shorter-wavelength exploration targets. The traditional, textbook solution, for example in gravity modeling, is to remove the regional signal and model the residual. We argue that in many cases, the regional gravity component needs to be modeled because interfering long-wavelength signals may cancel each other out, resulting in very different models. This is especially important where laterally extensive (> 100 km) sources may lie beneath each other; the conventional practice of using vertical exaggeration in geophysical imaging and omitting depth scales on geological cross-section sketches, are two practices that often obscure the importance of deeper contributions. A note of caution should be applied to the use of geophysical inversion methods, especially as applied to gravity data: it is impossible to resolve laterally extensive sources lying beneath each other with inversion alone. Using a sequential method of forward modeling deep sources, and subtracting their contribution, and then applying inversion may be fruitful (Finn et al., 2015). As exploration targets become deeper, the whole crust and even upper mantle contributions are incorporated in target generation. This is a fruitful area for cooperation between academia and industry; examples in southern Africa of crustal scale projects include the Kaapvaal Project (James et al., 2001; Fouch et al., 2004); SAMTEX (Jones et al., 2009; Evans et al., 2011), and the AfricaArray program (Nyblade et al., 2011). All of these crustal scale projects have resulted in models that have contributed to exploration programs; the data are freely available and continue to generate new ideas.

In this review, we have examined the Bushveld Complex in some detail, to explore how large-scale conceptual models have changed and influenced exploration efforts. A fully integrated approach is needed as the exploration environment becomes more challenging; here we support the development of 4D modeling; linking the geometry observed today to the processes that created it. A 4D model of the Bushveld Complex, which examines the 3D geometry through time in a tectonic context, would allow testing of emplacement models and postulated deformation events ensuring they are consistent with large-scale geological observations, rheological response, and age constraints. The rigorous integration of many data sets, including geophysical, mapping, age dating, and structural data, will require innovative data management, integration, and visualization.

While these various models and data all support a model of the Bushveld Complex that is mostly connected at depth, they also suggest a complex 3D structure (Cole, 2018; Finn et al., 2015). The improved resolution of these models has identified a number of potential exploration targets, such as the feeder zone west of Polokwane, the westward extent (~200 km) of the flat Platreef in the Northern Lobe, and the detailed geometry of the connection between the Western Lobe and Eastern Lobe. These targets are, however, all poorly resolved with current technology and potential field modeling. The use of reflection seismic data as an exploration tool is becoming more common, and there is an urgent need for a reflection seismic campaign on which to base further brownfields exploration in the Bushveld Complex. This will allow a better assessment of this major resource. Although targets are likely to be deep, there is an urgent need to determine the 3D structure of the Bushveld Complex with enough lead time to develop the technologies needed to extract these deeper deposits in the future.

REFERENCES

Ashwal, L. D., Webb, S. J., & Knoper, M. W. (2005). Magmatic stratigraphy in the Bushveld Northern Lobe: continuous geophysical and mineralogical data from the 2950 m Bellevue drillcore. *S. Afr. Jour. Geol.*, *108*(2), 199–232.

Campbell, G. (1990). The seismic revolution in gold and platinum prospecting. *South African Geophysical Association Year Book*, 37–45.

Campbell, G. (2007). The Phosiri Prospect: An Example of 2D Seismic Reflection Mapping in the Bushveld. *Proceedings 10th International South African Geophysical Association of South Africa (SAGA) Biennial Conference and Exhibition.* Wild Coast, South Africa

Campbell, G. (2011). Exploration geophysics of the Bushveld Complex in South Africa. *The Leading Edge*, *30*(6), 622–638.

Cawthorn, R. G., & Webb, S. J. (2001). Connectivity between the western and eastern limbs of the Bushveld Complex *Tectonophysics*, *330*, 195–209.

Cawthorn, R. G., Cooper, G. R. J., & Webb, S. J. (1998). Connectivity between the western and eastern limbs of the Bushveld Complex. *S. Afr. Jour. Geol.*, *101*(4), 291–298.

Cole, J. (2018). Three Dimensional Geometry of the Bushveld Complex Derived from Potential Field Modeling. PhD dissertation, School of Geoscience, University of the Witwatersrand.

Cole, J., Webb, S. J., & Finn, C. A. (2014). Gravity models of the Bushveld Complex: Have we come full circle? *Journal of African Earth Sciences*, *92*, 97–118.

Cole, J., Webb, S. J., & Finn, C. A. (2017). A 3D Potential field model of the Bushveld. South African Geophysical Association, 15th Biennial Conference, Cape Town (10–13 September).

Cole, P. (2017). PyGMI: Python Geophysical Modeling and Interpretation. Computer program, Council for Geoscience. Available at: http://patrick-cole.github.io/pygmi/ (Accessed, 3 August 2018).

Council for Geoscience. (1997). Gravity Map of South Africa, 1:1,000,000 Map Series.

Cousins, C. A. (1959). The structure of the mafic portion of the Bushveld Igneous Complex. *Trans. Geol. Soc. S. Afr.*, *62*, 174–189.

Delph, J. R., & Porter, R. C. (2015). Crustal structure beneath southern Africa: Insight into how tectonic events affect the Mohorovičić discontinuity. *Geophys. Jour. Int.*, *200*, 254–264.

Du Plessis, A., & Levitt, J. G. (1987). On the structure of the Rustenburg layered suite: Insight from seismic reflection data. In *Indaba on the Tectonic Setting of Layered Intrusives*, Geological Society of South Africa, Geological Survey of South Africa, University of Pretoria, Institute for Geological Research.

du Toit, A. L. (1954). *The Geology of South Africa*. Oliver and Boyd, Edinburgh.

Eales, H. V., Botha, W. J., Hattingh, P. J., de Klerk, W. J., Maier, W. D., & Odgers, A. T. R. (1993). The mafic rocks of the Bushveld Complex: A review of emplacement and crystallization history, and mineralization, in the light of recent data. *Jour. Afr. E. Sci.*, *16*, 121–142.

Ernst, R. E. (2014). *Large Igneous Provinces*. Cambridge.

Evans, R. L., Jones, A. G., Garcia, X., Muller, M., Hamilton, M., Evans, S., Fourie, C. J. S., et al. (2011). Electrical lithosphere beneath the Kaapvaal craton, southern Africa. *Journal of Geophysical Research*, *116*, B04105. doi:10.1029/2010JB007883.

Finn, C. A., Bedrosian, P. A., Cole, J., Khoza, T. D., & Webb, S. J. (2015). Mapping the 3D extent of the Northern Lobe of the Bushveld layered mafic intrusion from geophysical data. *Precambrian Research*, *268*, 279–294.

Fouch, M. J., James, D. E., VanDecar, J. C. Lee, S. V. D., & Group, K. S. (2004). Mantle seismic structure beneath the Kaapvaal and Zimbabwe Cratons. *S. Afr. Jour. Geol.*, *107*, 33–44.

Hales, A. L., & Gough, D. I. (1962). The gravity survey of the Republic of South Africa. *Handbook 3. Part II: Isostatic Anomalies and Crustal Structure*. Government Printer, Pretoria.

Hall, A. L. (1932). The Bushveld igneous complex of the central Transvaal. *Geological Survey of South Africa Memoir*, *28*.

Hattingh, P. J. (1980). The structure of the Bushveld Complex in the Groblersdal-Lydenburg-Belfast area of the Eastern Transvaal as interpreted from a regional gravity survey. *Transactions of the Geological Society of South Africa*, *83*, 125–133.

James, D. E., Fouch, M. J., VanDecar, J. C., van der Lee, S., & Group, K. S. (2001). Tectospheric structure beneath southern Africa. *Geophys. Res. Let.*, *28*, 2485–2488.

Jones, A. G., Evans, R. L., Muller, M. R., Hamilton, M. P., Miensopust, M. P., Garcia, X., Cole, P., et al. (2009). Area selection for diamonds using magnetotellurics: Examples from southern Africa. *Lithos*, *112S*, 83–92.

Kgaswane, E. M., Nyblade, A. A., Julia, J., Dirks, P. H. G. M., Durrheim, R. J., & Pasyanos, M. E. (2009). Shear wave velocity structure of the lower crust in southern Africa: Evidence for compositional heterogeneity within Archaean and Proterozoic terrains. *Journal Geophysical Research*, *114*, B12304, 1–19. doi: 10.1029/2008JB006217.

Kgaswane, E. M., Nyblade, A. A., Durrheim, R. J., Julià, J. P., Dirks, H. G. M., & Webb, S. J. (2012). Shear wave velocity structure of the Bushveld Complex, South Africa. *Tectonophysics*, *554*, 83–104.

Khoza, T. D., Jones, A. G., Muller, M. R., Evans, R. L., Webb, S. J., Miensopust, M. P., et al. (2013). Tectonic model of the Limpopo belt: Constraints from magnetotelluric data. *Precambrian Research*, *226*, 143–156.

Kinloch, E. D. (1982). The regional trends in platinum group mineralogy of the Critical Zone of the Bushveld Complex, South Africa. *Econ. Geol.*, *77*, 1328–1347.

Lee, C. A. (1996). A review of mineralization in the Bushveld Complex and some other layered mafic intrusions. In R.G. Cawthorn (Ed.), *Layered Intrusions, Developments in Petrology* (pp. 103–145). *15*, Elsevier.

Maré, L. P., & Tabane, L. R. (2004). Physical properties of South African rocks. *South African Geophysical Atlas*, *4*. Council for Geoscience, 40.

van der Merwe, M. J. (1976). The layered sequence of the Potgietersrus limb of the Bushveld Complex. *Econ. Geol.*, *71*, 1337–1351.

Meyer, R., & Beer, J. H. D. (1987). Structure of the Bushveld Complex from resistivity measurements. *Nature*, *325* (6105), 610–612. doi:10.1038/325610a0.

Molengraaff, G. A. F. (1901). Géologie de la République Sud Africaine du Transvaal. In *Bulletin de La Société Géologique de France*, *4* Série Tome. English translation by J. H. Ronaldson; published in 1904 by Esson and Perkins, Johannesburg, with addition and alterations by the author, 13–92.

Molyneux, T. G., & Klinkert, P. S. (1978). A structural interpretation of part of the eastern mafic lobe of the Bushveld Complex and its surrounds. *Trans. Geol. Soc. S. Afr.*, *81*, 359–368.

Nair, S. K., Gao, S. S., Liu, K. H., & Silver, P. G. (2006). Southern African crustal evolution and composition: Constraints from receiver function studies. *Journal of Geophysical Research*, *111*, B02304.

Nguuri, T. (2002). Crustal Structure of the Kaapvaal Craton and Surrounding Mobile Belts: Analysis of Teleseismic P-Waveforms and Surface Wave Inversions. PhD thesis, University of the Witwatersrand.

Nguuri, T. K., Gore, J.,. James, D. E, Webb, S. J., Wright, C., Zengeni, T. G., Gwavava, O., et al. (2001). Crustal structure beneath southern Africa and its implications for the formation and evolution of the Kaapvaal and Zimbabwe cratons. *Geophysical Research Letters*, *28*, 2501–2504.

Nyblade, A. A., Durrheim, R., Dirks, P., Graham, G., Gibson, R., & Webb, S. J. (2011). Geoscience initiative develops sustainable science in Africa, *EOS*, *92*, 161–162.

Odgers, A. T. R. (1998). Structure of the Southern Bushveld Complex as determined from region reflection seismics. *Southern African Geophysical Review*, *2*, 79–82.

Odgers, A. T. R., Hinds, R. C., & von Gruenewaldt, G. (1993). Interpretation of a seismic reflection survey across the southern Bushveld Complex. *South African Journal of Geology*, *96* (4), 205–212.

Sargeant, F. (2001). The Seismic Stratigraphy of the Bushveld Igneous Complex, South Africa. PhD dissertation, The University of Liverpool.

Scoon, R. N. (2002). A new occurrence of Merensky Reef on the flanks of the Zaaikloof Dome, Northeastern Bushveld Complex: Relationship between diapirism and magma replenishment. *Economic Geology*, *97*, 1037–1049.

Sharpe, M. R., & Snyman, J. A. (1980). A model for the emplacement of the eastern compartment of the Bushveld Complex. *Tectonophysics*, *65*, 85–110.

Sharpe, M. R., Bahat, D., & Von Gruenewaldt, G. (1981). The concentric elliptical structure of feeder sites to the Bushveld Complex and possible economic implications. *Geol. Soc. South Africa Trans.*, *84*, 239–244.

Stettler, E. H., Fourie, C. J. S., & Cole, P. (2000). Total magnetic field intensity map of the Republic of South Africa (in 4 panels). Council for Geoscience, Pretoria, South Africa.

Stettler, E. H., Fourie, C. J. S., Bühmann, J. R., Hattingh, E., Cole, P., Kleywegt, R. J., et al. (1999). Magnetics. *South African Geophysical Atlas.* Volume 2. Council for Geoscience, South Africa, 2, CD Rom.

Uken, R., & Watkeys, M. K. (1997a). Diapirism initiated by the Bushveld Complex, South Africa. *Geology*, *25*, 723–726.

Venter, C. P.,. du Plessis, J. G., Stettler, R. H., Potgieter, T. D., Kleywegt, R. J., Hattingh, E., et al. (1999). Gravity, *South African Geophysical Atlas*, Volume 1, Council for Geoscience, South Africa, 1.

Viljoen, M. J. (1999). The nature and origin of the Merensky Reef of the western Bushveld Complex based on geological facies and geophysical data. *S. Afr. Jour. Geol.*, *102*(3), 221–239.

Wagner, P. A. 1929). *The Platinum Deposits and Mines of South Africa.* Oliver and Boyd, Edinburgh.

Watts (2001). *Isostasy and Flexure of the Lithosphere.* Cambridge University Press, Cambridge.

Webb, S. J. (2009). The Use of Potential Field and Seismological Data to Analyze rhe Structure of the Lithosphere Beneath Southern Africa. PhD thesis, University of the Witwatersrand. http://wiredspace.wits.ac.za/handle/10539/8003.

Webb, S. J., Ashwal, L. D., & Cawthorn, R. G. (2011). Continuity between eastern and western Bushveld Complex, South Africa, confirmed by xenoliths from kimberlite. *Contributions to Mineralogy and Petrology*, *162*(1), 101–107.

Webb, S. J., Nguuri, T. K., Cawthorn, R. G., & James, D. E. (2004). Gravity modeling of Bushveld Complex connectivity supported by southern African seismic experiment results. *S. Afr. Jour. Geol.*, *107*(1/2), 207–218.

Weiss, O. (1940). Gravimetric and Earth-Magnetic measurements on the great dyke of Southern Rhodesia, *Trans. Geol. Soc. S. Afr.*, *43*, 143–152.

Youssof, M., Thybo, H., Artemieva, I. M., & Levander, A. (2013). Moho depth and crustal composition in Southern Africa. *Tectonophysics*, *609*, 267–287.

10

A Summary of Some Recent Developments in Potential Field Data Processing in South Africa with Mining and Exploration Applications

G. R. J. Cooper

ABSTRACT

Potential field data sets are widely used in South Africa, both for exploration and mining. Hence, considerable research has taken place to improve the way in which the data are processed so as to extract the maximum information from them. The research areas fall into different categories: image processing filters, such as sunshading or the Tilt Angle, data transformations, such as pole reduction or vertical continuation, and semiautomatic interpretation techniques, such as Euler Deconvolution or the Tilt-Depth method. Recent developments in these areas are summarized and discussed.

10.1. INTRODUCTION

South Africa is fortunate to posses many mineral deposits of considerable economic importance, the most significant of which are the Witwatersrand basin (source of much of the world's gold), and the Bushveld igneous complex (where 80% of the world's platinum originates; Eales, 2001). The use of geophysical methods in exploration was recognized as early as the 1920s (De Beer, 2015, p.54). The Geological Survey of South Africa (now the Council for Geoscience) acquired regional aeromagnetic (mostly at 1 km line spacing, with some at 200 m line spacing) and gravity data (more than 120,000 stations) over the whole country from the 1950s to the 1990s, and various mining and exploration companies commissioned higher resolution surveys over specific areas of interest.

As computer technology became ever cheaper and more powerful, digital processing and storage of geophysical data replaced hardcopy maps, and old pen and paper rules of thumb methods of estimating the depths of geologically simple bodies (such as dykes and contacts) were replaced by sophisticated mathematical techniques. The display of data on a computer screen allowed the manipulation of the image in many ways, for example to enhance weak (but geologically significant) anomalies. There is a large amount of well-known transformations of potential field data possible (e.g., vertical continuation, or pole reduction; Baranov, 1957) and these methods are all employed routinely. However, under certain circumstances, they do not work well and therefore had to be modified. Finally, because large volumes of aeromagnetic data, which can be acquired rapidly using modern equipment, aids to interpretation were necessary. Semiautomatic interpretation methods provide an initial estimate of source parameters (such as location and depth, sometimes also dip and susceptibility), which can then be refined by detailed modeling and inversion. Recent research on several such methods are discussed below.

School of Geosciences, University of the Witwatersrand, Johannesburg, South Africa

10.2. IMAGE PROCESSING TECHNIQUES

10.2.1. Sunshading

Sunshading is a commonly used image enhancement technique that produces an amplitude balanced horizontal derivative of the data. A "sun" is considered to be illuminating the data from a particular azimuth and

Ore Deposits: Origin, Exploration, and Exploitation, Geophysical Monograph 242, First Edition.
Edited by Sophie Decrée and Laurence Robb.

elevation, and the reflected light is calculated. Different reflectance models can be used. If the surface is a Lambertian reflector, then it reflects all light incident upon it and reflects the light equally in all directions. The reflectance is given by (Horn, 1982):

$$R = \frac{1 + p_0 p + q_0 q}{\sqrt{1 + p^2 + q^2} \cdot \sqrt{1 + p_0^2 + q_0^2}} \tag{10.1}$$

where $p_0 = -\cos \Phi \tan \theta$, $q_0 = -\sin \Phi \tan \theta$, θ is the sun elevation (measured from the vertical), and Φ is the azimuth (measured anticlockwise from east). The p and q are two orthogonal horizontal gradients of the data. Illuminating the data from an azimuth φ diminishes features with that azimuth and enhances features at 90° to φ. The method is primarily used to enhance linear features such as faults or contacts that have a particular azimuth. Cooper (2003) modified the algorithm so that instead of the geophysicist choosing the filter azimuth, a central location in the data was chosen. Equation (10.1) then was applied in a radial manner so that the filter azimuth became a variable. The azimuth used at the pixel (i,j) is then calculated from

$$\Phi_{i,j} = \tan^{-1}(\Delta y_{i,j} / \Delta x_{i,j}) \tag{10.2}$$

where Δx and Δy are the distances EW and NS between the chosen center location and the location of each pixel in the image. Thus Φ is a radius vector of any circle centered on the chosen point. The result was to enhance near-circular features of any radius (such as meteorite impact craters or kimberlite pipes) that were centered on the chosen point. Figure 10.1 shows an application of the method to gravity data over the Witwatersrand basin and Bushveld complex, South Africa.

Figure 10.1b shows the effect of sunshading the data from the South; features with azimuths that lie near EW have been enhanced, but those with near NS azimuths have been diminished. Figure 10.1c shows the result of applying radial sunshading centered on the Vredefort dome meteorite impact structure. Almost the whole outline of the dome has been enhanced, as has the Witwatersrand basin that surrounds it. If the filter center is moved to the center of the Pilanesberg complex (Fig.10.1d), then both it and the eastern limb of the Bushveld complex are now enhanced.

10.2.2. The Tilt Angle

Geophysical data sets often contain anomalies of comparatively small amplitude, which are nevertheless of geological importance. Various Automatic Gain Control (AGC) methods can be employed to make them appear more prominent. The simplest AGC method involves dividing each data value by an average value computed in a moving window (Rajogopalan & Milligan, 1995). However the result is strongly dependent on the window size, and the shapes of anomalies are distorted. The Tilt Angle (Miller & Singh, 1994) is an amplitude balanced first vertical derivative of the data, which can produce excellent results (see Fig.10.2b). It is defined as

$$T = \tan^{-1}\left(\frac{\frac{\partial f}{\partial z}}{\sqrt{\left(\left(\frac{\partial f}{\partial x}\right)^2 + \left(\frac{\partial f}{\partial y}\right)^2 \right)}} \right) \tag{10.3}$$

where f is the potential field data.

Vertical continuation is a mathematical technique that takes measured potential field data and computes from it the data that would have been measured at a different altitude (Blakely, 1995, p. 320). For example, data measured by an aircraft can be downward continued to the Earth's surface to enable it to be compared with ground surveys. Downward continuation enhances detail but is very sensitive to noise, while upward continuation is a commonly used smoothing technique for noisy data. In the frequency domain it is given by,

$$A'(u,v) = A(u,v)e^{-z\sqrt{u^2+v^2}} \tag{10.4}$$

where $A(u,v)$ is the Fourier coefficient of the potential field anomaly at east-west and north-south frequencies u and v, and z is the difference in elevation between the altitude at which the data were measured and the new altitude. While the Tilt Angle cannot be vertically continued using equation (10.2) because it is not a harmonic function, it naturally can be applied to data from any altitude. Unfortunately, because it is based on a ratio of derivatives, it is very sensitive to noise and the Tilt Angle of downward continued data is frequently unusable. Cooper (2016) noticed that when applied to magnetic data from vertically magnetized, vertically dipping contacts, the Tilt Angle from contacts at different depths only differed by a constant inside the arctan function, that is,

$$T_\alpha = \tan^{-1}\left(\frac{\alpha \frac{\partial f}{\partial z}}{\sqrt{\left(\left(\frac{\partial f}{\partial x}\right)^2 + \left(\frac{\partial f}{\partial y}\right)^2 \right)}} \right) \tag{10.5}$$

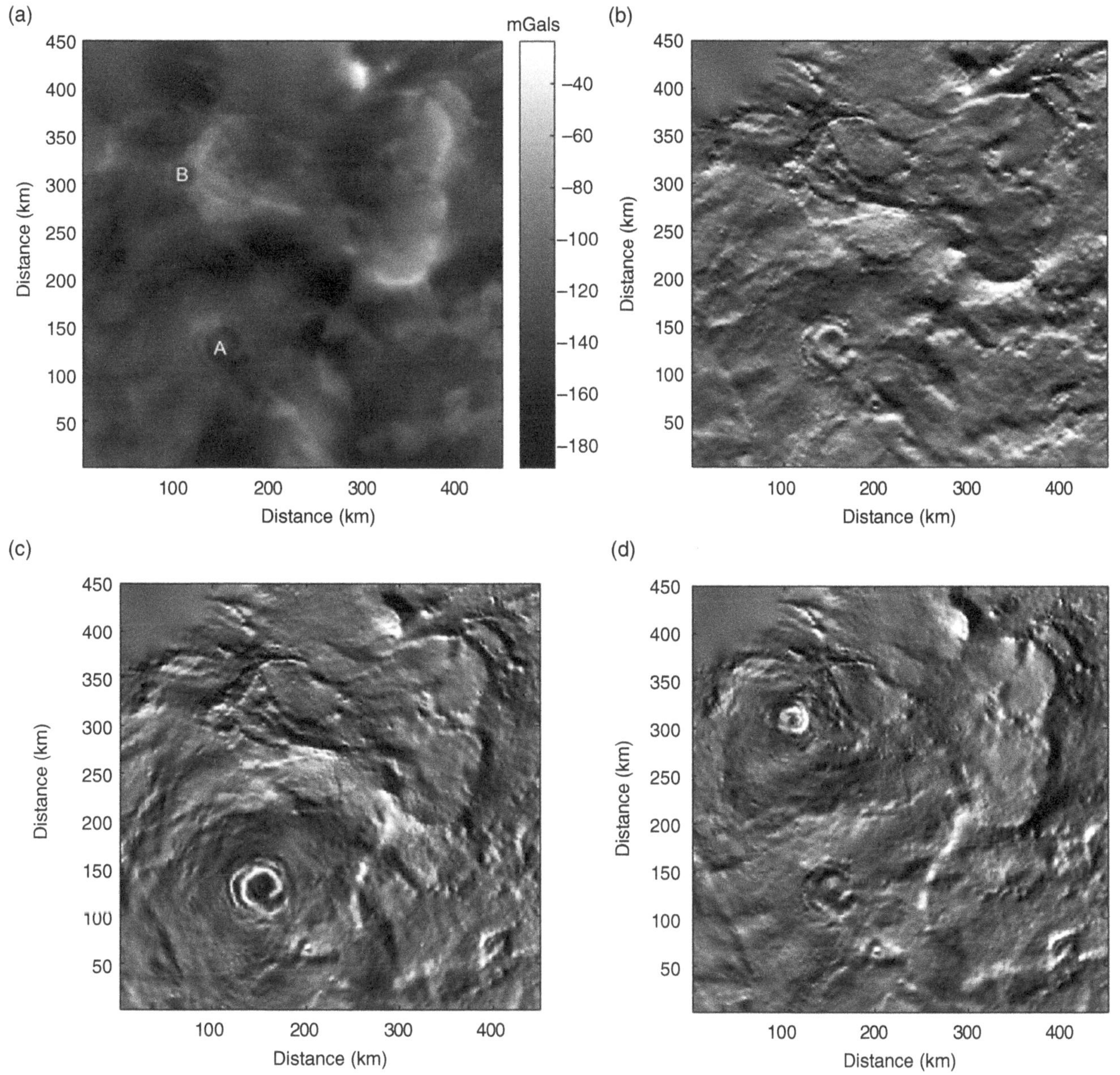

Figure 10.1 (a) Gravity data over the Witwatersrand basin (lower left) Bushveld Complex (upper middle) of South Africa. The grid interval is 1 km. (b) Gravity data from (a) sunshaded from the south using an elevation angle of 10°. (c) Gravity data from (a) sunshaded radially from the center of the Vredefort dome, marked as A in (a) using an elevation angle of 10°. (d) Gravity data from (a) sunshaded radially from the center of the Pilanesberg complex, marked as B in (a), using an elevation angle of 10°.

So the Tilt Angle over a contact at a depth z can be transformed into that from a contact at a depth z/α by multiplying the vertical derivative term by α. In addition to allowing improved Tilt Angle images to be produced (see Fig.10.2), this reduces interference problems when the Tilt-Depth semiautomatic interpretation method is used (see section 10.3.1). Equation (10.4) downward continues the anomalies from all sources by a fixed distance, but equation (10.5) downward continues the Tilt Angle from magnetic contacts by a fixed factor (α), which means that it is impossible to downward continue the data to or past the source. This makes the method much more stable than equation (10.4), although of course it is not as general in its application.

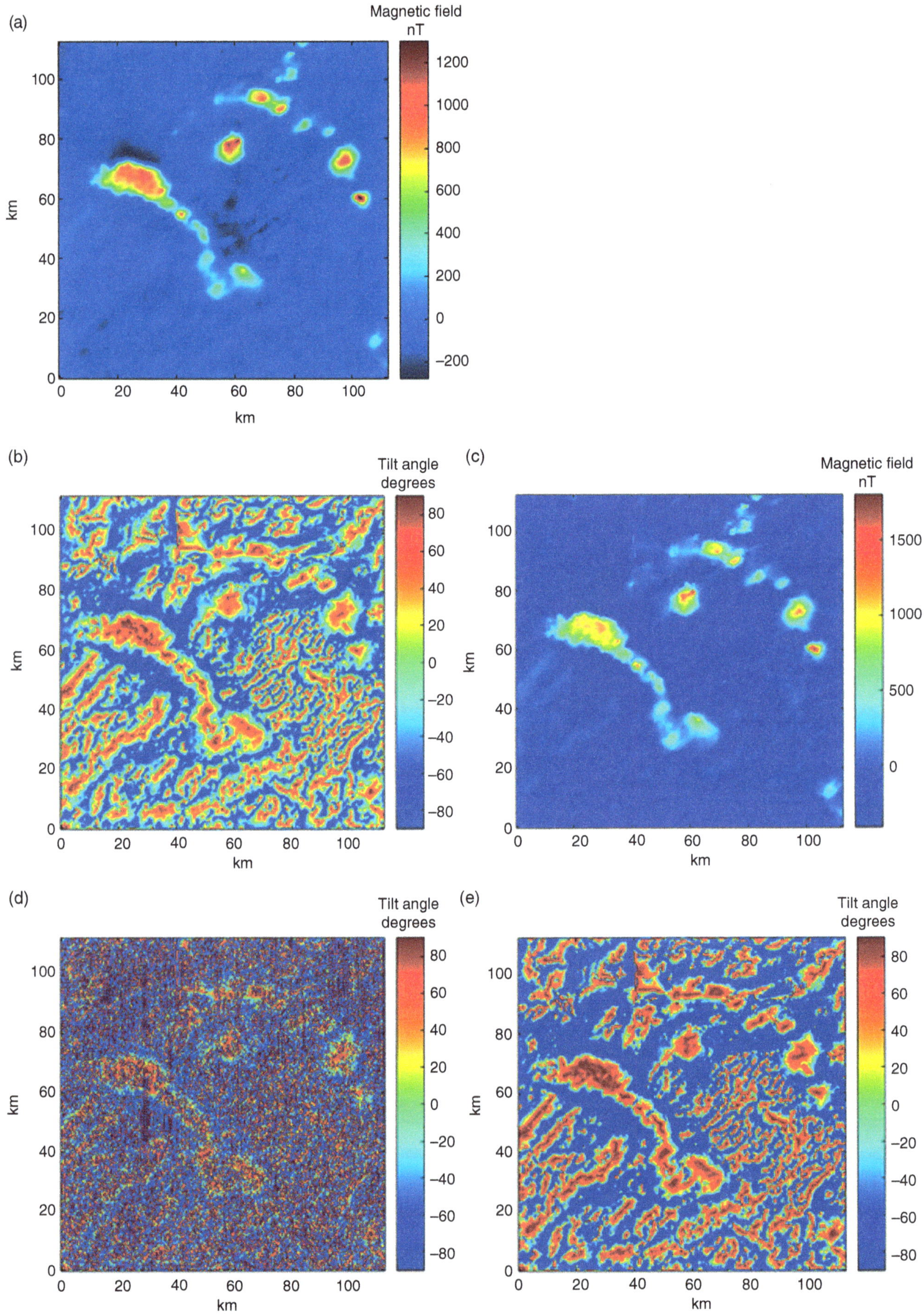

Figure 10.2 (a) Pole reduced aeromagnetic data set from Southern Africa. The grid spacing is 250.0 m. The flight line direction is north-south, the flight height is 100 m, and the line spacing is 1 km. (b) Tilt Angle T of the data from (a). (c) Data from (a) downward continued by 250.0 m (1 sample interval). (d) Tilt Angle T of the data from (c). (e) Tilt Angle T_α of the data from Figure 10.1a, α = 2 (from Cooper, 2016).

10.3. SEMIAUTOMATIC INTERPRETATION TECHNIQUES

10.3.1. The Tilt-Depth and Contact-Depth Methods

Salem et al. (2007) substituted the analytic expressions for the magnetic anomaly over a contact into the Tilt Angle, equation (10.3). When the contact was vertically dipping and vertically magnetized, the expression simplified to

$$T = \tan^{-1}\left(\frac{\Delta x}{\Delta z}\right) \tag{10.6}$$

where Δx and Δz are the horizontal and vertical distances from the current measurement point to the contact. Hence, when $T=45°$, $\Delta x=\Delta z$, and when $T=-45°$, $\Delta x=-\Delta z$. So the depth to the contact could be obtained by measuring (half) the distance between the ±45° contours of the Tilt Angle. The location of the contact is taken as the zero value of the Tilt Angle. Figure 10.3 shows the application of the method to data from southern Africa, and some problems with the practical application of the method become immediately apparent. First, when the geology is complex, it is often not clear which distances should be measured. Second, the method is very computer intensive because it involves measuring the distance between every point on every contour and every point on every other contour. To avoid both these problems, Cooper (2014a) differentiated the contents of the arctan function in equation (10.6), immediately giving the reciprocal of the depth without any need for measuring distances between contours, that is,

$$R = \frac{\frac{\partial f}{\partial z}}{\frac{\partial f}{\partial x}}\Big|_{contact} = \frac{\Delta x}{\Delta z}, \frac{\partial R}{\partial x} = \frac{1}{\Delta z} \tag{10.7}$$

Figure 10.3 compares the Tilt-Depth and Contact-depth methods. The data set was upward continued by 1.25 km prior to the application of the semiautomatic interpretation methods, to reduce noise. Determining the depth to the contacts in the data using the Tilt-Depth method involves measuring the distance between the yellow and white contours of the Tilt angle, which are overlain on the aeromagnetic data. Also overlain on the figure are the depths obtained from equation (10.7), which have been plotted only when $T=0$. The indicated depths are similar to those that would be obtained from measuring (half) the distance between adjacent Tilt angle contours; the depths increase as the contours get farther apart, and decrease when the contours become closer together (note that the upward continuation distance must be subtracted after measuring the distance between Tilt angle contours; this has already been done for the contact-depth method). The depths and positions are obtained directly, without any of the complexities and ambiguities involved in trying to determine which point on which contour is associated with the "correct" point on another contour. Unfortunately, there is no useful geological information available for the area covered by this aeromagnetic data, and the actual depths of the structures that are visible in the aeromagnetic data are unknown, but the fact that both methods give similar depths gives confidence in the results.

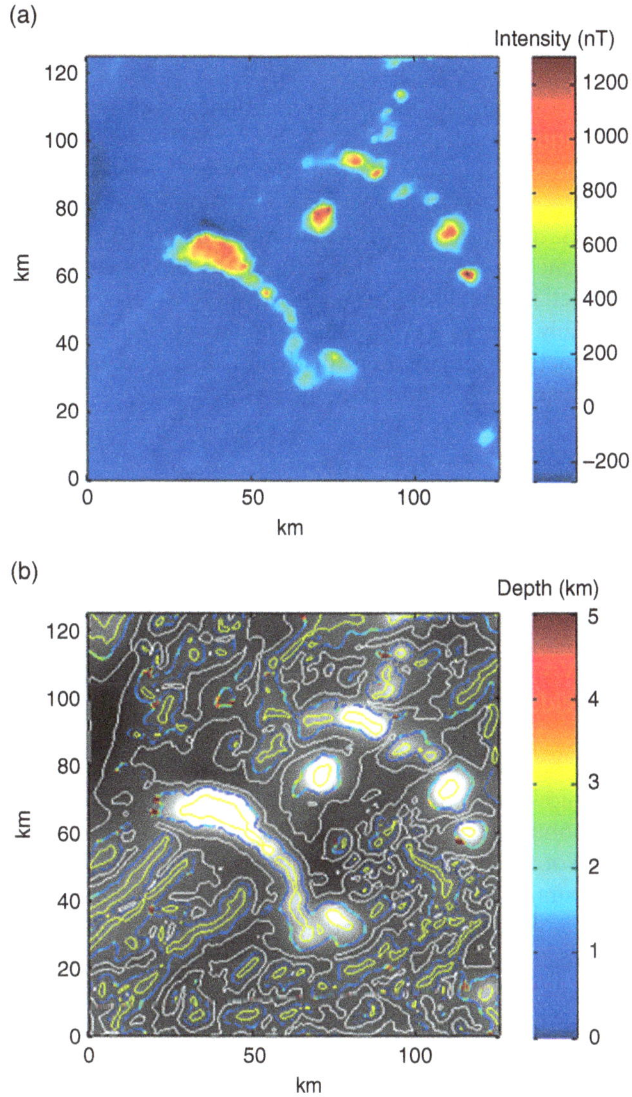

Figure 10.3 (a) Pole-reduced aeromagnetic data from Southern Africa. The grid interval is 0.25 km. (b) Depths obtained from the contact-depth method are shown in color. Overlain are the ± 45° contours of the Tilt angle, which are plotted in yellow and white. In both cases, the data were upward continued by 1.25 km prior to the application of the method (from Cooper, 2014a).

10.3.2. Euler Deconvolution and Source-Distance Methods

Euler deconvolution is a semiautomatic interpretation technique that obtains the horizontal and vertical distances to the source from the current measurement point. Unlike methods such as Tilt Depth and Contact Depth, it can be applied to anomalies from a variety of different models, and is not affected by dip or remanent magnetization. These powerful advantages have made it arguably the most commonly semiautomatic interpretation method for potential field data. Euler's equation is (Thomson, 1982; Reid et al, 1990):

$$\Delta x \frac{\partial f}{\partial x} + \Delta z \frac{\partial f}{\partial z} = -N.f \tag{10.8}$$

where the known quantities are the field strength f, its horizontal and vertical gradients, and the structural index N. N is determined by the source type, for example, for magnetic data, $N=0$ for a contact; $N=1$ for a dyke; and $N=3$ for a dipole. When applied to derivatives of the field, the structural index is increased by the order of the derivative. The unknown quantities are Δx and Δz. Equation (10.8) is solved by the least-squares inversion of a moving window of data, and produces one solution per data point. The location of potential field sources is indicated by clusters of solutions, which are usually found around the upper corners of source bodies. While the structural index can also be solved for, in practice it is usually chosen by the geophysicist. Different values can be used, and the correct one chosen on the basis that it gave the tightest clusters of solutions. Cooper (2014b) recast equation (10.8) from its Cartesian form into a radial coordinate system. For sources whose amplitude decreases with distance r, as $1/r^N$, equation (10.8) becomes

$$r = \frac{Nf}{As_1} \tag{10.9}$$

where As_1 is the first order analytic signal amplitude given by (Nabighian, 1972)

$$As_1 = \sqrt{\left(\frac{\partial f}{\partial x}\right)^2 + \left(\frac{\partial f}{\partial z}\right)^2}$$

The magnetic anomaly from sources such as dykes or contacts does not have the form $1/r^N$, but their analytic signal amplitude does, so equation (10.9) becomes (Cooper, 2014c)

$$r = \frac{(N+1)As_1}{As_2} \tag{10.10}$$

where (Hsu, 1996):

$$As_2 = \sqrt{\left(\frac{\partial^2 f}{\partial z \partial x}\right)^2 + \left(\frac{\partial^2 f}{\partial z^2}\right)^2} \tag{10.11}$$

and, in general

$$r = \frac{(N+\alpha)As_\alpha}{As_{\alpha+1}} \tag{10.12}$$

When the method is used, the negative of r is plotted. At the horizontal position when $-r$ reaches a maximum (i.e., becomes closest to zero), then $r=\Delta z$ (the depth of the source), and this position is the horizontal location of the source. Unlike Euler deconvolution, if an incorrect structural index is used the horizontal location of the source will still be correct (although the depth will be incorrect with both methods). Figure 10.4 shows an application of the method to aeromagnetic data from the Bushveld Igneous complex, South Africa.

The eastern Bushveld igneous complex, South Africa, is crossed by many dykes, which affect the platinum mining that takes place in the area. The dykes are believed to be of Precambrian age (Letts et al., 2003), and many of them are remanently magnetized. While some of the dykes outcrop, others are deeper and many are weathered. The mapping of the dykes helps the mines in the area to plan their mining activities. Figure 10.4a shows an aeromagnetic data set from the region. The dykes are clearly visible and most trend from the southwest to the northeast. Figure 10.4b is a plot of the distances to the dykes, calculated using equation (10.12) with $\alpha=0$. To enable the correlation between the aeromagnetic data and the depths to be more clearly visible, the aeromagnetic data have been overlain on the 3D surface of the depths. The dykes can be seen to be mostly shallow, although some appear to get deeper in different parts of the region. Figure 10.6c shows a profile that was abstracted from the aeromagnetic data, and Figure 10.6d plots the distances to the sources that were calculated using equation (10.12) and also solutions from Euler deconvolution, calculated using equation (10.8). The general agreement is good, although the spread in the Euler solutions often makes clusters hard to see.

10.3.3. Using Wavelets to Obtain the Source Location and Depth

Wavelets analyze data sets on multiple scales, as a function of time (or position). The transform uses a wavelet Ψ as the basis of the analysis. To be admissible as a wavelet, a function must have a zero mean. The continuous

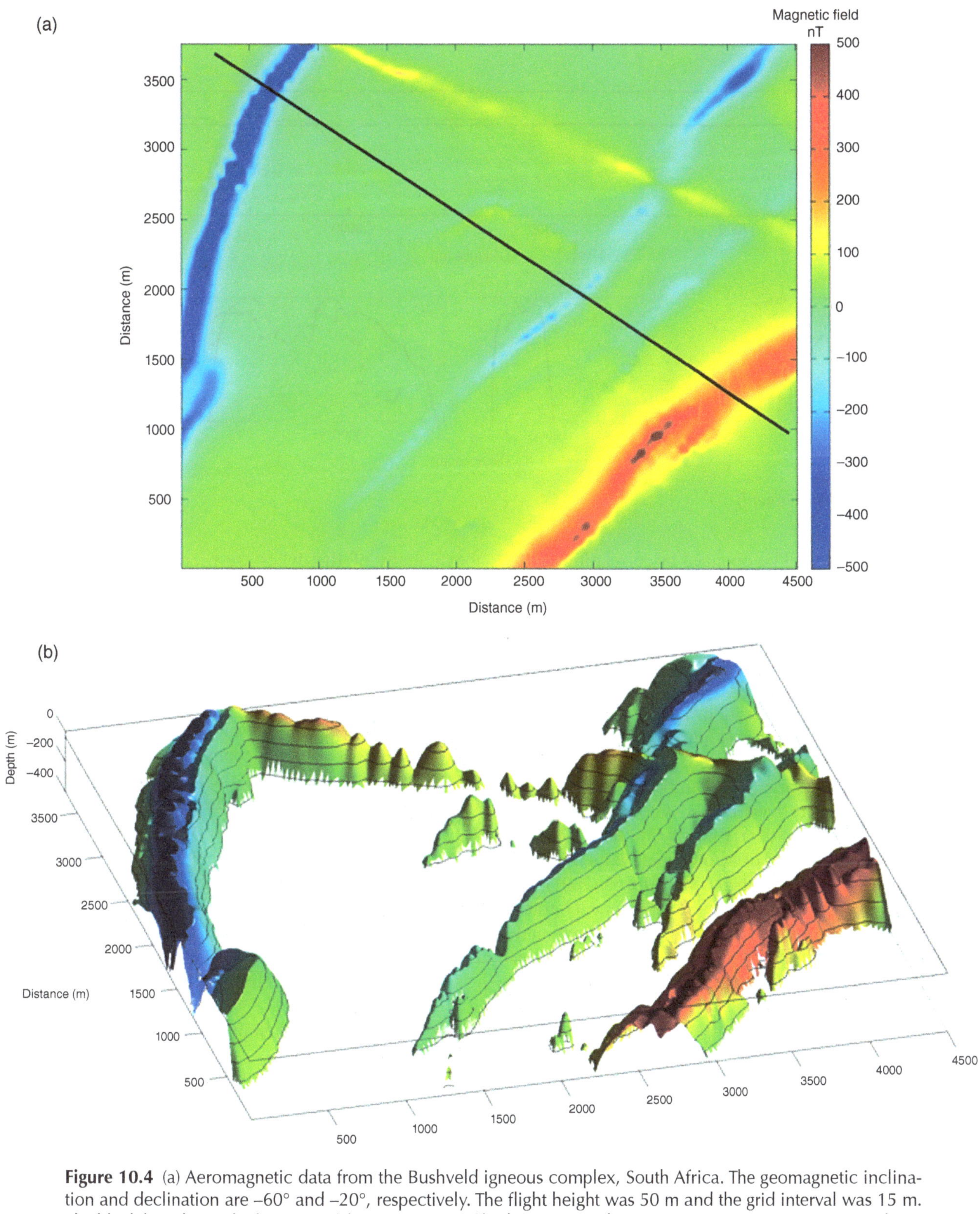

Figure 10.4 (a) Aeromagnetic data from the Bushveld igneous complex, South Africa. The geomagnetic inclination and declination are –60° and –20°, respectively. The flight height was 50 m and the grid interval was 15 m. The black line shows the location of the magnetic profile shown in (c). (b) Distances to magnetic sources as determined by equation (10.12) with $N = 1$ and $\alpha = 0$, after the data were upward continued by 15 m. Overlain is the aeromagnetic data from (a). Contours at depths of 50, 150, 250, and 350 m are plotted in black. Distances have been corrected by the flight height of 50 m and the upward continuation height of 15 m. (c) Aeromagnetic data from the profile is marked in black in (a). The sampling interval was 15 m. (d) Distances to dykes (solid black line) overlain with solutions from Euler deconvolution (black + symbols). The SI used with the Euler deconvolution was 1 and the window size was 11 points. The flight height of 50 m was subtracted from all depth estimates.

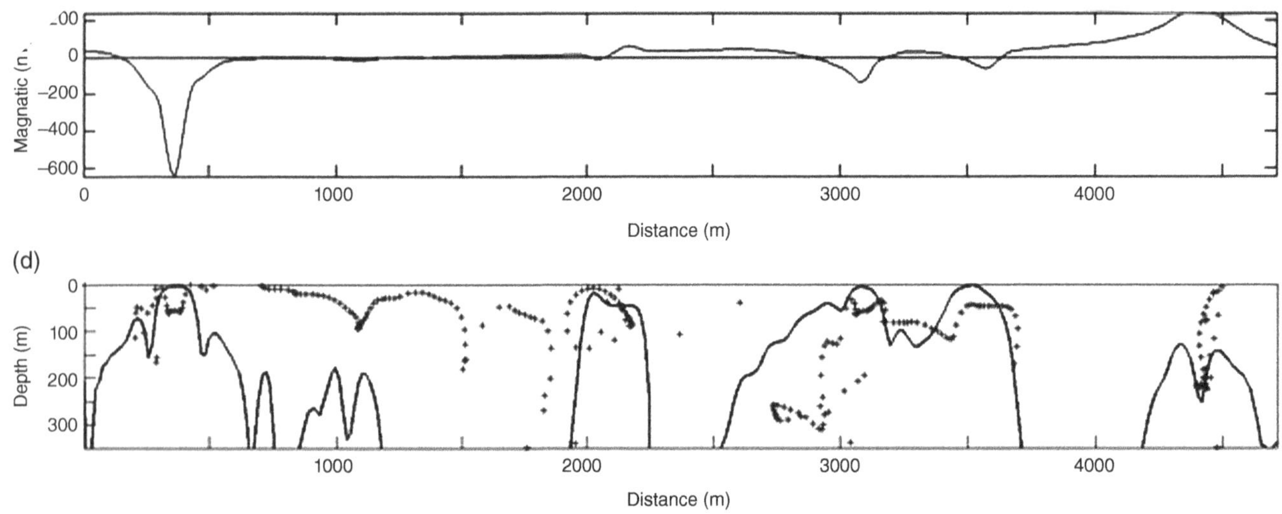

Figure 10.4 (Continued)

wavelet transform (CWT) is obtained by correlating scaled versions of the wavelet with the data (Mallat, 1998, 5),

$$Wf(u,s)=\int_{-\infty}^{\infty} f(x)\frac{1}{\sqrt{s}}\Psi^{*}\left(\frac{x-u}{s}\right)dt \qquad (10.13)$$

where s is scale and u is position. Unlike Fourier transforms, which have a fixed basis function (the sine wave), the wavelet transform can be used with many different wavelets. The Haar, Morlet, and Daubechies wavelets are in common use. There have been many applications of wavelets to potential field data. Ridsdill-Smith and Dentith (1999) denoised derivatives of aeromagnetic data using the wavelet transform, as did Leblanc and Morris (2001). De Oliveira Lyrio et al. (2004) used it to denoise gravity gradiometry data. Fedi and Quarta (1998) performed regional-residual separation using the wavelet transform. Valee et al. (2004) used complex wavelets to estimate source depth from a ratio of the first and second order wavelet coefficients. Cooper (2006) introduced a particularly simple method of determining the location and depth of potential field sources using the wavelet transform. The wavelet used was based on the horizontal derivative of the target model, for example the gravity response of a horizontal cylinder or the analytic signal amplitude of the magnetic anomaly from a thin dyke or a contact. The wavelet analysis is then applied not to the data, but to its horizontal derivative. Different orders of derivative can be used. For the thin dyke, the wavelet used is

$$\Psi_{Dyke}=\frac{x}{(x^2+1)^2} \qquad (10.14)$$

Figure 10.5 shows an application to magnetic data from the Bushveld complex, South Africa. As discussed above, many of the dykes are permanently magnetized and their presence hinders platinum mining in the area. Figure 10.5b shows the result of computing the CWT of the first horizontal derivative of the magnetic data using a wavelet based on the first horizontal derivative of the analytic signal amplitude of the anomaly from a thin dyke. Large CWT coefficients (shown in red) occur when the correlation between the wavelet and the data is high. However, unlike Euler deconvolution, the CWT is affected by the amplitude of the magnetic anomalies, so that the response from the larger amplitude anomalies will tend to be greater than those from weaker anomalies. This makes the method less sensitive to noise than Euler deconvolution or the Source-Distance semiautomatic interpretation approaches.

10.4. POTENTIAL FIELD PROCESSING TECHNIQUES

10.4.1. Differential Pole Reduction

Due to the dipolar nature of the geomagnetic field, magnetic anomalies located anywhere other than at the magnetic poles are asymmetric even when the magnetic source distribution is symmetrical, which complicates interpretation. Pole reduction (RTP) takes magnetic anomalies and changes their asymmetric form to the symmetric form, which would have been observed had the causative magnetic bodies lain at the magnetic poles. The frequency domain operator is (Baranov, 1957):

$$A'(u,v)=\frac{A(u,v)}{\left(\sin\theta+i\cos\theta\sin(\phi+\alpha)\right)^2} \qquad (10.15)$$

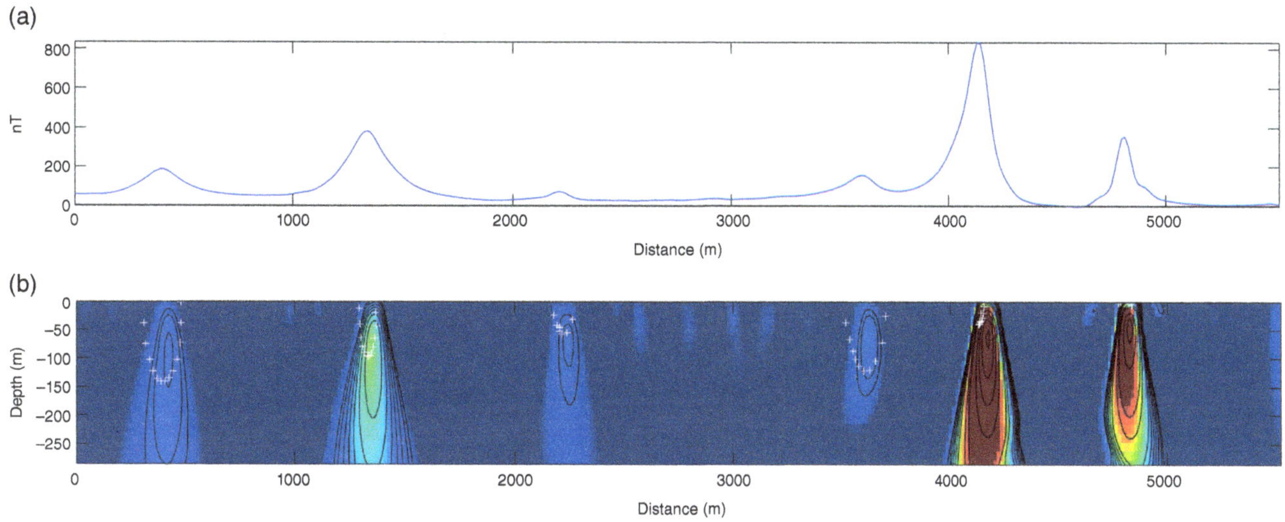

Figure 10.5 (a) Aeromagnetic data from the Bushveld igneous complex, South Africa. The geomagnetic inclination and declination are –60° and –20°, respectively. The flight height is 50 m and the sample interval is 15 m. (b) Continuous wavelet transform of the first horizontal derivative of the analytic signal amplitude of the data shown in (a) computed with equation (10.13) using the wavelet given by equation (10.14). Overlain are solutions from Euler deconvolution (white + symbols). The SI used with the Euler deconvolution was 1 and the window size was 11 points.

where $A(u,v)$ is the Fourier amplitude at frequencies (u,v), θ and Φ are the geomagnetic inclination and declination respectively, and α is $\tan^{-1}(v/u)$. One problem of the method is that θ and Φ must remain constant throughout the area of application of the filter. This restriction is acceptable for small surveys, but becomes a problem when regional aeromagnetic maps are being produced. One solution is to consider the variations of the field parameters as perturbations about the average field values of the region. Arkani-Hamed (1988) used this idea in the frequency domain in an iterative manner. However, the method is rarely used in practice because of the large data storage requirements of the algorithm and its demanding computational requirements (Swain, 2000). An alternative approach is to apply the perturbations in the space domain (Cooper & Cowan, 2005), that is,

$$RTP_{\text{var}} = RTP_{mean} + \Delta inc\frac{\partial RTP}{\partial inc} + 0.5\Delta inc^2\frac{\partial^2 RTP}{\partial inc^2} + \Delta dec\frac{\partial RTP}{\partial dec} + 0.5\Delta dec^2\frac{\partial^2 RTP}{\partial dec^2} + \ldots \tag{10.16}$$

where RTP_{mean} is the data set reduced to the pole using the average field inclination and declination of the area. Δinc is the difference between the inclination at a given point and the average inclination, and Δdec is computed similarly. The derivatives are computed in the space domain by differencing.

The algorithm was applied to the regional aeromagnetic grid of the Northern Territory, Australia (Fig. 10.6a). The magnetic field inclination ranges from –57° in the south to –33° in the north. Previous RTP of this data set was performed by Geoscience Australia by breaking the data set into a series of 300 x 300 km overlapping panels, using the mean geomagnetic inclination and declination of each panel, retaining only the central 100 x 100 km portion and then stitching the panels together. This worked well for the vertical magnetic gradient but was less successful for the total magnetic intensity. This method has the disadvantage that long wavelength features (comparable in size or greater than the panel size) are not correctly reduced to the pole. Figure 10.6 shows the result of conventional RTP as well as differential RTP. Although the differences are not large in amplitude, significant changes in anomaly shapes can be seen (Cooper & Cowan, 2005).

10.4.2. Stable Downward Continuation

As discussed in section 10.2.2 above, vertical continuation is a mathematical technique that takes measured potential field data and computes from it the data that would have been measured at a different altitude (Blakely, 1995, p. 320). While upward continuation is a stable, smoothing process, downward continuation (which enhances detail) is unstable and is very sensitive to noise. In addition, any problems in the implementation of the

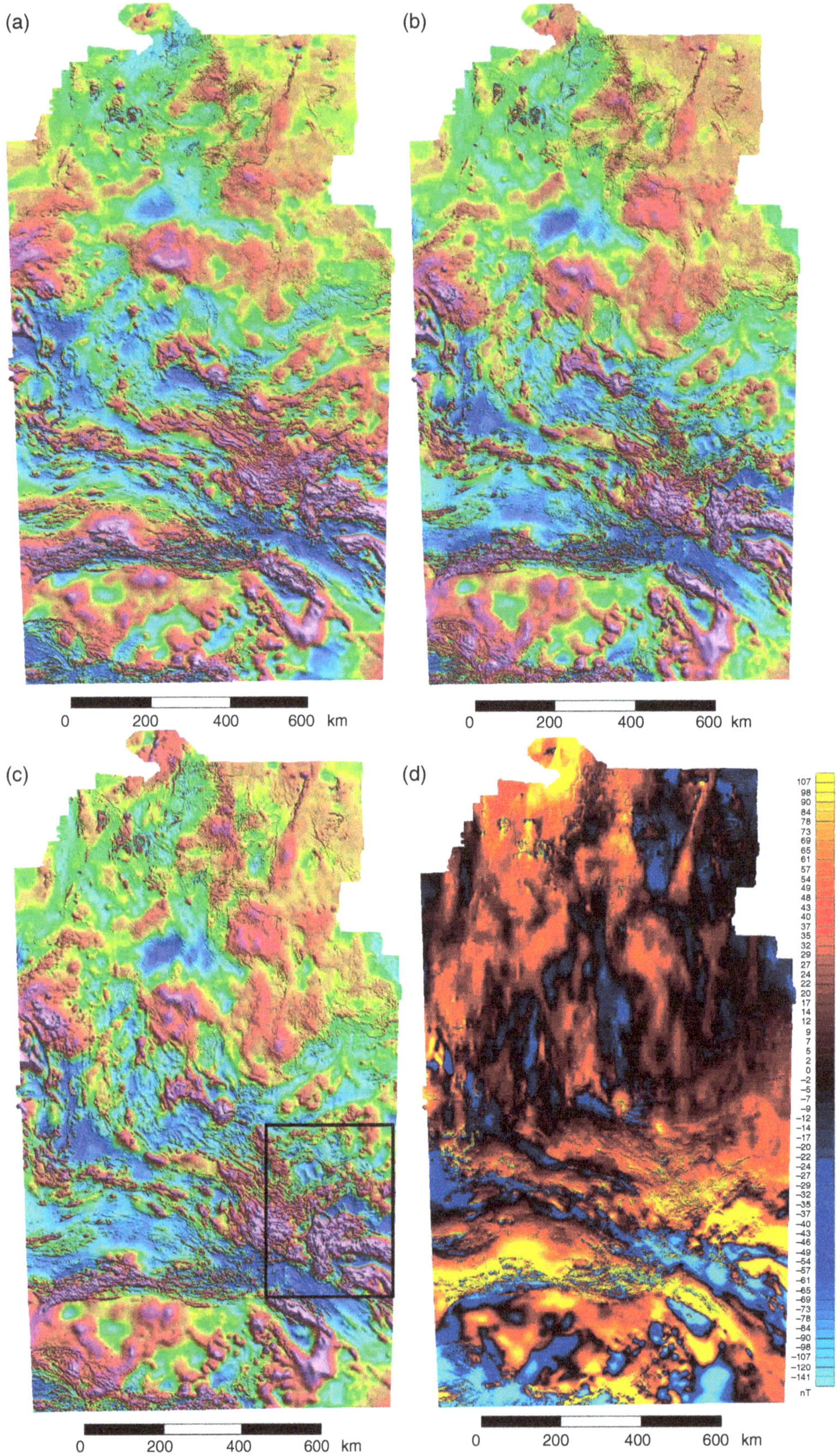

Figure 10.6 (a) Aeromagnetic data from the Northern Territory, Australia. (b) Data from (a) after the application of standard RTP. (c) Data from (a) after the application of differential RTP. (d) Difference between data in (b) and (c) (from Cooper and Cowan, 2005).

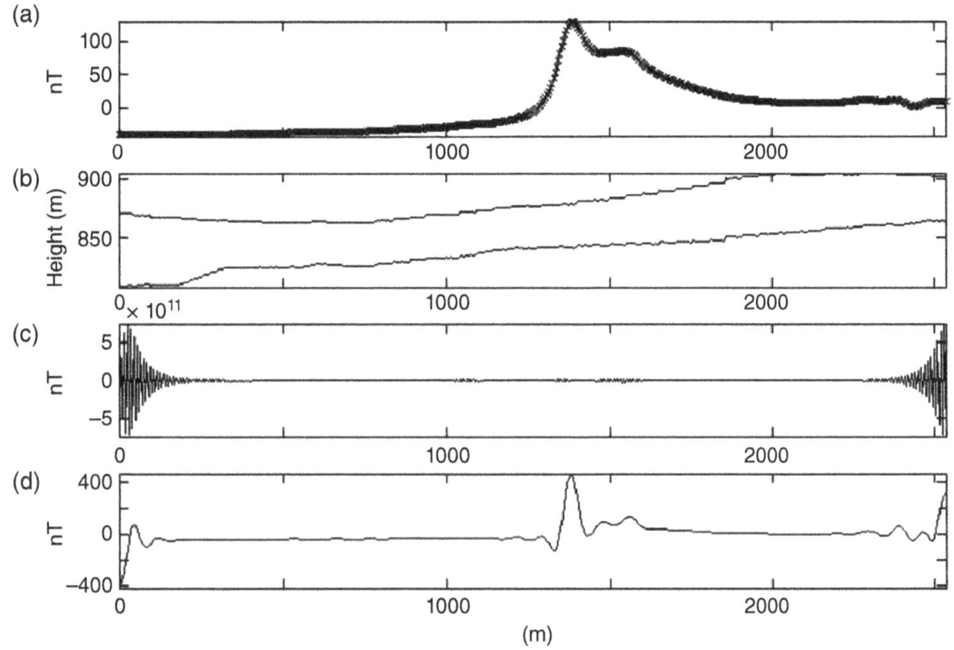

Figure 10.7 (a) Aeromagnetic data from the Bushveld igneous complex (x symbols, interpolated to 5 m sample interval) overlain with upward continued downward continued data (solid line). (b) Terrain and flight path of the aircraft. (c) Data downward continued 50 m (10 sample intervals) using the standard Fourier method. Note the scale on the *y*-axis. (d) Data downward continued 50 m (10 sample intervals) using the inversion method, with $k = 0.1$.

Fourier transform (such as edge-padding and tapering) are amplified by the downward continuation process. To avoid these problems, the downward continuation problem was reformulated. The downward continued field was considered as that when upward continued became the original potential field. The procedure was to start with an initial estimate of the downward continued field, then modify it using least-squares inversion until, when upward continued, the original field was reconstructed (Cooper, 2004). Specifically,

$$f_{downward} = \left(A^T A + kI\right)^{-1} A^T e \tag{10.17}$$

where e is the misfit between the original data and the upward continued data. A is the gradient matrix, I is the identity matrix, and the constant k is a damping factor. A is calculated by a numerical differencing operation. The inversion process is even determined since each data point is a parameter. Figure 10.7 shows the downward continuation of an aeromagnetic data set from the Bushveld Complex, South Africa. Using the standard method, equation (10.4), the downward continuation has diverged and the profile is dominated by edge effects with an amplitude of 10^{11} nT. Using equation (10.17), however, produced a stable result, with the individual anomalies on the profile being much more clearly distinguishable than in the original data. The disadvantage of the method is that it is computationally intensive when applied to large grids. Zhang et al. (2013) stated that "the downward continuation by the least-squares inversion (Cooper, 2004) is one of the most robust downward continuation methods."

10.5. CONCLUSIONS

A brief summary of some of the techniques that have been developed in the 21st century in South Africa for the processing and interpretation of potential field data has been presented. The techniques have a wide range of applications and some have been implemented in commercial geophysical software.

ACKNOWLEDGMENTS

The reviewers and editors are thanked for their constructive comments, which improved this manuscript. Thanks to the National Research Foundation (Pretoria) for funding this project. Gordon Chunnet of the Angle Platinum corporation is thanked for the use of the aeromagnetic data shown in Figures 10.4, 10.5, and 10.7. The Council for Geoscience (Pretoria) is thanked for the use of the gravity data shown in Figure 10.1. Geoscience Australia is thanked for the aeromagnetic data shown in Figure 10.6.

REFERENCES

Arkani-Hamed, J. (1988). Differential reduction to the pole of regional magnetic anomalies. *Geophysics, 53* (12), 1592–1600.

Blakely, R. J. (1995). *Potential Theory in Gravity and Magnetic Applications*, Cambridge University Press.

Baranov, V. (1957). A new method for interpretation of aeromagnetic maps: Pseudo-gravimetric anomalies. *Geophysics, 22* (2), 359–383.

Cooper, G. R. J. (2003). Feature detection using sunshading. *Computers & Geosciences, 29* (8), 941–948.

Cooper, G. R. J. (2004). The stable downward continuation of potential field data. *Exploration Geophysics, 35*(4), 260–265.

Cooper, G. R. J., & Cowan, D. R. (2005). Differential reduction to the pole. *Computers & Geosciences, 31*(8), 989–999.

Cooper, G. R. J. (2006). Interpreting potential field data using continuous wavelet transforms of their horizontal derivatives. *Computers & Geosciences, 32* (7), 984–992.

Cooper, G. R. J. (2014a). The automatic determination of the location, depth, and dip of contacts from aeromagnetic data. *Geophysics, 79* (3), J35–J41.

Cooper, G. R. J., (2014b). Euler deconvolution in a radial coordinate system. *Geophysical Prospecting, 62*(5), 1169–1179.

Cooper, G. R. J. (2014c). The automatic determination of the location and depth of contacts and dykes from aeromagnetic data. *Pure and Applied Geophysics, 171*, 2417–2423.

Cooper, G. R. J. (2016). The downward continuation of the tilt-angle. *Near Surface Geophysics, 14* (5), 385–390.

De Beer, J. H. (2015). *The History of Geophysics in Southern Africa*. Sun Press.

Eales, H. V. (2001). A first introduction to the geology of the Bushveld complex and those aspects of geology that relate to it. Council for Geoscience, Pretoria.

de Oliveira Lyrio, J. C. S., Tenorio, L., & Li,Y. (2004). Efficient automatic denoising of gravity gradiometry data. *Geophysics, 69* (3), 772–782.

Fedi, M., &. Quarta, T. (1998). Wavelet analysis for the regional-residual and local separation of potential field anomalies. *Geophysical Prospecting, 46*, 507–525.

Horn, B. K. P. (1982). Hill shading and the reflectance map. *Geo-Processing, 2*, 65–146.

Hsu, S-K, Sibuet, J-C., & Shyu, C-T (1996). High-resolution detection of geologic boundaries from potential field anomalies: An enhanced analytic signal technique. *Geophysics, 61*, 373–386.

Leblanc, G. E., & Morris, W. A. (2001). Denoising of aeromagnetic data via the wavelet transform. *Geophysics, 66*, 1793–1804.

Letts, S., Torsvik, T. H, Webb, S. J., & Ashwal, L. D. (2003). Palaeomagnetism of Mafic Dykes from the Eastern Bushveld Complex (South Africa). 8th SAGA Biennial Technical Meeting and Exhibition, Pilanesberg, South Africa.

Mallat, S. (1998). *A Wavelet Tour of Signal Processing*. Academic Press, New York.

Miller, H. G., & V. Singh (1994). Potential field tilt: A new concept for location of potential field sources. *Journal of Applied Geophysics, 32*, 213–217.

Nabighian, M. N. (1972). The analytical signal of 2D magnetic bodies with polygonal cross-section: Its properties and use for automated anomaly interpretation. *Geophysics, 39*, 85–92.

Rajogopalan, S., & Milligan, P. (1995). Image enhancement of aeromagnetic data using automatic gain control. *Exploration Geophysics, 25*, 175–178.

Reid, A. B., Allsop, J. M., Granser, H. Millet, A. J., & Somerton, I. W. (1990). Magnetic interpretation in three dimensions using Euler deconvolution. *Geophysics, 55*, 80–91.

Ridsdill-Smith, T. A., & Dentith, M. C. (1999). The wavelet transform in aeromagnetic processing. *Geophysics, 64*, 1003–1013.

Salem, A., Williams, S., Fairhead, J. D., Ravat, D., & Smith, R. S. (2007). Tilt-depth method: A simple depth estimation method using first-order magnetic derivatives. *The Leading Edge,* October, 1502–1505.

Swain, C. J. (2000). Reduction to the pole of regional magnetic data with variable field direction, and its stabilisation at low inclinations. *Exploration Geophysics, 31*, 78–83.

Thomson, D. T. (1982). EULDPTH: A new technique for making computer-assisted depth estimates from magnetic data. *Geophysics, 47* (1), 31–37.

Valee, M. A., Keating, P., Smith, R. S., & St-Hilaire, C. (2004). Estimating depth and model type using the continuous wavelet transform of magnetic data. *Geophysics, 69*, 191–199.

Zhang, H., Ravat, D., & Hu, X. Y. (2013). An improved and stable downward continuation of potential field data: The truncated Taylor series iterative downward continuation method. *Geophysics, 78* (5), J75–J86.

11

3D Reflection Seismic Imaging for Gold and Platinum Exploration, Mine Development, and Safety: Case Studies from the Witwatersrand Basin and Bushveld Complex (South Africa)

M. S. Manzi, E. J. Hunt, and R. J. Durrheim

ABSTRACT

The discovery of giant mineral and metal deposits needed to sustain long-term global growth is a great challenge. An integrated exploration approach involving geological mapping, high-resolution 3D seismic imaging, and drilling coupled with borehole geophysics will likely yield the best results. Seismic methods, in particular, have become, and will continue to be, an important tool to help unravel structures hosting mineral deposits at great depth. This chapter demonstrates how the reflection seismic method (mainly 3D seismic) has been successfully used to explore some of the world's largest gold and platinum deposits and map the gross structural architecture that controlled the formation of these ore bodies, support mine planning, and contribute to safety. Three case studies from the Witwatersrand Basin and Bushveld Complex in South Africa are presented, demonstrating the importance of reprocessing the legacy data using new seismic imaging techniques for high-resolution structural mapping. We further discuss the pitfalls with conventional 2D and 3D seismic surveys compared with broadband 3D seismic surveys for mapping both shallow (<500 m) and deep targets (>500 m) in hard rock environments. We argue that future deep exploration programs should use high-resolution 3D broadband seismic surveys.

11.1. INTRODUCTION

The reflection seismic method has been widely used in the oil industry since the 1960s to locate oil and gas reservoirs in "soft" sedimentary rocks. Over the past few decades, the method has been developed and actively used in "hard rock" metallogenic provinces worldwide for deep-crustal studies, mineral exploration, mine planning, and safety (Milkereit et al., 1996, 2000; Pretorius et al., 2000; Trickett et al., 2004; Malehmir & Bellefleur, 2009; Malehmir et al., 2012). For example, in the last 20 years, Geosciences Australia has acquired numerous regional 2D seismic lines (more than 15,000 km) and a few 3D seismic surveys across the Yilgarn Craton to provide insights into the crustal-scale mineral-bearing fluid pathways and possible source rocks (Urosevic et al., 2005).

The discovery of giant mineral and metal deposits needed to sustain long-term global growth is a great challenge. However, the answers are likely to be found using a state-of-the-art exploration approach involving a combination of improved geophysical surveys together with drilling, field geological mapping, and physical property measurements. The reflection seismic method has proven to be the only surface geophysical method that provides a high-resolution image of the subsurface and information about structural and lithological relationships that control mineral deposits at depths. In structurally complex areas, such as the Witwatersrand goldfields and Bushveld Complex in South Africa, only a 3D seismic survey can fully resolve the

University of the Witwatersrand, School of Geosciences, Johannesburg, South Africa

Ore Deposits: Origin, Exploration, and Exploitation, Geophysical Monograph 242, First Edition.
Edited by Sophie Decrée and Laurence Robb.

complex 3D architecture of the ore body and geological structures, such as faults and dykes.

Although 3D seismic surveys are expensive relative to linear 2D seismic surveys, aeromagnetic surveys, and traditional exploration drillings, they offer great advantage for processing and interpretation of the seismic data, thus providing a more detailed understanding of the complex 3D structural networks of the ore body and the tectonics of the area. Generally, 3D seismic surveys are carried out only at previously well-explored sites with known economic ore deposits (such as an existing mining site), where exploration drilling and 2D seismic surveys have already been conducted. The 2D seismic surveys, on the other hand, continue to be used at an initial exploration stage, to map the regional structural framework of the target area and to evaluate the acquisition parameters for a potential follow-up 3D seismic survey. There are, however, many challenges faced in adapting a conventional seismic method that was developed primarily for hydrocarbon exploration to hard rock mineral exploration. These include, but are not limited to, low mapping resolution for shallow and deep targets due to: (1) the low bandwidth of the vibroseis sweeps source used for acquisition, typically four and, at most, five octaves; (2) a low density of source and receiver positions; and (3) a large bin size, shot offset and low nominal fold.

Over the last few years, there has been a proliferation of seismic broadband solutions, which employ specific combinations of equipment, acquisition, and processing techniques, that can be applied in both hard rock and soft rock situations to improve the imaging resolution (Duval, 2012; Denis et al., 2013). The best broadband results to date have come from solutions that have used a high-density of receiver and source arrays and extended the seismic bandwidth to six octaves. These studies show that the data acquired with high-source and receiver density, small bin size, long offset, wide azimuth, and broadband vibrators are less subject to tuning effects (Widess, 1973) and allow better detection of small faults and thin beds that cannot be resolved by conventional band-limited seismic surveys (Duval, 2012; Denis et al., 2013). These developments have made the technique increasingly effective to the point that it has become well-established in the oil and gas industry. Following the successful application of broadband reflection seismic technology for oil and gas exploration, one of South Africa's largest gold mining companies successfully acquired South Africa's first-ever land broadband 3D seismic data in one of their prospect areas to evaluate its potential in hard rock mining. However, despite these developments, 3D broadband reflection seismic is underutilized in the hard rock environments and remains an unfamiliar technique to many geoscientists in hard rock exploration and mining.

This chapter demonstrates how the reflection seismic method (mainly 3D seismic) has been used to explore for deep mineral deposits (mainly gold- and platinum-bearing horizons) and map the gross structural architectures that control the formation of these deposits, focusing on case studies from the Witwatersrand Basin and Bushveld Complex in South Africa. We further discuss the pitfalls with conventional 3D seismic surveys compared with broadband 3D seismic surveys for mapping both shallow (<500 m) and deep targets (>500 m) in hard rock environments. We argue that for future deep exploration programs to be successful, 3D broadband seismic surveys should be utilized.

11.2. DEEP MINING AND REFLECTION SEISMIC METHODS IN SOUTH AFRICA

Gold-bearing quartz pebble conglomerates (known as reefs) were first discovered near modern-day Johannesburg in 1886, which significantly contributed to the economic growth of South Africa. The initial mining was associated with outcrops around the perimeter of the Witwatersrand Basin. As mining proceeded, some of the ore bodies were found to persist to great depths (>1 km) beneath the cover rocks, with significant challenges to continue exploration and extraction. More than 500 exploration diamond drill holes were sunk in the Witwatersrand goldfields, but most were concentrated close to the known deposits and did not provide a comprehensive understanding of the extent of the gold-bearing horizons (Manzi et al., 2013a). The economic potential of the Bushveld Complex was discovered much later, as the platinum deposit, which later became known as the Merensky Reef, was only located in 1924 (Cawthorn, 1999). The low economic value of platinum over the following decades limited exploration, but the increasing requirements for platinum from the 1950s onward resulted in a significant expansion of mines around the Bushveld Complex by the 1970s (Cawthorn, 1999). Following this, most of the exploration was concentrated around the major mines in the Bushveld Complex (Cawthorn, 2010).

In the 1930s, potential field methods (gravity and magnetics) were successfully used to locate some of the gold-bearing strata concealed beneath cover rocks. These methods, however, were unable to image the structure of deeply buried strata in detail. Consequently, this led to the first application of the reflection seismic surveys in the 1980s to explore for gold in the Witwatersrand Basin. Since then, hundreds of 2D seismic surveys and tens of 3D seismic surveys have been acquired across the Witwatersrand Basin and the Bushveld Complex to map their structure, as well as delineate the deep-seated gold- or platinum-bearing horizons and other subsurface structures (e.g., faults, potholes, and dykes) (Pretorius et al., 1989;

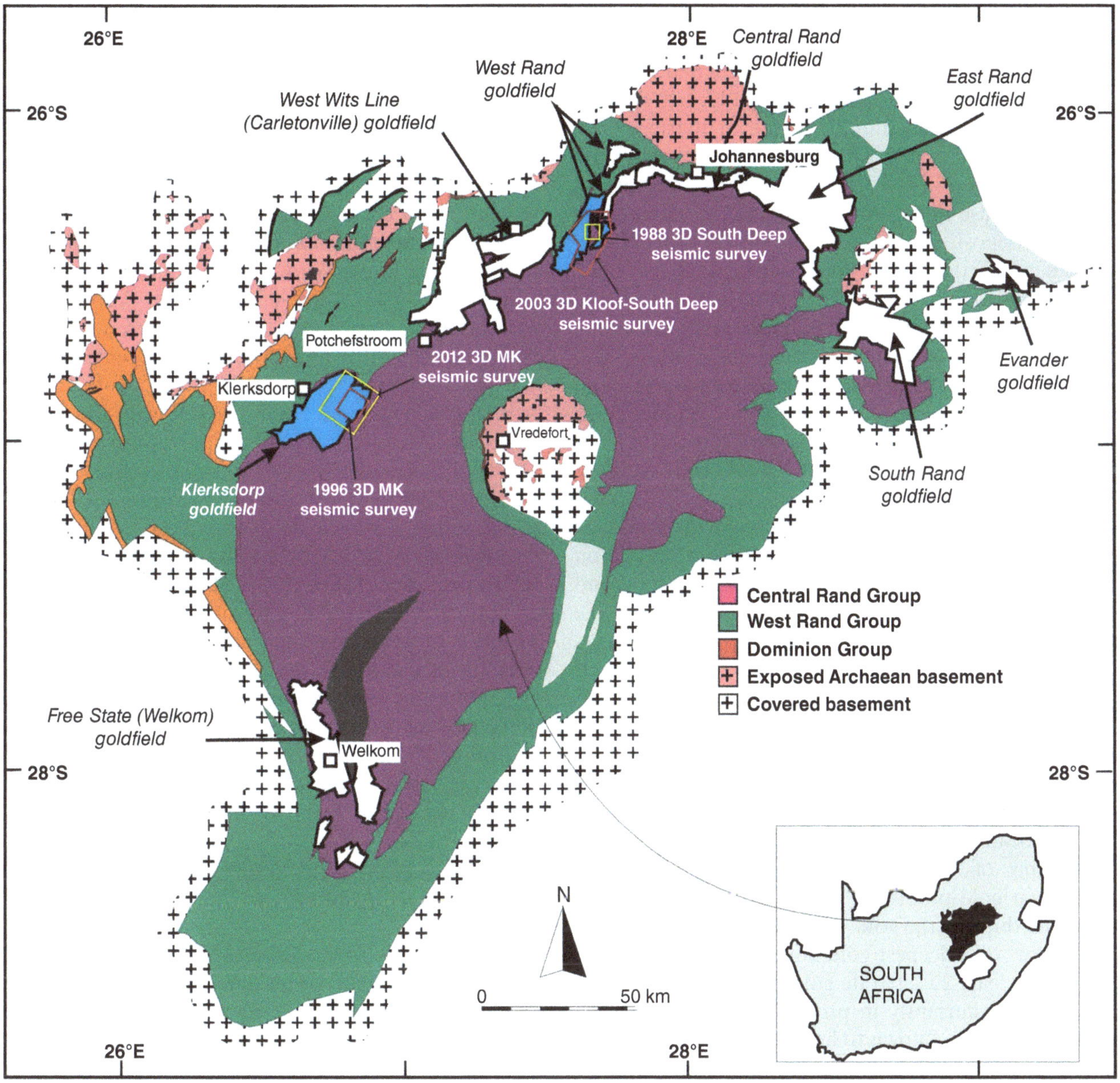

Figure 11.1 Map of the West Rand and Klerksdorp goldfields of the Witwatersrand Basin in South Africa, showing the location of 3D seismic reflection surveys covering Kloof, South Deep, and Moab Khotsong (MK) mines (modified after Manzi et al., 2013a).

Campbell & Crotty, 1990; Pretorius et al., 2003; Malehmir et al., 2012; Malehmir et al., 2014). The discovery of various goldfields beneath cover in the Witwatersrand Basin (Fig. 11.1) was the result of the integration of these various geological and geophysical methods. South Africa, to our knowledge, has acquired more hard rock seismic data than the rest of the world (Manzi et al., 2014).

Most mineral deposits have favorable physical properties to be targeted using various geophysical methods, but many of these methods do not have sufficient sensitivity and resolution at great depth (>500 m). The ability of the boundary between two geological entities to generate a strong reflection depends on the contrast in the acoustic impedance (product of velocity and density), the dimensions and geometry of the targets. The resolution of the seismic imaging depends on the seismic wavelength, which in turn depends on the seismic velocity of the rocks and the frequency of the seismic pulse. Typical seismic wavelengths for waves propagating in the hard rock environment are on the order of 50–100 m. Therefore, targets located at depths greater than approximately 500 m cannot be resolved vertically and horizontally if they are less than about 20 m thick and 350 m wide, respectively (Malehmir et al., 2012). However, advanced acquisition methods, processing and interpretation techniques can be used to enhance the detection of geological features below the seismic resolution limits (i.e., a quarter of the dominant wavelength). The latest developments are particularly interesting to South Africa's deep mining industry because they support the quest to locate these deeper ore deposits, as well as any subtle geological structures ahead of the mining face that might affect mine planning and safety. Furthermore, it is shown in several studies (Malehmir et al., 2012; Malehmir et al., 2014) that the success of the reflection seismic surveys depends on the site conditions (e.g., accessibility, topography, and thickness of the weathered layer) and the geology (e.g., rock properties, steepness of dips, and geological complexities such as tight folding). In the following sections, we present three case studies from the Witwatersrand Basin and the Bushveld Complex showing how 3D seismic surveys can significantly improve interpretation in geologically complex hard rock environments, and how they have been used to complement and guide the drilling programs.

11.3. WITWATERSRAND BASIN

11.3.1. Gold Deposits

The Mesoarchaean Witwatersrand Basin of South Africa is located in the center of the Kaapvaal Craton in the Republic of South Africa and is one of the world's best known gold-producing provinces. It has produced about one-third of the gold ever mined, worldwide, and hosts the world's deepest gold mines with depths down to 4 km. At least 20 separate gold ore bodies have been mined, or are currently in production throughout the basin. The basin is known to host substantial resources, many of which are currently located at great depths. These may become attractive to exploit in the future, depending on the gold price and technological developments.

Witwatersrand gold is found in fluvial conglomerate beds and is usually associated with pyrite mineralization and locally with uraninite mineralization. In most cases, the conglomerate layers are only a few decimeters to several meters thick, but in a few instances the conglomerate layers are stacked on top of each other, forming massive, gently dipping tabular ore bodies that are tens of meters thick. These deposits are found at several stratigraphic levels and extend for many kilometers along strike and dip; thus, mining takes place at different levels in a mining district or mine (Krapež,1985; Jolley et al., 2004; Manzi et al., 2013a). The basin is, in some parts, overlain by outliers of Karoo Supergroup shales and sandstones on the surface, which are underlain by Pretoria Group sediments and the Chuniespoort Group dolomites of the Proterozoic Transvaal Supergroup. The dolomites overlie the Klipriviersberg Group volcanic rocks of the Ventersdorp Supergroup, which in turn cap the gold-bearing Ventersdorp Contact Reef (VCR) and sediments of the Central Rand Group of the Witwatersrand Supergroup (Fig. 11.2). The latter two are the principal hosts of the economic ore bodies (Dankert & Hein, 2010; Manzi et al., 2013a).

Several authors have proposed different models for the origin of gold in the Witwatersrand Basin. The main genetic models for the gold-uranium ores of the Witwatersrand Supergroup postulate that: (1) they are paleoplacers modified by hydrothermal processes (Frimmel et al., 2005; Minter, 2006); or (2) they are of entirely hydrothermal origin (Law & Phillips, 2005), with the uranium and organic matter infiltrating the basin synchronously (Law & Phillips, 2005) or in separate events (Barnicoat et al., 1997). Recently, it also has been proposed that the gold is entirely syngenetic, having been precipitated by algal mats from surface waters and shallow seas (Frimmel & Hennigh, 2015; Heinrich, 2015). The models make different predictions regarding the origin and distribution of the ore, with significant implications for exploration strategies, the estimation of ore reserves, and mining layouts. The 3D seismic method can be used to image major and minor fault networks that cut the gold-bearing horizons, thus providing constraints on the timing of faulting and fracturing that may have played a significant role in transporting ore-bearing fluids (Malehmir et al., 2014).

11.3.2. Case Study 1: 3D Seismic Surveys of Kloof and South Deep Gold Mines

The case study from the Kloof and South Deep gold mines in the Witwatersrand Basin represents pioneering work that demonstrated the effectiveness of the 3D reflection seismic method for deep gold exploration. Kloof gold mine, now Kloof-Driefontein Complex (KDC) East, is a well-established intermediate to ultra-deep-level gold mine that is accessed from surface through vertical shaft systems to different depth levels. It is operated by Sibanye-Stillwater. The mine exploits two primary ore bodies, namely the Ventersdorp Contact Reef (VCR) located at the base of the Ventersdorp Supergroup, and the Middelvlei Reef (MR) near the base of the Central Rand Group (Fig. 11.2; Manzi et al., 2013b). Gold mineralization in Kloof gold mine is considered by Muntean et al. (2005) to be the result of structurally controlled fluid flows, with permeability developed largely within microfracture networks localized within conglomeratic horizons. It is suggested that alteration was associated with hydrothermal fluid flow, and gold was deposited due to highly efficient reduction of the fluids by hydrocarbon ("carbon") (Muntean et al., 2005). The mine has

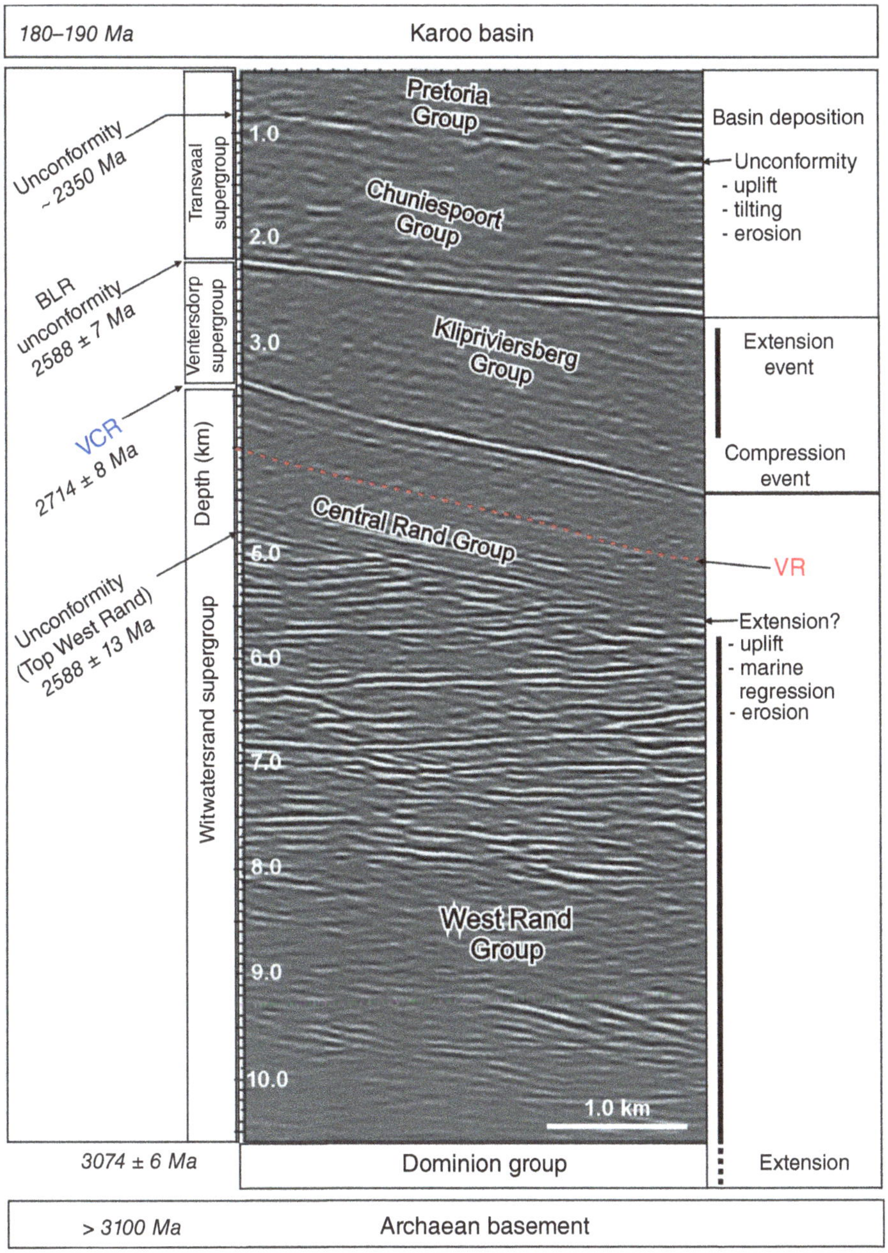

Figure 11.2 Generalized seismic stratigraphy and tectonic events of the Witwatersrand goldfields, with dates and ages from Dankert and Hein (2010). VCR: Ventersdorp Contact Reef; VR: Vaal Reef; BLR: Black Reef.

an estimated reserve of 44.1 Mt for a Life of Mine (LoM) head-grade of 13.9 g/t gold. These reserves are planned for depletion by 2027 at an average production rate of 170,000 t/month. The deepest mining level of the ore body is at 3.35 km below surface and the total gold production is averaged at 1667 kg/month at an average yield grade of 6.0 g/t (Manzi et al., 2013b).

The South Deep gold mine, which is owned and operated by Gold Fields Limited, is an intermediate to ultradeep level mine comprising two shaft systems. The mine is currently one of the world's premier gold ore bodies, with some 78 million ounces of resources and 29 million ounces of reserves (Manzi et al., 2012a; Manzi et al., 2013b). It exploits the largest known gold deposits in the Archaean Witwatersrand Basin, that is, the VCR and Upper Elsburg reefs (multistacked conglomerate units) of the Central Rand Group at depths between 2 and 3.5 km below the surface (Manzi et al., 2013b).

In the 1980s, a series of high-resolution, 2D seismic profiles were acquired to map the ore body and large-scale structures in the goldfields (Pretorius et al., 2003). However, 2D seismic surveys could not resolve the structural complexity of the South Deep ore body. In 1987, the first hard rock 3D seismic survey in South Africa was acquired to aid in mine planning and deep exploration at the South Deep project area (Campbell & Crotty, 1990). In 2003, Gold Fields Ltd. conducted a high-resolution 3D seismic reflection survey, targeting the down-dip portion of the auriferous VCR in the Kloof and South Deep gold mines (Fig. 11.1). The 2003 survey was designed to overlap with the 1987 3D South Deep seismic survey. These surveys were merged and processed as a single entity of final depth-converted prestack time migrated (PSTM) seismic data to produce a single seismic cube (Manzi et al., 2012a).

Figure 11.3 shows a merged 3D seismic volume demonstrating successful imaging of the VCR ore body across the two mines. The VCR ore body is typically less than 2 m thick and is not directly imaged seismically. The reason for strong reflection of the VCR is that it occurs at the interface between the overlying high-velocity (6400 m/s), dense (2.90 g/cm^3) basaltic lavas of the Ventersdorp Supergroup and the underlying low-velocity (5750 m/s), less-dense (2.67 g/cm^3) quartzite units of the Central Rand Group (Fig. 11.2) (Pretorius et al. 2000; Manzi et al., 2015).

To improve the resolution, the legacy data were reprocessed using the latest processing algorithms to image the continuity of the ore body and steeply dipping structures. Figure 11.4 shows the seismic section after the reprocessing of the data using modern seismic imaging techniques. This example demonstrates the benefit of reprocessing legacy seismic data for better delineation of the VCR ore body and steeply dipping faults at great depths.

In the depth structure map, which is shown in Figure 11.5a, the VCR ore body exhibits a variable strike and dip across the mines, probably due to folding, warping, and faulting. What also made the reprocessing of the 3D seismic data remarkable was the ability of new imaging techniques (Manzi et al., 2012a) to resolve a previously unknown 250 m wide and 1.0 km long discrete ore body block (VCR) preserved between two fault segments in a major fault zone (West Rand Fault, 2 km maximum throw), which is shown in Figure 11.5a, b. The identification of this new VCR block was confirmed by underground mapping. Establishment of the continuation of the ore body assisted in improving and enhancing the resource calculation, financial valuation of the ore body, and life-of-mine planning. The major extensional fault, the West Rand Fault (WRF), is believed to have played some role in the formation or transportation of the gold in the West Rand goldfields, and its kinematics have been the subject of detailed geological studies (Dankert & Hein, 2010; Manzi et al.,

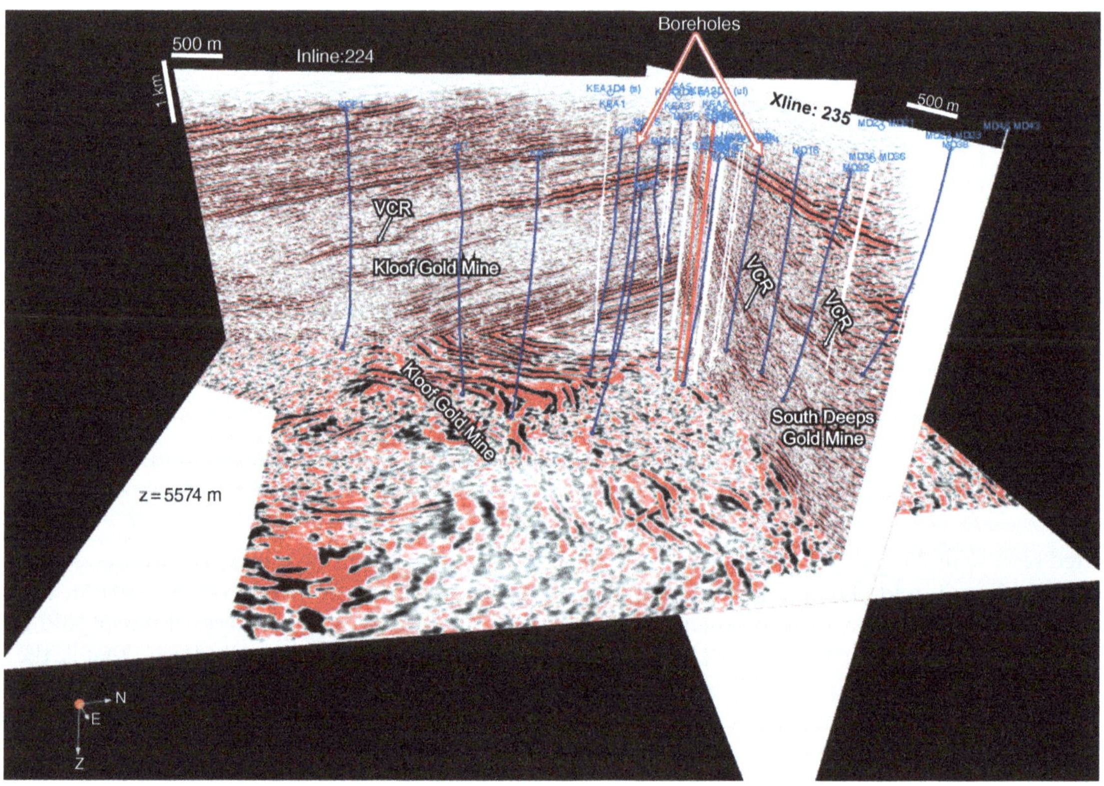

Figure 11.3 Merged 3D depth-converted prestack time migrated seismic cube (amplitude display) covering Kloof and South Deep gold mines, showing the location of exploration boreholes and the successful seismic imaging of the Ventersdorp Contact Reef (VCR) ore body across the mines.

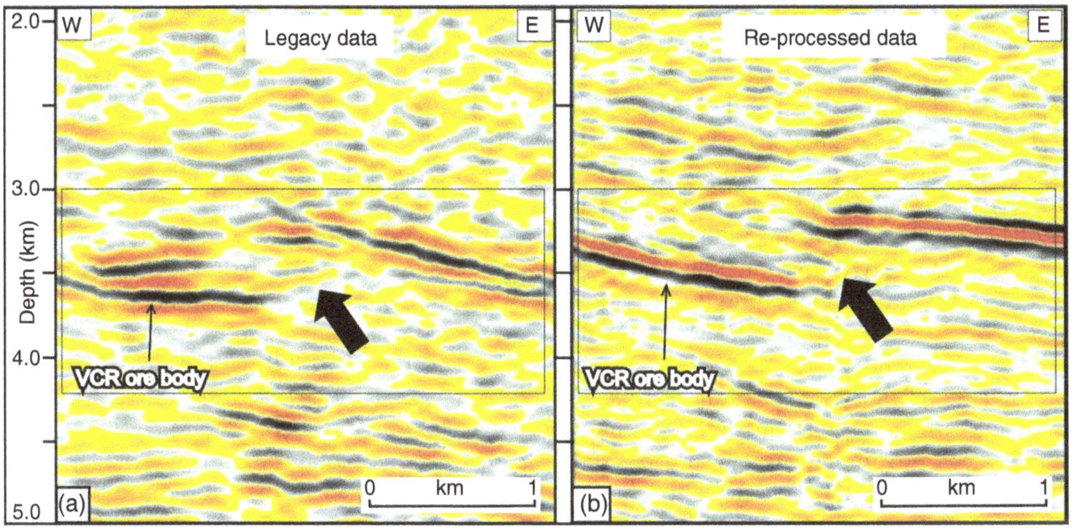

Figure 11.4 Comparison of legacy conventional data and reprocessed seismic data. (a) Legacy 1994 conventional data set processed using old imaging techniques (e.g., 3D poststack finite difference migration). (b) Reprocessed seismic data using 3D prestack Kirchhoff time migration. Note that the new imaging algorithms show better resolution relative to the poststack depth migration that was originally applied to the same data, shown in (a). The new image shows better imaging of the fault's offset of the Ventersdorp Contact Reef (VCR) ore body.

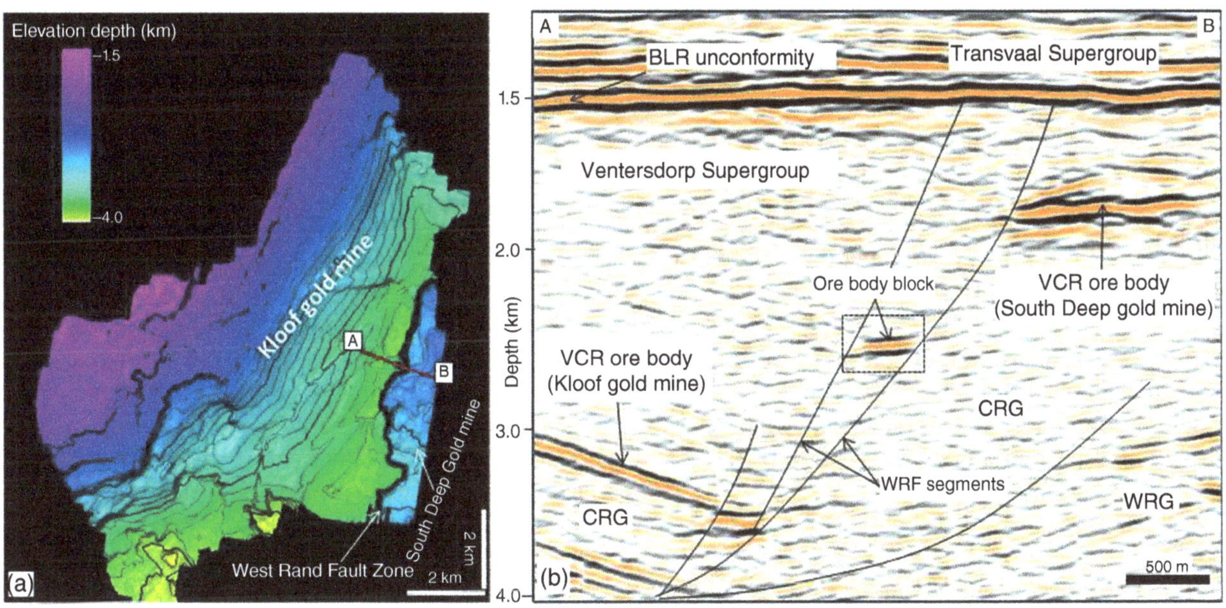

Figure 11.5 (a) The depth structure map (contour interval: 150 m) of the Ventersdorp Contact Reef (VCR) ore body showing the location of the West Rand Fault (WRF) Zone, which separates the Kloof and South Deep gold mines to the west and east, respectively. (b) Seismic section (amplitude display) across the West Rand Fault (with a throw of 1.5 km in the north of study area) showing a seismically imaged VCR ore body block preserved between segments of the steeply dipping West Rand Fault. CRG: Central Rand Group; WRG: West Rand Group.

2013a). In the seismic section, the WRF is north-south trending; west-dipping (65°–70°), and broadly listric in form with a maximum throw of 2 km. The fault cuts and displaces the West Rand and Central Rand Group strata (ca. 2.99–2.85 Ga) and the Ventersdorp Supergroup (ca. 2.71–2.63 Ga), and forms a décollement horizon at the top of the West Rand Group (Manzi et al., 2013a). The timing of this extensional faulting is constrained to the Neoarchaean, between 2.63 Ga and 2.59 Ga, because it does not breach the base of Palaeoproterozoic Black

Reef Formation of the Transvaal Supergroup (ca. 2.58–2.20 Ga) (Fig. 11.5). Results from the seismic data show that the area is structurally complex as it underwent compressional deformation during the Umzawami Event (ca. 2.73 Ga), which terminated deposition within the Witwatersrand Basin. This was followed by the major extensional deformation of the Hlukana-Platberg event (2.70–2.64 Ga), where extension and normal faulting accompanied the extrusion of the 2.71 Ga Klipriviersberg Group lavas, culminating in the deposition of the Platberg Group sediments (Manzi et al., 2013a).

New developments in data interpretation techniques, such as 3D seismic attributes, have made it possible to extract new information from old surveys (Manzi et al., 2012a, 2012b). In particular, the edge-detection seismic attribute, a combination of dip- and dip-azimuth attributes (Manzi et al., 2012a, 2012b), clearly images the faults that cannot be normally resolved on seismic sections, emphasizing the high interconnectivity and crosscutting relationships between faults that offset the ore body (Fig. 11.6a). This information provides a better understanding of the interrelationship between fault activity and the distribution of ore body blocks, and the relative chronology of tectonic events. The successful high-resolution seismic imaging of minor faults by advanced seismic attributes has paved the way for multidisciplinary projects to study small-scale structures that (1) make the reefs impractical to mine, (2) reduce the stability of the mining hanging wall in underground workings, and (3) may act as conduits to transport combustible gas to the mining levels (Manzi et al., 2012b; Mkhabela & Manzi, 2017).

Figure 11.6b shows the interpreted 3D geometry of an ore body model incorporating complex fault and dyke architectures derived from seismic, borehole, and underground mapping data. These studies have improved the quality of the ore body models in the South Deep and Kloof gold mines.

11.3.3. Case Study 2: 3D Conventional vs. Broadband Seismic Survey, Moab Khotsong Mine

In this case study, we show how the acquisition of the 3D broadband reflection seismic data has optimized the imaging of subtle geological features and target horizons at different depth intervals, thus improving the mine planning and reducing mining risks.

Moab Khotsong gold mine, located in the Klerksdorp goldfields (Fig. 11.1) in the western part of the Witwatersrand Basin, started its operations in 2003, making it the youngest of the South African deep level gold mines. The mine, acquired by Harmony Gold Mining Company Ltd. in 2018, was opened and operated by AngloGold Ashanti to exploit ore blocks adjacent and contiguous to the existing AGA mining areas. It is an intermediate to ultradeep level mine with workings accessed through three vertical shaft systems. The mine exploits the Vaal Reef ore body (VR~1 m thick) at depths between 2.5 km and 3.8 km below surface (Pretorius et al., 2000). The VR constitutes about 95% of the ore reserve and is the most economic gold-bearing conglomerate in the Klerksdorp goldfield. It is a narrow, tabular deposit situated near the middle of the Central Rand Group of the Witwatersrand Supergroup (Fig. 11.2). The structurally and stratigraphically controlled mineralization of the VR is developed over a lateral dimension in excess of 7 km × 12 km, with an average dip of ~14° to the south-southeast (Pretorius et al., 1989; 2000). It consists of a single bed of coarse-grained quartzite or clast-supported conglomerate, which is typically < 20 cm thick and can sometimes occur as a single line of pebbles. Much of the gold is concentrated in a thin (2 mm to a few cm) carbonaceous layer that intermittently follows this well-defined stratigraphic horizon on the mine scale, but locally transgresses through the conglomerate, indicating that the carbonaceous material was mobile. The ore body is crosscut, displaced, and compartmentalized by several suites of faults, dykes, and sills, spaced tens to a few hundred meters apart, thus making mining difficult (Pretorius et al., 1989; 2000).

In 1996, AngloGold Ashanti Ltd. (AGA) conducted a high-resolution 3D seismic reflection survey over Moab Khotsong mine, their most structurally complex property. The objective of the survey was to image the structure of the Vaal Reef (VR) auriferous conglomerate and define the complex geological features that bound its blocks at the mining levels (2.0–3.8 km depths). Since the Vaal Reef cannot be imaged directly by the reflection seismic method (Pretorius et al., 2003), it was envisaged that its structure would be mapped by imaging reflective marker horizons above and below it. The survey successfully imaged the entire stratigraphy from surface through the Ventersdorp Supergroup to the base of the Witwatersrand Basin (~11 km in depth). However, it was unable to map the thin beds, shallow and deeply buried strata at a resolution sufficient for designing the final geometry of the deep gold mine. The survey also did not achieve the goal of better defining the structural complexity, that is, crosscutting relationship between faults, and the continuity and connectivity of the faults, in the area.

In 2012, AngloGold Ashanti (AGA) commissioned a test 3D seismic survey over a prospect area of 35 km^2 covering the Moab Khotsong mining area. This was to

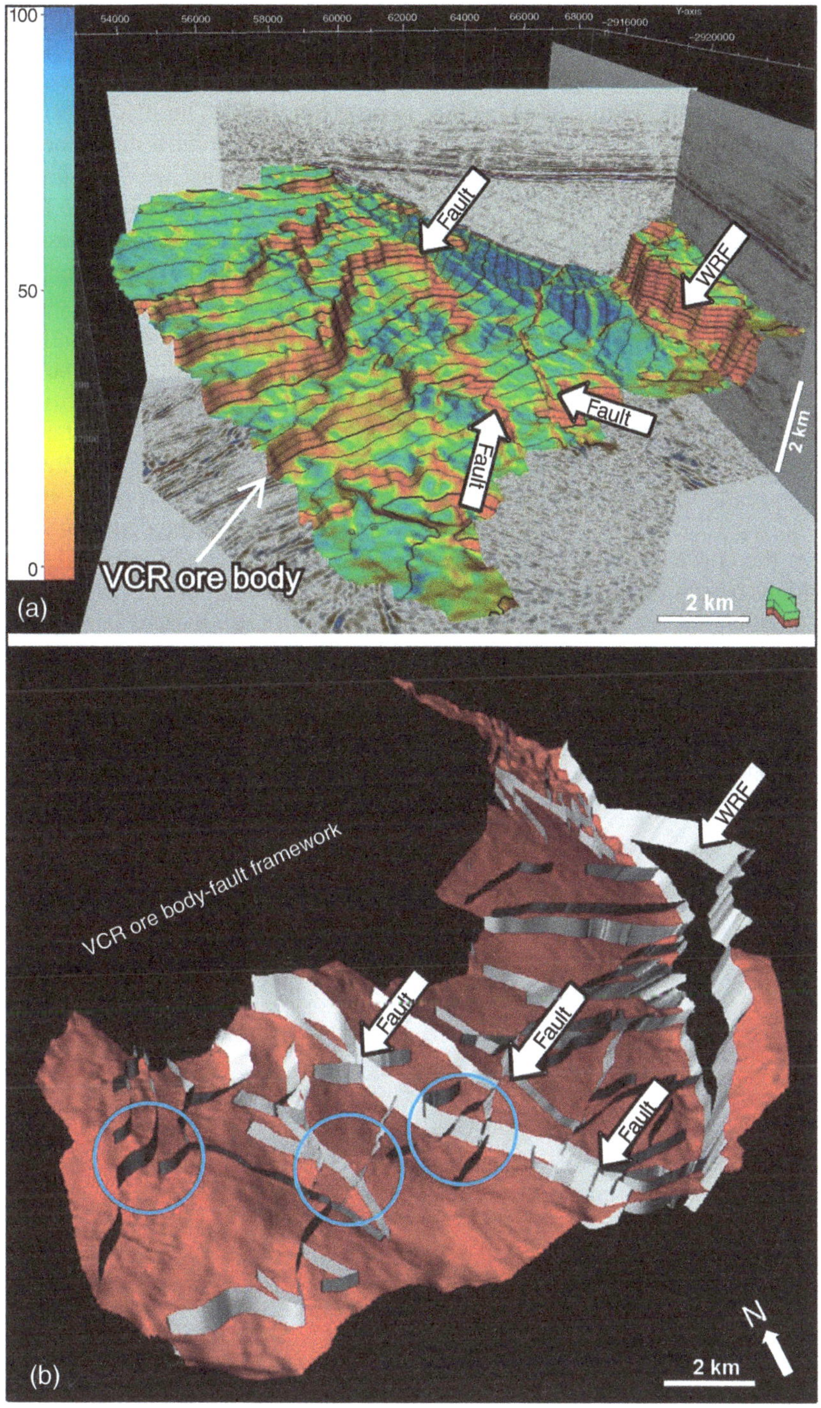

Figure 11.6 (a) Ventersdorp Contact Reef (VCR) horizon-fault framework defined from the edge-detection seismic attribute (color bar is given in percentages). The edge-detection attribute shows the successful imaging of faults. The attribute enhances the detection of the faults' continuity, bifurcation, and crosscutting relationships. The edge-attribute map also shows the West Rand Fault (WRF ~2 km maximum throw) that crosscuts the VCR ore body. (b) Final ore body model with incorporated seismically mapped faults, which can be integrated into data mine packages for mine planning and design purposes. Note the crosscutting relationship between faults that is often difficult to see through drilling and/or underground mapping.

improve structural imaging of the area and, at the same time, evaluate the value of dense 3D broadband seismic survey (i.e., a wider band of frequencies is recorded compared with conventional seismic acquisition) for the mining industry. The survey was designed to overlap with the 1996 legacy survey to enable comparison. This test was the first-ever onshore 3D broadband reflection seismic survey in South Africa. It presented an opportunity to illustrate how new onshore acquisition techniques, dense geometries, and the latest processing can improve land seismic imaging for mine planning and designs. The 3D seismic data have been the subject of several studies (e.g., Ogasawara et al., 2016) given the unique nature of the complex geology. Here, we illustrate the superiority of the broadband 3D seismic data in delineating and resolving key deep structures compared with the conventional 3D seismic survey.

Figure 11.7 compares the conventional data acquired in 1996 and the new high-density broadband data set. Both data show good imaging of the major prominent seismic horizons: the Black Reef (2 m thick, at 1 km depth) and the Ventersdorp Contact Reef (1.5 m thick; ~3 km depth), which lies 500 m above the narrow Vaal Reef ore body (Fig. 11.2). The Black Reef horizon is characterized by strong seismic reflections due to a significant acoustic impedance contrast between the overlying dolomites of the Chuniespoort Group of the Transvaal Supergroup and the underlying lower-velocity, less-dense Ventersdorp lavas (Fig. 11.2; Manzi et al., 2013a).

As evident in Figure 11.7, it is clear that the broadband data provided much better imaging of both the shallow and deeper targets when compared to the 1996 data. This is mainly due the expanded frequency range (3 to 160 Hz, or almost six octaves) of the broadband data when compared with that of the 1996 data (10 to 90 Hz, or about three octaves). The results demonstrate that the expectations from the dense acquisition, a small bin size, large offset, and high nominal fold were met, with outstanding image quality at all depths from shallow to deep targets. With this level of data quality, it was also possible

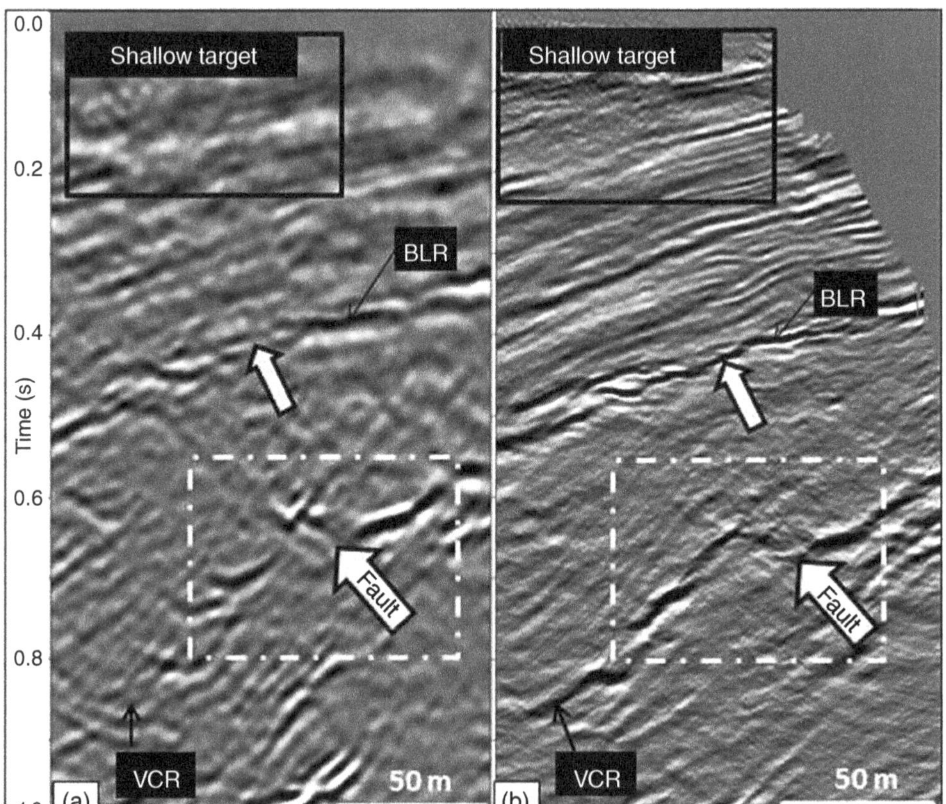

Figure 11.7 Comparison of post-stack migrated images for (a) legacy 1996 conventional data set, 10–90 Hz; (b) broadband (full bandwidth) high-density dataset, 3 –160 Hz. It is clear that the broadband data have provided much better imaging of the shallow targets (square), deeper targets (Black Reef (BLR) and Ventersdorp Contact Reef (VCR) ore horizons), complex faulting and steeply dipping structures (white box) when compared with the 1996 data. The improvement in the imaging is mainly due to the high-frequency content of the data (up to 160 Hz) introduced during acquisition and the new seismic imaging tools used during processing.

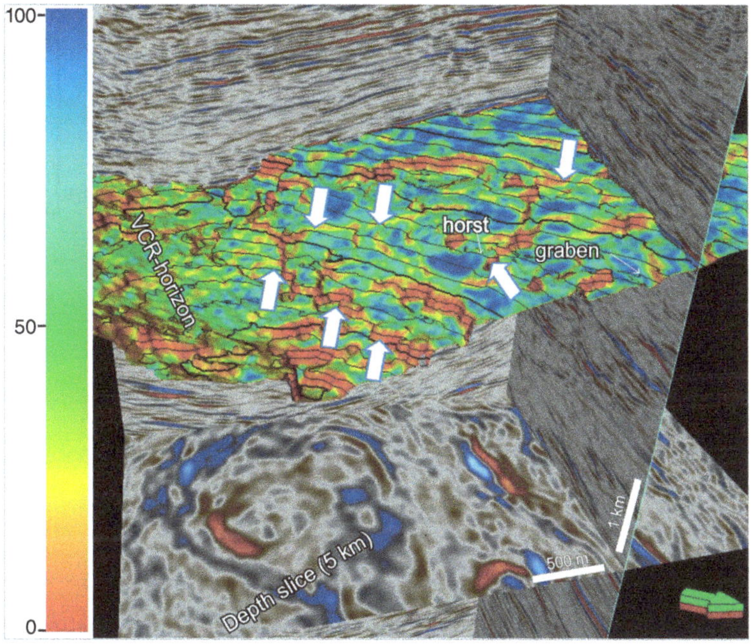

Figure 11.8 Three-dimensional visualization of combined 3D VCR edge-detection surface (color bar for edge-detection attribute is given in %) and modeled faults. The map shows the successful imaging of seismic horizons at shallow and deep levels. The Ventersdorp Contact Reef (VCR) ore body, which is situated approximately 500 m above the Vaal Reef ore body, is offset by major and minor normal faults (white arrows) with variable displacements and dips.

to significantly improve the structural interpretation of the ore body using seismic attributes (Manzi et al., 2015). This allowed for successful imaging of faults and fracture zones that offset the ore bodies with vertical displacements as small as 10 m (Fig. 11.8). Successful imaging of thin stratigraphic units and minor faults is attributed to a sharp seismic wavelet and lack of side lobes, resulting from the six-octave broadband source. The absence of interference in the side lobes of the seismic wavelet, in particular, makes it easier to resolve thin layers with low impedance contrasts. Furthermore, the usage of high frequencies (up 160 Hz) and dense acquisition design provided better sampling of the near surface variations and improved the surface-consistent processing (statics, deconvolution). This, in turn, improved the mapping resolution of the thin shallow targets.

It is therefore clear from these examples that the full structural complexity of the area cannot be solely delineated by conventional 3D seismic data. In particular, the advanced seismic attributes and the processing of the data using broadband technology have revealed a complex degree of faulting at great depths, which was not known prior to this study. For example, as shown in Figure 11.8, many faults are resolved as multifault segments that bound unmined blocks (grabens and horsts) leading to the discovery and delineation of resources in structurally complex areas of the mines. The existence of these major faults within the volume has since been confirmed through exploration drill holes and underground mapping. The amount of vertical movement on these faults also has major implications for mining as it can eliminate the entire ore body mining level along with affecting siting of shafts and exploration boreholes. Furthermore, the quality of these data has paved the way for a multidisciplinary International Continental Drilling Program (ICDP)-funded project "Drilling into Seismogenic Zones of M2.0–M5.5 Earthquakes in South African Gold Mines (DSeis)" in the Klerksdorp region to study geological structures that may have hosted the 2014 M5.5 Orkney earthquake that took place below the mining horizon (~4 km) (Ogasawara et al., 2016). This information could be used to assess and mitigate the risks posed by seismicity as mining proceeds to greater depths.

11.4. BUSHVELD COMPLEX

11.4.1. Platinum Deposits

The platinum deposits of South Africa are associated with the ~2.06 Ga Bushveld Complex (Fig. 11.9a), which is widely known as the world's largest igneous complex with an estimated minimum area of 64,000 km^2

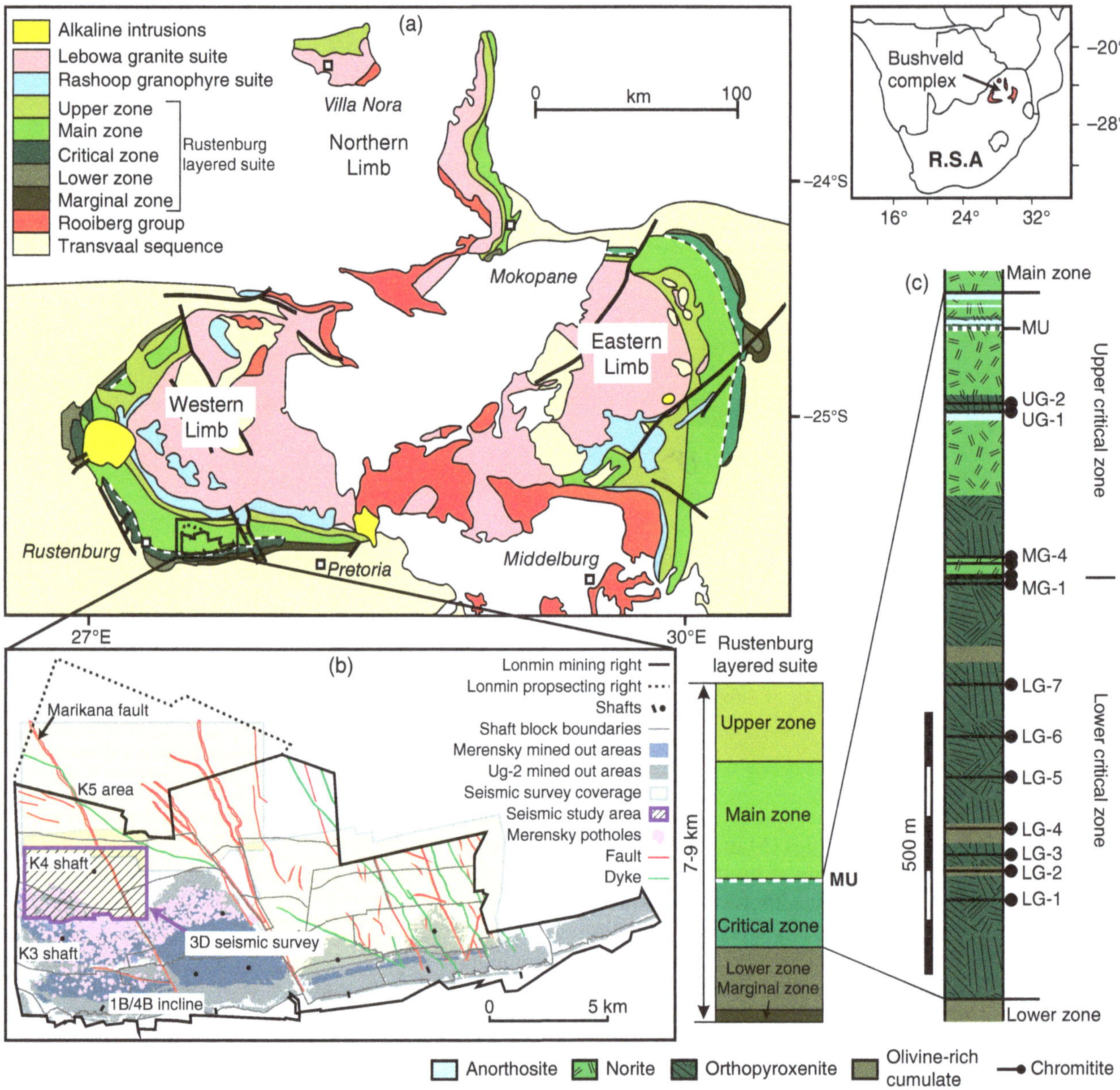

Figure 11.9 (a) Sketch map of the Bushveld Complex indicating location of Lonmin Marikana, modified from *Webb et al.* (2004). (b) Map of the Lonmin Marikana properties indicating locations of: the field study area in the K3 shaft; the overlapping seismic study area; workings on the UG-2 (grey) and Merensky (blue) Units; and known Merensky potholes (pink) in the worked areas. Note Karee mine consists of the 1B/4B, K3 and K4 workings. Adapted from Lonmin Platinum (2016). (c) Schematic stratigraphic column of the Critical Zone indicating the locations of the chromitites and the Merensky Unit (MU) (adapted from Eales & Costin, 2012).

(Carr et al., 1994). The potential of the Bushveld Complex for hosting platinum has been known since the early twentieth century, with the first economically significant platinum deposit being discovered by Hans Merensky and associates in 1924 (Cawthorn, 1999). The platinum is associated with the sequence of rocks known as the Rustenberg Layered Suite (2055.91 ± 0.26 Ma; Zeh et al., 2015) with the bulk of the deposits hosted within the Critical Zone (Fig. 11.9b, c). This principally comprises orthopyroxenites and chromitites within the Lower Critical Zone with the appearance of norites and anorthosites within the Upper Critical Zone. The economic resources, platinum-group elements (PGEs), chromium and vanadium, are found in stratiform horizons, referred to in mining terminology as "reefs." The bulk of these resources are hosted in chromitites (>60% Cr-spinel) with

Figure 11.10 Open pit mine on the UG-2 horizon with potholes marked by black outlines (photo courtesy of Lonmin Platinum).

the principal layers in the Western Limb of the Bushveld Complex being the Lower Group (LG-1 to −7), Middle Group (MG-1 to −4), and the Upper Group (UG-1 and −2) chromitites, with the LG-6, MG-1, MG-4, and UG-2 chromitites being economically viable to mine (Naldrett et al., 2009a,b). Additional economically viable PGE resources are hosted within the feldspathic pyroxenites of the Merensky Unit (MU, Fig. 11.9c) and the Platreef in the Northern Limb (Naldrett et al., 2009a).

Exploitation of these resources is affected by geological complexities, which includes faults, dykes and more complex structures such as iron-rich ultramafic pegmatite (IRUP) bodies and potholes. Large-scale faults (throw > 25 m, Fig. 11.9b) are well constrained by surface and underground mapping, while dykes (Fig. 11.9b) or IRUP bodies can be distinguished by their surface expressions, underground mapping and magnetic signatures (e.g., Cole et al., 2013; Hoffman & Plumb, 2014). Potholes are transgressive features where one of the typically stratiform units cuts down into the underlying units, so that the ore body lies at a lower stratigraphic level than expected, developing circular or ovoid pits (Figs. 11.9b and 11.10). Determination of pothole locations ahead of mining is currently unreliable, and statistical models are used for mine planning purposes (Hoffmann, 2010). These geological features, combined, are estimated to result in geological losses, on average, of 15% for the UG-2 unit and 14% for the Merensky Unit in the Marikana Region (Lonmin Platinum, 2016), with up to 30% losses in spatially restricted areas (Hoffmann, 2010). Thus, additional tools beyond exploration drilling and geological mapping are required to adequately characterize the ore bodies during mine development.

11.4.2. Case Study 3: Integration of 3D Seismic Survey Data with Geological Field Data, Karee Mine of the Bushveld Complex

Through this case study, we show that seismic methods are capable not only of detecting platinum ore bodies associated with thin (<5 m) stratiform rock horizons at depths below 200 m, but also of helping the mining industry to optimize mine planning in a cost-effective manner.

The Karee Mine is part of the Lonmin Marikana properties. The mine exploits two main PGE-, chromium-, the UG-2 and Merenskey reefs, at depths of 1300 m below surface. These are accessed through one inclined shaft and two vertical shafts. The Marikana properties in total have 3PGE + Au (platinum, palladium, rhodium, and gold) resources of 115.0 million ounces and reserves of 30.1 million ounces with average resource grades at 4.91 g/t (3PGE + Au) and average reserve grades at 4.06 g/t (3PGE + Au) (Lonmin Platinum, 2016). The economic life of the Karee mine shafts varies with the different generations, the 1B/4B inclined shaft is expected to produce until 2033; the K3 vertical shaft until 2037; and the K4 vertical shaft is expected to start producing in 2020 and continue until 2081. Across the Marikana underground operations 230,000 tonnes of rocks from the Merensky Unit area extracted per month, while 830,000 tonnes are extracted from the UG-2 Unit (Lonmin Platinum, 2016).

The term *UG-2 Reef* is used to refer to the massive chromitite (>60% Cr-spinel), which attains a thickness of ~1 m at Karee Mine. The typical dip of the UG-2 Reef is 8–10° toward the north and it has a sharp, often cuspate boundary to the footwall, which comprises either a medium-grained orthopyroxenite (>90% orthopyroxene) or a medium-grained poikilitic anorthosite (orthopyroxene oikocrysts up to 5 cm). The hanging wall is in sharp, planar contact with a medium-grained orthopyroxenite, which typically contains two "marker" chromitites, each up to 10 cm thick. Toward the west of the Karee Mine, the UG-2 Reef is "split" by a band of medium-grained orthopyroxenite (up to 0.6 m) into lower and upper chromitites. The UG-2 Reef is continuous with local variations in dip and dip azimuth across the study area (Hunt et al., 2018).

The term *Merensky Reef (MR)* at Karee Mine is used to refer to a package of pyroxenite with up to three thin (<5 cm) chromitites. It dips 8–10° to the north and is principally composed of a medium- to coarse-grained feldspathic orthopyroxenite (70% primocryst orthopyroxene), but has been subdivided into six facies by Lonmin according to the total thickness of the reef; presence or absence of the basal coarse-grained feldspathic orthopyroxenite (50% primocryst orthopyroxene), also referred to as a "pegmatite"; and number of chromitites (cf. Hunt et al., 2018). The reef thickness varies from 0.5 to 2.0 m, of which the coarse-grained feldspathic orthopyroxenite forms < 0.5 m when present.

Initially, statistical methods, 2D seismic surveys and potential field methods were used in the Karee Mine property to estimate the location of potholes and faults. However, these methods could not estimate the actual spatial distribution of these structures, leading to estimates of up to 30% of the reefs being considered an economic loss at the ore body level (e.g., Hoffmann, 2010). The initial positive results of the 3D seismic surveys in the Witwatersrand Basin for gold exploration sparked the interest of the platinum industry in the Bushveld Complex. In 1999, the Karee Mine conducted a high-resolution 3D reflection seismic survey for mine planning and deep mineral exploration purposes (Fig. 11.9b), which was acquired by Compagnie Générale de Géophysique (CGG). The details about the acquisition and processing of the data are discussed by Larroque et al. (2002). The main objective of the 3D seismic survey was to delineate the Merensky and UG-2 reefs, and to image faults, potholes, and fracture zones critical for geotechnical planning of the mine. The dominant frequency obtained during 3D seismic acquisition over the Karee region was 65 Hz, with a 6000 m/s mean velocity value observed during processing. These parameters provide a dominant wavelength of 92.3 m. Therefore, for one-fourth and one-eighth dominant wavelength Widess criteria (Widess, 1973), the achievable resolution limits are 23 m and 12 m, respectively. These suggest that the seismic method cannot resolve the top and bottom of the horizons with thickness less than 23 or 12 m. However, their detection can be enhanced using volume and horizon-based seismic attribute analysis (Manzi et al., 2012b; Manzi et al., 2015).

The Merensky and UG-2 reefs rarely exceed 2 m in thickness and are too thin to be directly resolved by surface reflection seismic method. Instead, it is the interface between their hanging wall and footwall rooks that is imaged due to large acoustic impedance (product of the bulk density × seismic velocity) contrasts and significant reflection coefficients (RC ≥ 6 %) at these interfaces. Figure 11.11 shows seismic sections extracted from the Karee Mine 3D seismic volume, providing successful seismic imaging of the Merensky Reef (MR), UG-2 Reef, and structures (faults, IRUPs and potholes). As evident in Figures 11.11a,b, the MR is characterized by a strong seismic response, but is less prominent than the UG-2 Reef, due to a lesser acoustic impedance contrast between the Cr-spinel-bearing feldspathic orthopyroxenites (~6400 m/s seismic velocity and a ~2.9 g/cm^3 density) and the less-dense anorthosites and leuconorites (~5700 m/s seismic velocity and 2.7 g/cm^3 density) (cf. Hunt et al., 2018). The UG-2, on the other hand, exhibits a strong seismic reflection character due to a major acoustic impedance contrast between the more-dense chromitite (~6400 m/s velocity and density up to 4.8 g/cm^3) and the less-dense pyroxenite, norite, or anorthosite (~5700 m/s velocity and densities between 2.7 and 2.9 g/cm^3). Therefore, the large acoustic impedance contrast is controlled by the bulk density variations within the target strata, rather than by seismic velocity, as there is no significant contrast in the seismic velocities. Continuity of both the UG-2 and Merensky reefs is confirmed by the seismic data, although both display variations in their strike and dip across the study area (Fig. 11.11).

To enhance the detection of faults below typical seismic resolutions (throw of < 30 m) and small-scale potholes (diameter of < 100 m), we applied the 3D seismic attributes to the UG-2 horizon. Attribute analysis allows for detection of small-scale features that would be difficult to resolve with traditional seismic techniques. Figure 11.12 shows improved mapping of faults and potholes by the Root Mean Square amplitude attribute (Manzi et al., 2014)

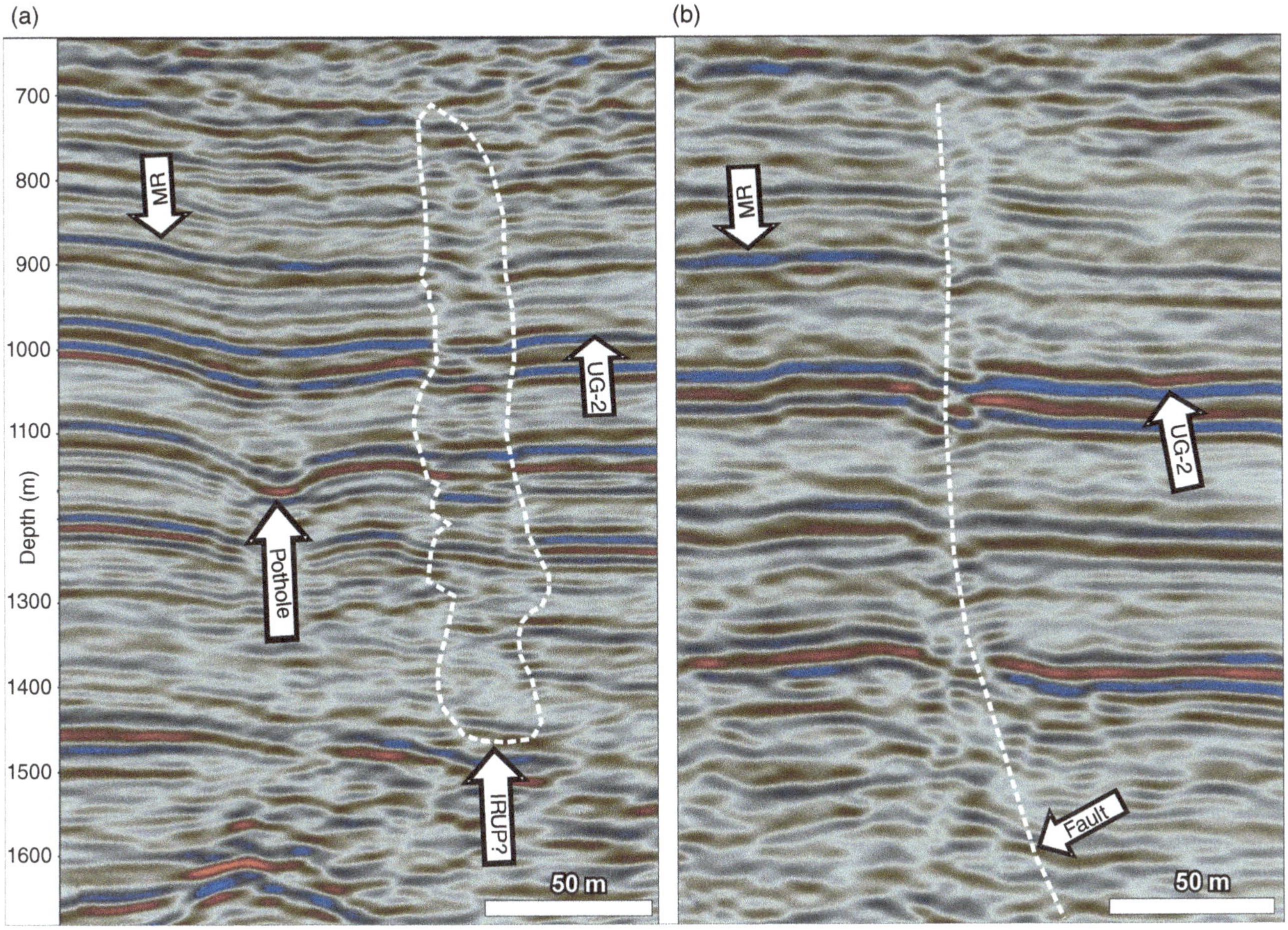

Figure 11.11 (a) Seismic section (amplitude display) extracted from the seismic cube, showing the seismic reflections associated with the Merensky Reef (MR), chromitite (UG-2), and interpreted pothole and iron-rich ultramafic pegmatite (IRUP) structures. (b) Seismic section (amplitude display) extracted from the seismic cube, showing the seismic reflections associated with the Merensky Reef (MR), chromitite (UG-2), and interpreted fault. Note, MR and UG-2 horizons are confirmed by drilling.

in 3D space. The most dominant feature imaged by the seismic method in the area is the Marikana Fault (F1), which is a north-northwest trending, normal fault (~500 m maximum throw) that steeply dips at ~75–80° to the west. The depth positions and apparent dips of the Marikana Fault have been confirmed by underground mapping and drilling (Lonmin Platinum, 2016). However, its timing of activity and its relationships with the formation of the ore bodies have not been fully studied. The identification of F1 and its variations in dip angles and vertical throws is important, as it is likely to cause weak ground conditions during mining. In addition, the constraints on fault orientations and timing of activity across the mining areas are significant and will need to be studied and be factored into future mine planning and development.

Furthermore, edge-detection attribute reveals some faults and potholes that are not resolvable on seismic sections and amplitude attribute displays (Fig. 11.12). For example, seismic attributes indicate the occurrence of randomly distributed, elliptical potholes (Fig. 11.13a). Integration of these attributes with underground observations (Fig. 11.13b) confirm that many meter-scale geological features are present and contribute to the complexity of the area (see Fig. 11.10) (Hunt et al., 2018). This allows for the interpretation of seismically small-scale faults from subtle offsets of the reflectors and for association of potholes with faults (Fig. 11.13c,d), despite their apparent random distribution (Fig. 11.13a).

Our integrated results further suggest that potholes are not associated with the regional faults, for example, the Marikana Fault (F1) (Figs. 11.9b and 11.13a); instead they are associated with much more subtle, small-scale features. The association of potholes with brittle faults is considered to be related to a structural control on their formation (e.g., Carr et al., 1994; Carr et al., 1999).

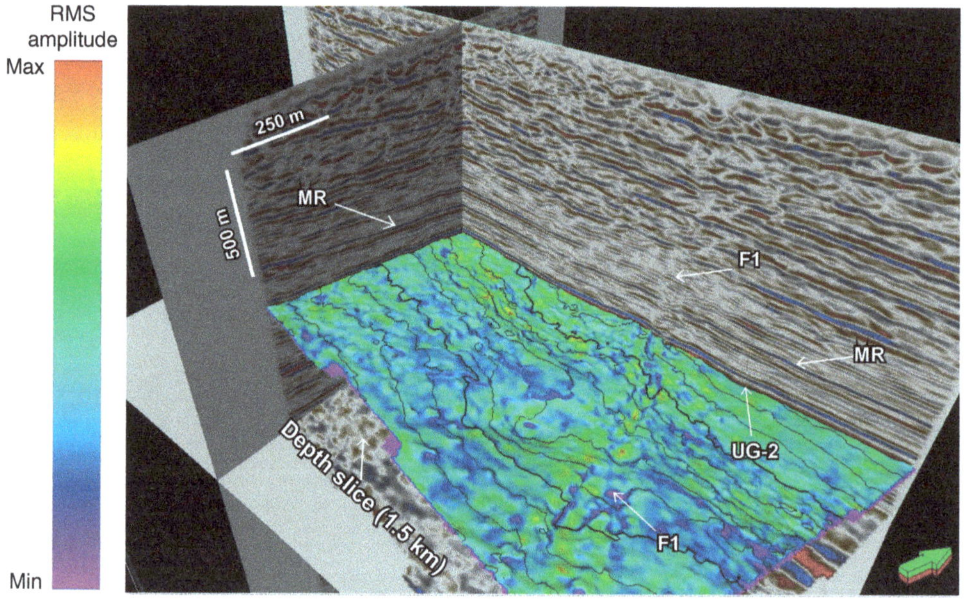

Figure 11.12 3D visualization of the 3D seismic data showing the successful imaging of the Merensky Reef (MR), chromitite (UG-2) and its associated seismically mapped Marikana fault (F1) on the Root Mean Square (RMS) amplitude attribute display.

However, as evident in the seismic sections, not all potholes occur in association with brittle faulting. Thus, instead the deformation associated with their development is suggested to have occurred while the rocks were still cooling and thus underwent ductile deformation. The small scale of some of the potholes (Fig. 11.13e) indicates that this deformation may have been very localized, potentially affecting areas only a few tens of square meters in size.

The third principal geological complexity in the study area is the IRUP bodies. While their near surface expressions are visible in magnetic data, bodies at depth are not visible using the magnetic method (Trickett et al., 2004). At depth, IRUPs can be detected within seismic data as they result in scattering of the acoustic energy, so can be recognized from diffuse portions of the reflectors (Trickett et al., 2009). Pipelike IRUP bodies are interpreted in the region from seismic sections, but are not expressed in the attribute analysis (Fig. 11.13e,f). This indicates that comprehensive analysis of the 3D seismic data is important for detecting all the geological complexity within the ore bodies and is vital for any mine planning purposes.

While the seismic data cannot show the small-scale complexity of potholes that is visible in the field, use of attribute analysis techniques allows for the interpretation of pothole locations. Combining these data with those collected during the normal operation of the mine workings, such as underground mapping and exploration drilling, provides a holistic overview of the geology of the area and can contribute considerably to accurate resource estimation and development planning. The data have provided the motivation for a large multidisciplinary project in the Bushveld Complex to understand the large-scale magma chamber dynamics during the crystallization of layered igneous rocks and the economic resources they can host. Hunt et al. (2018) present the more detailed results of this work with a focus on the integration of underground mapping data and geochemistry to provide constraints on the existing models.

11.5. DISCUSSION AND CONCLUSIONS

Mineral exploration at depth requires a systematic multidisciplinary approach that includes detailed underground mapping, physical property studies, 2D and 3D seismic surveys followed by downhole sonic measurements. We have illustrated, through case studies from hard rock environments in South Africa, that deep mineral exploration using low-quality legacy 2D seismic data and conventional 3D seismic surveys may provide misleading interpretations. We have demonstrated that the reprocessing of the legacy seismic data using new sophisticated imaging techniques and acquisition of new 3D seismic data using a broadband technology may provide better images of the subsurface structures. The case study from Kloof and South Deep gold mines clearly illustrates the risk in relying on the interpretation of old seismic data, and the value of reprocessing the data for better evaluation of the ore block resources.

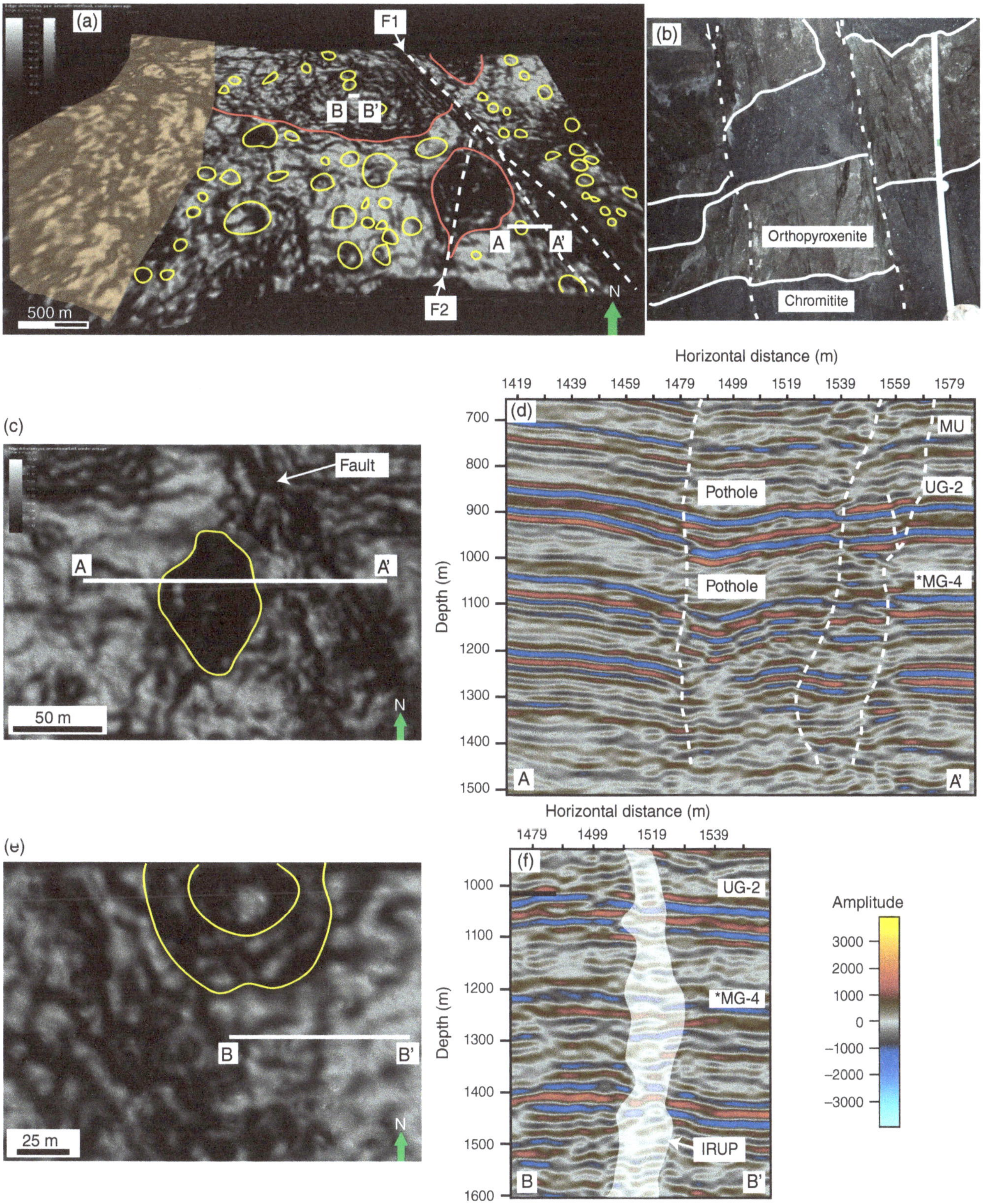

Figure 11.13 (a) Interpreted edge-detection map computed for the UG-2 horizon combining dip and dip azimuth attributes. Elliptical structures, interpreted as potholes, are outlined in yellow; large-scale (>500 m diameter) depressions are outlined in red; faults (F1, Marikana Fault, and F2) are marked by dashed white lines. Note the left-hand side of the area (shaded in orange) has not been interpreted due to a low signal-to-noise ratio as the area is at the edge of the survey. Location of seismic area is shown in Figure 11.9. (b) Underground exposure of UG-2 Unit (K3# 15cW 59 RSE) affected by brittle faults (dashed lines), scale bar is 1 m in length. (c) Enlarged area marked by A-A′ in (a) with central pothole outlined in yellow. (d) Seismic section from line A-A′ indicates both UG-2 and interpreted MG-4 horizons are potholed, in association with faulting. (e) Enlarged area marked by B-B′ in (a) with complex pothole outlined in yellow. (f) Seismic section from line B-B′ with IRUP pipe structure.

Most gold mines in the Witwatersrand Basin are approaching great depths; in some cases the reefs are heavily faulted and closely spaced and have poor seismic contrasts with their host rocks, implying that they cannot be directly detected, or are at best poorly resolved, by a conventional 3D seismic survey. In addition, the structural complexities at these mines make it difficult for traditional exploration drilling and underground mapping methods to locate significant ore body blocks within the fault zones. An example from the Moab Khotsong gold mine demonstrates that the use of the 3D broadband seismic method is an ideal future technique for hard rock mineral exploration as it is shown to produce high-resolution images of structures at great depths.

A case study from the Bushveld Complex demonstrates the impact of a high-resolution 3D seismic survey on the development program of a platinum mine, and shows that reflection seismology can play an important role in determining the structural position of thin, layered ore bodies and image potholes and faults. In the Bushveld Complex, the data have also been used to study the mechanisms by which the potholes are developed, possibly providing some insights in unraveling complex magmatic processes. The techniques applied within this study are not only of interest to the academic community, but also to the mining industries, that is, they can benefit planning operations by providing better resource estimation, and better-informed siting of future shafts. Thus, future mine development plans should consider acquiring high-resolution 3D reflection seismic data when assessing the potential to extend the Life of Mine and reduce the costs.

ACKNOWLEDGMENTS

We are thankful to numerous people for their contributions to this chapter. We thank the National Research Foundation (NRF) of South Africa, DST-NRF CIMERA, Shell South Africa Ltd., Compagnie Générale de Géophysique (CGG), and Thuthuka National Research Funding for financial support. We thank Schlumberger for providing licenses for the Petrel software package, which was used for seismic interpretation. The authors gratefully acknowledge support from Lonmin Platinum for access to the seismic data set from the Karee region. Dennis Hoffmann and the anonymous reviewer are thanked for their valuable input to the chapter.

REFERENCES

Barnicoat, A. C., Henderson, I. H. C., Knipe, R. J., Yardley, B. W. D., Napier, R. W., Fox, N. P. C., et al. (1997). Hydrothermal gold mineralization in the Witwatersrand Basin, *Nature*, *386*, 820–824.

Campbell, G., & Crotty, J. H. (1990.). 3-D seismic mapping for mine planning purposes at the South Deep Prospect. In D. A. J. Ross-Watt & P. D. K. Robinson (Eds.), SAIMM Symposium Series S10, 2 (pp. 569–597). Proceedings International Deep Mining Conference.

Carr, H. W., Groves, D. I., & Cawthorne, R. G. (1994). The Importance of synmagmatic deformation in the formation of Merensky Reef potholes in the Bushveld Complex. *Economic Geology*, *89*, 1398–1410.

Carr, H. W., Kruger, F. J., Groves, D. I., & Cawthorn, R. G. (1999). The petrogenesis of Merenksy Reef potholes at the Western Platinum Mine, Bushveld Complex: Sr-istopic evidence for synmagmatic deformation. *Mineralium Deposita*, *34*, 335–347.

Cawthorn, R. G. (1999). The discovery of the platiniferous Merensky Reef in 1924. *South African Journal of Geology*, *102*(3), 178–183.

Cawthorn, R. G. (2010). The platinum group element deposits of the Bushveld Complex in South Africa. *Platinum Metals Review*, *54*(4), 205–215.

Cole, J., Finn, C. A., & Webb, S. J. (2013). Overview of the magnetic signatures of the Palaeoproterozoic Rustenberg Layered Suite, Bushveld Complex, South Africa. *Precambrian Research*, *236*, 193–213.

Dankert, B. T., & Hein, K. A. A. (2010). Evaluating the structural character and tectonic history of the Witwatersrand Basin. *Precambrian Research*, *177*, 1–22.

Denis M., Brem, V., Pradalie, F., Moinet, F., Retailleau, M., Langlois, J., & Bai. B., et al. (2013). Can land broadband seismic be as good as marine broadband?, *Leading Edge*, *32*, 1382–1388.

Duval, G. (2012). How broadband can unlock the remaining hydrocarbon potential of the North Sea. *First Break*, *30*, 85–91.

Eales, H. V., & Costin, G. (2012). Custally contaminated komatiite: Primary source of the chromitites and marginal, lower and critical zone magmas in a staging chamber beneath the Bushveld Complex. *Economic Geology*, *107*(4), 645–665.

Frimmel, H., & Hennigh, Q. (2015). First whiffs of atmospheric oxygen triggered onset of crustal gold cycle. *Mineralium Deposita*, *50*, 5–23.

Frimmel, H. E., Groves, D. I., Kirk, J., Ruitz, J., Chesley, J., & Minter, W. E. L. (2005). The formation and preservation of the Witwatersrand goldfields, the world's largest gold province. *Economic Geology*, 769–797 (100th Anniversary Volume).

Heinrich, C. A. (2015). Witwatersrand gold deposits formed by volcanic rain, anoxic rivers and Archaean life. *Nature Geoscience*, *8*, 206–209.

Hoffmann, D. (2010). Statistical size analysis of potholes: An attempt to estimate geological losses ahead of mining at Lonmin's Marikana mining district. Paper presented at the 4th International Platinum Conference, Platinum in transition "Boom or Bust." The Southern African Institute of Mining and Metallurgy.

Hoffmann, D., & Plumb, S. (2014). Predicting the probability of iron-rich ultramafic pegmatite (IRUP) in the Merenksy Reef at Lonmin's Karee Mine, in The 6th International Platinum Conference, Platinum-Metal for the Future. The Southern African Institute of Mining and Metallurgy.

Hunt, E. J., Latypov, R., & Horváth, P. (2018). The Merensky Cyclic Unit, Bushveld Complex, South Africa: reality or myth? *Minerals*, *8*, 144, doi: 10.3390/min8040144.

Krapež, B. (1985). The Ventersdorp Contact placer: A gold-pyrite placer of stream and debris-flow origins from the Archaean Witwatersrand Basin of South Africa, *Sedimentology*, *32*, 223–234.

Jolley S. J., Freeman, S. R., Barnicoat, A. C., Phillips, G. M., Knipe, R. J., Pather, A., et al. (2004). Structural controls on Witwatersrand Gold mineralization. *Journal of Structural Geology*, *26*, 1026–1086.

Law, J. D. M., & Phillips, G. M. (2005). Hydrothermal replacement model for Witwatersrand Gold. In J. W. Hedenquist, J. F. H. Thompson, R. J. Goldfarb, & J. P. Richards (Eds.), *One Hundreth Anniversary Volume* (pp. 799–811). Society of Economic Geologists, Inc., Littleton, Colorado,

Larroque, M., Postel, J. J. P., Slabbert, M., & Duweke, W. (2002). How 3D seismic can help enhance mining. *First Break*, *20*(7), 472–475.

Lonmin Platinum (2016). Mineral Resource and Mineral Reserve Statement Report, Johannesburg.

Malehmir, A., & Bellefleur, G. (2009). 3D seismic reflection imaging of volcanic-hosted massive sulfide deposits: Insights from reprocessing Halfmile Lake data, New Brunswick, Canada. *Geophysics*, *74*(6), 209–219.

Malehmir, A., Durrheim, R., Bellefleur, G., Urosevic, M., Juhlin, C., White, D. J., et al. (2012). Seismic methods in mineral exploration and mine planning: A general overview of past and present case histories and a look into the future. *Geophysics*, *77*(5), 173–190.

Malehmir A., Koivisto, E., Manzi, M., Cheraghi, S., Durrheim, R., Bellefleur, R., et al. (2014). A review of reflection seismic investigations in three major metallogenic regions: The Kevitsa Ni-Cu-PGE district (Finland), Witwatersrand goldfields (South Africa), and the Bathurst Mining Camp (Canada). *Ore Geology Reviews*, *56*, 423–441.

Manzi, M. S. D., Cooper, G., Malehmir, A., Durrheim, R. J., & Nkosi, Z. (2015). Integrated interpretation of 3D seismic data to enhance the detection of the gold-bearing reef:Mponeng Gold mine, Witwatersrand Basin (South Africa). *Geophysical Prospecting*, *63*(4), 881–902.

Manzi, M. S. D., Durrheim, R. J., Hein, K. A. A., & King, N. (2012b). 3D edge-detection seismic attributes used to map potential conduits for water and methane in deep gold mines in the Witwatersrand basin, South Africa. *Geophysics*, *77*(5), 133–147.

Manzi M. S. D., Gibson, M. A. S., Hein, K. A. A., King, N., & Durrheim, R. J. (2012a). Application of 3D seismic techniques to evaluate ore resources in the West Wits Line goldfield and portions of the West Rand goldfield, South Africa. *Geophysics*, *77*, 163–171.

Manzi, M. S. D., Hein, K. A. A., Durrheim, R. J., & King, N. (2013b). Seismic attribute analysis to enhance detection of thin gold-bearing reefs: South Deep gold mine, Witwatersrand Basin, South Africa. *Journal of Applied Geophysics*, *98*, 212–228.

Manzi, M. S. D., Hein, K. A. A., Durrheim, R. J., & King, N. (2014). The Ventersdorp Contact Reef model in the Kloof Gold Mine as derived from 3D seismics, geological mapping and exploration borehole datasets. *International Journal of Rock Mechanics and Mining Sciences*, *66*, 97–113.

Manzi, M. S. D., Hein, K. A. A., King, N., & Durrheim, R. J. (2013a). Neoarchaean tectonic history of the Witwatersrand Basin and Ventersdorp Supergroup: New constraints from high-resolution 3D seismic reflection data. *Tectonophysics*, *590*, 94–105.

Milkereit, B., Berrer, E. K., King, A. R., Watts, A. H., Roberts, B., Adams, E., et al. (2000). Development of 3-D seismic exploration technology for deep nickel-copper deposits: A case history from the Sudbury basin, Canada. *Geophysics*, *65*, 1890–1899.

Milkereit, B., Eaton, D., Wu, J., Perron, G., Salisbury, M. H., Berrer, E. K., et al. (1996). Seismic imaging of massive sulphide deposits; Part II, Reflection seismic profiling. *Economic Geology*, *91*(5), 829–834.

Minter, W. E. L. (2006). The sedimentary setting of Witwatersrand placer mineral deposits in an Archaean atmosphere. Special volume on the evolutionof the Earth's atmosphere. In S. E. Kessler & H. Ohmoto (Eds.), *Evolution of Early Earth's Atmosphere, Hydrosphere and Biosphere: Constraints from Ore Deposits. Pardee Symposium* (pp. 105–119). Geological Society of America Memoirs, 198.

Mkhabela, M., & Manzi, M. S. D. (2017). Detection of potential methane gas pathways in deep South African gold mines. *Journal of Geophysics and Engineering*, *14*, 960–974.

Muntean, J. L., Frimmel, H. E., Phillips, N., Law, J., & Myers, R. (2005). Controversies on the origin of world-class gold deposits. Part II. Witwatersrand gold deposits. *SEG Newsletter*, *60*, 7–19.

Naldrett, A. J., Kinnaird, J., Wilson, A., Yudovskaya, M., McQuade, S., Chunnett, G., & Stanley, C. (2009b). Chromite composition and PGE content of Bushveld chromitites: Part 1: The Lower and Middle Groups. *Applied Earth Science (Trans. Inst. Min. Metall. B)*, *118*(3/4), 131–161.

Naldrett, A. J., Wilson, A., Kinnaird, J., & Chunnett, G. (2009a). PGE metal tenor and metal ratios within and below the Merensky Reef, Bushveld Complex: Implications for Its genesis. *Journal of Petrology*, *50*(4), 625–659.

Ogasawara, H., Yabe, Y., Ito, T., van Aswegen, G., Cichowicz, A., & Durrheim, R. (2016). Drilling into seismogenic zones of M2.0–M5.5 earthquakes in deep South African gold mines (DSeis). European Geosciences Union, General Assembly 2016, Vienna, Austria. ID EPSC2016-2057, 17–22 April 2016.

Pretorius, C. C., Muller, M. R., Larroque, M., & Wilkins, C. (2003). A review of 16 years of hardrock seismic of the Kaapvaal Craton. In D. W. Eaton, B Milkereit., & M. H. Salisbury (Eds.), *Hardrock Seismic Exploration, Geophysical Developments*, *10* (pp. 247–268). Society of Exploration Geophysicists.

Pretorius, C. C., Trewick, F. W., Fourie, A., & Irons, C. (1989). Application of 3-D seismics to mine planning at Vaal Reefs gold mine, number 10 shaft, Republic of South Africa. *Geophysics*, *77*, 1862–1870.

Pretorius, C. C., Trewick, W. F., Fourie, A., & Irons, C. (2000). Application of 3-D seismics to mine planning at Vaal Reefs gold mine, number 10 shaft, Republic of South Africa. *Geophysics*, *65*, 1862–1870.

Trickett, J. C., Düweke, W., Linzer, L., Kock, S., & Tootal, K. (2009). From Aspirin to "The Bin" in a Mere 85 Years, in 11th SAGA Biennial Technical Meeting and Exhibition, 303–310, Swaziland.

Trickett, J. C., Düweke, W. A., & Kock, S. (2004). Three-dimensional reflection seismics: Worth its weight in platinum. *The Journal of The Southern African Institute of Mining and Metallurgy*, *105*, 357–363.

Urosevic, M., Stoltz, E., & Massey, S. (2005). Seismic Exploration for Gold in a Hard Rock Environment –Yilgarn Craton, Western Australia: 67th Meeting, EAGE, Expanded Abstracts, G009.

Webb, S. J., Cawthorn, R. G., Nguuri, T., & James, D. (2004). Gravity modeling of Bushveld Complex connectivity supported by Southern African Seismic Experiment results. *South African Journal of Geology*, *107*(1–2), 207–218.

Widess, M. B. (1973). How thin is a thin bed?, *Geophysics*, *38*(6), 1176–1180.

Zeh, A., Ovtcharova, M., Wilson, A. H., & Schaltegger, U. (2015). The Busvheld Complex was emplaced and cooled in less than one million years, results of zirconology, and geotectonic implications. *Earth and Planetary Science Letters*, *418*, 103–114.

INDEX

Ore Deposits: Origin, Exploration, and Exploitation, Geophysical Monograph 242, First Edition.
Edited by Sophie Decrée and Laurence Robb.